Red Tides
Biology, Environmental Science, and Toxicology

Red Tides
Biology, Environmental Science, and Toxicology

Proceedings of the First International Symposium on Red Tides held November 10–14, 1987, in Takamatsu, Kagawa Prefecture, Japan

Editors:

Tomotoshi Okaichi
Faculty of Agriculture
Kagawa University
Miki-cho, Kagawa Prefecture, Japan

Donald M. Anderson
Biology Department
Woods Hole Oceanographic Institution
Woods Hole, Massachusetts

Takahisa Nemoto
Ocean Research Institute
University of Tokyo
Nakano-ku, Tokyo, Japan

Elsevier
New York • Amsterdam • London

Elsevier Science Publishing Co., Inc.
655 Avenue of the Americas, New York, New York 10010

Sole distributors outside the United States and Canada:

Elsevier Science Publishers B.V.
P.O. Box 211, 1000AE Amsterdam, the Netherlands

Library of Congress Cataloging in Publication Data

International Symposium on Red Tides (1st : 1987 : Takamatsu-shi, Japan)
Red tides.

"Proceedings of the International Symposium on Red Tides held November 10-14, 1987, in Takamatsu, Kagawa Prefecture, Japan."
Includes index.
1. Red tide—Congresses. I. Okaichi, Tomotoshi, 1929- . II. Anderson, Donald M. (Donald Mark) III. Nemoto, Takahisa, 1930- . IV. Title.
SH177.R4I58 1987 589.4'104522 88-30224
ISBN 0-444-01343-1

Current printing (last digit)
10 9 8 7 6 5 4 3 2 1

Manufactured in the United States of America

CONTENTS

III ENVIRONMENTAL SCIENCE AND BIOLOGY

IV BIOCHEMISTRY

SYMPOSIUM SPONSORS

Kagawa University

Ocean Research Institute,
University of Tokyo

Ministry of Education
Science and Culture
Government of Japan

Kagawa Prefecture

Takamatsu City

SYMPOSIUM CO-SPONSORS

Intergovernmental Oceanographic
Commission (IOC)

Scientific Committee on Oceanic
Research (SCOR)

International Association for
Biological Oceanography (IABO)

IOC Regional Committee for the
Western Pacific (WESTPAC)

Bureau of Environment
Government of Japan

Bureau of Fisheries
Government of Japan

The Japanese Society of
Scientific Fisheries

The Oceanographical Society
of Japan

The Plankton Society of Japan

Osaka Prefecture

Osaka 21st Century Association

Kagawa Prefecture Federation of
Fishery Cooperative Associations

SYMPOSIUM HOST

Organizing Committe of the International
Symposium on Red Tides

EDITORIAL COMMITTEE

INTERNATIONAL ADVISORY COMMITTEE

Donald M. Anderson
Woods Hole Oceanographic Institution
U.S.A.

Vagn Kr. Hansen
Danish Institute of Fisheries and Marine Research
Denmark

Yuzuru Shimizu
University of Rhode Island
U.S.A.

Suraphol Sudara
Chulalongkorn University
Thailand

F. J. R. Taylor
University of British Columbia
Canada

LOCAL ORGANIZING COMMITTEE

T. Okaichi,(Chairman)	Kagawa University
T. Nemoto,(Vice-Chairman)	University of Tokyo
A. Fujita	Kagawa Prefecture
K. Hashimoto	University of Toyko
R. Hirano	University of Toyko
S. lizuka	Nagasaki University
H. Inoue	Kagawa University
T. Kubo	Takamatsu City
S. Montani	Kagawa University
T. H. Murakami	Kagawa Medical School
Y. Nakanishi	Kagawa University
T. Ochi	Kagawa University
C. Ono	Akashiwo Research Institute of Kagawa
K. Okutani	Kagawa University
T. Sasaki	Kagawa University
A. Shirota	Nansei Regional Fisheries Research Laboratory
H. Tanaka	Kagawa University
M. Terazaki	University of Tokyo
M. Watanabe	National Institute for Environmental Studies
T. Yanagi	Ehime University

PREFACE

Research on red tides and toxic dinoflagellates has expanded significantly over the past 15 years. During that time, a series of International Conferences on Toxic Dinoflagelletes was initiated, with meetings held in 1974, 1978 and 1985. Despite this growing research activity and scientific interest, some areas of research and some areas of the world were not adequately represented by these conferences since they were generally restricted to dinoflagellates and all were convened in North America. Although the toxic dinoflagellate conferences will continue, with the fourth in the series scheduled for June, 1989 in Sweden, there was a need to hold a more general meeting that would include all bloom-forming organisms - animal and plant, toxic and non-toxic. To this end, the First International Symposium on Red Tides was convened in Takamatsu, Kagawa Prefecture, Japan in November, 1987. Twenty-seven countries were represented at the Symposium which was a unique opportunity for individuals from the many disciplines that fall under the "red tide" umbrella to meet each other and interact scientifically and socially. Red tides are clearly interdisciplinary in nature, and the symposium was purposefully held in one large room (i.e. without concurrent sessions) so that interactions across disciplines could take place.

The program of the Symposium consisted of keynote lectures and both oral and poster contributed communications. The subjects covered were:

1. Regional Problems - case reports or reviews of red tides in specific areas;
2. Biology - taxonomy, physiology, ecology, and biochemistry;
3. Environmental Science - The physical, chemical, and biological structure of the marine environment, remote sensing, and numerical simulations;
4. Toxicology - toxic species, assay methods, and chemistry of red tide toxins.

Despite the complications inherent in bringing such a diverse assemblage of scientists, regulators, and industry officials together, the Symposium proceeded smoothly and was a technical and, we feel, a social success. The global distribution of this particular scientific community then became painfully apparent when it came to collecting over 100 manuscripts and reviewing them for scientific content and English grammar, all within a few short months of the meeting, as required by our publisher.

The editors and publisher believe that one of the values of a Proceedings Volume is that it provides an up-to-date picture of the status of a field of research that is difficult to obtain from the general scientific literature. Due to the large number of manuscript submissions, strict page limits had to be imposed that in turn restricted the depth and detail of the contributions, but we believe that it is important for all of those involved in this diverse field to have their efforts read and understood by their colleagues. This is a special problem for those who do not typically publish their results in English, and we hope that our efforts will allow the international communication that was so visible during the Symposium to continue. Those reading this volume will recognize that not all manuscripts have perfect word structure and usage; the final product is a compromise between time constraints and editorial propriety. Some authors will find their papers to have been changed significantly prior to publication since many were edited extensively and some were re-typed completely after we received them in "final form".

This Symposium would not have been possible without the interest and dedication of the participants and of certain Japanese and International organizations. The Intergovernmental Oceanographic Commission (IOC), consulting with WESTPAC, was the primary force behind the initial plans, with the proposal being supported by the Ministry of Education, Science and Culture, Kagawa Prefecture, and Takamatsu City. The organizers and editors wish to acknowledge Dr. Peter Bjornsen, Dr. Koichi Okutani, Ms. Tomoko Watanabe and Ms. Akemi Nakai for assisting with the organization of the Symposium. We thank the numerous volunteers, student assistants and chairpersons who were so efficient and courteous throughout the meeting, Dr. Theodore Smayda should be acknowledged for his willingness to take the time to assimilate all of the scientific contributions and to write an Overview for this book. We are also grateful to the workshop lecturers, Drs. F.J.R. Taylor, Kazumi Matsuoka and Yasuwo Fukuyo on Biology and Taxonomy, and Drs. Takeshi Yasumoto, Yasukatsu Oshima, Sachio Nishio and Michio Murata on Toxicology. The Editors gratefully acknowledge the efforts of the Editorial Committee in revising manuscripts, sometimes on exceedingly short notice, and Mr. Allan Ross of Elsevier for his enthusiasm and willingness to publish the Proceedings. Finally, we would like to express our sincere gratitude to the Symposium sponsors for providing the financial support that made it all possible.

Tomotoshi Okaichi
Kagawa University

Donald M. Anderson
Woods Hole Oceanographic Institution

Takahisa Nemoto
University of Tokyo

Red Tides

Biology, Environmental Science, and Toxicology

SYMPOSIUM LOGO

Mrs. Fusa Miyauchi (1883-1986) produced beautiful paper-mache toys in Takamatsu City. One of her dolls, known as the "Taimochi Ebisu" (the Japanese folk God of fishermen who always carries a lucky "Tai" or sea bream) was selected as the design for the 1959 New Year's commemorative stamp by the Japanese Post Office. Miss Akemi Nakai, Kagawa University, used this doll and the red tide dinoflagellate *Gymnodinium* to make the Symposium logo. This logo represents our wish for the extinction of red tides and an improvement of fish catch throughout the world.

I KEYNOTE ADDRESSES AND OVERVIEW

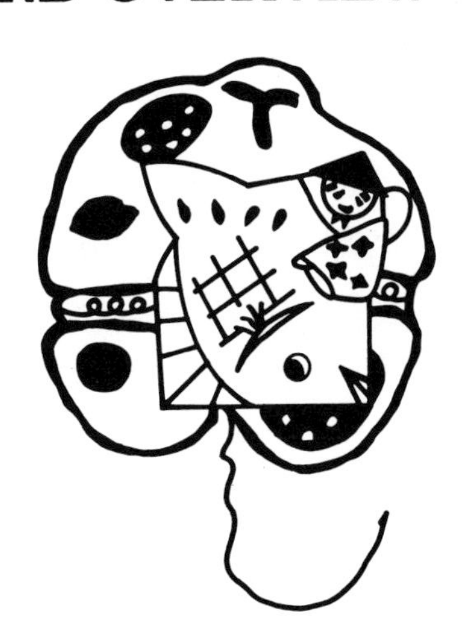

RECENT PROGRESS OF RED TIDE STUDIES IN JAPAN: AN OVERVIEW

HIDEO IWASAKI
Faculty of Bioresources, Mie University, Edobashi 2-80, Tsu 514, Japan

ABSTRACT

The red tide problem in Japan is introduced, followed by an overview of recent progress of red tide studies in the following areas: 1) Taxonomy of red tide flagellates, 2) initiation of blooms, 3) nutrients, growth promoters, and environment, 4) vertical migration of the causative organisms as nutrient support and a concentration mechanism, 5) population growth and interaction of organisms, and 6) prediction of red tide occurrence.

INTRODUCTION

In Japan, the definition of red tide is still vague. Historically, a seawater discoloration, irrespective of the causative organisms, has been called a red tide. Phytoplankton blooms, even diatom blooms have often been included in the definition of a red tide by some workers. So, the Japanese meaning of red tide seems to be different from the American's or the Westerner's. Since 1900, when the first published report occurred, red tide has often been observed mainly in the cultural grounds of pearl oyster (*Pinctada martensii)*. In the Seto Inland Sea, red tide outbreaks accompanied by damage to fisheries increased conspicuously in the 1960's. Therefore, the primary concern was concentrated on the damage to fisheries, although the appearance of paralytic shellfish poisoning was reported in 1961. To clear up the causative mechanisms of red tides, a research group was organized under the sponsorship of Fisheries Agency in 1966. The program of study was intended to clarify (1) the relation between the formation of low dissolved oxygen water mass in the bottom layer (anoxic condition) and the occurrence of red tide, (2) the influence of water pollution or industrial wastes on red tide occurrence, and (3) physiological characteristics of the causative organisms. The results[1] contributed to the development of a profile of red tides, and significantly influenced subsequent studies in Japan. The main problems faced in these studies were: (1) where the seed beds were in relation to the existence of the cysts, and (2) identification of unfamiliar or unknown organisms.

Since the frequent occurrences of red tide at that time had attracted public attention, studies on red tide were started in each research field, and carried out actively. Biological studies were focused on the following genera because of their economic and social effects, Raphidophyceae: *Heterosigma, Chattonella;* Dinophyceae: *Gymnodinium, Prorocentrum, Protogonyaulax,* and *Dinophysis*.

TAXONOMY

As mentioned above, the appearance of unfamiliar or unknown red tide flagellates induced confusion of identification, because there was no workable identification system owing to the lack of detailed information about their ultrastructure. The organisms in question, therefore, were reexamined.

Chattonella antiqua was first found in Hiroshima Bay, the Seto Inland Sea, in 1969, and described under *Hemieutreptia antiqua* by Hada[2]. Takano[3] reported the morphological similarity between *Hemieutreptia antiqua* and *Hornellia marina*. Since then, it has been called "*Hornellia*" in Japan.

Published 1989 by Elsevier Science Publishing Co., Inc.
RED TIDES: BIOLOGY, ENVIRONMENTAL SCIENCE, AND TOXICOLOGY
Okaichi, Anderson, and Nemoto, Editors

Since Hollande and Enjumet[4] had already asserted that *Hornellia marina* was the same species as *Chattonella subsalsa*, the taxonomic position of this organism has been unclear. Ono and Takano[5] and Hara and Chihara[6] agreed that *Hemieutreptia* is synonymous with *Chattonella* and described it under the new combined name of *C. antiqua*, because it differs from *C. subsalsa* in cell size and chloroplast structure. For the smaller form, Hara and Chihara[6] proposed the new, combined name of *Chattonella marina* (Subrahmanyan) Hara and Chihara.

Heterosigma akashiwo was first isolated in Bingo-Nada, the Seto Inland Sea, and Gokasho Bay in Kii peninsula in 1966. The genus *Heterosigma* was proposed by Hada to encompass *Entomosigma akashiwo*[7] Hada and a new species, *Heterosigma inlandica*[8], which the author considered to be unarmoured dinoflagellates. Hara and Chihara[9] reexamined these taxa using cultured material, including type cultures of both *H. akashiwo* and *H. inlandica* as well as specimens referred to as *Olisthodiscus luteus*, and maintained in CCAP and UTEX by light and electron microscopy. They concluded that these algae can be accommodated as a single species, *H. akashiwo*, because of their similarity in appearance and ultrastructure.

In the Seto Inland Sea there was another unknown microalga, tentatively called *Botryococcus* sp., which was often found to be one of the main causes of red tide harmful to mariculture. Toriumi and Takano[10] proposed this organism as longed to a new genus and species *Fibrocapsa japonica* Toriumi and Takano.

INITIATION OF BLOOMS

In relation to the initiation of blooms, several investigations have been carried out. Cysts of *Protogonyaulax tamarensis* were found in Tanabe Bay, Kii peninsula, Ise • Mikawa Bays, and Ofunato, Okirai Bays in the northeastern part of Japan. This suggests that this organism is distributed widely in the coastal waters of Japan. Fukuyo[11] evaluated the (1) vertical distribution of *Protogonyaulax* cells, (2) the numbers of benthic cysts per unit area, and (3) the germination rate of the cysts in Ofunato Bay from January, 1980 to June, 1981. He suggested the high possibility that the cysts play an important role as "seed populations", and that the appearance of swimming cells was controlled by environmental conditions, such as temperature, rather than the resting period. He described the planozygotes, which formed only in the early stage of planozygote formation and could transform to hypnozygotes; planozygotes were formed during exponential growth.

Studies on the cysts of Raphidophyceae have made slow progress because of their fragility. Research on the cysts has been tried in two different ways: laboratory cultures and field surveys. Furuki *et al.*[12] observed the morphological changes of *Chattonella antiqua* occurring under various environmental conditions, and described the conditions which induced their transformation to globular form and recovery. Appearance of vegetative cells of *Chattonella* in seawater incubated with bottom sediment samples suggested that these organisms have a benthic dormant stage, such as a cyst, during the course of their life cycle. From surveys using the extinction dilution method, Imai *et al.*[13,14] studied the distribution of the dormant cells in Harima-Nada and Suō-Nada located on the east and west side of the Seto Inland Sea, and confirmed that 1) the dormant cells accumulated in the restricted area, and 2) vegetative cells regularly occurred in the surface when bottom temperature rose to about 20°C. They reported that it is essential to find the seed beds first and then to monitor the rise of bottom water temperature to predict the appearance of cells in the early stage of the *Chattonella* red-tide outbreak. Imai and Itoh[15] also investigated the effect of storage temperature on the dormancy and maturation of the dormant cells of *Chattonella*. They summarized the annual life cycle as follows: (1) vegetative cells in early summer originate from germination of dormant cells in sediments; (2) they form overwintering dormant cells during the summer season; (3) dormant cells spend a duration of post dormancy, an enforced one due to low temperature, until early summer. This explanation seems to be well suited to the temperature regime in temperate sea such as the Seto Inland Sea. Finally, they[16] found the cysts of *Chattonella* and described their morphological characteristics.

NUTRIENTS, GROWTH PROMOTERS, ENVIRONMENT

The physiological characteristics of the primary red tide flagellates appearing in the coastal waters of Japan were already summarized, and shown that each species has specific preferences and tolerances to the environments[17]. Recently, interesting studies have been conducted. Takahashi and Fukazawa[18], in three semi-continuous culture experiments using natural populations, observed two types of responses for macro- and micro-nutrient dependency. The former group includes *Skeletonema costatum, Thalassiosira* sp. and *Eutreptiella* sp., whose growth was mainly stimulated by macro-nutrients. *H. akashiwo* was in the latter group which was stimulated by micro-nutrients. *Gymnodinium* sp. showed an intermediate type. Each flagellate showed characteristic growth responses under different nutrient conditions. *H. akashiwo* maintained a high growth rate even under poor macro-nutrients, and a further growth enhancement was obtained by moderate concentration of ammonia, manganese and vitamin B_{12}. *Gymnodinium* sp. growth was stimulated by macro-nutrients, but not as significantly as the diatoms, and also by iron and vitamin B_{12}. The nutrient requirements of *Prorocentrum micans* were clarified by Iwasaki and Iwasa[19]. Enrichment of the seawater (Ise Bay) with nitrate, phosphate, and vitamin B_{12} enhanced its growth. Moreover, the addition of chelated iron or acid extract (at pH 3.0) of marine mud greatly increased the growth, though no enhancement was observed following addition of nitrogen and phosphorus. They suggested that vitamin B_{12}, organometallic compounds, and organic substances play a significant role in the blooming of this organism in Ise Bay. The same result was obtained in another strain of *P. micans* by Yamochi[20] using Osaka Bay water. In the experiment using AGP method, Yamochi[20,21] reported that the addition of chelated iron to filtered seawater (Osaka Bay) enhanced the growth of *H. akashiwo,* but not the addition of nitrogen, phosphorus or vitamin B_{12}. Enrichment with a combination of thiamine, vitamin B_{12}, and Fe-EDTA increased by 15-fold the yield of *Eutreptiella* sp. Neither nitrogen nor phosphorus was effective unless sufficient vitamins, Fe-EDTA and Mn-EDTA were added, whereas the growth yield of *Chattonella marina* was promoted primarily by addition of Fe-EDTA or phosphorus. Nitrogen was effective only after enrichment with Fe-EDTA and phosphorus. The primary importance of Fe-EDTA was detected both in seawater collected in the early phase and at the climax of *C. marina* red-tide. He suggested that, in the case of eutrophic waters, chelated iron is one of the crucial factors triggering red tide outbreaks. Secondary nutrients, which stimulate the growth of red tide flagellates in coupling with chelated iron, are different from species-to-species.

Regarding iron uptake, Okaichi *et al.*[22] using radioactive iron EDTA-^{59}Fe, calculated the half saturation constants for uptake by *C. antiqua, Prorocentrum minimum* and *H. akashiwo*. The values were almost the same. However, the time required to reach the maximum rate of uptake was the shortest in *C. antiqua*.

Although the role of the vitamin B group in red tide formation has been discussed by many workers, knowledge of its distribution and dynamics in Japanese coastal waters is limited. Recently, vitamin concentration of seawater was measured in Harima-Nada[23], Ise Bay[24], and elsewhere[25]. Nishizima and Hata[23] measured vitamin B_{12}, thiamine, and biotin concentrations in seawater and bottom sediments at 24 stations in Harima-Nada in early summer and autumn. They showed that the distribution of vitamins corresponded with the distribution of heterotrophic bacteria, and was higher in eutrophic areas. From the seasonal variation of vitamins, they suggested that many phytoplankton in the waters consumed vitamin B_{12} for growth and excreted both thiamine, and biotin during the growing season. They also examined vitamin dynamics in the water, and found that 1) the vitamin load from polluted inland water was an important source into the sea; 2) release rates of vitamins into the overlying water from the bottom sediments increased with increasing temperature: 3) both bacterial and phytoplankton communities may play the most important role among the agents supplying and removing vitamin B_{12} in eutrophic waters. Furuki *et al.*[26] also showed that *C. antiqua* grew well in coexistence with bacteria in vitamin B_{12}-free medium. Nishizima[25] calculated vitamin B_{12} half saturation constants for growth, photosynthesis and uptake of several red tide flagellates. *C. antiqua* required a cell vitamin B_{12}-quota of about 8.5 ng/*l* to maximize its growth rate. Each red tide flagellate required about the same vitamin B_{12} concentration (10-20 ng/*l*). Consequently, he suggested that the concentration of vitamin B_{12} in the water was one of the most important factors controlling the outbreak of red tides in the coastal waters.

The relationship between organic or trace metal pollution originating in fish cultivation and the occurrence of red tides has stimulated discussion. Seawater collected near fish farms promoted the growth of *Gymnodinium nagasakiense*. Growth was also enhanced greatly by the addition of extracts from mackerel meat and yellowtail feces, while seawater enriched with inorganic nitrogen and phosphorus, and nitrogen, phosphorus plus iron did not promote growth. Growth of *C. antiqua* was enhanced only after addition of chelated iron to seawater, among various nutrients. From these results, Nishimura[27] concluded that dissolved organic matter at the fish culture site contributes greatly to the occurrence of *G. nagasakiense* red tide.

The macro-nutrient requirement of *C. antiqua* was reexamined in a nitrogen- or phosphorus-limited semi-continuous culture system by Nakamura[28]. He calculated the half saturation constants of nitrate, ammonium, and phosphate for uptake and growth. The results showed that these values were unexpectedly low in contrast to the cell volume. Comparisons of the half saturation constants with nutrient concentrations in the Seto Inland Sea in summer suggested that phosphate is one of the controlling factors of population growth of *C. antiqua*.

VERTICAL MIGRATION

Flagellates are well known to undertake vertical migrations. Such migrations have the great advantage of permitting the utilization of the nutrients in a water column of 10-20 meters. Hirano[29] has pointed out the importance of vertical migration in the development mechanism of red tide from the serial observations of the vertical distribution of *Gymnodinium* in a tide pool. Recently, more detailed observations of vertical migration have been made almost concurrently by several workers in the laboratory and field. Concerning *H. akashiwo* and *C. antiqua*, Watanabe *et al.*[30] showed that the timing of descent and ascent did not coincide exactly with the onset or end of light phase, and nitrate and phosphate uptake were less affected by light conditions in these species than in coastal diatoms. From the combination of diurnal vertical migration in thermally stratified water and the ability to take up nutrients at night, they described that these species have increased ecological advantages over coastal diatoms. Hatano *et al.*[31] observed the speed of migration of *H. akashiwo* and its adaptation to photoperiodic rhythm using a cylinder (D 4 x L 30 cm). The downward movement commenced at a half an hour before the light at a speed of 3.5 cm/h. Under continuous illumination or darkness, no vertical migration occurred, but a homogeneous distribution was thus obvious. The vertical migration was changed to a new rhythm at a 6L : 6D photoperiod. It took 3 to 5 days to acclimate to a new migration regime.

In field observations at two fishing ports (about 3 and 8 m in depth) in Osaka Bay, Yamochi and Abe[32] obtained interesting experimental results to prove the ecological significance of diel vertical migration of *H. akashiwo*.

They observed diel changes in the vertical profile of cell numbers together with several environmental factors at intervals of 0.5 to 3.0 hours when *H. akashiwo* was predominant. They also prepared dialysis bags filled with red water of the organism. One was suspended at the surface, another close to the bottom, and a third one was exposed to the surface in the daytime and the bottom at night. The organism migrated toward the surface early in the morning at a speed of 1.0 to 1.3 m/h. Downward migration was found in the afternoon, and more than 3,000 cells/m*l* aggregated in the bottom layer. *H. akashiwo* could cross steep temperature and salinity gradients (6.5°C and 5.7 ‰ S) during its diel vertical migration. High values of particulate organic carbon and nitrogen concentration were obtained in dialysis bags suspended *in situ* at an identical layer with high cell concentration, while the values for surface and bottom bags were comparatively low. These results show that the organism migrated toward the surface to carry out photosynthesis effectively, and to the bottom to utilize nutrients efficiently.

POPULATION GROWTH AND INTERACTION OF ORGANISMS

Population density and growth rate of *G. nagasakiense* were measured by Iizuka[33] in Ōmura Bay in two ways. Cell number was counted in water samples collected from different depth, and in dialysis tubes filled with *in situ* seawater containing the organism and suspended at 2.5, 7.5, and 12.5 meters for one to two days. The population growth started 11 days before reaching the density of a red tide. The growth rate was highest at the surface layer, $\mu_2 = 1.05$ on average, and lowest at the bottom. The growth rate at the middle layer was about 60 per cent of the surface; however, high cell density was maintained for a week after the disappearance at the surface. He also discussed the carrying capacity from the information of the maximum cell density of each organism and the maximum chlorophyll-*a* content in different waters.

Honjō[34] described the competitive relation between *S. costatum* and *H. akashiwo* in phytoplankton succession, and suggested that *H. akashiwo* produced an allelopathic substance inhibiting the growth of *S. costatum* at a concentration of 10^7 cells/*l* of *H. akashiwo*. The interrelation among phytoplankton, bacteria, and *Chattonella* during the process of development of the *Chattonella* bloom was investigated by Furuki *et al*[35]. in Harima-Nada. The results showed that suitable communities and characteristics of bacteria were required for the growth of *Chattonella*. In mixed culture of the bacteria and *Chattonella* in vitamin-deficient medium, they grew well when they could live together symbiotically, and showed no growth when cultured separately. Ishio *et al*[36]. succeeded in isolating a new species *Vibrio algoinfesta*, from the bottom sediments of Hakata Bay which produces dinoflagellate growth inhibitor (DGI). Since then, they have been trying to isolate the substance, determine the chemical structure, synthesize the substance chemically, and produce it in high yield.

Concerning grazing pressure, based on a feeding experiment with *Acartia clausi*, Tsuda and Nemoto[37] suggested that it could detect the biomass peak and alter its feeding behavior rapidly within at most 6 hours. Itoh and Imai[38] measured the grazing rate of five copepods at different concentration of *C. marina*. The grazing rate of *Calanus sinicus* increased with increasing cell concentration up to an average maximal rate of 380 cells/animal/h. They suggested that the feeding activity of these copepods on *Chattonella* presumably influences the initiation or termination of *Chattonella* red-tide. Uye[39] investigated the grazing of *C. antiqua* for five species of copepods. Ingestion rate increased linearly with increasing cell concentration untill a maximum level was reached, beyond which the rate was constant. This cell concentration was higher for larger copepods. The weight-specific maximum ingestion rates were higher in the small species. He calculated grazing pressure of the natural copepod community in Harima-Nada by itegration of the laboratory-determined feeding rates and field measurement of zooplankton biomass. The daily removal rate was 3.4 to 30.8 % (mean 12.8 %) of *C. antiqua* biomass at 20 cells/m*l* and decreased to 0.4 - 4.3 % (mean 1.8 %) at 500 cells/m*l*. From these results, he concluded that the grazing pressure by the copepod community is important at the initial stage of the red tide.

PREDICTION

Several attempts have been made to explore the possibility of predicting red tide occurrences. From discriminant analysis based on water temperature and salinity of surface water at Akashi strait from 1971 to 1981, Aoyama *et al*[40]. found prominent differences in the mean deviations from average daily values during May occurred between the years of red tide occurrence and years of non-occurrence in Harima-Nada. Both water temperature and salinity were higher in the years of red-tide occurrence; the probability of classificatory error was 0.022 in the discriminant function. They presented a formula predicting the date of the occurrence of *Chattonella* red-tide based on the cumulative value of daily water temperature from Februaty 1. Ōuchi[41] also tried to predict red tide (phytoplankton bloom) by discriminant analysis based on water quality data of surface water in the northeastern part of Hiroshima Bay. All the data (N = 526) obtained from 1976 to 1982 were divided into two groups, a non-red tide group and a red tide group, according to a critical value of 250 µg/*l* for the PON concentration. He used six variables — WT, SAL, TDP(total dissolved phosphorus), DIN, DON and PON. When a station is classified into the non-red tide group and its mis-classification probability becomes more than 0.25 or less than 0.05, there is a great tendency for red tide to occur around that station. The red tide actually occurred at a rate of 52 per cent. When a station belonging to the red tide group has a mis-classification probability of more than 0.17, there is a great tendency for red tide to disappear. Ōuchi considered the method to have great advantage in judging red tide occurrences, because only one mis-classification probability was used instead of six variables.

REFERENCES

1. T.Hanaoka, H.Irie, F.Uyeno, T.Okaichi and H.Iwasaki, "The cause of Red-tide in Neritic Waters", Nippon Suisan Shigen-hogo Kyōkai, Tokyo, pp. 1-105, (1972) (in Japanese)
2. Y.Hada, *Bull. plankt. Soc. Jap.*, 20, 112, (1974)
3. H. Takano, "Kankyō to Seibutsu-shihyō 2, Suikai-hen" (Environment and Biological indices - Hydrosphere), Kyōritsu Shuppan, Tokyo, 234-242, (1974) (in Japanese)
4. A.Hollande and M.Enjumet, *Bull. Trav. Publ. Stat. Aquic Pêche Castiglione N.S.*, 8, 273, (1956)
5. C.Ono and H. Takano, *Bull. Tokai Reg. Fish. Res. Lab.*, 102, 93, (1980)
6. Y.Hara and M.Chihara, *Jap. J. Phycol.*, 30, 47, (1982)
7. Y.Hada, *Bull. Suzugamine Women's Coll. Nat. Sci.*, 13, 1, (1967)
8. Y.Hada, *Bull. Suzugamine Women's Coll. Nat. Sci.*, 14, 1, (1968)
9. Y.Hara and M.Chihara, *Bot. Mag. Tokyo*, 100, 151, (1987)
10. S.Toriumi and H.Takano, *Bull. Tokai Reg. Fish. Res. Lab.*, 76, 25, (1973)
11. Y.Fukuyo, "A Thesis of Degree" Tokyo Univ., (1982)
12. M.Furuki, H.Kitamura and T.Tsukamoto, *Bull. Plankt. Soc. Jap.*, 28, 43, (1981)
13. I.Imai, K.Itoh and M.Anraku, *Bull. Plankt. Soc. Jap.*, 31, 35, (1984)
14. I.Imai, K.Itoh, K.Terada and M.Kamizono, *Bull. Jap. Soc. Sci. Fish.*, 52, 1665, (1986)
15. I.Imai and K.Itoh, *Mar. Biol.*, 94, 287, (1987)
16. I.Imai and K.Itoh, *Bull. Plankt. Soc. Jap.*, 33, 61, (1986)
17. H.Iwasaki, in "Biochemistry and Physiology of Protozoa (2nd Ed.), Vol.1" Academic Press, New York, 357, (1979)

18. M.Takahashi and N.Fukazawa, *Mar. Biol.*, 70, 268, (1982)
19. H.Iwasaki and K.Iwasa, *Bull. Fac. Fish. Mie Univ.*, 9, 49, (1982)
20. S.Yamochi, *Bull. Plankt. Soc. Jap.*, 31, 97, (1984)
21. S.Yamochi, *J. Oceanogr. Soc. Jap.*, 39, 310, (1983)
22. T. Okaichi, Report of Co-ope. Res. Minist. Educ. Sci. Cul., 194, (1987) (in Japanese)
23. T.Nishizima and Y.Hata, *Bull. Jap. Soc. Sci. Fish.*, 52, 1533, (1986)
24. H.Noda, Report of Gen. Res. Minist. Educ., 29, (1982) (in Japanese)
25. T.Nishizima, *Memor. Fac. Agri. Kochi Univ.*, 43, 1, (1986)
26. M.Furuki and K.Kubo, *Hakkōkogaku*, 62, 181, (1984)
27. A.Nishimura, *Bull. Plankt. Soc. Jap.*, 29, 1, (1982)
28. Y.Nakamura, *J. Oceanogr. Soc. Jap.*, 41, 381, (1985)
29. R.Hirano, *Inform Bull. Plankt. Jap., Commenmor. number of Dr.Matsue's 60th Birth.*, 25, (1967)
30. M.M.Watanabe, *Jap. J. Phycol. (Sourui)*, 31, 161, (1983)
31. S.Hatano, Y.Hara and M.Takahashi, *Jap. J. Phycol.(Sourui)*, 31, 263,(1983)
32. S.Yamochi and T.Abe, *Mar. Biol.*, 83, 255, (1984)
33. S.Iizuka, Report of Co-ope, Res., Minist. Educ. Sci. Cul., 91, (1987) (in Japanese)
34. T.Honjō, M.Asakawa and M.Yokote, Nippon Suisan Gakkai, Shūki-taikai Kōen-youshishū, 99, (1985) (in Japanese)
35. M.Furuki, M.Moriguchi and H.Kitamura, *Hakkōkogaku*, 63, 61, (1985)
36. S.Ishio, M.Kumehara and H.Kitamaru, Nippon Suisan Gakkai, Shūki-taikai Kōen-youshishū, 42, (1984) (in Japanese)
37. A.Tsuda and T.Nemoto, *Bull. Plankt. Soc. Jap.*, 31, 7. (1984)
38. K.Itoh and I.Imai, *Nansei Reg. Fish. Res. Lab.*, 20, 115, (1986)
39. S.Uye, *Mar. Biol.*, 92, 35, (1986)
40. K.Aoyama, Y.Konishi, Y.Iwata and H.Hara, Akashio Taisaku Gijutsu Kaihatsu-shiken Hōkokusho (Fisheries Agency), 1-199, (1984) (in Japanese)
41. Ōuchi, *Bull. Jap. Soc. Sci. Fish.*, 50, 1647, (1984)

TOXIC ALGAL BLOOMS AND RED TIDES: A GLOBAL PERSPECTIVE

DONALD M. ANDERSON
Woods Hole Oceanographic Institution, Woods Hole, MA 02543 USA

ABSTRACT

The literature on toxic algal blooms and red tides documents a global increase in the frequency, magnitude, and geographic extent of these events over the last two decades. Some of this increase is undoubtedly a result of the increased awareness and analytical capabilities of the scientific community, but a strong correlation between the number of red tides and the degree of coastal pollution or utilization of coastal waters for aquaculture argue that there are other contributing factors. It also appears likely that toxic algal species have spread within regions over spatial scales of hundreds of kilometers, moving with major water currents and storms. Long distance transport of species across oceans may have occurred as well, but the evidence is not conclusive and the hypothesis controversial.

BACKGROUND

Red tides, both toxic and non-toxic, have occurred throughout recorded history, but in recent years, there has been a global increase in the number of these events. This trend is most easily seen in the number of countries represented at international conferences on the subject. In 1974, the First International Conference on Toxic Dinoflagellate Blooms [1] had participants from 3 countries; the second conference in 1978 [2] represented 17 countries, and the third in the series in 1985 [3] had 22 participating countries. Now at this First International Symposium on Red Tides we have participants from 27 countries. At each meeting, outbreaks were reported for the first time at a variety of locations. In 1974, the major new concern was the massive 1972 New England red tide; in 1978, Paralytic Shellfish Poisoning (PSP) caused by Spanish mussels was recorded, as was a mysterious gastro-intestinal illness caused by Dutch mussels from clean, unpolluted waters; in 1985, toxic events were reported for the first time in the Faroe Islands, in Argentina, in Thailand and in Newfoundland Canada, and the gastro-intestinal illness first publicized in 1978 was shown to be of dinoflagellate origin and termed Diarrhetic Shellfish Poisoning (DSP). At the 1987 Symposium on Red Tides, more new outbreaks have been described, including brown tides in Rhode Island and New York that kill shellfish and submerged aquatic vegetation, or PSP in shellfish from Tasmania, Taiwan, Guatemala, Korea, Hong Kong, and Venezuela.

This sequence obviously neglects other events reported in the open literature (especially those concerning toxicity or mortalities associated with non-dinoflagellates) and is biased against contributions from Southeast Asia because of its geographic distance from the North American conferences, yet the overall impression is unmistakably one of a major expansion of the geographic extent, frequency, magnitude, and species complexity of red tides throughout the world. The obvious question is whether this expansion is truly occurring, or is instead a result of other factors such as the increased number of scientists working in the field or

RED TIDES: BIOLOGY, ENVIRONMENTAL SCIENCE, AND TOXICOLOGY
Okaichi, Anderson, and Nemoto, Editors

the increased sophistication and accuracy of analytical equipment. There is no easy answer to this question, but it is worthwhile to scrutinize the data more closely with these concerns in mind. Restricting our attention to toxic blooms because they are most reliably documented in the literature or in unrecorded folklore, the number of first occurrences or possible spreading events from the last two decades can be grouped into three categories. This compilation is by no means comprehensive, but does include examples of a variety of different phenomena.

Dispersal Within a Region

Included in this category are those events where there is reason to believe that a toxic algal species was present within a region from which it then expanded its geographic range on spatial scales of hundreds of kilometers. The best example is the 1972 New England red tide [1], which caused PSP along the southern New England coastline where there had been no previous records of shellfish toxicity. Every year since that event, PSP has occurred in that region, often in historically-popular clamming areas. Based on health records and local folklore as well as more sophisticated scientific analysis using isozyme electrophoresis [4], this is one case where it can be said with surety that spreading did occur in 1972 from established populations that had long caused toxicity in northeastern Canada and Maine. Similarly, in late 1975, a bloom of Pyrodinium bahamense in Papua, New Guinea (where toxic red tides were common events) was followed in early 1976 by the first reported PSP episode in Sabah and Brunei [5]. It has been suggested that this transport was via the Southern Equatorial Current, but a scarcity of plankton records and the uncertainties of interviews with inhabitants make it difficult to prove that this was a true spreading event. Nevertheless, a PSP outbreak in the Philippines in 1983 may represent further dispersal of P. bahamense within the region.

Introduction to a Region

This category includes events where the causative algal species may have been introduced to a region from far away - even across oceans. This is by far the most controversial category, since it is easier to argue that a species was present in a region at low concentrations and escaped detection than to prove that it was never there at all. One possible example of long-distance spreading involves Gyrodinium aureolum, a dinoflagellate which was not observed in an extensive series of plankton counts in the North Sea until 1966 [6]. Within two years, that species was blooming and causing fish kills that are a serious problem to the western European fish farming industry to this day. In northwest Spain, episodes of PSP began unexpectedly in 1976 in an area with an extensive mussel farming industry. The causative organism was Gymnodinium catenatum, which subsequently bloomed in 1985, 1986, and 1987. Phytoplankton counts dating back many years in that region and a long history of shellfish consumption and culture both suggest that G. catenatum is an introduced species. In 1985, this same species was linked to PSP in Tasmania, Australia for the first time, and as in Spain, there is speculation that the species was introduced. Plankton records suggest that G. catenatum was present in 1980, but apparently not before [7]. Originally described in 1942 in Mexico, this species has been reported in only four other widely-scattered locations - Argentina, Spain, Tasmania, and Japan.

Another new and disturbing event is the 1987 outbreak of PSP in Champerico, Guatemala that killed 26 people [8]. As usual, historical evidence for the complete absence of past toxicity in that region is not conclusive, but again there has been speculation that the problem is introduced. Here the important observation is that the causative organism is P. bahamense, a dinoflagellate reported in tropical waters of the western hemisphere but never directly associated with toxicity. Part of the uncertainty here is that there are two strains of P. bahamense, the toxic variety compressa and the non-toxic variety bahamense. Early reports of this species in Central and South America have typically specified variety bahamense or have not provided sufficient detail for a proper designation. In Champerico, P. bahamense var. compressa caused the PSP, so there is concern that the the toxic variety that has caused fatalities and economic hardship in Southeast Asia has now crossed the Pacific. Just as we saw an apparent dispersal of this organism between Papua, New Guinea and Sabah, Brunei, and the Philippines in the last ten years, will there now be a spreading of PSP in Central America and the Caribbean? It is a disturbing prospect.

Unusual Conditions

In this category, a toxic species may have been present in an area for years, but a unique set of conditions favorable for growth or accumulation allowed that species to fluorish at one point in time. A clear example is the 1982 outbreak of PSP in Newfoundland, Canada. Since this island is near areas in northeastern Canada that have a long history of PSP, shellfish in Newfoundland have been tested since the 1950's. Results were always negative until 1982 when PSP reached dangerous levels immediately after two weeks of warm, sunny weather - a rare event in that region. A reasonable conclusion [9] is that the causative organism (Protogonyaulax tamarensis) had been in those waters for years, but conditions had never been right for the cells to bloom. In 1983, an outbreak of PSP occurred in the Gulf of Thailand, caused by a species of dinoflagellate new to the growing list of toxin producers - Protogonyaulax cohorticula. The episode was sudden and unexpected, and has not recurred.

Perhaps the most important new occurrence among the list of unusual events is the "brown tide" that first appeared in parts of the northeastern United States in 1985 [10]. The organism is a previoulsy undescribed chrysophyte named Aureococcus anorexefferens. Its blooms are so dense that shellfish either stop filtering or retain very little food and starve to death. The blooms also block the sunlight and thus destroy eelgrass beds, an important habitat for scallops and other marine organisms. Aureococcus anorexefferens is such a tiny (2 µm) featureless organism that it is difficult to say whether it was ever present prior to the 1985 blooms. It has since bloomed in 1986 and 1987 as well and has devastated the multi-million dollar scallop resource on Long Island. Although this organism might have spread from elsewhere, it seems more likely that it was a minor component of the local phytoplankton assemblage that fluorished under unusual conditions. The only record of a similar occurrence is an unpublished report from France where oysters died of starvation during a bloom of a tiny, unnamed picoplankter that turned the water brown for several months [11].

Contributing Factors

There are a number of possible explanations for the increased number of reports of red tides or toxicity episodes throughout the world during the last 20 years. These can be summarized as follows:

1) The scientific community is better informed and more alert for the signs and symptoms of plankton blooms and toxicity episodes. Analytical capabilities have improved considerably as well, making it much easier to detect toxins. The 1985 PSP outbreak in Tasmania is an interesting issue in this context, since toxicity was detected shortly after a major marine

laboratory moved to the island from the mainland of Australia. Gymnodinium catenatum may still have been a recent introduction to the area during the years prior to that discovery, but there is no doubt that the prompt diagnosis of PSP and the isolation of the causative organism was in part due to the increased scientific scrutiny of local waters.

2) The use of coastal waters for shellfish and finfish farming can amplify an area's existing toxicity potential. Not only are the products of marine farms monitored and assayed very carefully, but the nutrient enrichment of local waters from the food and excretion of the cultured animals could stimulate red tide phytoplankton. The introduction of a shellfish or finfish resource could also make an existing toxic organism more noticeable. For example, over the years there had been sporadic reports of illnesses among humans eating Anadara and other shellfish from Balete Bay in the Philippines, but one year after the green mussel was first successfully cultured in the bay, PSP was detected [12]. Mussels are well-known for their ability to accumulate PSP toxins more rapidly than other filter feeders

3) Even though old plankton records do not document the existence of certain species, collection and preservation techniques were often inadequate and the resulting species identifications should be considered incomplete. The use of plankton nets with large mesh sizes or the preservation of samples with 10% formalin as was common in the past would have biased observations against smaller species or those that deform or are destroyed by strong fixatives (a particularly serious problem with naked or fragile flagellates such as G. aureolum or G. catenatum).

Dispersal Mechanisms

Although the above explanations can account for some new reports of toxicity, there is no doubt that certain spreading events are genuine. The dispersal of a species within a region by currents or other major water movements is not only possible but probable, the 1972 New England red tide being a noteworthy example. But how can we explain the introduction of a species to a region thousands of kilometers from its source? In this context, it must be stressed that many of our toxic red tide species form resting stages (e.g. Protogonyaulax species, G. catenatum, Chatonella antiqua, P. bahamense) and that these resistant cells can remain fully viable under harsh conditions. One possible transport mechanism is in the ballast water of ships. Cells taken into the ballast in one port could either survive as motile cells for short distances or as cysts for longer trips before being discharged into the water at the ship's destination. An excellent review of the importance of this mechanism in the biogeography of many marine animals is given by Carlton [13], and there is every reason to believe that toxic algal species have travelled in similar fashion. It is also important to note that since cysts are rapidly buried below the sediment surface where many do not germinate until they are resuspended, the introduction of a species to an area is often followed by years of recurrent outbreaks. It is easy for cyst-forming species to be introduced to a region, but their eventual disappearance is a long, gradual process.

The other side of the spreading or dispersal argument deserves discussion as well - namely that in an evolutionary sense, phytoplankton have had ample time to reach and inhabit suitable environmental niches throughout the world and thus that recent spreading is unlikely. Taylor [14] argues that similar environments in different oceans typically have the same general plankton assemblage. Whereas marine animals or other less primitive organisms may need man's inadvertant assistance to disperse to new regions, phytoplankton species may have already achieved a stable distribution.

Although these two points of view on spreading are difficult to reconcile in hindsight, there are ways to compare geographically-distant populations. For example, electrophoretic analysis of _P. tamarensis_ isolates from the east coast of the US and Canada revealed a genetically homogeneous population, consistent with the hypothesized recent dispersal from a common source during the 1972 New England red tide [4]. A similar analysis of isolates from the west coast showed far more diversity among strains [15], as would be expected given that region's long history of PSP. Although similar approaches would be highly informative when applied to isolates of _G. catenatum_ or _P. bahamense_ from distant locations to examine the spreading hypothesis, the truly definitive information requires molecular taxonomy using RNA sequencing or cloned DNA probes. Another method for studying species dispersal is based on the ability of both _P. bahamense_ and _G. catenatum_ to form resting cysts with resistant cell walls. This means that if one of these species was present in a region hundreds or thousands of years ago, the cyst walls would still be present in the sediment. A coring study that examined sediments below the well-mixed zone where bioturbation activity is high could thus determine whether a species was newly introduced to a region or not.

Long-Term Trends

In addition to the spreading mechanisms and contributing factors discussed above, we cannot ignore the possibility that long-term trends in pollution or natural environmental parameters can be major factors in that increase. It is now firmly established that there is a direct correlation between the number of red tides and the extent of coastal pollution, measured either as the chemical oxygen demand of effluents as in Japan [16] or the population density in a watershed as in Hong Kong [17]. The reason that some noxious or toxic phytoplankton dominate other species in polluted waters remains a mystery, and it is not clear whether this correlation holds for PSP-producing species such as _P. tamarensis_ or _P. bahamense_, but the fact remains that man's activities in the coastal zone can directly affect the incidence of red tides. There may, however, also be natural environmental trends that affect the growth and dominance of certain phytoplankton. For example, a long-term dataset from the central north Pacific Ocean documents a statistically-significant doubling in the chlorophyll content of the mixed layer over the last 20 years [18], presumably due to a decrease in sea surface temperature. In coastal waters, data from Narragansett Bay, Rhode Island since 1959 [19] document: 1) an increase in the annual average water temperature by almost one degree Celsius; 2) a 10% decrease in the annual average wind speed; 3) a corresponding decrease in the depth of light penetration in the water column; and 4) an increase in the abundance of non-diatom phytoplankton, especially since the late 1970's. These data suggest that global environmental trends may be superimposed on effects due to pollution, spreading, or other factors. To detect such trends, we need long-term datasets similar to those described above. One useful parameter in such monitoring efforts would be shellfish toxicity since regular assays at the same stations can reveal important trends (such as the surprising increases in PSP toxicity in eastern Canadian shellfish over the last several decades [20]). Toxin monitoring efforts could be included in a global "Mussel Watch" program similar to that established in the US to detect trends in trace metal and organic pollutants.

At a practical level, there is little we can do to prevent the spreading of species by currents and major water movements, nor can we expect to regulate ship ballast or other accidental dispersal mechanisms. However, it is possible to control pollutants, especially when the increasing political power of aquaculture interests are brought to bear on the problem, but here again, the economic pressures on under-developed countries may preclude expensive pollution control strategies. What is needed is a way to help countries anticipate and manage problems from algal blooms when they arise. Existing techniques make it relatively easy to conduct surveys on a regional level to find the cysts or motile stages of dangerous organisms, a precaution that would be most useful in areas where fisheries developments are being planned or where there is an existing industry worth protecting. The technology and expertise for such studies exists, but is often not found in the countries most needful of the information. We clearly need a degree of formal international cooperation and training that does not exist at present. Several international working groups on red tides have been established in recent years, and attempts are underway at the Intergovernmental Oceanographic Commission to coordinate red tide training and information exchange even further. These are worthwhile efforts, but we must do more. The red tide problem is a global issue and should be formally recognized as such. If we continue to conduct our research in relative isolation, communicating mainly through the literature or at conferences, we may miss the opportunity to document and respond to immensely important trends or spreading events.

Acknowledgements

Research supported in part by the National Science Foundation (OCE86-14210) and by the Office of Sea Grant, National Oceanic and Atmospheric Administration through a grant to the Woods Hole Oceanographic Institution (NA86AA-D-SG090, R/B-76). Contribution No. 6669 from the Woods Hole Oceanographic Institution.

REFERENCES

1. V.R. LoCicero. Toxic Dinoflagellate Blooms, Elsevier, New York. 541 pp. (1974).
2. D.L. Taylor and H.H. Seliger. Toxic Dinoflagellate Blooms, Elsevier, New York. 505 pp. (1979).
3. D.M. Anderson, A.W. White and D.G. Baden. Toxic Dinoflagellates, Elsevier, New York, 561 pp. (1985).
4. B.A. Hayhome, D.M. Anderson, D.M. Kulis and D.J. Whitten. Submitted manuscript.
5. J.L. MacLean. In: Toxic Red Tides and Shellfish Toxicity in Southeast Asia, A.W. White, M. Anraku and K.K. Hooi, eds. Southeast Asian Fisheries Development Center. 1984. p. 92.
6. G. Boalch, personal communication.
7. A.W. White. In: The Impact of Toxic Algae on Mariculture. In press.
8. F. Rosales-Loessener. This volume. 1988.
9. D.R.L. White and A.W. White. In: Toxic Dinoflagellates. Elsevier, New York. p. 511 (1985).
10. T.J. Smayda. This volume. 1988.
11. P. Lassus. Personal communication.
12. R.Q. Gacutan, M.Y. Tabbu, E.J. Aujero and F. Icatlo. Mar. Biol. 87, 223 (1985).
13. J.T. Carlton. Oceanogr. Mar. Biol. Ann. Rev. 23, 313 (1985).
14. F.J.R. Taylor. In: The Biology of Dinoflagellates. Blackwell Scientific, Oxford. p. 399 (1987).
15. A.D. Cembella and F.J.R. Taylor. Biochem. System Ecol. 14, 311 (1986).
16. M. Murakawa. Oceanus 30, 55, (1987).
17. K. Lam. This volume. 1988.
18. E.L. Venrick, J.A. McGowan, D.R. Cayan and T.L. Hayward. Science. 238, 70 (1987).
19. T.J. Smayda, unpublished data.
20. A.W. White. Can. Tech. Rep. of Fish and Aquat. Sci. No. 1064. Biological
Stn., St. Andrews. 1982.

TOXICOLOGY AND PHARMACOLOGY OF RED TIDES: AN OVERVIEW

Y. SHIMIZU
Department of Pharmacognosy and Environmental Health Sciences,
College of Pharmacy, The University of Rhode Island
Kingston, Rhode Island, U. S. A. 02881

ABSTRACT

Recent progress in pharmacological and biochemical studies of the well-known red tide organisms, *Gonyaulax (=Protogonyaulax, Alexandrium)* spp. and *Gymnodinium breve (=Ptychodiscus brevis)* will be reviewed. Discussion will also be presented as to the toxicity of other deleterious red tide organisms which cause serious economic and environmental problems.

INTRODUCTION

There are now many microalgal organisms known to produce toxins. The list seems to be growing constantly. However, not all of them form the blooms which show the general appearance of "red tides". In this paper, only the toxins of selected organisms which form classical red tides with immense deleterious effects will be subjects of discussion.

Red Tide Toxins to Cause Paralytic Shellfish Poisonings

Paralytic shellfish poisoning (PSP) is the first health hazard found to be related to the red tides. The problem is ubiquitous and found in both temperate and tropical waters. The dinoflagellate, *Gonyaulax catenella* was first recognized as the causative organism. Now, more than a half dozen organisms including a fresh-water blue-green alga are know to produce the related toxins.

Saxitoxin, which was isolated in a large quantity from Alaska butter clam, *Saxidomus giganteus* in 1957, was the sole known toxin responsible for the poisoning. In 1975, four new toxins, gonyautoxin-I, -II, -III, -IV were isolated from the soft-shell clam, *Mya arenaria* exposed to a massive red tide caused by *Gonyaulax tamarensis* in Massachusetts. Since then, more than a dozen new toxins have been isolated from various biological specimens and causative organisms and their structures established. This finding of toxin heterogeneity opened a new dimension of PSP research.

First, observed differences in the specific toxicity and bioavailability of the toxins cast serious questions about the validity of the on-going PSP montioring program which depends on the mouse bioassay using saxitoxin as the standard. Of particular concern are the cryptic toxins bearing a sulfate group on the carbamoyl group. This group of compounds typically exhibits toxicity less than one-tenth in the mouse ip test or isolated nerve system. The sulfate moiety is easily hydrolyzed with a dilute acid solution resulting in the production of the normal toxins. The behavior of the toxins in the gastro-intestinal system, especially interactions with the intestinal flora should be investigated.

Variations of toxin profile in different PSP samples seem to necessitate a more precise assay method which will provide precise toxin contents and toxicity calculated from the toxicity of individual toxins. Such a system can be easily developed by adding a simple computer program

Published 1989 by Elsevier Science Publishing Co., Inc.
RED TIDES: BIOLOGY, ENVIRONMENTAL SCIENCE, AND TOXICOLOGY
Okaichi, Anderson, and Nemoto, Editors

to the HPLC system developed by Sullivan.

The most important contribution of the new PSP toxins has taken place in the electrophysiological study of sodium channels. Tetrodotoxin and saxitoxin had been known to be specific blockers of sodium channels and used as important tools to investigate the excitable membranes. When the new toxins were discovered in 1975, it was immediately recognized that those compounds could be valuable molecular probes for pharmacology research. The structure variations could provide information regarding the toxin receptors and the topology surrounding the sodium channels. The electrophysiology and receptor binding studies of the new toxins have been carried out by such researchers as Narahashi, Strichartz and Kao. Meanwhile, physicochemical properties of the toxin molecules pertinent to the interpretation of pharmacological experiments have been elucidated.

The results confirmed the imidazoline quanidinium group as an active site as assumed earlier by Hille and other researchers. The results also suggested a need for a revision of the classical toxin-receptor interaction model proposed by Hille in 1975, which depicts the toxin molecule plugged into the channel by ion pairing and other interactions. The analysis of observed activities of toxins having various fuctionalities clearly excludes such a model. Alternate models have been proposed in which the toxin molecule interacts with the surface or shallow cavity near the channel either by noncovalent or covalent bonding. The exact nature of the binding is still a subject of intensive discussion.

Another very important aspect of the toxin-sodium channel research is the use of the toxins as molecular probes to differentiate types of sodium channels. Sodium channels play a major role in the transmission of nerve impulse. The presence of different channels in different tissues had been recognized earlier. The recent structural specificity of the toxins is expected to be an important parameter in such studies. The recognition of channel heterogeneity and varied function will provide informatin critical for the understanding of disease states and drug action.

Two other important developments seen in the PSP research are the elucidation of the biosynthetic pathway and the preparation of antibodies of saxitoxin. The author's group has recently shown the molecular origin of PSP toxins and some elaborate mechanisms involved in the molecular building. Kodama's group reported that an endosymbiotic bacterium is the actual toxigenic organism. A similar discovery is reported by Yasumoto's group.

Past attemps to prepare antibodies for PSP toxins were troubled with low titer and specificity. Recent reports, however, seem to be more promising. Especially, Davio's antibody gave protection to mice from saxitoxin. The actual use of the antisera for assay and therapeutic purposes still requires further investigation of corss-reactivity and adversary reactions. One obvious possibility will be the selection of the most desirable antibody by the monoclonal method.

Gymnodinium breve Red Tide

This red tide in the Gulf of Mexico continues to be a serious menace causing large scale fish kills, threatening the red fish and oyster resources important to the local economy. Also, once a red tide occurs, the resort beaches are often devasted by rotting, dead fish.

The organism was brought into culture by Wilson and as far as this author knows, all in-depth toxin studies have been carried out using this

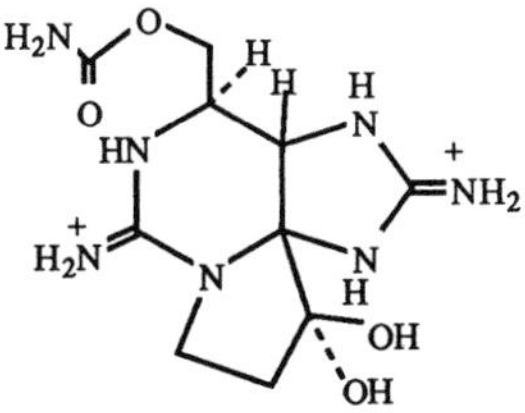

Saxitoxin

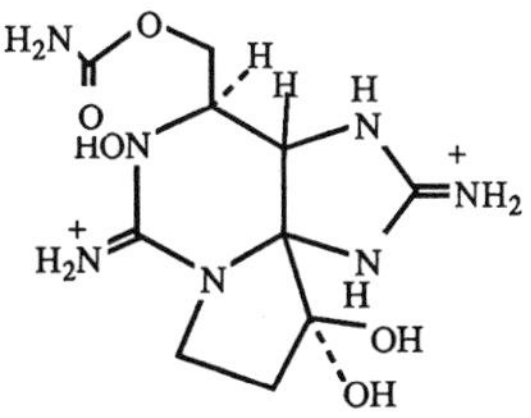

Neosaxitoxin

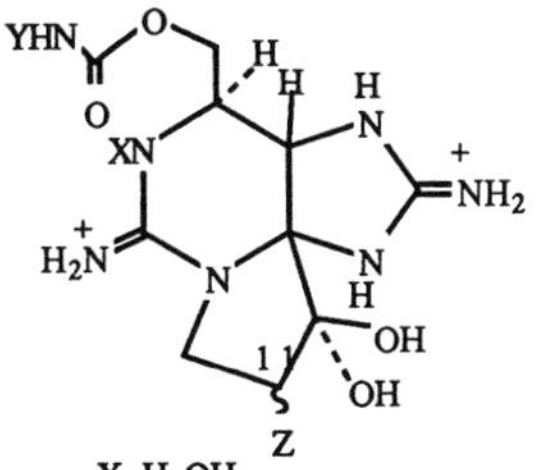

X=H, OH
Y= H, SO_3H
Z=H, OSO_3H

ALL COMBINATIONS OF X,Y,Z AND 11-EPIMERS.

Gonyautoxins

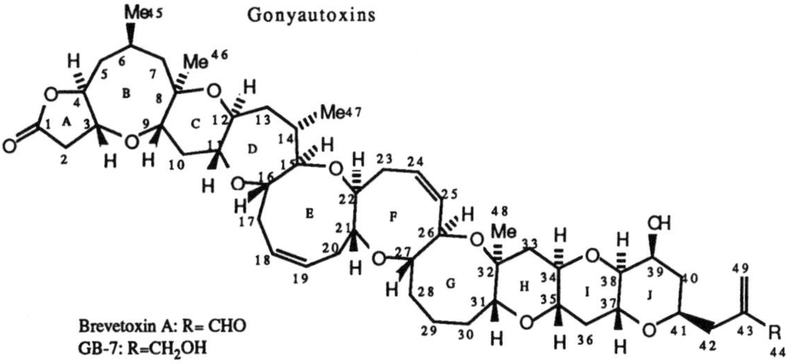

Brevetoxin A: R= CHO
GB-7: R=CH_2OH

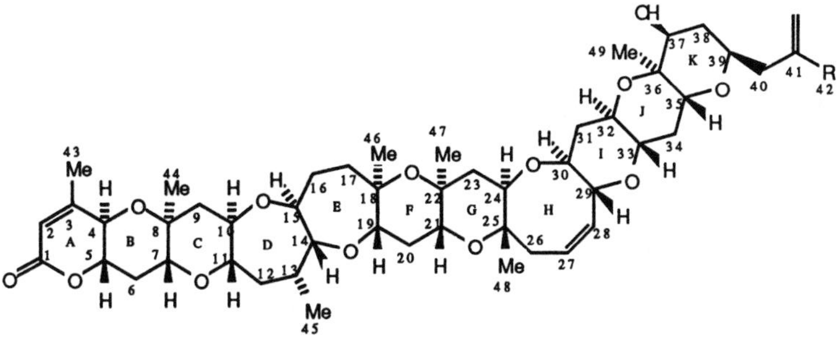

Brevetoxin B: R=CHO
GB-3: R=CH_2OH
GB-6: R=CHO, 27,28beta-epoxide

Wilson strain. Thus there is no assurance that the exact toxin profile of the wild cells matches that reported from the cultured cells. This is one important aspect of *G. breve* toxin research which needs further investigation.

The earlier stage of the toxin research was marred by confusions. The first reliable report is that of Alam and Trieff, who named two different toxic bands on preparative TLC as T_1 and T_2 for toxin-1 and toxin-2. T_2 was further purified to a chromatographically and spectoscopically pure compound and reported by the author's group in 1974. From the high-field NMR, MS, IR and UV spectra, the toxin, specified as GB-2 toxin (*G. breve* toxin-2) to avoid confusion with trichothecene toxins, was determined to be a polycyclic ether of molecular weight ~800 having an aldehyde group, a conjugated carbonyl, a terminal methylene and several methyl groups. The identical compound was re-isolated and reported under different code names by two groups in 1978 at the 2nd International Conference on Toxic Dinoflagellates. In 1980, the X-ray determined structure of GB-2 was reported under the name, brevetoxin-B. In the same report, the toxin, which corresponds to GB-1 toxin, was referred to as brevetoxin-A. Confusion over the nomenclature of the *G. breve* toxins was further intensified by the renaming of the organism to *Ptychodiscus brevis*, which is now questioned by most taxonomists. Although GB-1 and GB-2 are the first given names, the author's group agreed to use brevetoxin-A and -B, and also suggests the use of abbreviations, GBTX-A, -B, -3 etc. The previously used BTX is confused with the prominent neurotoxin, batrachotoxin.

A total of eight toxins have been isolated and their structures determined. All of them are potent ichthyotoxins and active at concentrations of ca. 10^9 µg/ml. The pharmacology of brevetoxin-A, -B and GB-3 toxin have been studied by several groups. The toxins are selective depolarizers of sodium channels, like veratridine or sea anemone toxins. It is speculated that they penetrate into the membrane and act on the hydrophobic portion around the gating control region of the channel. The presence of conjugated aldehyde is not a requirement for the activity, and the shape of the toxin molecule seems to play the key role. In that respect, our finding that the sheet-like structure of the toxin can be twisted at the eight-membered ring may be an important factor for the toxin binding. The presence of an eight-membered ring,whose conformation changes with a very small energy barrier (ca. 2 Kcal), is a common structural feature in both brevetoxin-A and brevetoxin-B series.

Gyrodinium aureolum Red Tide

This oganism is known to form large red tides along the British and Scandinavian coasts. The red tide also causes massive fish kills and damage to fish aquaculture. The situation resembles of *G. breve*, a taxonomically close species. Attempts to pinpoint the toxic component in the organism, however, have so far failed. In collaborative research with Tangen, the author's group examined the extracts of cells cultured in both Norway and Rhode Island. We could detect neither toxin nor ichthyotoxicity in the extracts. This result is very puzzling in view of the extreme virulence shown by the organism in the field.

Brown Tide

This new species in red tides poses probably the most serious threat to the northeast US coast. The widespread bloom caused the devastation of many resources in the region. For example, it wiped out the wild *Mytilus edulis* population and caused 40% mortality at a mussel aquaculture facility in Rhode Island. The details of the brown tide are reviewed by Smayda in this conference.

The toxic symptoms in the mussels were the thining of the flesh, opening of the shell and death. The toxin of the causative organism, which is now identified as *Aureococcus anorexefferens*, is not known. However, in our attempt to identify the toxin in the wild cells, we observed the release of an unusually large amount of sulfide from the dying organisms. Therefore, sulfides derived from the ingested organisms might have been the direct cause of the shellfish deaths. If so, it is comparable to deaths caused by sulfate reducing bacteria.

Conclusion

Of the major red tides which produce serious health and economic impact, good progress has been made with PSP and *Gymnodinium breve* toxins. The isolated toxins have also proved to be important tools in pharmacological research. However, almost nothing is known about the cause of toxicity of *Gyrodinium aureolum* and *Aureococcus* sp. despite efforts by many researchers. The virulence of these organisms against marine organisms is unparalleled with that of other red tides and difficult to explain by simple oxygen deficiency or physical disturbances. Chloromonads represented by *Heterosigma akashiwo*, which is the most formidable fish-killer in the Inland Sea of Japan, may be also added to this list of the most-wanted. The problems are not easy to solve in view of the long history of the research. What is needed here is a close collaboration of scientists in different disciplines, such as pathology, biochemistry and organic chemistry.

Acknowledgement

Support for the research by NIH grants and travel fund by the Naito Foundation are greatly appreciated.

HOMAGE TO THE INTERNATIONAL SYMPOSIUM ON RED TIDES: THE SCIENTIFIC COMING OF AGE OF RESEARCH ON AKASHIWO; ALGAL BLOOMS; FLOS-AQUAE; TSVETENIE VODY; WASSERBLÜTE

SMAYDA, T. J., Graduate School of Oceanography, University of Rhode Island, Kingston, RI 02881 USA

Professor Nemoto kindly asked last night if I would try to summarize the scientific product of our Symposium and provide some perspective. I was reluctant to accept, since I knew the extent to which I selected from among the > 125 papers and posters for concentration. Hence I feared that my synthesis would overly reflect my primary research interests at the expense of a balanced overview. Professor Nemoto prevailed. Now, with renewed hesitancy I offer the following overview, not as a _rapporteur_ of this remarkable International Symposium on Red Tides, but as the openly acknowledged and biased focus of a physiological ecologist. To those whose work or research areas I may have neglected or collectivized into anonymity because of the limitations of space and/or my limited qualifications to comment on such work, I apologize for this unintended slight. Others undoubtedly would have different impressions or emphasize different aspects of the Symposium.

First, a basic feeling: this is the most extraordinary phytoplankton meeting that I have ever attended. The vivid scientific ferment evident in the numerous, excellent presentations collectively (may I add finally) have brought research on nuisance blooms in the sea into the mainstream of phytoplankton ecology. Their anecdotal treatment and relegation as scientifically curious, enigmatic, and episodic blooms of rogue species, which may discolor seawater and are sometimes accompanied by commercial and human loss, are approaches of the past. Papers ranging over the gamut of organismal, toxicological and ecological approaches have been presented; from taxonomy to molecular biology, to nutritional requirements, to toxins, to modeling; even global dynamics. The presentation of high quality research; the application of new techniques; a focus on ecological first principles and holistic approaches, and application of sound research procedures were commonplace during this Symposium - a sure sign of the coming-of-age of a discipline. Collectively, these scientific presentations yielded a most remarkable scientific spinoff of the International Symposium on Red Tides - the gelling of a discipline. This notable achievement is particularly due to the impressive organizational skills, dedicated attention to detail and personal qualities of Professors Tomotoshi Okaichi and Takahisa Nemoto. We collectively owe them much gratitude, recognition and a mighty thank you for their efforts. They, by assembling us, made both this Symposium and its spinoff possible, a dual event which should be acclaimed by future historians of Biological Oceanography and readers of the Proceedings of this Symposium.

Now, some scientific impressions and comments:

I. TAXONOMY

The extent to which dinoflagellates are under- and over-classified, and the identity and taxonomic utility of conservative morphological features were addressed by Drs. Gaines, Taylor and Steidinger. Their treatment of these basic uncertainties put into clearer perspective the basis of the present, widespread taxonomic confusion hampering our efforts, including such unresolved issues as: Are _Gyrodinium aureolum_ and _Gymnodinium nagasakiense_ conspecific? Are _Olisthodiscus_ and _Heterosigma_ the same taxon? Is _Aureococcus anorexefferens_ really a _Pelagococcus_? And is it _Alexandrium tamarense_ or _Protogonyaulax tamarensis_? The need to supplement classical (morphologi-

RED TIDES: BIOLOGY, ENVIRONMENTAL SCIENCE, AND TOXICOLOGY
Okaichi, Anderson, and Nemoto, Editors

cal) criteria with new approaches is both clearly evident and affirmed by the provocative analyses of isozyme patterns by Drs. Sako, Uchida and Ishida. They not only showed the potential value of their approach to resolve taxonomic problems, but demonstrated the high, though variable, degree of genetic polymorphism that occurs within dinoflagellate taxa. Such clonal variation recalls Dr. Shimizu's observations on progressive toxin attenuation in more southerly clones of Protogonyaulax tamarensis and Braarud's [1] demonstration of physiological clones in dinoflagellates. The extent to which ecomorphs generally characterize bloom species, similar to that established for the diatoms Thalassiosira gravida - Thalassiosira rotula [2] and the dinoflagellate genera Dinophysis and Peridinium [3,4], must also be further defined. We must take greater cognizance of the considerable evidence that physiological, genetic and morphological variations characterize phytoplankton species generally, including nuisance bloom taxa. Elucidation of such clonal variations is fundamental for a proper taxonomic and ecological understanding.

The growing reliance on minute, morphological elements and anticipated molecular and genetic approaches for proper taxonomic identification increasingly require highly specialized skills, which poses other problems. That is, toxicologists and ecologists must increasingly depend upon taxonomic specialists for proper clonal characterizations and species identification; including identification of obscure nannoplanktonic representatives whose blooms appear to be more commonplace. Hence, a system of archiving material and clonal culture maintenance to ensure proper identification, including hindsight identifications, by taxonomic specialists seems called for.

II. DISTRIBUTION, DISPERSAL, OCCURRENCE

A fundamental revision in our scientific approach to red-tides and toxic algal blooms is required, based on Dr. Anderson's excellent review and convincing conclusions. (I shall use red-tides as an expression of nuisance algal blooms generally.) There is a globally significant increase in their frequency of occurrence accompanied by regional spreading and involving more species. Differences in bloom species between regions are merely secondary manifestations of the basic phenomenon: the global increase in such blooms. More than 25 papers reporting such bloom events in the coastal waters of all continents, ranging from boreal to tropical regions and involving taxa from several phylogenetic groups, were presented at this Symposium. They provide additional support for Dr. Anderson's conclusion which now begs the question: What is triggering these blooms? One possibility, of course, is that new species' introductions are filling "open niches" leading to their blooms. The explosive growth and persistence of the diatoms Biddulphia sinensis and Coscinodiscus nobilis upon introduction into the North Sea [5] and of Rhizosolenia calcar-avis into the Caspian Sea [6] are well known. The spreading of the toxic Gymnodinium catenatum reported here by Dr. Hallegraeff and coworkers may be such an example. However, as Dr. Anderson pointed out, many of the causative bloom species are often normal components of the local community. This has focused attention on the potential role of anthropogenic factors as the triggering mechanisms. Dr. Lam's provocative data set from Tolo Harbour, Hong Kong, revealed a long-term increase in nitrogen and phosphorus levels accompanied by increased dinoflagellate abundance and frequency of red-tide outbreaks. This implicit concept of an anthropogenic trigger seems to be the favored notion. Specifically, it is thought to involve factor(s) accompanying nutrient enrichment, i.e., a nutrient and/or water quality trigger.

I am not ready to embrace this view, despite the widespread, provocative evidence. It is striking to me that shellfish and finfish aquacultural activities are often followed by blooms of both benign and toxic algal species. I recall the 1976 Ceratium tripos bloom off New York which resulted in the anoxic die-off of finfish and shellfish [7]. Ceratium is frequently ungrazed because of its large size. Let us also remember that aquacultural activities augment and modify the local grazing structure, as do benthic and pelagic fishery operations. I suggest that the attendant, altered grazing structure and pathways have not as yet been ruled out as triggers of such algal blooms. The occurrence of differential grazing effects on red-tide producers were reported on by Drs. Iizuka, Kim, Kishi and co-workers. Recall that an altered grazing structure can also result from the selective effects of nutrient enrichment on the composition and size of phytoplankton species, and even on the grazers themselves. In this way, the effect of nutrients on blooms may be indirect in primarily influencing nutrition of the grazers, and not directly, primarily or solely due to nutritional stimulation of the bloom phytoplankters themselves.

Regardless of the specific interplay between nutrients and grazing as bloom triggers, the focus on local anthropogenic effects should be tempered; even broadened. To what extent does the global epidemic of algal blooms, notably dinoflagellate blooms, reflect natural, long-term cyclical patterns in biotic and environmental variability analogous to the Russell Cycle, well known for the North Sea [8], or even longer time periods (> 1000 years)? Is there a linkage with subtle, global climatological changes ("greenhouse" effect; El Nino events) which are becoming stimulatory to algal bloom development? At issue is the extent to which this global increase in blooms is a manifestation of natural, planetary cycles which are presently convergent with, and further augmented by local anthropogenic stimuli. It is essential to resolve this issue. Experimental evaluation of anthropogenic effects is possible; validation that natural, cyclical events are primarily responsible for these blooms is not tractable experimentally. Our ability to forecast and control such blooms depends upon resolution of the roles of anthropogenic and cyclical (= climatological, geological epoch) events as bloom determinants.

III. BLOOM REGULATION

Numerous papers evaluated the specific triggering events and factors regulating subsequent bloom development. We learned of the need to distinguish between the seeding of bloom species and their actual blooming. For those species producing a cyst, the bloom trigger is a two-step process. The first is a morphogenetic event: excystment producing pelagic vegetative cells; the second, triggering of vegetative cell growth. Dr. Imai and co-workers demonstrated the role of temperature on germination/dormancy of the ichthyotoxic rhaphidophycean, Chattonella.

Encystment leading to renewed dormancy probably requires a different environmental trigger and can be quite involved, as Drs. Inoue and Fukuyo have shown for the phagotrophic dinoflagellate Protoperidinium leonis. Collectively, the various reports convincingly established the importance of life cycle as a major factor in bloom events of over-wintering species. The discovery by Anderson and Keafer (9) that excystment of some dinoflagellates may be under endogenous control further suggests that this morphogenetic step and subsequent seeding result from careful environmental monitoring by the cysts, including cuing reactions, rather than from fortuitous, stochastic events.

Bloom species which do not have a dormant stage, such as the dinoflagellate genera Ceratium and Prorocentrum, obviously don't depend upon morphogenetic triggers. Their mechanisms of overwintering and seeding consequently must differ. Dr. Franks and Dr. Anderson showed how physical conditions could accumulate such blooms and provide seed populations. Taken together with previous well known reports by Dr. Holligan and co-workers, we must now accept that both "physical-forcing" and excystment mechanisms provide seed populations for bloom development. There is not a common mechanism.

An impressive set of observations from different regions on different species vividly showed the astonishing degree of complexity of bloom dynamics, its regulation, and our need to revise some basic paradigms. The actual blooming of a species, once seeded, must now be treated as a separate event most likely triggered by factors other than those regulating the seeding itself. Hence, bloom phenomena are minimally two-trigger events. The data also collectively suggest that chemical factors having both nutritional and water quality effects are significant growth regulators. Moreover, two basic nutrient forms and patterns of regulation must be acknowledged: regulation by macro-nutrients and regulation by micro-nutrients. Macro-nutrients in the form of nitrogen and phosphorus appear to regulate biomass or environmental carrying capacity, i.e. are yield-regulating. They may also regulate growth rates and sexuality, as shown by Dr. Tomas et al. Evidence that the intensity and duration of a bloom are linked to macro-nutrient availability and/or buildup was presented, for example, for Tolo Harbour (Dr. Lam) and the Seto Inland Sea (Dr. Nakamura et al. and Professor Okaichi's group). Micro-nutrients, in contrast, appear to regulate species occurrences, growth rates, competitive ability and perhaps community structure. Professor Okaichi and his excellent group drew our attention to the role of iron; Dr. Ishimaru to selenium; Dr. Fuse and co-workers to iodide and iodate; Drs. Nishijima and Hata to vitamin B_{12} and Dr. Graneli to effects of complexation compounds such as humic acids. Collectively, these results suggest the riskiness of applying Liebig's Law of the Minimum to algal bloom events. Multiple, concurrent nutrient regulation of algal blooms in terms of biomass, species growth rates, competition and community structure seems more probable.

Our collective focus on searching only for a growth stimulant(s) of algal bloom events seems too restrictive to me; this approach rules much of our conceptual thinking and research protocol. The many reports of bloom developments presented here and elsewhere suggest, however, the co-occurrence of a simultaneous repression of species that ordinarily would then be in bloom, or expected to take advantage of increased nutrients in regions where the latter stimulate blooms. Are unusual algal blooms, then, usually accompanied by species-specific growth repressors (= chemical) and/or abundance repressors (= grazers) in addition to the presumed triggers (= stimulants) of the bloom species? Future studies need to evaluate the growth characteristics of the non-blooming components of the phytoplankton community using such a stimulant-repressor paradigm: are these non-blooming taxa replaced? merely out-competed? Future studies must also evaluate species interaction effects along the lines of Pratt's [10] _Olisthodiscus luteus_ - _Skeletonema costatum_ co-action responses and Dr. Iwasaki's [11] important studies.

Clearly, unraveling the environmental regulation of species' blooms is heavily dependent upon autecological studies. Dr. Iwasaki presented an impressive overview, rooted strongly in experimental autecological principals, of our contemporary knowledge of bloom species in Japanese coastal waters. The great need for experimental autecological studies on bloom-producing species implicit in his review is increasingly recognized; scientific progress in this area was very evident. Drs. Watanabe, Hirayama and co-workers at Tsukuba and Nagasaki focused our attention on the role of phosphorus - that neglected nutrient. Dr. Tomas et al. reported on the effects of temperature-nutrient interactions on growth rate; their results convincingly demonstrating the need for such experimentation. The applications and refinement of newer autecological techniques were likewise impressive. Dr. Clarice Yentsch outlined the applicability of flow-cytometry and cell-sorting as a new, sensitive and highly promising autecological technique and approach. The *in situ* species-specific productivity measurements by Han and co-workers at the Ocean Research Institute represent a major advance in field techniques.

Phototaxic and migratory behavior were the autecological aspects that received the most attention. Biophysical aspects such as the influence of temperature, cell size and chain-length on swimming speed, and the effect of cell membrane charge on motility were discussed by Drs. Kamykowski, Fraga and Miyagi. The effects of inorganic nitrogen species on motility were demonstrated by Dr. Nakamura and co-workers. Drs. Takahashi and Watanabe and co-workers presented provocative contributions on diel migratory patterns of *Heterosigma akashiwo* and their affect on growth rate and phosphorus uptake. These and other reports clearly indicate that swimming behavior is more than a response to light; it serves several fundamental ecological roles. It influences retention within an area; it is a mechanism of depth keeping; it is sensitive to the nutrient field; it facilitates accumulation. These, in turn, influence seeding and bloom dynamics. Motility, *per se*, however, does not appear to be a mode of nutrient uptake [12].

There is seemingly a paradox associated with the phototaxic and migratory behavior of motile bloom-species. That is, their aggregative tendencies might be expected to aggravate their competition for nutrients (whose concentrations influence flagellate motility). Thus, what is the selective advantage of such a behaviorism in the nutrient-limited aquatic milieu? Permit me a wild speculation! The very high sensitivity of dinoflagellates (relative to diatoms) to trace metals, particularly Cu, is well-known. (The production of metallothionein as a cellular-detoxification mechanism, and whose induction in a diatom was reported by Dr. Maita and co-workers at this Symposium, occurs in certain organisms.) I recall a paper written by Professor Ishida [13] some years ago in which he documents the secretion of various volatile sulfur compounds, including H_2S, from dinoflagellates, compounds capable of complexing trace metals such as Cu. I recall also the stimulatory effects of sulfide pulp wastes on the bloom species *Eutreptiella* sp. and *Prorocentrum triestinum* reported by Okaichi and Yagyu [14], and Iizuka and Nakashima's [15] observations that sodium sulfide stimulated the photosynthesis of red-tide dinoflagellates. Sulphur-containing compounds appear to function as growth regulators of bloom-forming flagellates! Is it possible that aggregation of dinoflagellates, for example, resulting from phototaxis or pattern-swimming, or even physically forced, is an environmental (= habitat) detoxification strategy? That is, does the aggregation of a critical number of growing cells, which may secrete trace metal complexation substances, initially and then through increased population density progressively increase the concentrations of such complexing compounds in the microhabitats surrounding the cells? Do these then detoxify and/or poise water-quality triggering both initial growth and allowing subsequent sustained growth? Motility itself could provide the mechanism of finding nutrient micro-patches and also facilitate nutrient diffusion into the nutrient free zone surrounding the cells. Thus, nutrient acquisition and competitive uptake may be less of a problem than trace metal toxicity. As Professor Okaichi and Yagyu [14] pointed out, the chelating activity of such sulphides may be too weak or variable, however, reducing their metal detoxification potential. The fact remains that sulphides appear to stimulate dinoflagellate and euglenid growth [14] and dinoflagellates secrete such compounds [13]. Whatever the exact mechanism of this proposed auto-stimulation of growth, the commonly observed cellular aggregation of flagellates possibly represents a mechanism facilitating a chemically-induced environmental improvement beneficial at both cellular and population levels.

IV. SURVEILLANCE, MONITORING, MODELING

Algal blooms are usually detected after the fact - signaled by water discoloration, sometimes die-offs. Rarely are field analyses in progress at the time of their inception to assess associated triggering factors. Our predictive capability is, thus, very limited. Several papers presented applications of remote sensing techniques based on pigment levels, including use of a novel combination of satellites, aircraft and balloons proposed by Dr. Charles Yentsch. Routine, effective aerial surveillance of bloom development over the requisite short time scales and large spatial coverage now seems imminent.

There is also a need for biological surveillance, an area in which much remains to be done. Monitoring for dinoflagellate cyst occurrences prior to the establishment of maricultural activities was one approach suggested at the Symposium. There is also need to develop bioassay techniques to test the growth-supporting potential of the water column for indigenous red-tide producers. While most of the species are now available in culture, an appropriate bioassay protocol needs development. The papers by Drs. Okamoto and Hirano on enrichment bioassays and Dr. Kogure and co-workers on tissue culture assays hopefully will stimulate such development.

Predictability of bloom outbreaks has high priority, particularly in aquacultural areas. At this stage of our knowledge, our best hopes for predictability are for localized areas. The development of predictive capacity depends upon refinement of phenomenological techniques such as aerial and biological assay surveillance and, where suitable data sets exist, seeking and applying statistical correlations between previous bloom events and environmental parameters. This latter approach ideally should be carried out in combination with monitoring buoys, such as described by Dr. Hiiro and co-workers. Dr. Iwasaki reported the effective use of discriminant analysis techniques employed by Drs. Aoyama, Ôuchi and co-workers to predict red-tide outbreaks of *Chattonella* in Harima-Nada and Hiroshima Bay based on temperature, salinity and nutrient deviations. Dr. Sekiguchi and co-workers reported on similar long-term environmental correlates and toxic *Protogonyaulax tamarensis* blooms in Ofunato Bay. These studies should encourage similar efforts elsewhere, where data sets allow.

Ultimately, however, effective modeling of algal blooms will provide greater predictive capability. A major, exciting development at this Symposium was the numerous modeling efforts presented by Japanese scientists. The inappropriate and unsuccessful attempts to use a diatom template to model red-tide and other non-diatomaceous blooms in the sea have finally been abandoned, and new insights reached. The keen awareness of the role of life cycles and migration (where appropriate), increased availability of quantitative data on the growth and nutritional kinetics of red-tide species and other autecological data have led to this success. The summer blooms of *Chattonella antiqua* in the Seto Inland Sea have been evaluated by the chemical environments model of Dr. Nakamura et al.; by the physical environment/migration model of Dr. Uchiyama et al.; by the numerical simulation model of Drs. Kishi and Ikeda. Dr. Yamochi presented a bloom outbreak mechanisms model for *Heterosigma akashiwo* in Osaka Bay. The Nakamura et al. model revealed vertical water-column differences in the growth-supporting capacity of nutrients, and confirmed the Uchiyama and Kishi results, which showed the need to incorporate motility and migration into the models. A weak effect of grazing was suggested by Kishi and Ikeda's model.

These pioneering efforts further challenge the common perception that nuisance blooms are fundamentally regulated by single factors and/or mechanisms. An interactive, multifactorial approach seems called for. The development of such models increases the need for experimentally-based kinetic constants on key bloom species and their rates of nutrient uptake, growth, photosynthesis and grazing. An issue presently unresolved, and requiring modeling and autecological experimentation, is whether bloom outbreaks result from fast (= explosive) growth or represent the accumulation of slow growing, ungrazed populations subject to physical entrainment and concentration. Dr. Iizuka's and co-workers excellent, continuing efforts on in situ population growth of Gymnodinium nagasakiense in Omura Bay suggest red-tide development follows a period of fairly rapid growth (1 doubling/day) and low grazing pressure. Dr. Furnas reported high flagellate and dinoflagellate growth rates, up to 1.5 to 2 doublings/day; large ceratians reached 0.7 doubling/day (Tomas et al.); Prorocentrum minimum reached 1.4 doublings/day (Okamoto and Hirano); Chattonella antiqua $\sim$ 1.0 doubling/day (Ono; Nakamura et al.). Collectively, these growth rates considerably exceed the slow generation time of 72 hours (0.3 doubling/day) usually reported for cultured dinoflagellates. Do such bloom species grow faster in situ than our common perception, at least at the time of the critical triggering event(s)? We clearly need to provide modelers with more information on growth rates. They, in turn, must incorporate within their models variable growth rates to reflect inter- and intra-specific variability and combine these with suitable, experimentally derived grazing terms.

V. TOXIN ECOLOGY AND ECOSYSTEM DYNAMICS

Dr. Shimizu's excellent overview prepared us for the impressive papers on the pharmacological and biochemical aspects of toxin production. Since I am not a toxicologist, I must restrict my comments to some general impressions from my vantage point as a physiological ecologist. These will undoubtedly seem naive to the experts and certainly inadequate to put into perspective their toxicological contributions. The diversity, complexity, widespread phylogenetic occurrence and diverse toxic effects of the toxins were unexpected. We learned from Dr. Baden et al. about the eight toxins in Ptychodiscus brevis and competitor brevetoxins; about ichthyotoxin occurrences and their neurotoxic, hemolytic and hemagglutinative components from Drs. Mitsui, Onoue, Oshima and their colleagues. A bacterial source of saxitoxin in Protogonyaulax tamarensis was reported by Dr. Kodama et al., which stimulated much discussion. A unique, non-toxic mechanism of fish kills was posited by Dr. Jenkinson: increased viscosity resulting from long-chain polysaccharides secreted during algal blooms may cause high shear stress on gill epithelia and damage tissue. Enigmas remain, such as the uncertainty whether Gyrodinium aureolum and Aureococcus anorexefferens produce toxins. These various studies clearly demonstrate that deleterious effects of nuisance algal blooms on upper trophic levels can come about from diverse toxic principles or non-pharmacological stresses, such as anoxia, anorexia and possibly the Jenkinson effect. Moreover, there are numerous pathways and mechanisms by, and through which nuisance blooms may negatively impact trophic components. Our earlier, simpler conceptual models must be upgraded.

The increased focus of toxicologists on the environmental conditions influencing toxin production represented a major contribution, also helpful to ecologists and modelers. Dr. Mitsui and co-workers showed hemolytic toxin production in the cyanobacterium Synechococcus sp. is dependent upon temperature, salinity and irradiance. Temperature and irradiance influence saxitoxin production in several dinoflagellates, leading Dr. Ogata and co-workers to suggest a linkage with photosynthesis. Growth-phase also influences toxin production, as shown by Drs. Mitsui, Baden and co-workers. The significance of these observations is their demonstration that toxin production is a dynamic process subject to environmental regulation by the same parameters which regulate organismal growth. The relevance of this important finding to assessing the potential for, and the consequences of ecosystem disruption accompanying nuisance algal blooms remains to be clarified. Perhaps some key insights into potential bloom control procedures may ultimately be found in this line of research.

We must now recognize that nuisance algal blooms can not be treated as isolated events unique to red-tide producers or triggered by special factors. Rather, such bloom phenomena are manifestations of a basic characteristic of phytoplanktonic cells and their ecology. They are also under regulation by the same first principles which generally regulate phytoplankton growth in the sea. Further, we must begin to focus attention on the effects of nuisance blooms on the total ecosystem to supplement the impressive strides evident in the taxonomic, autecological, toxicological and modeling efforts presented at our Symposium. Field, experimentally-oriented studies on several trophic levels during, pre- and post-blooms are essential. The evidence that blooms of *Aureococcus anorexefferens* negatively impacted zooplankton grazing and benthic feeding in Narragansett Bay; the well-known induction of fish kills by ichthyotoxins; nekton and benthic die-offs due to anoxia; fish larvae mortality during toxic dinoflagellate blooms, as reported by Dr. White; the antagonistic effects of the bacterium *Vibrio algoinfesta* on flagellate bloom species reported by Dr. Ishio and colleagues. These observations clearly reveal that nuisance algal cells and their blooms frequently impact other trophic levels, can cause considerable ecosystem dysfunction, and may alter trophic processes. We need to describe and evaluate these effects. This is one area where research on nuisance algal blooms is still lagging. Obviously, such research is costly, requires concurrent studies on many processes and trophic groups, and is interdisciplinary. This Symposium has led me to conclude, however, that we now have the critical mass of expertise and international momentum to tackle such an ecosystem-based research effort. Should we think about selecting one or more areas with a high incidence of nuisance algal blooms for concentrated study? Regions where scientists representing the different requisite skills may pool their efforts in carrying out joint-research in a multi-disciplinary, international effort. That I suggest this, reaffirms my belief that the International Symposium On Red Tides at Takamatsu enabled the scientific coming-of-age of research on red-tides, and indeed fulfilled the wishes of our distinguished conveners Professor Okaichi and Nemoto, who wrote: "We deeply expect that the Symposium will open up new approaches for red tide studies". It has; let us collectively build upon this progress.

REFERENCES

1. T. Braarud, in: Oceanography, M. Sears, ed. (American Association Advancement Science. Washington, Publ. No. 67, 1961); pp. 271-298.
2. E.E. Syvertsen, Nova Hedwigia, Beih. 54, 99 (1977).
3. O. Paulsen, Kgl. Danske Vid. Selsk. Biologiske Skrifter 6(4), 1 (1949).
4. I. Solum, Nytt Mag. Bot. 10, 5 (1962).
5. G.T. Boalch, Br. Phycol. J. 22, 225 (1987).
6. L. Zenkevitch, Biology of the Seas of the SSSR. (Interscience Publ., N.Y., 1963).
7. T.C. Malone, NOAA Technical Report NMFS Circular 410, 1 (1978).
8. D.H. Cushing, Climate and Fisheries (Academic Press, N.Y., 1982).
9. D.M.Anderson and B.A.Keafer, Nature (Lond.) 325, 616 (1987).
10. D.M. Pratt, Limnol. Oceanogr. 11, 447 (1966).
11. H. Iwasaki in: Biochemistry and Physiology of Protozoa (2nd Ed.), Vol. 1, M. Levandowsky and S.H. Hutner, eds. (Academic Press, N.Y. 1979) pp. 357-393.
12. J. Gavis, J. Mar. Res. 34, 161 (1976).
13. Y. Ishida, Mem. Coll. Agric., Kyoto Univ. 94, 47 (1968).
14. T. Okaichi and A. Yagyu, Bull. Plankt. Soc. Japan 16, 126 (1969).
15. S. Iizuka and T. Nakashima, Bull. Plankt. Soc. Japan 22, 27 (1975).

II RED TIDES IN REGIONS

HISTORICAL TRENDS IN THE RED TIDE PHENOMENON IN THE RIAS BAJAS OF NORTHWEST SPAIN

T.WYATT,* AND B. REGUERA**
*Instituto de Investigaciones Marinas, Muelle de Bouzas, Vigo, Spain
**Instituto Espanol de Oceanografia, Aptdo.1552, Vigo, Spain

ABSTRACT

Recent studies of the nutrient kinetics of phytoplankton cultures indicate that distinct strategies exist in relation to different nitrogen substrates. We examine the changing status of the Rias Bajas of Spain in relation to four sources of nutrient enrichment. These are urban drainage, agricultural runoff, mussel culture, and forest fires. Two independent hydrographic mechanisms, an intermittently closed vertical circulation cell, and internal waves, provide mechanisms whereby this nutrient input can be retained in the rias and pumped into the surface layer. The change from nontoxic to toxic blooms which took place in 1976 is discussed against this background. It is suggested that toxin production may provide a mechanism for eliminating excess nitrogen from the water column.

INTRODUCTION

Red tides have been recognized as a more or less annual event in Galician waters, at least since 1918 [24]. It is only however since 1976 that toxic blooms are known to have occurred. The PSP problem in Galicia is associated with *Protogonyaulax tamarensis* Lebour and *Gymnodinium catenatum* Graham. From the data available, both appear to be newcomers to the phytoplankton of the Galician rias, as does *Olisthodiscus luteus* Carter, found in the Ria de Pontevedra in 1980 [12], and probably in Ria de Vigo the same year [4]. None of these organisms was sufficiently conspicuous to be recorded in the Galician flora prior to the 1970s.

Sobrino Buhigas [24] identified *Gonyaulax polyedra* Stein as the dominant species in red patches in the Ria de Pontevedra. He also drew attention to the hypothesis of Odon de Buen that red tides inside the rias originated in coastal waters, and were subsequently carried into the rias by currents. The importance of such injections of oceanic waters into the rias was recognized more formally by Saiz et al. [22], and points to the need to consider the coastal water flora as well as that of the rias themselves. *P. tamarensis* ,*G. catenatum* and *O. luteus* are not found in the only species lists available for Galician coastal waters [27,10].

For the Ria de Vigo, we have several species lists from the 1940s and 1950s [17,18,19,9,28,29]. There were notable red tides in that period due to different species of *Gonyaulax* (*G. spinifera*, *G. polyedra*, and *G. diacantha*), and to *Mesodinium*. But none of these studies provide any evidence that *P. tamarensis* or *G. catenatum* occurred in the ria then.

In summary, there may have been an important change in the environment of the rias between the 1950s and 1970s, which has prevented some algae from forming red tides, and allowed others to do so. Since the cultivation of mussels was well developed by 1976, it seems likely that had toxic dinoflagellates occurred in large numbers in earlier decades, they would have caused public health problems.

We can therefore tentatively date the proposed change in the environment to 1975 or 1976, or alternatively suggest that some progressive change then reached a stage which induced a sudden alteration in the local phytoplankton.

RED TIDES: BIOLOGY, ENVIRONMENTAL SCIENCE, AND TOXICOLOGY
Okaichi, Anderson, and Nemoto, Editors

PHYTOPLANKTON CHANGES ELSEWHERE

In 1966, a massive bloom of Gyrodinium aureolum, previously unknown in Europe, occurred in Oslofjord [25]. Since then, it has been widespread in northern European waters. The first outbreak of PSP in Europe occurred in 1968 in NE England. There were significant trends in diatoms (decrease) and microflagellates (increase) in the North Sea between 1966 and 1973 [21], which have been linked to climate change [7]. In these years too the well known Russell Cycle reversed [5]. Thus the phytoplankton changes in Galician rias may reflect a widespread climatic phenomenon. But the particular changes which have taken place in Galicia require a local cause.

NUTRIENT STATUS OF RIAS BAJAS

We resume briefly four new sources of nutrient enrichment in the rias:

i) Mussel production is most intensive in Ria de Arosa (2500 rafts in 300 km^2). Using figures of Bayne [1,2], we estimate the ammonium input from this source at 1 ugl^{-1} daily for Arosa, and about half that for Ria de Vigo, for a water layer 10 m deep.

ii) The use of agricultural fertilizers is increasing in the region, but the quantities are still relatively small, and we do not regard them as a significant new input.

iii) The current discharge rate of sewage from Province of Pontevedra is of the order of 50,000 kg BOD, and 425 x 10^3 kg of solids daily. These figures do not include industrial wastes. Both the population and the pattern of discharge are changing. The population is increasing, and septic tanks are being replaced on a large scale by urban drainage systems. None of these wastes are treated at present. From these discharges we can expect a 10% decrease in oxygen and a 10% increase in nitrates. Domestic sewage also contains about 50 mgl^{-1} of ammonia nitrogen, and 20 mgl^{-1} of organic nitrogen.

iv) The number of hectares of forest burned in Galicia prior to 1971 was quite small. Since then destruction has become substantial. Following rain, each hectare burned can loose between 25 and 50 t of soil, most of which reaches the sea. Soil extract (Erdschreiber's medium), with its nutrients and chelators, is the classical medium for algal culture! The sterilization of soils also reduces the time lag between rainfall and runoff, so that stratification of the water column may increase. The fine silt content will also depress light penetration.

NUTRIENT AVAILABILITY IN RIAS

Some of the nutrient input listed may be flushed from the rias by intrusion of coastal water and by upwelling pulses, but two mechanisms ensure that its residence time may often allow bloom development. Fraga [11] has identified an intermittent circulation pattern which prevents exchange of nutrients with adjacent waters. Nutrient pumping by baroclinically generated internal waves also occurs.

DISCUSSION

We have suggested that the Rias Bajas have become significantly hypertrophic in recent decades. The forest fire data indicate that a rather sudden change occurred in the 1970s, but we can not show that it was of overriding importance. Urban drainage must also be having a major impact on local waters. We conclude that the appearance of toxic blooms, or more generally, the change in phytoplankton in the Rias Bajas, is a response to hypertrophication.

We end with some speculation about the role which toxic blooms may play in nutrient enriched waters. It has been argued that the chemical composition of the sea and atmosphere is regulated by the biosphere [16]. This hypothesis led to a realization of the important part played by phytoplankton in transfering sulphur (in dimethyl form) from the sea to the land. The slope sink hypothesis [30,31] provides another potential regulatory mechanism, for the removal of carbon from the ocean in response to anthropogenic enrichment.

In this context we may probe the role of toxic dinoflagellates which synthesize saxitoxin and related compounds. This synthesis diverts arginine, methionine, and energy, from growth, and results in a stable compound with a remarkably high nitrogen content. Red tides are often linked to stable hydrographic conditions when nutrients are scarce and toxin production might then be seen as a nitrogen storage mechanism [15]. We should thus expect toxins to be depleted during ambient nutrient scarcity, but this is not generally observed.

We know that some dinoflagellates can store inorganic nitrogen [8] but this ability is not related to toxicity. We also know that ammonia is inhibitory at high concentrations [26], and that toxin production is stimulated by phosphorus deficiency [3]. Vegetative cells of <u>Gonyaulax</u> may contain as much as 24 pg/cell of SXT and derivatives [23] -about 4 parts in 10^{5} by weight- and resting cysts even more [6]. Is the production of toxin, paradoxically, a detoxification mechanism? Does the production of saxitoxin render inert excess quantities of nitrogen, and hence provide its producers with a "competitive" advantage in hypertrophic waters? If so, one might for example seek a link between the frequency of toxic blooms and the relative inputs of nitrogen and phosphorus.

Finally, we can also suggest that the production of saxitoxin is intended to literally bury nitrogen. The evidence of sedimentary sterols indicates that dinoflagellates make an important contribution to the removal of carbon from the biosphere [6]. The very high levels of saxitoxin in some cysts [6] -an unnecessarily expensive protective measure in view of their sporopollenin walls- would then be validated in Gaian terms. Mankind has increased the biospheric nitrogen flux by about 30%. If toxic red tides are fulfilling the role suggested here, then the price of vigilance vis a vis our consumption of seafoods becomes more acceptable.

ACKNOWLEDGEMENTS

We thank Dr.Diaz Fierro, University of Santiago de Compostela, for making the forest fire data available.

REFERENCES

1. B.L. Bayne, J. mar. biol. Ass. U.K. <u>53</u>, 39-58 (1973).
2. B.L. Bayne, R.J. Thompson and J. Widdows in: Marine Mussels, B.L. Bayne, ed. (Cambridge U.P. 1976).
3. G.L. Boyer, J.J. Sullivan, R.J. Andersen, P.J. Harrison, and F.J.R. Taylor in: Toxic Dinoflagellates, D.M. Anderson, A.W. White, and D.G. Baden, eds. (Elsevier, New York 1985) pp.281-286.
4. M.J. Campos, S. Fraga, J. Marino, and F.J. Sanchez, ICES L:27 (mimeo) (1982)
5. D.H. Cushing, Climate and Fisheries (Academic Press 1982) 373pp.
6. B. Dale, C.M. Yentsch, and J.W. Hurst, Science (NY) <u>201</u>, 1223-1225 (1978).
7. R.R. Dickson and P.C. Reid, J. Plank. Res. <u>5</u>, 441-455 (1982).
8. Q. Dortsch, J.R. Clayton, S.S. Thorensen, and S.I. Ahmed, Mar. Biol. <u>81</u>, 237-250 (1984).
9. M. Duran, F. Saiz, M. Lopez-Benito, and R. Margalef, Inv. Pesq. <u>4</u>, 67-95 (1956).

10. M. Estrada, J. Plank. Res. 6, 414-434 (1984).
11. F. Fraga, in press (CSIC Madrid).
12. F. Gomez Figueiras, Ph.D.Thesis, (Universidad de Barcelona, 1985).
13. C. Lee, J.N. Farrington, and R.B. Gagosian, Geochim. Cosmochim. Acta 43, 35-46 (1979).
14. C. Lee, R.B. Gagosian, and J.N. Farrington, Organic Geochem. 2, 103-113 (1980).
15. A.R. Loeblich in: Dinoflagellates, D.L.Spector, ed. (Academic Press 1984) pp. 299-342.
16. J.E.Lovelock, ed. Gaia-a new look at life on earth- (Oxford U.P. 1979).
17. R. Margalef, P. Inst. Biol. Apl. 9, 83-118 (1952).
18. R. Margalef, Inv. Pesq. 5, 113-134 (1956).
19. R. Margalef, M. Duran, y F. Saiz, Inv.Pesq. 2, 85-129 (1955).
20. S. Mori, Y. Nakamura, M.M. Watanabe, S. Yamohi, and M. Watanabe, Res. Rep. Natl. Inst. Envir. Studies 30, 71-86 (1982).
21. P.C. Reid, Rapp. Proces Verb. Cons. int. Explor. Mer 172, 384-389 (1978).
22. F. Saiz, M. Lopez-Benito, y E. Anadon, Inv. Pesq. 8, 29-88 (1957).
23. R.J. Schmidt, and A.R. Loeblich, J. mar. biol. Ass. U.K. 59, 479-487 (1979).
24. R. Sobrino Buhigas, Mem. R. Soc. Esp. Hist. Nat. 10, 407-458 (1928).
25. K. Tangen, Sarsia 63, 123-133 (1977).
26. W.H. Thomas, J. Hastings, and M. Fujita, Mar. envir. Res. 3, 291-296 (1980).
27. M. Varela, Bol. Inst. Esp. Oceanogr. 7, 191-222 (1982).
28. F. Vives, y M. Lopez-Benito, Inv. Pesq. 10, 45-106 (1957).
29. F. Vives, y M. Lopez-Benito, Inv. Pesq. 13, 87-125 (1958).
30. J.J. Walsh, G.T. Rowe, R.L. Iverson, and C.P. McRoy, Nature 291, 196-201 (1981).
31. J.J. Walsh, E.T. Premuzic, J.S. Gaffney, G.T. Rowe, G. Harbottle, W.L. Balsam, P.R. Petzer, and S.A. Macko, Deep Sea Res.32,853-883, (1985).

STUDIES ON RED TIDE PHENOMENA IN KOREAN COASTAL WATERS

Park, J. S., H. G. Kim and S. G. Lee
National Fisheries Research and Development Agency, Pusan 606, Korea

ABSTRACT

The red tides in Jinhae Bay (the most active red tide area in Korean waters) occur from April to October with maximum blooms in July every year. Twenty-six species of causative organisms were identified, of which 16 species were flagellates. Among the causative organisms, diatoms such as *Skeletonema costatum*, *Chaetoceros curvisetus* and *Nitzschia seriata* were the most outstanding species until the first half of the 1970s. Since then, however flagellates such as *Prorocentrum micans*, *P. minimum*, *Heterosigma akashiwo*, *Gymnodinium nagasakiense*, *G. sanguineum* and *Protogonyaulax fraterculа* were found to be the main causative organisms. Therefore, red tides could be characterized by a change from diatoms to dinoflagellates through the years.

Recent outbreaks of red tides have increasingly been large scale and widespread with long periods and densities. Red tides in Masan Bay, mainly by *Prorocentrum micans* and *Heterosigma akashiwo*, occurred regularly every year from April to October in high densities (more than 10^4 cells/ml), due to increasing eutrophication by organic pollutants from land.

INTRODUCTION

Outbreaks of red tides in Korean neritic waters have remarkably increased in the last decade and have caused severe damage to cultured shellfish and other living organisms.

Most of the 104 red tides recorded in Korean waters during the period from 1972 to 1979 appeared in Jinhae Bay and its vicinity (by Park (1)). Since the unprecedented large scale red tides of 1981 in and around Jinhae Bay, more frequent and persistent heavy red tides have occurred from April to October every year. Such red tides can have a severe effect on the economy of a region due to detrimental effects on fisheries and mariculture.

The present paper deals mainly with studies on periodic change in red tide occurrence and succession of the causative organisms using samples collected from a comprehensive red tide research and monitoring programme.

MATERIALS AND METHODS

Plankton samples of red tide organisms and environmental parameters such as nutrient salts, dissolved oxygen, chemical oxygen demand, salinity and temperature were collected from about 100 key stations (Fig. 1) established in shellfish growing areas and fishing grounds. In addition to the key stations, more sampling collections were added for large scale red tide outbreaks.

RESULTS

1. Succession of red tide organisms

Twenty-six species of red tide organisms in Korean neritic waters were identified: 16 flagellates, 9 diatoms and 1 ciliate. The main ones were *Heterosigma akashiwo*, *Prorocentrum micans*, *P. minimum*, *P. triestinum*,

RED TIDES: BIOLOGY, ENVIRONMENTAL SCIENCE, AND TOXICOLOGY
Okaichi, Anderson, and Nemoto, Editors

Gymnodinium nagasakiense, G. sanguineum, Eutreptiella gymnastica, Protogonyaulax fratercula, Noctiluca miliaris, Skeletonema costatum, Chaetoceros curvisetus, Nitzschia seriata, Thalassiosira allenii and Th. conferta.

Two categories of causative organisms of Jinhae Bay are distinguished: one is the organisms (mostly Prorocentrum species and Heterosigma akashiwo from May to July and Skeletonema costatum after August) forming the red tides for Masan Bay and Haengam Bay, and the other is the organisms (mostly chaetoceros curvisetus, Skeletonema costatum, Nitzschia and Thalassiosira species) creating the red tides for the other 6 subareas in the bay. Another distinctive feature in the other 6 subareas of the bay is that the large scale red tides caused either by Heterosigma or Prorocentrum do not resemble those of Masan Bay. However, Protogonyaulax fratercula and P. tamarensis populations increased year by year.

Until 1977, the dominant organisms in Masan Bay were diatoms such as Skeletonema costatum, Chaetoceros and Nitzschia (from park & Kim (2) ; Park (1)). Since then, the dominant organisms have been dinoflagellates such as Prorocentrum micans, P. minimum, P. triestinum, Gymnodinium nagasakiense and Heterosigma akashiwo.

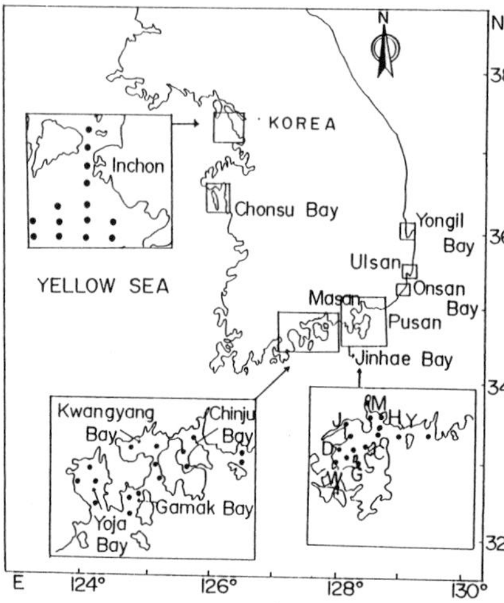

FIG. 1. Map showing a location of red tide sampling areas in Korean coastal waters. Solid circles in boxes indicate sampling stations. 8 subareas are divided in Jinhae Bay (M: Masan Bay, J: Jindong Bay, D: Dangdong Bay, H: Haengam Bay, Y: Yongwon Bay, C: Chilchondo area, W: Wonmun Bay, G: Gajodo area).

Other red tide region along the coast (besides Jinhae Bay) proves to have its own characteristic red tide organisms. They are Noctiluca miliaris for Inchon Bay ; Scrippsiella trochoidea and Prorocentrum micans for Ulsan Bay ; Skeletonema costatum, Thalassiosira sp., Prorocentrum micans and Heterosigma akashiwo for Onsan Bay ; Chaetoceros sp., and Noctiluca miliaris for Yoja Bay; Cochlodinium for Chinju Bay ; Prorocentrum sp., Heterosigma akashiwo and Gymnodinium sp. for Yongil Bay ; Skeletonema costatum for Kwangyang Bay ; Noctiluca miliaris and Prorocentrum micans for Gamak Bay.

2. Duration and scale of red tide occurrences

A comparion of periodic change of phytoplankton abundance from 1981 to 1986 associated with red tide phenomena among 8 subareas (Fig. 1) in the bay was made (Fig. 2). In 1981, heavy red tides of Gymnodinium nagasakiense occurred from July through September in the entire bay. The red tides of Jinhae Bay may be roughly divided into two types: large scale red tides with long persistency covering two out of 8 subareas such as Masan Bay and Haengam Bay, and small scale, sporadic red tides in the other 6 subareas.

Since 1982, the red tides of Masan Bay and Haengam Bay have habitually begun in April and lasted until September or October, showing the largest concentration of over 10,000 cells/ml for the period of May to July. The red tides of the 6 subareas occur sporadically from May to October becoming maximal in July in all the years and areas. Since 1982, red tides in other coastal regions such as Inchon Bay, Gamak Bay, Chinju Bay and Ulsan Bay appeared mostly from June through September.

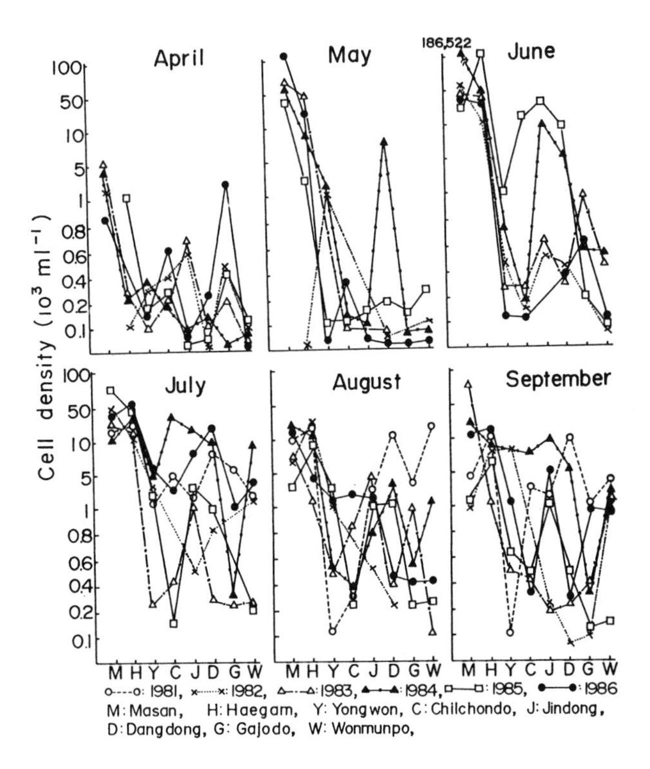

FIG. 2. Phytoplankton abundance as red tide formation by 8 different areas in Jinhae Bay from April to September, 1981 - 1986.

DISCUSSION

Recent outbreaks of red tides have been characterized by extensive area coverage and a succession from diatoms to dinoflagellates suspected of being harmful to living organisms.

According to Adachi and Kawai (3) and Adachi and Eguwa (4), the main causative organisms of red tides in Ise Bay, Japan were Noctiluca scintillans in 1974, Skeletonema costatum in 1975, changing into Scrippsiella trochoidea, Prorocentrum minimum and Olisthodiscus sp. in 1976. From 1977 to 1980, red tides were caused by Prorocentrum spp. in 1977-1978 and Skeletonema costatum, Chaetoceros, Nitzschia and Thalassiosira in 1979-1980. The organisms in Hakada Bay (5) were Skeletonema and Noctiluca in 1971, Prorocentrum spp. and Olisthodiscus sp. in 1972 and Gymnodinium type '65 in 1973.

Differences in red tide phenomena were related to environmental factors. Masan Bay, representing the largest red tides by flagellates contained the highest concentrations of nitrogen and phosphorus. In Gajodo area, showing minor red tides by diatoms, there were the lowest concentrations of nutrient salts with relatively high salinity and low temperature, as shown in Fig. 3.

However, although Gajodo area in August had sufficient basic nutrient salts (N and P) to produce dense blooms, it did not have large scale red tides like Masan Bay. Perhaps the massive red tides of Heterosigma and Prorocentrum in Masan Bay were induced by growth promoting substances, special organic compounds such as amines and amino acids, or metals like iron and manganese (5, 6), which are continuously supplied through the influx of domestic sewage and industrial waste.

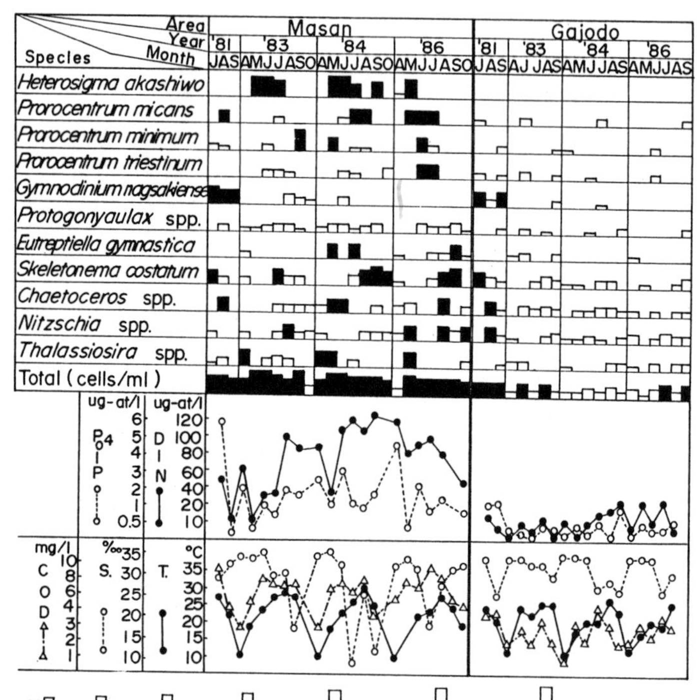

FIG. 3. Comparison of red tides and environmental factors between Masan Bay and Gajodo area, 1981 - 1986.

REFERENCES

1. J.S. Park, Bull. Fish. Res. Dev. Agency, 23, 7-157 (1980).
2. J.S. Park, and J.D. Kim, Bull. Fish. Res. Dev. Agency, 1, 65-79 (1967).
3. R. Adachi and H. Kawai, Rep. Envir. Sci. Mie Univ., 4, 123-136 (1979).
4. R. Adachi and S. Eguwa, Spec. Res. Rep. Envir. Sci., B 148-R 14-8, Minis. Educ., Japan (1982).
5. T. Honjo, bull. Tokai, Reg. Fish. Res. Lab., 79, 77-121 (1974).
6. H. Iwasaki, Bull. Pl. Soc. Japan, 19(2), 46-56 (1973).

OCCURRENCES OF RED TIDE IN THE GULF OF THAILAND

SUNEE SUVAPEPUN
Marine Fisheries Division, 89/1 Sapanpla, Yannawa,
Bangkok 10120, Thailand

ABSTRACT

This presentation summarizes the results of 7 year of investigations on red tide phenomena in the Gulf of Thailand. Red tides always occur between February and May. *Trichodesmium erythraeum* was recorded as the major species of phytoplankton usually found during red tides in offshore water. In estuaries and coastal waters the species responsible for red tides is *Noctiluca scintillans*. In certain places *Noctiluca* occurred together with *Ceratium furca* or with *Trichodesmium erythraeum*. In 1983 *Trichodesmium erythraeum* blooms caused extensive damage to fish farms on the east coast. No incident of toxic red tide has been detected in the Gulf of Thailand. A case of shellfish poisoning was reported in May 1983, but none has occurred since then.

INTRODUCTION

Red tide is common around the coast of the Gulf of Thailand. Most early observations were concerned with the impact on the fisheries industry [1] . In recent years, there has been increasing documentation of red tides, usually of monospecific blooms in coastal waters which produce harmless water discolouration that may be yellow-green, green, reddish brown or milky in colour depending on the organism involved [2, 3, 4, 5]. Piyakarnchana and Tamiyavanich [6] reported that the red tide phenomenon could become more serious because of increasing pollution from rivers in the Inner Gulf of Thailand. Excess nutrients and organic and inorganic pollutants from river run-off have periodically induced red tides resulting in the mass mortalities of cultured animals [7, 8, 9,]. The problem is therefore of considerable importance.

MATERIALS AND METHODS

The investigations were carried out in the coastal area of the Gulf of Thailand (Fig.1). The total area of the Gulf is about 350,000 square kilometers and corresponds to a coastline of 1,875 kilometers in length. It is bordered by the coasts of Vietnam, Kampuchea and Thailand. The Gulf is relatively shallow with an average depth of 45 m. Surface salinities are between 30.5 and 33 ppt, oxygen concentrations between 4 and 4.6 ml/l. There are two layers of water in the Gulf; low salinity water ($<$ 33 ppt.) which has been diluted by rainfall and run-off flows out of the Gulf at the surface, while highsalinity water (33-44 ppt.) from the South China Sea flows into the Gulf [10].

Monthly surveys for red tides were conducted between 1981 and 1987 by research vessels. Five-litre water samples were collected at three depths: subsurface, mid depth, and near bottom, using Van Dorn water sampler. Samples were filtered through nylon net of 20 microns mesh size then fixed and preserved in 10% neutralized formalin. At stations where red tides occurred, surface water samples were collected and preserved in Lugol's iodine for species identification and direct counting of algal cells or filaments. Temperature, pH, salinity, chlorophyll, dissolved nutrient and dissolved oxygen were measured at all stations indicated in Fig.1.

RED TIDES: BIOLOGY, ENVIRONMENTAL SCIENCE, AND TOXICOLOGY
Okaichi, Anderson, and Nemoto, Editors

RESULTS

The most common red tide organisms in the Gulf of Thailand were Trichodesmium erythraeum Ehr. and Noctiluca scintillans (Macartney) Kofoid that produce harmless water discolorations. At the start of Trichodesmium blooms, the water appears yellow-green in colour and changes to grey and reddish-brown later. Noctiluca blooms are always green due to the presence of prasinomonad symbionts (Pedinomonas noctilucae) within the cells. This dinoflagellate was restricted to coastal waters and occurred especially in the vicinity of river mouths and in sheltered bays. No toxic effects are known, but it often causes depletion of oxygen and high ammonia levels in shrimp farms which result in the reduction of normal yiels. Diatom blooms cause gelatinous white coloured water. Coscinodiscus jonesianus (Greville) Ostenfeld. Rhizosolenia styliformis Brightw. were the causative organisms, and usually occurred mixed with Hemidiscus, Chaetoceros and Bacteriastrum. In Tha-Chin River mouth, Noctiluca and Ceratium furca blooms in February 1981 produced two parallel streaks, one green and the other reddish-brown. Trichodesmium and Noctiluca occurred together in April 1987 at two stations on the west coast of the Gulf.

There were 43 occurrences of red tide along the coast of the Gulf of Thailand from 1981 to 1987 of which 21 were caused by Trichodesmium erythraeum, 17 by Noctiluca scintillans and the others by diatoms (Fig.2). (Red tide usually occurs between February and May, but in 1985, 11 red tides appeared throughout the year from January to December, mostly caused by Noctiluca scintillans. A large bloom of Trichodesmium erythraeum was detected in 1983 scattered over a wide area of approximately 7000 km^2 , spreading from the east coast to the Inner Gulf from May to June. The organisms accumulated at the sea surface and were driven by tide and wind into bands of yellow - green water several miles long. In June the red tide was driven to the shore and water became reddish-brown along a stretch of almost 27 km. When washed ashore T. erythraeum accumulated into reddish-brown foam patches in the surf of a sandy beach about 20 km long. The decomposition of masses of T. erythraeum caused respiratory irritation, but no harmful effects on fish were observed in the sea. However, this bloom caused extensive damage to fish farms along the east coast; there was an estimated loss of about U.S.$ 1.16 million [4]. This mass death of fish was caused by the decomposition of Trichodesmium in fish ponds that generated anoxic conditions.

It does not appear that these organisms produce toxins since neither extensive fish kills nor PSP are associated with them. None of the red tides observed appeared to be directly toxic, even in May 1983, when an outbreak of shellfish poisoning was reported at Pranburi River mouth, in the northwest of the Gulf. The source of this biotoxin has not been identified. The plankton community during the incident was dominated by Cyanobacteria and diatoms, of which Skeletonema coastatum, Thalassiosira spp, Chaetoceros spp. and Cyclotella sp. were the major species. Several species of dinoflagellates were found to be more abundant than usual. The most abundant species was Protoperidinium quinquecorne. Other species found in large numbers included Prorocentrum micans, Peridinium spp. abd Dinophysis spp. Small numbers of Protogonyaulax sp. appeared in the water samples during the occurrence [4].

DISCUSSION

Most red tides in the Gulf of Thailand are caused by Trichodesmium erythraeum and should be regarded as natural phenomena, unrelated to industrial pollution. Trichodesmium blooms were observed by Charernpol [1].before modern industry developed in the countries around the Gulf. This study also found no correlation between Trichodesmium blooms and nutrient concentrations in sea water. The blooms appear to be harmless events, but under exceptional conditions the algae were driven ashore by the wind, causing problems to

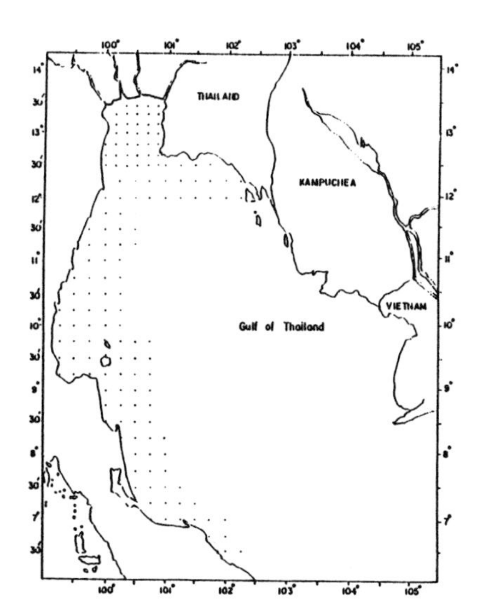

Figure 1. Location of sampling stations in the Gulf of Thailand.

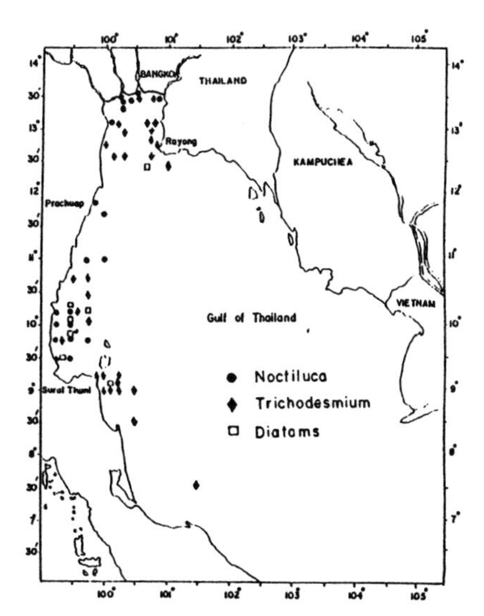

Figure 2. Areas affected by red tides during 1981-1987

coastal aquaculture activities.

Heavy blooms of Noctiluca deplete oxygen in the sheltered bays especially in the river mouth, and have had economic impacts on shrimp farms in the rivers and in the coastal area close to river outlets. Future studies should include details of the economic impact of red tide on aquaculture activities, and about what should be done to control it.

Protogonyaulax tamarensis was implicated as the causative organism for an outbreak of PSP in the Gulf of Thailand in 1983 [11]. However, Kodama et al [12] found that all strains of P. tamarensis collected from the Gulf of Thailand were non-toxic, so that some other organism must have been responsible for the shellfish poisoning on that occasion. There are still no firm conclusions about the causative organism of PSP in the Gulf of Thailand.

[Editors' note: see the paper by Fukuyo et al., in this volume.]

ACKNOWLEDGEMENT

I thank Dr. Veravat Hongskul Secretry - General of the Southeast Asian Fisheries Development Center (SEAFDEC) for his interest and valuable comments on the manuscript. The author's attendance at the International Symposium on Red Tides was possible due to the generous support of the SEAFDEC.

REFERENCES

1. S. Charernpol, IX Pacific Science Congress, Bangkok, Thailand, 18-30 November, (1957).
2. S. Chernbumroong and C. Tharnbupha, Marine Fisheries Laboratory Report No. 24/1, Bangkok, Thailand, (1981).
3. S. Suvapepun, Fisheries Gazette 35(6), 581-593 (1982).

4. S. Suvapepun, S. Chernbumroong and V. Wangcharernporn, 22nd conference, Fisheries Section, Kasetsart University, Bangkok, Thailand, (1984).
5. S. Suvapepun, Fisheries Conference, Department of Fisheries Bangkok, Thailand, (1985).
6. T. Piyakarnchana and S. Tamiyavanich, Journal of Aquatic Animal Diseases, 2(4), 207-215 (1979).
7. P. Menasveta, Water sources and pollution problems, Chulalongkorn University Press, Bangkok, (1982).
8. S. Tamiyavanich, Proceeding of the third Seminar on the Water Quality and the Quality of Living Resources in Thai Waters, NRCT, 481-486 (1984).
9. P. Rojanavipart, ASEAN-EEC Seminar on Marine Science, Manila, 12-16 April (1987).
10. M.K. Robinson, NAGA Report 3(1), 5-109 (1974).
11. S. Tamiyavanich, M. Kodama and Y. Fukoyo in: "Toxic Dinoflagellates" D.M. Anderson, A.W. White and D.G. Baden, eds. (Elsevier, New York 1985) pp. 521-524
12. M. Kodama, T. Ogata, Y. Fukuyo, T. Ishimaru, P. Pholpunthin, S. Wisessang, K. Saitanu, V. Panikiyakarn and T. Piyakarnchana. Nippon Suisan Gakkaishi 53(8), 1491 (1987).

RED TIDE: THE PHILIPPINE EXPERIENCE

Gonzales, C.L., J.A. Ordonez and A.M. Maala

Research Division, Bureau of Fisheries and Aquatic Resources
1184 Quezon Avenue, Quezon City, Philippines

ABSTRACT

An account of red tide occurrences in the Philippines is given here, emphasizing the results of the monitoring surveys conducted in 1984 up to 1987 in the red tide-affected areas.

INTRODUCTION

In June 1983, several cases of poisoning, attributed to the green mussel (*Perna viridis*), were reported in Catbalogan, Samar, Philippines, where government-initiated mussel culture projects are located. The poisoning was later found to be caused by a toxic dinoflagellate, identified as *Pyrodinium bahamense* var. *compressa* (1).

As a result of the incident, the government immediately imposed a ban on the gathering, selling and transportation of all kinds of shellfish from the area in order to prevent further intoxication and loss of lives.

Several days after the incident, red tide occurred first in Maqueda Bay and later in the neighboring Samar Sea and Carigara Bay (Fig. 1). Except in Sorsogon, which is about 90 nm northwest of Maqueda Bay, no other sightings were reported at that time. However, 7 cases of paralytic shellfish poisoning (PSP) - 6 attributed to *Amusium pleuronectes* (Asian moon scallop) and 1 to *P. viridis* - were reported in Northern Panay and in Mati, Davao Oriental, respectively. The incident resulted in great economic loss (estimated at about ₱10 million, approx. U.S. $500,000 in 1983) to the mussel culture projects in Maqueda and Villareal Bays, not to mention the loss incurred by the fishing industry in general as a result of the red tide scare.

After the 1983 red tide occurrence, the Bureau of Fisheries and Aquatic Resources (BFAR) implemented a monitoring program in the area. For a period of 4 years, no significant biological activity of the toxic dinoflagellate was observed. However, in mid-1987 there was a recurrence of the red tide in Samar. Almost during the same period, PSP, also caused by *P. bahamense* var. *compressa*, was reported in Masinloc, some 400 nm from Maqueda Bay.

MATERIALS AND METHODS

Plankton samples analyzed for this study were taken from vertical plankton net hauls made in Maqueda Bay, as well as in Masinloc, Zambales, during the recurrence of red tide in the country in 1987.

Counts of plankton organisms were made from a 1-ml aliquot placed in a Sedgewick-Rafter counting cell. These counts were the basis of the numerical estimates of plankton population, taking into consideration the mouth diameter of the net, sampling depth, and volume of plankton sample.

Fish and shellfish samples were also taken from the red tide-affected

RED TIDES: BIOLOGY, ENVIRONMENTAL SCIENCE, AND TOXICOLOGY
Okaichi, Anderson, and Nemoto, Editors

areas. Their stomach contents were examined and toxicity analyses, using the standard mouse bioassay test according to A.O.A.C. methods [2], were performed by the Bureau of Food and Drugs, Department of Health.

RESULTS AND DISCUSSION

Distribution and Relative Abundance of Pyrodinium Cells

The distribution and abundance of Pyrodinium during the 1983 red tide occurrence have been reported by Estudillo and Gonzales [3].

Results of surveys conducted in 1984 showed that the highest proportion of Pyrodinium in the total plankton population was recorded in September 1984. From March 1985 up to April 1987, Pyrodinium was from 0 to 9.84% of the total plankton collected.

In April 1987, Pyrodinium was present only at 2 stations, with 1 station having a density of 4 cells/l and the other with only 1 cell/l. The diatoms comprised 99% of the total plankton population, with Chaetoceros, Rhizosolenia and Lauderia as the major components of the samples. From the results of this particular survey, there was no indication that Pyrodinium would reappear in less than two months' time.

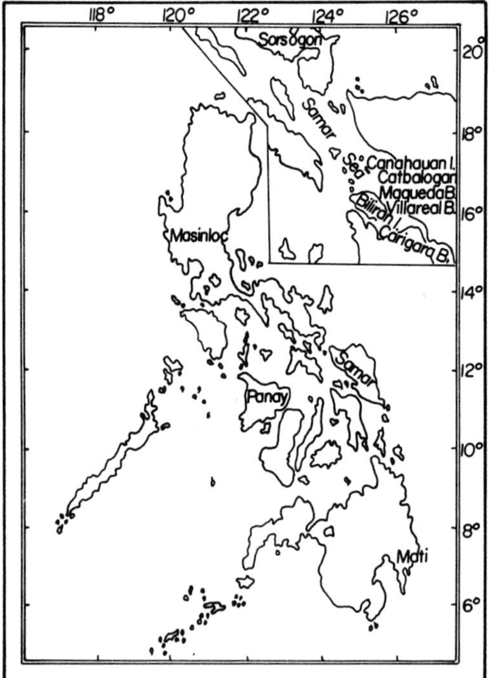

Fig. 1. Philippine map showing the red tide-affected areas.

Pyrodinium was present in all plankton samples collected in Maqueda, Villareal and Carigara Bays, and Samar Sea during the recurrence of red tide in June 1987. The denser population of the organism was observed in some parts of Carigara Bay and waters along the coasts of Western Samar. In Maqueda and Villareal Bays, the densities of Pyrodinium were less than 10,000 cells/l, except at the northern part where the values ranged from 14,857 cells/l to 43,502 cells/l. A maximum count of 8.57 x 10^6 cells/l was recorded between Biliran and Canahauan Islands, while counts of more than 2 x 10^6 cells/l were observed in Carigara Bays. Red tide sightings in Samar Sea coincided with the areas of high Pyrodinium counts.

In Masinloc, Zambales, the concentration of Pyrodinium ranged from 9,015 cells/l to 42,265 cells/l during the peak of its bloom in June 1987. The density decreased considerably in July and started to dissipate in August. The entire coastal water of Zambales was affected by the toxic dinoflagellate but the areas of high density were observed in Masinloc.

Examination of Stomach Contents

Results of the examination of the gut contents of fish and invertebrate samples from Maqueda Bay and Samar Sea revealed that several species of fish and invertebrates contained toxic dinoflagellate.

The Indo-Pacific (*Rastrelliger brachysoma*) and striped mackerels (*Rastrelliger kanagurta*) were the most contaminated among the fish samples examined. Although other pelagic species also contained *Pyrodinium* cells, their number was limited to less than 100 cells per fish, with the exception of sardines (*Sardinella longiceps*) which, on two occassions, were found to contain more than 100 cells per fish.

Demersal fishes, on the other hand, had low concentrations of *Pyrodinium* cells. Except for slipmouth and siganid, all fishes containing toxic dinoflagellate had less than 30 cells in their guts.

Among the invertebrates, the highest density of *Pyrodinium* cells was consistently observed in the green mussels, oysters (*Crassostrea* spp.) and pen shells (*Pinna* sp.). Blue crab, shrimp and squid were the least affected.

Red Tide Intoxication

A total of 211 cases of PSP, 6 of which were fatal, were recorded in Samar from May 26 to August 7, 1987. This figure brings to 489 the total reported PSP cases in the area since the first occurrence of toxic red tide in 1983. The records of the Integrated Provincial Health Office in Catbalogan, Samar disclosed that 4 of the fatalities were children whose age ranged from 4 to 9 years. The most affected age groups were those between 5 and 9 years, 45 and 49 years, and 10 and 14 years, in that order. More males than females were affected by PSP: 1.06:1. About 67% of the total PSP cases reported in Samar was caused by green mussel and only 13% by fish. While fish were observed to be safe for human consumption (as long as it was fresh and washed thoroughly after all its entrails were removed, and was not cooked in vinegar), still they continued to cause PSP in Samar due to the preference by some local residents to cook and eat fish with its intestines and gills still intact. Twenty-nine cases were reported to have been caused by fish, while 13 were attributed to shrimp and crab.

PSP Toxin

The principal toxins of *Pyrodinium* are saxitoxin (STX) and neosatitoxin (neoSTX) which were reported to comprise 78% of the toxicity of the dinoflagellate in Palau. Saxitoxin is a dibasic salt that is very soluble in water and is the most potent neurotoxin found in dinoflagellate blooms. It was shown by Arafiles *et al*. [4] that the crude toxin from green mussels had a maximum toxicity at pH 3. This seems to confirm an earlier finding by Hashimoto (1979), as cited by Arafiles *et al*. [4], that the toxicity of STX is enhanced at an acid pH.

Harada *et al*. [5] found out that the specific toxicity of 2 of the *Pyrodinium* toxins, viz., gonyautoxin V (GTX_5) and gonyautoxin VI (GTX_6), increased by 15 times when treated with dilute hydrochloric acid. On acidification, GTX_5 was converted to STX while GTX_6 was converted to neoSTX.

The effect of acid on the toxin of *Pyrodinium* was illustrated in a case in Roxas City, in September 1983 when 5 families from the locality bought Asian moon scallops caught by a bottom trawler from northern Panay waters. The scallops were simply cooked either by boiling, broiling, or in vinegar.

None of those who ate boiled or broiled scallops manifested any PSP symptoms, while those who ate scallops cooked in vinegar suffered PSP with one fatality (i.e., a 5 1/2-year old boy).

Yentsch and Incze [6] reported that research efforts to develop an antidote for PSP toxins were relatively unsuccessful. In Samar, however, coconut milk is used as an antidote for all kinds of poisons, so during the first toxic red tide incident in the area, some victims of PSP were given a glassful of coconut milk. The victims claimed that they were relieved of the symptoms a few hours afterward. Gacutan [7] was able to show that the toxicity of crude toxin from red tide-contaminated mussels when reacted in vitro with coconut milk decreased by 68.5%.

Mussel Toxicity

The results of the toxicity analysis performed by the Bureau of Food and Drugs showed that green mussels in Maqueda Bay had a toxin content of 256 μg/100 g of shellfish meat during the third week of July 1987. The toxicity dropped to 182.6 μg/100 g after two weeks and decreased further to below the maximum tolerable limit for shellfish toxin set by the U.S. Food and Drug Administration at 400 MU/100 g (about 80 μg/100 g). Beginning mid-September, the toxicity of green mussel in Maqueda Bay ranged from 0 to 36 μg/100 g. In contrast, mussels outside the bay had higher toxicity at 107.8 μg/100 g during the same period.

ACKNOWLEDGMENT

The authors would like to thank Ms. Ma. Ethel G. Llana for the critical reading of the manuscript. Thanks are also due to the officers and staff of the BFAR, especially those in the Fisheries Research Division; the Bureau of Food and Drugs; the Department of Health; the Department of Agriculture Regional Office Nos. 3 and 8; and the local governments of Samar and Zambales for providing assistance during the investigation.

REFERENCES

1. R.A. Estudillo, Fish Newsl., BFAR 12, 2 (1983).
2. W. Horwitz, ed., Official Methods of Analysis of the Association of Official Analytical Chemists, 13th Ed., A.O.A.C., Wash. D.C. 298 (1980).
3. R.A. Estudillo and C.L. Gonzales in: Toxic red tides and shellfish toxicity in Southeast Asia, A.W. White, M. Anraku and K.K. Hooi, eds., SEAFDEC/IDRC, 52-79 (1984).
4. L. Arafiles, R. Hermes and J.B.T. Morales in: Toxic red tides and shellfish toxicity in Southeast Asia, A.W. White, M. Anraku and K.K. Hooi, eds., SEAFDEC/IDRC, 43-51 (1984).
6. C.M. Yentsch and L.S. Incze, Dev. Aquac. Fish. Sci., 7, 223-246 (1980).
7. R. Gacutan in: Proc. 1st Asian Fish. Forum, J.L. Maclean, L.B. Dizon and L.V. Hosillos, eds., Asian Fish. Soc., 311-313, (1986).

RED TIDES IN TOLO HARBOUR, HONG KONG

Lam, Catherine W.Y. and K.C. Ho
Environmental Protection Department, Hong Kong

ABSTRACT

Tolo Harbour, a 15km long landlocked inlet in the northeastern part of Hong Kong, has been severely affected by red tides over recent years. The frequency of occurrence has markedly increased since 1980 with 69 incidents recorded from 1980-1986. A corresponding increase in dinoflagellate numbers in the inlet was found. More than twenty causative organisms were identified. *Prorocentrum triestinum*, *P. dentatum*, *P. sigmoides*, *Gymnodinium* sp., *Noctiluca scintillans* and a variety of small flagellates occurred most frequently. Red tides were found throughout the whole year with peak occurrence in spring and autumn. The water quality, oceanography and meteorological conditions characterising red tide occurrences during the period from 1976-1986 were analysed and reported. The increasing red tides are a consequence of accelerated eutrophication in Tolo Harbour following intensive urban development in the catchment. The nutrient loadings of nitrogen and phosphorus have increased more than two-fold in the past decade. Concurrent increase in the nutrient levels in the inlet water has been found. The major impact of red tides at present is causing fishkills by anoxia.

INTRODUCTION

Tolo harbour is a sea inlet in the northeastern part of Hong Kong in the sub-tropical region (Fig. 1). The long and narrow marine bay which is 15km long and only 1km wide at its mouth is landlocked with slow tidal exchange. The catchment at the head of the bay has been undergoing intensive urban development since mid-1970's. There has also been a decline in water quality in the inner parts of the bay with frequent occurrence of red tides over recent years in the 1980's [1]. This paper analyses information on red tides which occurred in Tolo Harbour during the period from 1976-1986. Existing water quality monitoring data collected during the same period, together with available background oceanographic and meteorological information, were used to identify conditions which were conducive to red tide formation in this sub-tropical environment.

FIG. 1. Geographical location of Hong Kong and Tolo Harbour.

RED TIDES: BIOLOGY, ENVIRONMENTAL SCIENCE, AND TOXICOLOGY
Okaichi, Anderson, and Nemoto, Editors

EXPERIMENTAL

The water quality in Tolo Harbour has been monitored since 1976. Water samples were collected from fixed stations at biweekly intervals from three depths (surface, middle, bottom). A wide range of physical, chemical and biological determinants were analysed including phytoplankton identification and enumeration. Red tide occurrences which were detected during water sampling were recorded. Details, including the nature of the causative organisms and their concentrations, the location and extent of areas affected, and the occurrence of fishkills were noted.

RESULTS AND DISCUSSION

The number of red tide incidents has increased markedly in Tolo harbour in the 1980's (Fig. 2). Whilst there were only 4 red tide reports in the 1970's, a total of 69 incidents were recorded from 1980 to 1986. A corresponding increase in dinoflagellate numbers in the inner part of Tolo Harbour was also observed (Fig. 3).

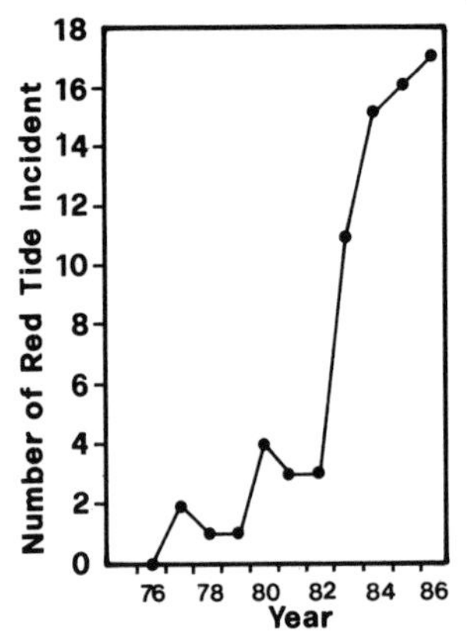

FIG. 2. Increase of red tide incidents in Tolo Harbour from 1976-1986.

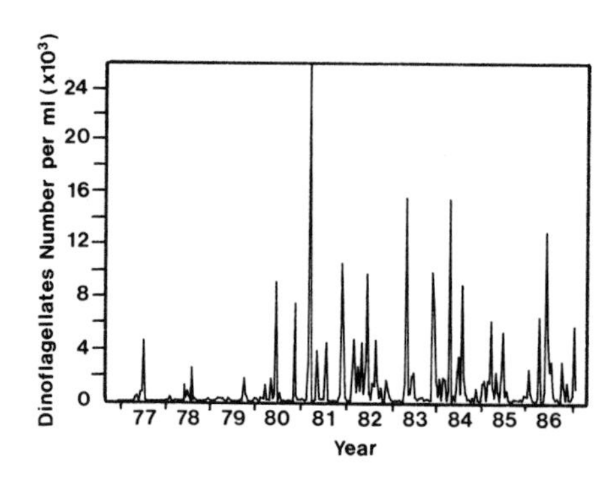

FIG. 3. Increase of dinoflagellate numbers in Tolo Harbour from 1976-1986.

Twenty major causative organisms were identified (Fig. 4). Among these, the most dominant organisms were Prorocentrum triestinum, P. dentatum, P. sigmoides, Noctiluca scintillans, Gymnodinium species and a wide range of small flagellates.

Red tides occurred throughout the whole year with high rates of occurrences during spring and autumn (Fig. 5). Nearly half of the incidents were recorded from March to May. On the other hand, incidents were rarely reported during the summer months of July and August. The spring peak relates to the additional nutrient supplies from increased river discharges during the first major flush and the resumption of warm temperatures after winter. During summer, the surface water is devoid of nutrients after the spring algal bloom. However, there is no replenishment from the nutrient-rich bottom water as the water body is vertically stratified with both thermal and salinity gradients. The condition becomes favourable again for algal growth at the advance of the autumn overturn.

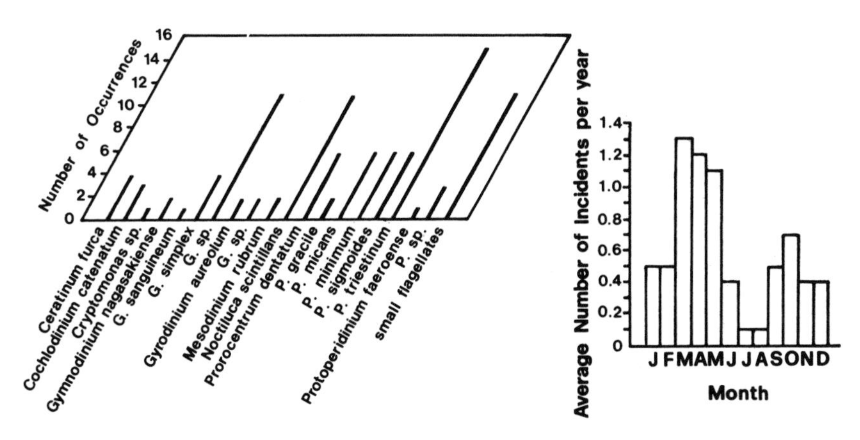

FIG. 4. Dominant red tide causative organisms in Tolo Harbour.

FIG. 5. Seasonal pattern of red tide occurrences in Tolo Harbour.

Table I. Environmental conditions associated with red tide occurrences in Tolo Harbour from 1976-1986.

Parameter	Range
Water temperature °C	16 - 30
Salinity %	28 - 32
pH	8.2 - 8.6
Wind : direction	NEE to E
Beaufort grade	3 - 4
Speed knots/hr	11 - 16
Relative humidity %	70 - 90
Global monthly bright sun-shine, hr	50 - 250
Inorganic nitrogen mg/l	0.05 - 0.2
Inorganic phosphorus mg/l	0.005 - 0.015
Dissolved oxygen mg/l	
surface water	7 - 11
bottom water	3 - 5
Current speed m/s	0.044 (average)

The environmental conditions which were associated with the majority of red tide occurrences are presented in Table I. Some of these conditions appear to be favourable for red tide formation in Tolo Harbour. The sub-tropical climate with mild temperature range and good sunshine allow a wide variety of red tide causative organisms to bloom throughout the year. The narrow ranges of salinity and pH do not introduce extreme conditions. The moderate to fresh north-easterly winds assist to aggregate red tide in the inner part of the inlet. The calm water with slow current movement also provides a stable environment for bloom development. Nutrient concentrations were recorded at rather low ranges during red tides. However, due to the cause-effect relationship of algal growth and nutrients, it would be difficult to define nutrient criteria levels for red tide initiation from the present data. It is noteworthy that there is a difference in oxygen status in the water column during red tides. Whilst the surface water is supersaturated, the bottom water tends to suffer from oxygen depletion. In fact, the impact of red tides so far has been causing fishkills mainly by anoxia.

During the past decade, the nutrient loadings into Tolo Harbour, arising mainly from livestock wastes and domestic sewage had increased more than two-fold in 1985 (2000kg N and 450kg P per day) as compared to 1976 (800kg N and 200kg P per day). A concurrent increase in nutrient concentrations in the water body was also observed (Table II). Coincidentally, the increase of red tide incidents corresponded to the growth of human population in the catchment (Fig. 6).

Table II. Annual median concentrations of inorganic nitrogen and phosphorus in inner Tolo Harbour from 1977-1986.

Year	Annual Median mg/l NO_3-N	PO_4-P
1977	0.005	0.008
1980	0.004	0.007
1983	0.026	0.026
1986	0.135	0.056

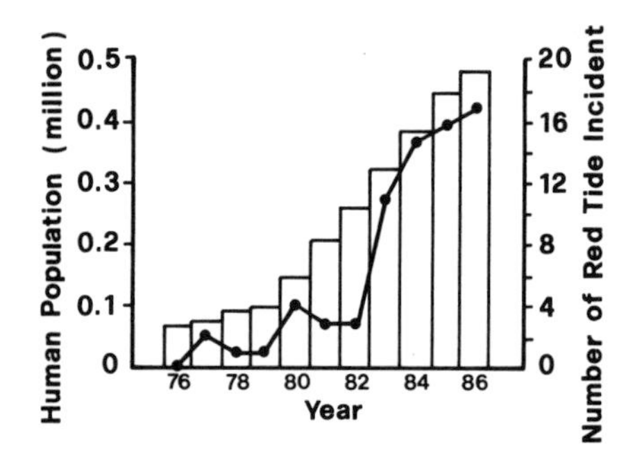

FIG. 6. Corresponding increase of red tide incidents with rising human population.

The increase of red tides in Tolo Harbour is therefore a consequence of accelerated eutrophication in the marine bay following intensive urban development in the catchment. Similar situation occurs in other areas of the world subject to eutrophication [2, 3]. The Government is vigilant to reduce red tide outbreaks and the implementation of a wide range of environment control measures is in progress to improve the eutrophic condition in Tolo Harbour.

ACKNOWLEDGEMENT

The Director of Environmental Protection, Hong Kong Government is acknowledged for his permission to publish this paper. We also wish to thank S.C. Cheung for technical assistance.

REFERENCES

1. P.R. Holmes and C.W.Y. Lam. Asian Marine Biology 2, 1 (1985).
2. J.L. MacLean and A.W. White in: Toxic Dinoflagellates, Proc. of the 3rd Intl. Conf. on Toxic Dinoflagellates, D.M. Anderdon, A.W. White and D.G. Baden, eds. (Elsevier North-Holland, New York 1985) pp.517-520.
3. F.J.R. Taylor in: The Biology of Dinoflagellates, F.J.R. Taylor ed. (Blackwell Oxford 1987) pp.399-502.

RED TIDES DUE TO NOCTILUCA SCINTILLANS (MACARTNEY) EHRENB. AND MASS MORTALITY OF FISH IN JAKARTA BAY

Q. Adnan

Center for Oceanological Research and Development-LIPI
Jakarta, Indonesia

ABSTRACT

In Jakarta Bay, red tides due to Noctiluca scintillans have been found throughout the year. Besides Noctiluca red tides, there have been blooms of diatoms dominated by Skeletonema costatum, Chaetoceros, Thalassiothrix and Coscinodiscus. Other dinoflagellates such as Ceratium, Dinophysis and Peridinium also caused red tides. Mass mortality of demersal fishes and benthic organisms was first reported in July, 1986 and was suspected to be related to the red tides which occurred in the nearby area.

INTRODUCTION

Jakarta Bay is an important industrial area, with two ports, oil terminals and an electric plant,and it is also important as a large fishing ground and a center for recreation. On the other hand, there are many rivers which bring much of the domestic, agricultural and industrial wastes into the bay. The river inflow thus influences the bay water, resulting in increased nutrients and growth of phytoplankton, which in turn affects other organisms. Therefore, continuous monitoring of plankton organisms as well as environmental factors is necessary.

In 1986 a mass mortality of demersal fishes occurred in Jakarta Bay. Red tides were suspected to be one of the causes of this phenomenon. The purpose of this communication is to report Noctiluca as the causative species of the most frequent red tides in the bay, the occurences of other phytoplankton groups and to discuss their relationships with the mass mortality of fish.

MATERIALS AND METHODS

Plankton surveys were carried out at a total of 12 areas (Fig. 1, A-L) in Jakarta Bay from 1975 to 1979. One to four stations were set in each area. In 1986-1987, Areas B, C and E were surveyed.

Samples were collected with a conical net (length: 120 cm; diameter: 31 cm; mesh size: 75 µm). A flowmeter was attached to the mouth of the net to determine the volume of water filtered. The net was towed horizontally at the surface for about five minutes. The sample was preserved in 4% formalin buffered with sodium tetraborate. Samples were examined in fractions using a chamber slide under a compound microscope.

RESULTS

Red tides due to Noctiluca scintillans (Macartney) Ehrenb. were common in Jakarta Bay. When red tides occurred the water always looked green, as if small green leaves had accumulated over a large area. The green color is due to the presence of a green flagellate, Pedinomonas noctilucae, in the vacuole of Noctiluca (1)

The occurrences of Noctiluca in Jakarta Bay are listed in Table 1. In November 1975, January 1976, May 1977 and July 1987, red tides occurred mostly on the east coast of the bay. In January 1978, red tides occurred along the entire coast of the bay, whereas in July 1979 it occurred on the middle to east coasts. In July 1986, when a mass mortality of demersal fish and benthic organisms occurred on the middle to the east coasts, red tides were observed at Station 3 in the Area E, close to the middle coast. High

RED TIDES: BIOLOGY, ENVIRONMENTAL SCIENCE, AND TOXICOLOGY
Okaichi, Anderson, and Nemoto, Editors

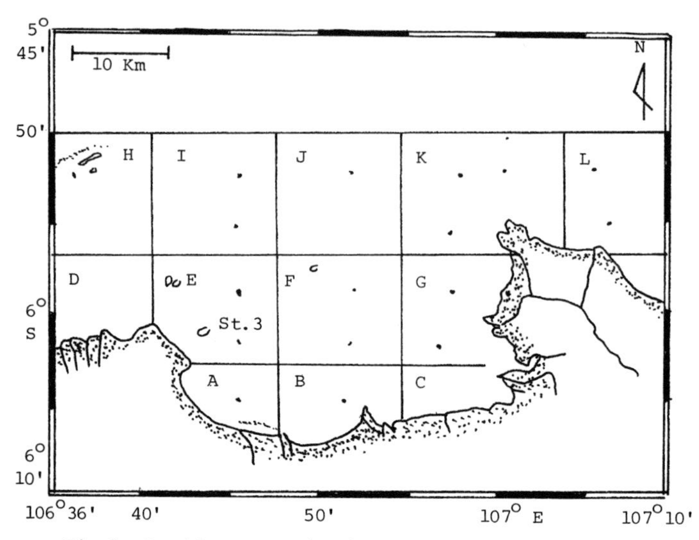

Fig. 1. Sampling areas (A-L) and stations (dots) in Jakarta Bay.

densities of Noctiluca have been observed mostly in Areas B, C and E with a maximum of 5.3 x 10^6 cells/m3 (Table 2). Other dinoflagellates which occurred in high densities were Ceratium (4 x 10^5), Dinophysis (8 x 10^5) and Peridinium (5 x 10^5).

Mass mortality of fish was first reported on 31 July by local fishermen, who, a week before, had found many fishes and benthic animals dead and floating with ruptured stomachs and black spots on the body (2). High concentrations of ammonia (9.21-9.49 ug atom N/l), nitrate (20.2-25.1 ug atom N/l) and phosphate (7.8-32.7 ug atom P/l) were observed at Stn. 3 on 5 August 1986, just after the mass mortality.

Besides Noctiluca red tides, diatom blooms have also been observed in Jakarta Bay, mostly dominated by Chaetoceros. The Chaetoceros bloom extended from the coast to offshore areas in November 1975, May 1976 and January 1977. In November 1977, it occurred together with Skeletonema in the middle offshore and with Coscinodiscus along the entire coast. In January and November 1978, the Chaetoceros bloom occurred in the eastern part, while in July 1979 it occurred with Thalassiothrix offshore. The Skeletonema bloom occurred mostly in the middle part of the bay. The Chaetoceros bloom looked white to brown and the Skeletonema bloom looked green.

TABLE 1. The occurrence of Noctiluca in Jakarta Bay. R: red tide. +: occurrence. -: absence. blank: no data.

Months	Area							
	A	B	C	D	E	F	G	H - K	L
Nov. '75	+	+	R	+	+	+	R	+	-
Jan. '76	-	-	-	-	-	-	-	-	-
May '76	+	+	R	+	+	+	+	+	+
Jan. '77	R	+	-	+	+	-	+	+	+
May '77	-	-	+	-	-	-	+	+	R
Nov. '77	+	+	+	+	+	-	+	+	+

TABLE 1. (continued)

Months	A	B	C	D	E	F	G	H - K	L
						Area			
Jan. '78	R	R	R	+	+	+	+	+	R
Nov. '78	-	R	+	-	-	+	-	+	+
July '79	+	R	R	+	+	+	+	+	+
July '86					R				
Feb. '87					+				
May '87		R	R						
June '87		-	-						
July '87		+	R						

TABEL 2. High densities of Noctiluca in Jakarta Bay (x 10^6 cells/m)

Month	Area	Density
May '76	C	2.1
Mat '77	C	2.1
Jan. '78	A	1.3
Jan. '78	C	1.0
July '79	C	1.15
July '86	E	0.6
July '87	C	5.3

DISCUSSION

According to the present survey, it seems that Noctiluca populations exist in Jakarta Bay throughout the year and cause red tides intermittently in accordance with changes of environmental condition. The high densities of Noctiluca and neritic diatoms in the coastal areas suggest that the inner part of the bay is under influence of eutrophication. At present, the cause of the high fish mortality is unknown. However, the high concentrations of Noctiluca and nutrients, especially ammonia, near the mortality sites on 5 August 1986 suggest that they are related. According to Aier (3, 4) a red tide of Noctiluca was toxic to fish and marine invertebrates. Halstead (4) and Morton and Twentyman (3) also reported Noctiluca to be toxic. According to Okaichi and Nishio (4), most of the toxicity of Noctiluca is due to vacuolar ammonia. Thus the possibility that Noctiluca caused the fish mortality cannot be ruled out. Establishing a red tide monitoring program and bioassay studies on the toxicity would be necessary for further understanding of the mechanisms of this mass mortality phenomenon.

ACKNOWLEDGEMENTS

The author thanks Prof. T. Nemoto and the JSPS for supporting of the author's attendance at the International Symposium on Red Tides. Dr. K. Romimohtarto is acknowledged for his guidance. Thanks are also to Drs. Y.Fukuyo, S.Nishida and Messrs. A.B. Sutomo, O.H. Arinardi, S.Thayeb and D.P.Praseno for their suggestions, and to Mr. Soedirdjo and Ms.Sugestiningsih for their help in processing data.

REFERENCES

1. B.M. Sweeney, J. Phycol. 14, 116, (1978)
2. "Kompas", a newspaper, Jakarta, 11th and 13th August, (1986)
3. B. Morton and P.R. Twentymen, Environmental Res., 4,544, (1971)
4. T. Okaichi and S. Nishio, Bull. Plankton Soc. Japan, 23,75,(1976)

RED TIDE DISCOLOURATION AND ITS IMPACT ON FISHERIES

Devassy, V.P.

National Institute of Oceanography
Dona Paula, Goa 403 004, India.

ABSTRACT

Incidents of intense red tides in the coastal waters of Goa, west coast of India (73°30'-73°50' E & 15°20'-15°40'N) during September 1973, May 1977 and February - April 1987 were studied. The red tides were caused by *Noctiluca scintillans* (Macartney), which ranged in concentration from 51.36 to 1004 x 10^6 cells m^{-3} during different seasons. Unlike the 1973 and 1977 red tides which imparted pink and orange colouration respectively, the 1987 one was green due to the presence of the flagellate *Pedinomonas noctilucae* (Subr.) Sweeney inside *Noctiluca*. This red tide extended over 600 km on the central west coast of India and prevailed intermittently for more than three months. Although the earlier two red tides did not provide any direct evidence of fish mortality, the 1987 one was followed by a substantial fall in fish catch. The prevailing environmental conditions and the probable impact on the fisheries during such red tides are discussed.

INTRODUCTION

The occurrence of red tides along the coasts of India has been fairly widespread. Hornell (1917) reported a *Noctiluca* red tide from the south west coast and concluded that it was not an active agent causing fish mortality, as Prasad (1953) also found from the east coast. However, according to Aiyar, (1936) oxygen deficiency resulted due to swarming and decay of *Noctiluca*, causing mortality of marine fauna in the east coast. On the south west coast, Bhimachar and George (1950) found that decaying *Noctiluca* caused mechanical obstruction to the movement of fish and its foulness was toxic to fish, causing fish to die or avoid such areas. Subrahmanyan (1954) described a flagellate viz. *Hornellia marina* occurring in the central west coast, which coloured the water green and accompanied mortality among fishes and crustaceans.

Only a few recent reports are available on toxic red tides, mortality of marine animals and shellfish poisoning, in India. One such report is from the east coast (Bhat, personal communication). Another report of paralytic shellfish poisoning and subsequent human death is by Karunasagar *et al.* (1984) from the west coast. In the present communication an attempt has been made to evaluate the occurrence of *Noctiluca* red tides and its impact on fisheries on the west coast of India.

RED TIDES: BIOLOGY, ENVIRONMENTAL SCIENCE, AND TOXICOLOGY
Okaichi, Anderson, and Nemoto, Editors

MATERIALS AND METHODS

Surface water samples were collected from between 73°30' - 73°50'E and 15°20' - 15°40'N. Noctiluca was counted and related environmental parameters, mainly dissolved oxygen and nutrients, were measured. Observations on Noctiluca red tides were spread over 3 different periods (September 1973, May 1977 and February 1987) when diverse environmental conditions prevailed.

RESULTS

During September 1973, a massive Noctiluca scintillans red tide was observed in the vicinity of a fertilizer factory on the central west coast of India. Incidentally, during this period a large scale fish mortality was reported from the region. On close observation it was noticed that an intense Noctiluca red tide occurred in the area. In the initial stages of the red tide, the Noctiluca count was 51 million m^{-3} and during the peak it was more than 1000 million m^{-3}. The red tide was fairly widespread, occupying several square km and had a pinkish brown colour. After a fortnight the red tide declined and the counts dropped to 0.05 million m^{-3} but it still retained the pinkish colour. During the period of study, dissolved oxygen at the surface varied from 4.1 to 6.0 ml l^{-1}, the inorganic nutrients (viz. PO_4-P and NO_3-N) varied from 2.05 to 4.59 and 0.43 to 0.94 µg at l^{-1} respectively.

The next Noctiluca red tide was observed during May 1977 following a Trichodesmium erythraeum bloom that ended at the end of March. This was followed by environmental disturbances and observations were discontinued for 3 weeks. When observations were resumed, the Trichodesmium red tide had disappeared and a mixed diatom bloom comprising largely of species of Chaetoceros had succeeded in its place. This was followed by a swarm of herbivores (viz. Evadne tergestina). Gradually Noctiluca populations increased and counts attained 640 million m^{-3}. This time, the bloom was a beautiful orange colour. Nutrients (viz. PO_4-P and NO_3-N) were relatively higher and varied from 0.3 to 30 and 0.7 to 2.3 µg at l^{-1} respectively. No fish mortality was associated with this Noctiluca red tide.

The third and most recent Noctiluca discolouration was observed during the first week of February 1987 in the same area. This time it gave a slimy green carpetted appearance to the coastal waters. This discolouration extended for about 600 km on the central west coast of India. During this time (February) the Noctiluca population varied from 2 to more than 51 million cells m^{-3}. Species of Ceratium also occurred abundantly along with Noctiluca. During the red tide peak, dissolved oxygen was 3 ml l^{-1}, PO_4-P 2.4 and NO_3-N 9.4 µg at l^{-1}. The size of N. scintillans collected varied between 650 and 760 µm.

Noctiluca harbouring Pedinomonas noctilucae (Subr.) Sweeney was responsible for imparting the deep green colouration to the water. Pedinomonas measured from 5 to 9.5 µm in length and 4.4 to 7.8 µm in breadth.

Fishing operations and fish catch were adversely affected during the peak discolouration period. Observations made on midwater trawls during January-May revealed the following : During January when no discolouration existed the fish catch varied from 60 to 200 kg h^{-1} . However, from February to April, when the green discolouration persisted the fish catch also dwindled, which varied from 0 to 22, 12 to 65 and 7 to 60 kg h^{-1} in these months. However, during May the discolouration disappeared and the fish catch recovered (15-200 kg h^{-1}).

DISCUSSION

Noctiluca red tides occurred during 3 different periods when diverse environmental conditions prevailed. It was rather difficult to attribute a particular reason for the fish mortality that occurred during September 1973. The presence of upwelled waters was considered to be the causative factor, the fertilizer factory effluents were also suspected. Yet another group claimed that fish mortality occurred due to red tide. During the red tide, PO_4-P and NO_3-N values were fairly high. A similar increase in nutrient values was observed by Prasad and Jayaraman (1954).

The most recent *Noctiluca* red tide imparted a green colour to the water due to the presence of the flagellate *Pedinomonas noctilucae* (Subr.) Sweeney in large numbers. Subrahmanyan (1954) measured the length of the flagellate as 5-6 µm and breadth 4.5 µm. Sweeney (1976) recorded the length of this flagellate as 2-6 µm.

The fishery of the central west coast was adversely affected when the green discolouration persisted. Bhimachar and George (1950) reported abrupt setback to the fishery on the south west coast of India due to the abundance of *N. scintillans*. These authors reported a steep fall in fish catch, especially the shoaling commercial population. Commercial fishes, particularly the shoaling fishes avoid *Noctiluca* red tide (Prasad and Jayaraman (1954). Subrahmanyan (1954) was of the opinion that mortality could occur due to *Pedinomonas* and not *Noctiluca*. Thus *Noctiluca* red tides along with others are on the increase due to more and more industrialization and the introduction of increasing quantities of effluents (Holmes and Lam, 1985). In view of this the occurrence of red tides assume greater importance to the fishery of the region.

ACKNOWLEDGEMENTS

The author is grateful to Dr. A.H. Parulekar, Head, Biology Division and to Mr. Sreekumaran Nair and Mr. Joaquim Goes for their help during the course of this investigation.

REFERENCES

1. J. Hornell, Bull. Madras Fish. Dept., **11**, 53 (1917)

2. R.R. Prasad, Proc. Indian Acad. Sci., **38 B**, 40 (1953)

3. R.G. Aiyar, Curr. Sci., 4, 488 (1936).

4. B.S. Bhimachar and P.C. George, Proc. Indian Acad. Sci., 13,339 (1950).

5. R. Subrahmanyan, Proc. Indian Acad. Sci., 39B, 118 (1954)

6. I. Karunasagar, H.S.V. Gowda, M. Subbaraj, M.N. Venugopal and I. Karunasagar, Curr. Sci., 53, 247 (1984)

7. R.R. Prasad and R. Jayaraman, Proc. Indian Acad. Sci., 40B,49 (1954)

8. B.M. Sweeney, Journ. Phycol., 6, 79 (1976)

9. P.R. Holmes and C.W.Y. Lam, Asian Mar. Biol., 2, 1 (1985)

INCIDENCE OF PSP AND DSP IN SHELLFISH ALONG THE COAST OF KARNATAKA STATE (INDIA)

INDRANI KARUNASAGAR, K. SEGAR AND I. KARUNASAGAR
Department of Fishery Microbiology, University of Agricultural Sciences
College of Fisheries, Mangalore 575 002, India

ABSTRACT

A monthly monitoring program was carried out for shellfish toxicity along the coast of Karnataka during 1984-86. Paralytic shellfish poison (PSP) was detected during April 1985 and during March-April 1986. Diarrhetic shellfish poison (DSP) was noticed sporadically and its appearance did not relate to season.

INTRODUCTION

Recent developments suggest that paralytic shellfish poisoning (PSP) and diarrhetic shellfish poisoning (DSP) may be more widely distributed than was thought earlier. Outbreaks of PSP in India [1], Philippines [2], Malaysia [3] and other Indo-Pacific countries [4] and Venezuela [5] have shown very convincingly that PSP is a problem of tropical/subtropical regions as well. Even DSP which was first detected in Japan [6] has been shown to be a problem in The Netherlands [7] and Chile [8]. Moreover, many cases of DSP might be missed because of the mild nature of the symptoms which are easily mistaken for bacterial food poisoning.

In India, following the 1983 outbreak of PSP [1], we undertook a monthly monitoring of shellfish harvesting areas along the 300 km coastline of Karnataka State. Results of the study during 1984-86 are presented here.

MATERIALS AND METHODS

Shellfish samples

Commercially important shellfish samples viz. clams (*Meretrix* sp, *Katelysia* sp, *Paphia* sp), oysters (*Crassostrea* sp) and mussels (*Perna* sp) were collected from the following estuaries: Kumble, Nethravathi, Mulki, Malpe/Udyavara, Kollur, Tadri and Kali. Samples were collected monthly depending upon availability.

Mouse bioassay for PSP and DSP

Presence of PSP in shellfish samples was tested by the mouse bioassay technique of AOAC [9]. Assay of DSP was carried out as described by Yasumoto *et al*. [6, 10].

RESULTS AND DISCUSSION

During the two year study (1984-86), PSP was detected on three occasions. A sample of oysters collected from Tadri estuary during April 1985 showed presence of PSP, but the levels were only 320 MU/100 g which is within the safety limit suggested by US FDA [11]. However, during the

RED TIDES: BIOLOGY, ENVIRONMENTAL SCIENCE, AND TOXICOLOGY
Okaichi, Anderson, and Nemoto, Editors

last week of March 1986, some shellfish showed presence of PSP, levels being highest in shellfish collected from Udyavara/Malpe area (1100-1200 MU/100 g) and less in other shellfish collected towards north (Tadri, Kali) and south (Mulki) of this estuary. Sommer and Meyer [12] estimated that sickness may result from shellfish containing about 1000 MU and to cause death, about 20,000 MU/100g are required. In our earlier study [1], we noted that shellfish containing 18,000 MU/100 g caused death of a boy and illness in several others. However, Prakash et al. [13] suggested that consumption of 1000 μg might bring about mild illness, 1900 μg moderate illness and 2000 μg severe illness. One MU is about 0.2 μg toxin (range being 0.16 μg-0.22 μg depending on strain of mice used [13], and by these standards, one should consume about 400 g of shellfish mentioned above (Udyavara/Malpe area) to get mild illness. No symptoms were reported by the local people at the time of sampling and the toxin levels declined to safe limits by a week's time.

These results suggest that a transient toxic dinoflagellate bloom might have occurred during March 1986. However, examination of water from shellfishing areas at the time of sampling the shellfish did not reveal the presence of any toxic dinoflagellates. The observation that the toxin levels were low and declined in a week's time further strengthens the view that the bloom was not very intense and was very transient. Perhaps the bloom occurred in Malpe/Udyavara area where toxicity was highest and spread to other areas. There are instances in literature when shellfish toxicities were noted even when dinoflagellate populations were small [14].

It is interesting to note that incidence of PSP along Karnataka coast is seasonal. The first outbreak occurred during April 1983 [1] and during this survey, toxicity was noticed during March-April 1985 and 1986.

Results in Fig.1 show that DSP was also present in several shellfish samples examined, however the incidence was not seasonal. The levels ranged from 0.37 to 1.5 MU/g hepatopancreas. Yasumoto et al. [15] reported that the minimum amount of toxin to induce symptoms in an adult was 12 mouse units based on analysis of left overs of patients' meals. By these standards, the toxin detected in our samples were low, but this should not lead to complacency and shellfish consumers and public health personnel must be alert to the problem. Moreover, the Japanese Government has introduced a regulation that the maximum allowance level of DSP in shellfish meat is 5 MU/100 g meat. This will certainly affect shellfish trade.

Studies carried out in Japan [6, 15] and The Netherlands [7] show that *Dinophysis* sp and *Prorocentrum* sp may be responsible for DSP in shellfish. Our studies (Karunasagar et al., this proceedings) also show the presence of these dinoflagellates in different shellfish harvesting areas along the coast of Karnataka. As with DSP, no seasonality is observed with these dinoflagellates.

ACKNOWLEDGEMENTS

The authors are thankful to Prof. H.P.C. Shetty for his pertinent suggestions during the course of the work. Thanks are due to Mr. M. Gilchrist of FDA, USA and to Dr. Yasumoto, Faculty of Agriculture, Tohoku University, Sendai, Japan for kindly providing reference PSP and DSP respectively. This project was supported financially by the Karnataka State Council for Science and Technology.

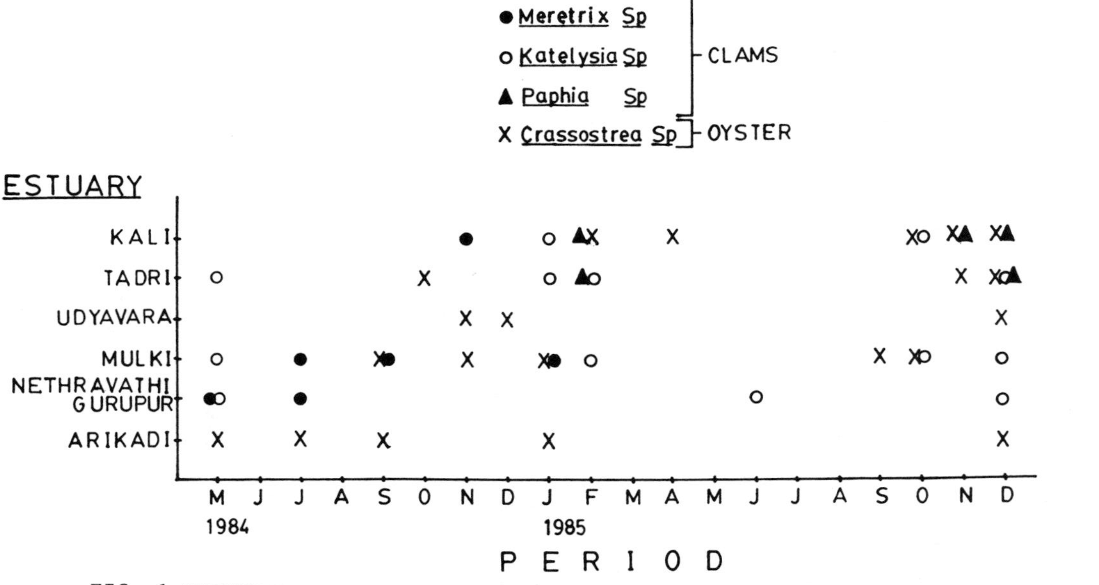

FIG. 1 INCIDENCE OF DSP IN DIFFERENT SHELLFISH SAMPLES ALONG KARNATAKA COAST DURING 1984-85.

REFERENCES

1. I. Karunasagar, H.S.V. Gowda, M. Subburaj, M.N. Venugopal and I. Karunasagar, Curr. Sci., 53, 247 (1984).
2. L.M. Arafiles, R. Hermes and J.B.T. Morales in: Toxic Red Tides and shellfish toxicity in Southeast Asia A.W. White, M. Ansaku and K. Hooi, eds. (SEAFDEC and IDRC 1984) p. 92.
3. A.A. Jotty in: Toxic Red Tides and Shellfish Toxicity in Southeast Asia A.W. White, M. Ansaku and K. Hooi, eds. (SEAFDEC and IDRC 1984) p.33.
4. J.L. Maclean in: Toxic Red Tides and Shellfish Toxicity in Southeast Asia A.W. White, M. Ansaku and K. Hooi,eds. (SEAFDEC and IDRC 1984) p. 92.
5. E. Ferraz-Reyes-Vasquer and I.B. Brezual in: Toxic Dinoflagellate Blooms D.L. Taylor, H.H. Seliger, eds. (Elsevier, North Holland 1979).
6. T. Yasumoto, Y. Oshima and M. Yamaguchi, Bull. Jpn. Soc. Sci. Fish. 44 (1978).
7. M. Kat in: Toxic Dinoflagellate Blooms D.L. Taylor and H.H. Seliger, eds. (Elsevier, North Holland 1979), 215.
8. L. Guzman, I. Canpodonico, Publ. Inst. Patagonia ser. Mon. 6 (1975).
9. W. Horwitz, Ed. In Official Methods of Analysis of Association of Official Analytical Chemists. 13th Ed. AOAC, Washington D.C. 298 (1980).
10. T. Yasumoto, V. Oshima, W. Sugawara, Y. Fuduyo, H. Oguri, T. Igarashi and N. Fujita, Bull. Jpn. Soc. Sci. Fish. 46, 1405 (1980).
11. Y. Hashimoto, Marine toxins and other bioactive marine metabolites, Japan Scientific Societies Press, Tokyo, 40 (1937).
12. H. Sommer and K.F. Meyer, A..M.A. Arch. Pathol. 24, 560)1937).
13. A. Prakash, J.C. Medcof, A.D. Tennant, Fish. Res. Bd. Can. Bull. 117, 15 (1971).
14. K. Tangen, Sarsia 68, 1 (1983).
15. T. Yasumoto, H. Murata, Y. Oshima, G.K. Matsumoto and J. Clardy in: Seafood Toxins E.P. Ragelis ed. (American Chemical Society, Washington D.C. 1984) p.207.

POTENTIALLY TOXIC DINOFLAGELLATES IN SHELLFISH HARVESTING AREAS ALONG THE COAST OF KARNATAKA STATE (INDIA)

I. Karunasagar, K. Segar and Indrani Karunasagar
Department of Fishery Microbiology, University of Agricultural Sciences
College of Fisheries, Mangalore 575 002, India

ABSTRACT

Monthly water samples were collected from major shellfish harvesting areas along the coast of Karnataka State (India) to study the presence of potentially toxic dinoflagellates during 1984-86. _Gonyaulax_ spp. was observed in two samples but the shellfish from these areas were not toxic suggesting that these may be non-toxic species. During April 1985 and March-April 1986, when low levels of PSP were detected in shellfish, no _Protogonyaulax_ or _Pyrodinium_ were observed in the water. Potentially DSP producing _Dinophysis_ and _Prorocentrum_ could be detected in the waters and acetone extracts of hepatopancreas from shellfish caused mouse death resembling that due to DSP.

INTRODUCTION

Filter feeding bivalve molluscs that have ingested toxic phytoplankton are known to cause shellfish poisoning and this is a recurring problem in many parts of the globe [1]. Toxic dinoflagellates have been often incriminated in the shellfish poisoning outbreaks. Toxic dinoflagellates belong to widely divergent genera like _Amphidinium_, _Dinophysis_, _Gambierdiscus_, _Gonyaulax_, _Gymnodinium_, _Ostreopsis_, _Prorocentrum_, _Protogonyaulax_, _Ptychodiscus_ and _Pyrodinium_ [2].

In India, there have been two outbreaks of paralytic shellfish poisoning (PSP), one on the east coast and the other on the west coast [3]. In the outbreak on the west coast, Karunasagar _et al_. (1984) detected PSP in incriminated shellfish and called for a study of the dinoflagellate population in the Indian water. Against this background, a study was undertaken to survey the major shellfish harvesting areas of Karnataka coast for the dinoflagellate populations and identify the potentially toxic genera which are incriminated in paralytic shellfish poisoning. The possibility of shellfish harbouring another type of toxin, diarrhetic shellfish toxin (DSP) due to ingestion of certain toxic dinoflagellates is recognised [4]. Identification of potential DSP causing dinoflagellates was also studied.

MATERIALS AND METHODS

Six sampling sites (Fig.1) were surveyed along the coast of Karnataka where there is commercial exploitation of clam and oyster beds, from April 1984 to June 1986. Due to heavy monsoon, sampling could not be carried out between June and August. Plankton samples were collected by towing a plankton net from a dug out canoe. The net had an opening of 50 cm x 50cm, a length of 1 m and was made of bolting silk cloth # 25. Towing time was maintained constant at 15 min. The formalin preserved samples were analysed after making up the volume to 500 ml and taking a subsample of 0.1 ml. Identification of dinoflagellates was carried out as per description of Wood [5].

RED TIDES: BIOLOGY, ENVIRONMENTAL SCIENCE, AND TOXICOLOGY
Okaichi, Anderson, and Nemoto, Editors

RESULTS AND DISCUSSION

Dinoflagellates observed in different sampling sites are shown in Table 2. Peridinium and Ceratium occurred most frequently followed by Dinophysis, Prorocentrum and Noctiluca. Gonyaulax, was observed only during May 1984 in three estuaries. Members of genus Gonyaulax/Protogonyaulax and Prorocentrum have been shown to produce toxins causing PSP while species of Dinophysis and Prorocentrum have been implicated in DSP [2]. However, shellfish harvested from estuaries showing Gonyaulax/Protogonyaulax did not show presence of PSP. This suggests that the species observed was non-toxic. Prakash and Sarma [6] noted a red tide off Cochin due to Gonyaulax polygramma which by mouse bioassay technique was found to be non-toxic.

Studies carried out in the Indo-Pacific region indicate that Pyrodinium bahamense var compressa is the dinoflagellate responsible for PSP [7]. In our study, however, we did not observe Pyrodinium sp. During the course of this study, PSP was detected during April 1985 in oysters from Tadri estuary and during March-April 1986 from a number of shellfish from different estuaries, but plankton samples do not reveal presence of any known toxic dinoflagellate. It is possible that the toxic dinoflagellate bloom was very weak, patchy and transient and therefore it was missed during the monthly sampling. Tangen [8] cited instances where outbreaks of PSP occurred when toxic dinoflagellates were observed only as fragments in the mussel gut and as subordinate component of plankton. In our study we did not examine the gut contents of toxic shellfish.

During October 1984, in Tadri estuary a red tide was observed which was due to Peridinium granii. However, no toxicity was observed in shellfish. There are earlier reports of reddish discolouration of waters along Indian coast and these have been attributed to Gonyaulax polygramme [6], Noctiluca miliaris [9], Cochlodinium sp [10], Peridinium sp [11] and Gymnodinium sp [12]. However, no shellfish toxicities have been reported to accompany these red discolourations.

Several species of Prorocentrum and Dinophysis are known to cause DSP and venurupin shellfish poisoning. P. minimum and P. lima have been implicated in Norwegian waters [7] and Prorocentrum sp accompanied shellfish poisoning in Dutch waters [13]. Potentially toxic species of Dinophysis reported so far are D. acuminata, D. acuta, D. norvegica and D. fortii [4, 8, 13, 14]. In our study, we observed D. caudata in many shellfish harvesting areas and acetone edtracts of hepatopancreas from a number of shellfish samples showed death in mice similar to that of DSP. Further chemical characterisation of these extracts is needed but at this stage it can be speculated that D. caudata might also produce DSP. Neither the incidence of DSP nor of Dinophysis caudata was found to be seasonal.

ACKNOWLEDGEMENTS

The authors thank Prof. H.P.C. Shetty for his useful suggestions during the course of the work and Karnataka State Council for Science and Technology for financial support.

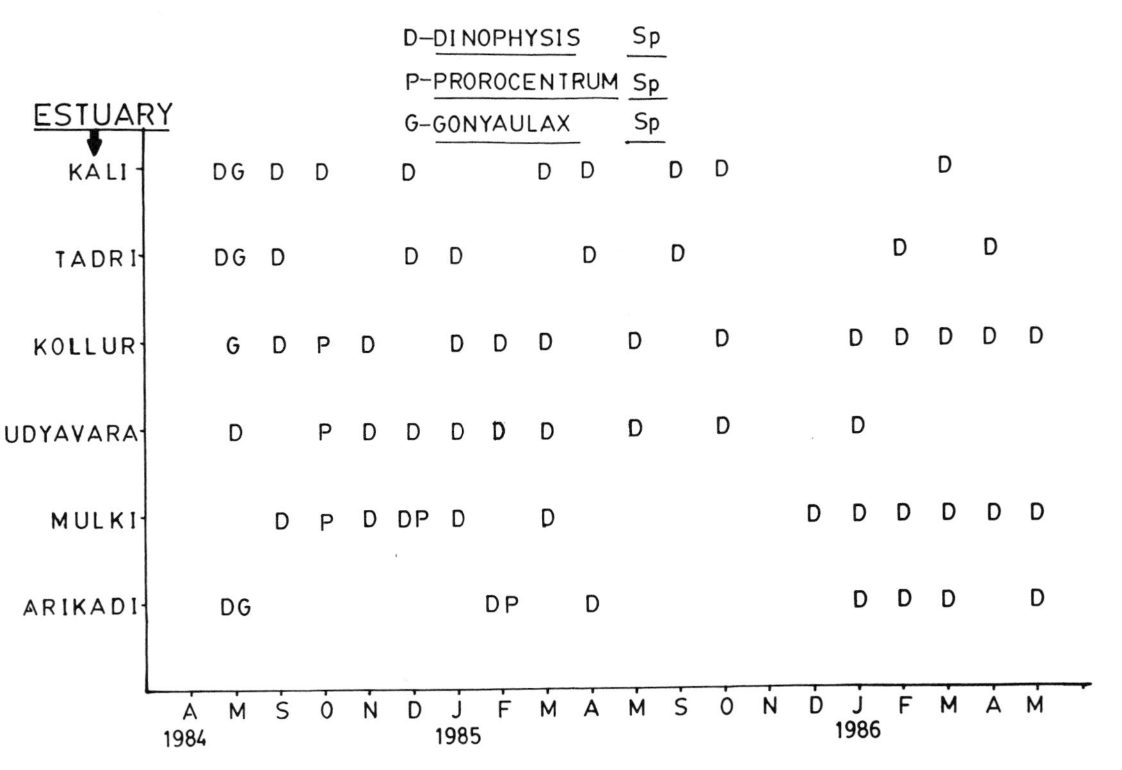

FIGURE 1. INCIDENCE OF POTENTIALLY TOXIC DINOFLAGELLATES ALONG KARNATAKA COAST DURING 1984–86.

REFERENCES

1. R.J. Schmidt and A.R. Loeblich, J. Mar. Biol. Assn., U.K., 59, 479 (1979).
2. K.A. Steidinger, Prog. Phyc. Res. 2, 147 (1983).
3. I. Karunasagar, H.S.V. Gowda, M. Subburaj, M.N. Venugopal and I. Karunasagar, Curr. Sci. 53, 247 (1984).
4. T. Yasumoto, Y. Oshima, W. Sugaware, Y. Fukuyo, H. Oguri, T. Igarashi and N. Fujita, Bull. Jpn. Soc. Sci. Fish. 46, 1405 (1980).
5. E.J.F. Wood, Aust. J. Mar. Freshwater Res. 5, 171 (1954).
6. A. Prakash and A.H.V. Sarma, Curr. Sci. 33, 168 (1964).
7. J.L. Maclean, in: Toxic Red Tides and Shellfish Toxicity in Southeast Asia (SEAFDEC and IDRC 1984) 92.
8. K. Tangen, Sarsia 68, 1 (1983).
9. B.S. Bhimachar and P.C. George, Proc. Ind. Acad. Sci. 31, 339 (1950).
10. J. Hornell and M.R. Naidu, Madras Fish. Bull. 17, 129 (1925).
11. J. Hornell, Madras Fish. Bull. 11, 53 (1917).
12. R. Subramanyan, Proc. Ind. Acad. Sci. 50, 43 (1959).
13. M. Kat, Antonie Van Leeuwenhoek 49, 417 (1983).
14. T. Yasumoto, Y. Oshima and M. Yamaguchi, Bull. Jpn. Soc. Sci. Fish. 44, 1249 (1978).

PERIODIC BLOOMS OF THE SILICOFLAGELLATE *DICTYOCHA PERLAEVIS* IN THE SUBTROPICAL INLET, KANEOHE BAY, HAWAII, U.S.A.

Taguchi, Satoru* and Edward A. Laws**
*Hawaii Institute of Geophysics and **Department of Oceanography
University of Hawaii at Manoa, 1000 Pope Road, Honolulu Hawaii 96822, U.S.A.

ABSTRACT

The silicoflagellate *Dictyocha perlaevis* was collected weekly for a period of one year at a station in the subtropical inlet of Kaneohe Bay, Oahu, Hawaii, U.S.A. Cell density was usually less than 5 cells$\cdot l^{-1}$ with maximum cell density of 47 cells$\cdot l^{-1}$. Periodic blooms occurred 11 times during the year. Median cell volume of 3.9×10^{4} μm^{3} was associated with a chlorophyll a (chl a) content of 49 pg Chla$\cdot$cell^{-1} during the bloom period. The median cell volume was not significantly different during the bloom and non-bloom periods, but the cellular chl a content was only 39 pgChla$\cdot$cell^{-1} during the non-bloom period. Median cellular photosynthetic rates of 0.052 and 0.043 pgC$\cdot$cell^{-1}h^{-1} were determined using a single cell isolation technique during the bloom and non-bloom period, respectively. The temporal pattern of ^{14}C incorporation indicated that *D. perlaevis* tended to accumulate storage products during non-bloom period and intermediate low molecular weight compounds at the expense of storage products during the bloom period. However, the percentage of ^{14}C allocated to protein was similar during bloom and non-bloom periods. These photosynthetic measurements suggest that *D. perlaevis* has a high photosynthetic activity associated with a high cellular chl a content during bloom periods. Relative growth rates were, however, similar during the bloom and non-bloom period and averaged 43±9% of nutrient saturated values as determined by the percentage of ^{14}C allocated to protein. The occurrence of blooms, therefore, seems to be associated with high nutrient-saturated growth rates as a result of favorable light and temperature conditions or other factors and not to a change in the degree of nutrient limitation.

INTRODUCTION

Silicoflagellates have chromatophores and are photosynthetic. They are ubiquitous in the world's ocean. Species of the genus *Dictyocha* have been reported as causing red tides (1). Before the sewage diversion of 1977, the frequent occurrence of red tides was sighted in Kaneohe Bay (2), possibly caused by *Dictyocha perlaevis*. Van Valkenburg and Norris (3) have reported the only successful effort to grow *Dictyocha fibula* in culture, and most information about the growth rate as a function of temperature and salinity was obtained as a result of that single study. Since then no physiological study has been done due to unsuccessful subsequent culture efforts until Taguchi and Laws (4) utilized a single-cell isolation technique to study *Dictyocha perlaevis*. *Dictyocha perlaevis* Frenquelli is one of the silicoflagellates reported from Kaneohe Bay, a subtropical inlet in Hawaii. We report here results of a seasonal study with the single-cell isolation technique to determine inorganic carbon assimilation and incorporation

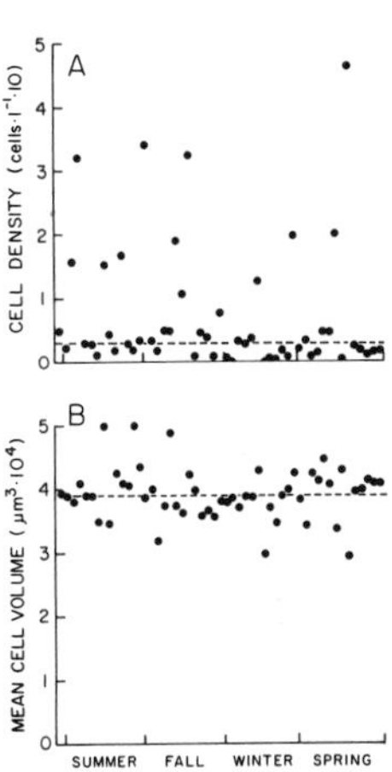

Fig. 1A: Cell density
B: Cell volume. Dashed line indicates median

RED TIDES: BIOLOGY, ENVIRONMENTAL SCIENCE, AND TOXICOLOGY
Okaichi, Anderson, and Nemoto, Editors

into major polymers by natural populations of *Dictyocha perlaevis*.

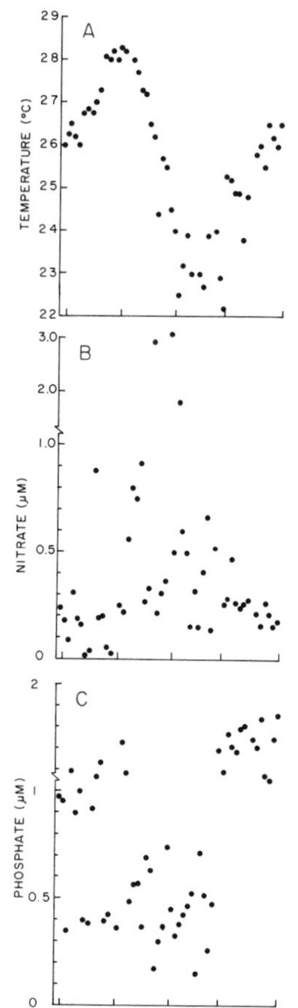

Fig. 2. A: Temperature B: Nitrate C: Phosphate

MATERIALS AND METHODS

A 35-μm mesh plankton net with a 30 cm diameter opening was towed vertically from 2 m above the bottom to the surface weekly from June 1983 to June 1984 in Kaneohe Bay, Hawaii. Photosynthetic measurements were made with a method described by Taguchi and Laws (4). Aliquots of net samples were preserved in Formalin solution at 2% final concentration. Mean cell dimensions were determined from microscopic measurements of 50 or more cells made on a Zeiss inverted microscope. Volumes were calculated on the assumption that the cells were oval shaped. Environmental data including photosynthetic active radiation (PAR), water temperature, silicate, nitrate, and phosphorus concentration were collected as described by Taguchi and Laws (5).

RESULTS

The seasonal occurrence of *Dictyocha perlaevis* is shown in Fig. 1. Cell density was usually less than 5 cells$\cdot l^{-1}$ with maximum cell density of 47 cells $\cdot l^{-1}$. Water temperature, PAR (data not shown), and nitrate concentration showed clear seasonality while phosphorus concentration showed only a moderate seasonality (Fig. 2). Blooms of *D. perlaevis* occurred 11 times during the course of the year but the timing of the bloom bore no clear relationship to environmental factors.

The median cell density was 23 cells$\cdot l^{-1}$ (95% confidence range: 13-34) and the median cell volume was 3.9×10^4 μm^3 (95% confidence range: $3.3\text{-}4.3 \times 10^4$) at the peak of blooms. The chl a content was 49 pgChla$\cdot$cell^{-1} (95% confidence range: 37-63) and the chl a concentration therefore 1.0 ng$\cdot l^{-1}$ (95% confidence range: 0.44-1.9). The median contribution of *D. perlaevis* to chl a concentration of phytoplankton cells larger than 35μm in diameter was 6.4% (95% confidence range: 3.3-62). During the non-bloom period the median cell volume was 3.9×10^4 μm^3 (95% confidence range: $3.7\text{-}4.0 \times 10^4$) and the chl a content was 39 pgChla$\cdot$cell^{-1} (95% confidence range: 23-48).

Cellular photosynthetic rate ranged from 0.0031 to 0.13 ngC$\cdot$cell^{-1}h^{-1} with a median of 0.043 ngC$\cdot$cell^{-1}h^{-1} (95% confidence range: 0.039-0.059) (Fig. 3). Photosynthetic rate normalized to cellular chl a ranged from 0.22 to 3.3 pgC$\cdot$[pgChla]$^{-1}$h^{-1} (95% confidence range: 0.89-1.3) (Fig. 3). Although there was no apparent relationship between photosynthetic rate and the occurrence of blooms, both photosynthetic rates normalized to either cell numbers or cellular chl a showed a significant seasonality; cellular photosynthetic rates in summer were about 3.3 times higher than in winter while summer photosynthetic rates per chl a were about 1.5 times higher than in winter (Table 1).

Time-series experiments of ^{14}C incorporation into major polymers show that during non-bloom periods, the incorporation of ^{14}C into protein is higher during photoperiod than at night. Cells maintain a relatively constant incorporation rate over time during bloom periods. The former condition is similar to that observed during the lagphase of growth (7). The latter condition is similar to that observed during exponential growth (8).

A total of eight experiments were conducted to study carbon partitioning patterns (Table 2). No significant difference of % ^{14}C allocated in protein was observed between bloom and non-bloom period. Both PNA and INT showed significant differences between bloom and non-bloom periods. During bloom periods % ^{14}C allocated in PNA reached 47±9%, while % ^{14}C allocated in INT reached 32±7% during non-bloom period. Although the difference in % ^{14}C allocated to lipids was marginal, lipid synthesis seemed to be higher during non-bloom periods.

Table 1. Summary of primary production, chl a concentration due to *D. perlaevis* and its photosynthetic rate normalized to cell and cellular chl a content during 12-month study in Kaneohe Bay. n is the number of data points obtained from weekly sampling. Values listed are median values. Numbers in parentheses are 95% confidence intervals (6).

Season	Primary Production ($ngC \cdot l^{-1}h^{-1}$)	Chlorophyll a ($pgChl\underline{a} \cdot l^{-1}$)	Photosynthetic Rate ($ngC \cdot cell^{-1}h^{-1}$)	Photosynthetic Rate ($pgC[pgChl\underline{a}]^{-1}h^{-1}$)
Summer	0.24	160	0.058	1.2
n=14	(0.11-0.74)	(77-830)	(0.040-0.091)	(0.81-2.0)
Fall	0.23	270	0.047	0.96
n=13	(0.033-0.93)	(67-1300)	(0.013-0.086)	(0.53-1.6)
Winter	0.031	46	0.018	0.80
n=14	(0.0060-0.13)	(10-170)	(0.011-0.043)	(0.56-1.0)
Spring	0.11	99	0.058	1.0
n=14	(0.070-0.24)	(74-200)	(0.040-0.092)	(0.95-2.1)

DISCUSSION

In the previous study (5) a distinct seasonality was noted for most environmental factors. However, the amplitudes were lower than for typical temperate waters. When biomass and photosynthetic rates of *D. perlaevis* are examined (Table 1), both tend to be low in winter compared to summer. Under these predictable environmental conditions and seasonal changes in photosynthetic rate, multiple blooms of *D. perlaevis* throughout the four seasons are not expected. Although different species vary independently in the total phytoplankton populations, one might expect a direct relation between physiological activity but not necessarily biomass and environmental factors. The contribution of *D. perlaevis* to the chl a concentration of phytoplankton cells larger than 35 μm in diameter ranged from 1 to 99% even at the peaks of bloom. The contribution of this net phytoplankton population to total chl a was higher than 25%. The observed multiple blooms cannot be associated directly with photosynthetic characteristics of *D. perlaevis*, and must be caused by unknown factors. Photosynthetic rates seemed to be controlled by environmental factors such as temperature and/or light. It is interesting that a reverse association is observed between cellular photosynthetic rates and nitrate concentration. This general trend also exists when photosynthetic rate per chl a is calculated (Fig. 3). This ratio is indicative of nutritional conditions of phytoplankton (9). These observations may suggest that photosynthetic rate of *D. perlaevis* is not limited by nitrate but rather by some other nutrient.

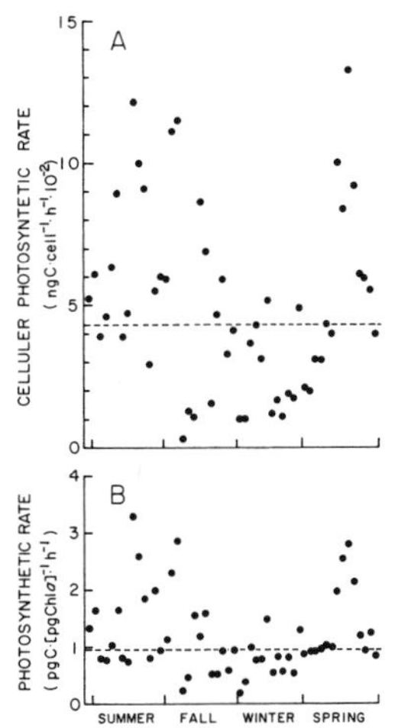

Fig. 3. Photosynthetic rate normalized by cell (A) and cellular chlorophyll a (B). Dashed line indicates median.

Characteristic differences in the pattern of ^{14}C incorporation into major polymers was observed between bloom and non-bloom periods. The results indicate that *D. perlaevis* accumulates storage products during non-bloom periods. *D. perlaevis* may enhance INT synthesis at the expense of storage products during bloom conditions. Relative growth rates calculated from % ^{14}C protein (8), however, do not show much difference between bloom and non-bloom conditions because the change of INT is cancelled by the change in storage products (Table 2). The calculated relative growth rate is 43±9% of the maximum nutrient saturated rate. This result indicates that blooms are associated with changes in the nutrient-saturated growth rate and not with changes in the degree of nutrient limitation.

Although the purpose of this study was to identify environmental factors associated with blooms of *D. perlaevis*, the conditions responsible for the occurrence of these blooms remain unknown. Follow-up studies aimed at identifying the causes of the apparent changes in nutrient-saturated growth rates could prove revealing.

Table 2. Percent protein, polysaccharides plus nucleic acids, lipids, intermediate low molecular weight compounds of total $^{14}CO_2$ assimilation at the end of 24 hr incubation with 12h:12h light dark cycle. B and N indicate bloom and non-bloom conditions, respectively. x±SD indicates mean and one standard deviation.

Experiment	Protein	Polysaccharides plus Nucleic Acids (PNA)	Lipids	Intermediate Low Molecular Weight Compounds (INT)
B	26	23	12	39
B	32	20	10	38
N	39	55	2	4
N	39	38	4	19
B	31	38	8	23
N	22	48	10	19
B	30	21	21	28
B	35	24	19	31
x±SD for N	33±10	47±9	5±4	14±9
x±SD for B	29±3	25±7	14±6	32±7

ACKNOWLEDGEMENTS

We thank J. Finn for his patient assistance in the field and J.A. Hirata for her nutrient analyses. This research was supported by the University of Hawaii Research and Training Revolving Fund. Contribution No. 740 from the Hawaii Institute of Marine Biology.

REFERENCES

1. R. Adachi, H. Irie, I. Umezaki, M. Saito, H. Takano, M. Chihara, S. Toriumi, Y. Fukuyo and K. Matsuoka, Red Tide Manual I, 175 pp, (1975).
2. D.C. Cox, P.-f. Fan, K.E. Chave, R.I. Clutter, K.R. Gundersen, N.C. Burbank, Jr., L.S. Lau and J.R. Davidson, Estuarine Pollution in the State of Hawaii, Vol. 2, Kaneohe Bay Study, 444 pp., (1973).
3. S.D. van Valkenburg and R.E. Norris, J. Phycol., 6, 48 (1970).
4. S. Taguchi and E.A. Laws, Mar. Ecol. Prog. Ser., 23, 251 (1985).
5. S. Taguchi and E.A. Laws, J. Plankt. Res., in press (1987).
6. M.W. Tate and R.C. Clelland, Nonparametric and Shortcut Statistics, 150 pp., (1957).
7. S. Taguchi, D. Jones, J.A. Hirata and E.A. Laws, submitted to Bull, Plankt. Soc. Japan, (1987).
8. G. DiTullio and E.A. Laws, Mar. Ecol. Prog. Ser., 32, 123, (1986).
9. H. Curl and L.F. Small, Limnol. Oceanogr., 10, R67, (1965).

TOXIC AND NON-TOXIC DINOFLAGELLATE BLOOMS ON THE DUTCH COAST

Marie Kat

Netherlands Institute for Fishery Investigations, 1970 AB IJmuiden-The Netherlands

INTRODUCTION

The water of the Dutch coastal area is a mixture of water entering the North Sea in the south through the Straits of Dover, fresh water from the rivers Rhine, Meuse and Scheldt and British coastal water (figure 2). The seasonal occurrence and year-to-year frequency for the period 1973-1984 will be described of 9 dinoflagellates occurring in a 70 km wide area along the coast. Special attention has been paid to the presence of toxin producing dinoflagellates and the results of this programme for 1986 and 1987.

MATERIALS AND METHODS

For the North Sea, one liter of water from 1.5 m depth was sampled. In the Waddensea net hauls (mesh size 33 µm) of 60 liter were taken. From both samples a subsample representing respectively 100 ml to 2.5 l was counted in an Ütermöhl Counting Chamber. Toxicity of mussels by DSP has been identified by the rat bioassay (2).

DINOFLAGELLATE SPECIES ABUNDANCES

The bloom season and year-to-year frequency of 9 dinoflagellate species are shown in figure 1, with concentrations represented by varying gray tones.

The spring bloom of the unarmoured dinoflagellate *Gyrodinium spirale* (since 1973) gradually extended after 1976 to the summer period and an extremely massive bloom was observed in 1983.

The less abundant *Nematodinium armatum* bloomed in 1977 and 1978, with another dense bloom in 1986.

Ceratium fusus had dense blooms in the period 1973-1975, but the peak of the blooms shifted from July till September in those years. Bloom density drastically decreased after 1975, but a moderate return was observed in 1981 and 1984.

Ceratium lineatum, although much lower in cell numbers than *C. fusus* , has a somewhat similar pattern. Occurring for 4 months a year during the period 1973-1975, it was maximal in 1976, but practically disappeared after 1977.

The short bloom period of *Prorocentrum micans* in late summer (1973-1975) became spring and autumn blooms after 1978. After the massive bloom in 1981, bloom periods and cell numbers decreased.

Prorocentrum minimum was observed after 1975, during summer sometimes lasting into September and coinciding with blooms of *Dinphysis acuminata* .

Prorocentrum balticum, not observed before 1977, usually appeared in mid-summer slightly earlier than *P. minimum* but with some overlap. An extraordinary massive bloom was observed in June 1978.

RED TIDES: BIOLOGY, ENVIRONMENTAL SCIENCE, AND TOXICOLOGY
Okaichi, Anderson, and Nemoto, Editors

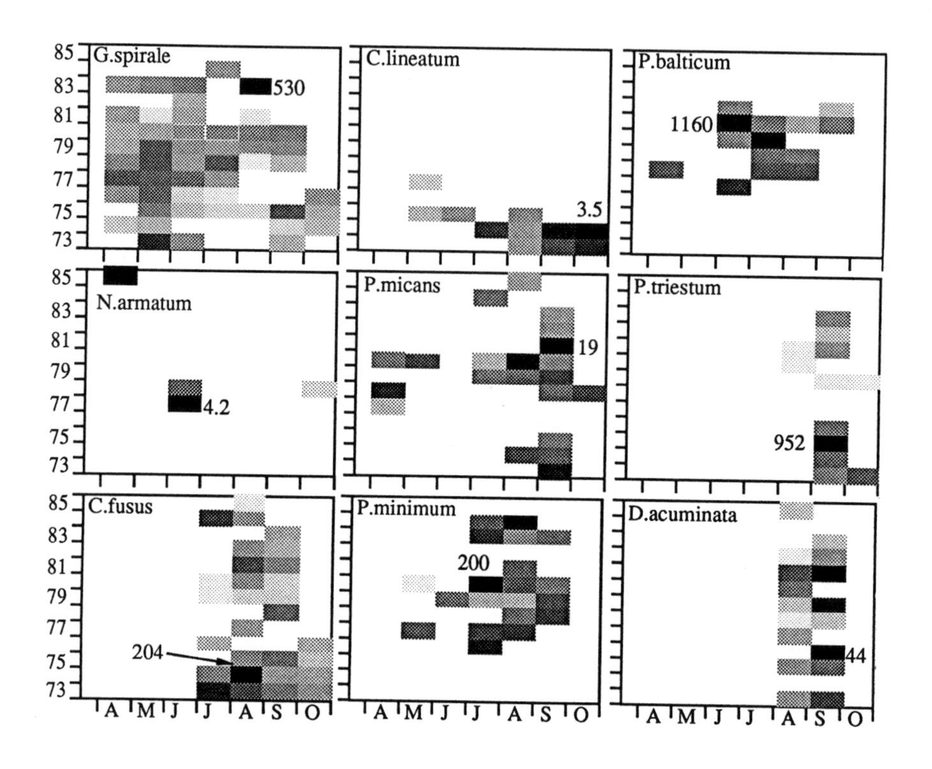

Figure 1 **Dinoflagellate species abundances.**
The logarithmic scale ranges from white (0.5 cells per ml) to completely black (the species maximum).

Prorocentrum triestinum was observed during September and October in four successive years 1973-1976, with a dense bloom in 1975. After 1981, this species was observed only in low concentrations.

Dinophysis acuminata occurrence is very seasonal. The blooms come in August but culminate mostly in September.

TOXICITY

Of all the dinoflagellate species described, only *D. acuminata* can be held responsible for the production of toxins, notably diarrheic shellfish poison (DSP). The spread of *D. acuminata* from the North Sea into the estuaries and the Waddensea

gave rise to DSP outbreaks in mussels from that area on several occasions (1,2,3,4 and 5).
The course of the 1986 DSP outbreak at a specific mussel site in the Waddensea (Meep) is depicted in figure 2. From 8 September till 20 October, about 80 *D. acuminata* per litre were present. This period was characterized by low wind speed and generally very calm and sunny weather. At the end of October, cell numbers decreased, and in the beginning of November *D. acuminata* disappeared. With the appearance of *D. acuminata*, mussels from the Meep area became slightly toxic. From 8-23 September, the toxicity slowly increased to a moderate level. During October, the continuing *D. acuminata* presence maintained a low level of toxicity in mussels at an average water temperature of 14 ^{o}C. Toxicity increased again in the last week of October, when the water temperature decreased from 12.5 to 9.5 ^{o}C. In spite of the disappearance of *D. acuminata* during November, the mussels remained moderately toxic. At the beginning of December, the mussels began to detoxify and by the end of December the mussels were completely free of DSP. During detoxification the water temperature decreased from approx. 8 to 6 ^{o}C.
The course of the DSP outbreak observed in 1987 in mussels from another site in the Dutch Waddensea (Oosterom) is depicted in figure 2.

During the second half of September, *D. acuminata* concentrations did not even exceed 30 cells per litre. Only at the end of September was a moderate level of DSP in mussels detected. Coinciding with a strong increase of toxicity in the mussels, the temperature dropped, and continued to decrease in October, while *D. acuminata* decreased to an average of 5 cells per litre. The toxicity level remained fairly constant until 26 October.

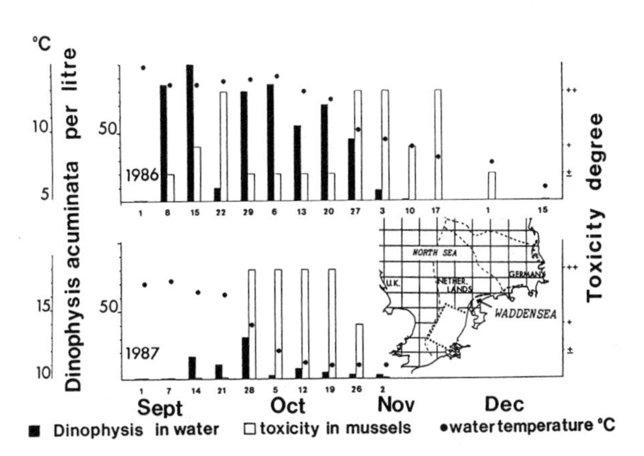

Figure 2 The course of the mussel toxicity 1986 and 1987.

Thereafter, a significant detoxification was observed. Temperature variations affect the metabolic activity of the mussels. The release of DSP from mussels is closely related to their metabolism and thus is likely to be delayed at lower temperatures. The results of the present work in the Dutch Waddensea as well as from the Bay of Vilaine France (6) and Scandinavia (7) are summarized in figure 3, which indicates that a *Dinophysis* "bloom" of 200 cells per litre at about 20 ^{o}C does not necessarily lead to diarrheic mussel poisoning. At temperatures of 13-14 ^{o}C however, *Dinophysis* concentrations of up to 50 cells per litre may lead to low toxicity in mussels. Concentrations of less than 30 cells per litre at approx. 10 ^{o}C can cause high toxicity.

CONCLUSIONS

Considering the occurrence of the 9 dinoflagellate bloom species, it can be concluded that for the period of the study 1973-1984:

a. There are variations in both year-to-year abundance and seasonal occurrence of most of the bloom species.
b. The cell numbers in the peak bloom of *C. fusus* and *P. triestinum* drastically decreased and *C. lineatum* even disappeared after 1975.
c. Blooms of *P. minimum* and *P. balticum* first appeared after 1975.
d. The toxic *D. acuminata* occurred very seasonally in August and September, sometimes leading to DSP outbreaks in the Waddensea and Easternscheldt. Even low cell numbers contributed to the continuing DSP content in mussels from that area.
e. Even low numbers of toxic species in the water of shellfish growing areas must be attended to, particularly during colder periods.

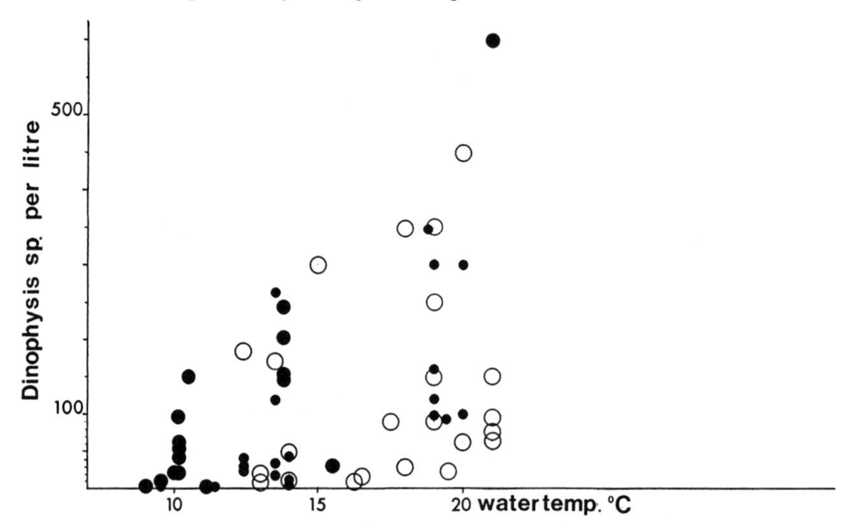

Figure 3 Dinophysis sp. cell concentrations and toxicity in relation to water temperatures.
O = no toxicity ● = low toxicity ● = moderate to high toxicity

REFERENCES

1. **Kat, M.** Sarsia 68: p. 81-84 (1983a).
2. **Kat, M.** Antonie van Leeuwenhoek 49, p. 417-427 (1983b).
3. **Kat, M.** In Toxic Dinoflagellates, D. Anderson, A. White and D. Baden (eds), p. 73-77, Elsevier Science Publishers (1985).
4. **Kat, M.** Visserij 3, May/June 1987 - Information Magazine for The Netherlands Fishery (in Dutch).
5. **Kat, M.** Rapp. P.-v. Réun. Cons. int. Explor. Mer, 187: 83-88 (1987).
6. **Lassus, P. and J.P. Berthome** Rapport Collectif DERO/MR et DRV/CSRU, IFREMER Nantes, France (1986).
7. **Lindahl, O. and M. Hageltorn** Proceedings of the XV Nordiska Veterinärkongressen Stockholm 28/7-1/8'86. Sveriges Veterinärförbund (1986).

Gymnodinium catenatum BLOOMS AND SHELLFISH TOXICITY IN SOUTHERN TASMANIA, AUSTRALIA

HALLEGRAEFF,G.M.[1],S.O.STANLEY[2], C.J.BOLCH[1] and S.I.BLACKBURN[1]

[1]CSIRO Marine Laboratories,GPO Box 1538, Hobart, Tas.7001, Australia

[2]Tasmanian Dept.of Sea Fisheries,Crayfish Pt, Taroona,Tas.7006,Australia

ABSTRACT

Dinoflagellate blooms of *Gymnodinium catenatum* (10^4 to 10^6 cells l^{-1}) in two southern Tasmanian estuaries in 1986 and 1987 resulted in PSP concentrations in mussels of up to 8350 µg/100 g. The dinoflagellate populations showed an unusually long growth season (December until June), at water temperatures of 12 to 18°C, sometimes following heavy rainfall events. Patterns of shellfish toxicity are discussed in relationship to dinoflagellate abundance, local hydrography, benthic cyst beds and shellfish feeding.

INTRODUCTION

Toxic dinoflagellates that can cause paralytic shellfish poisoning (PSP) in humans are well known from temperate waters of Europe, North America and Japan, but until recently none were reported from Australian waters (1). The conspicuous, chainforming dinoflagellate *Gymnodinium catenatum* Graham was first noticed in southern Tasmanian waters in 1980; it has produced seasonally recurrent blooms ever since. Dinoflagellate blooms were most extensive in 1986 when they caused two unambiguous human PSP poisonings and led to the temporary closure of up to 15 shellfish farms (2). *G.catenatum* is known to have caused previous human poisonings in Mexico and Spain (3,4), where this species tends to produce short-lived (2-4 wk) bloom events following coastal upwellings (4,5). In Tasmanian waters, the dinoflagellate blooms appear to be confined to two small estuaries near the main shipping port of Hobart (42°53'S, 147°19'E; Fig.1) and bloom events may last as long as 8 months. We describe here *G.catenatum* abundance and associated shellfish toxicity in southern Tasmanian waters in 1986 (high shellfish toxicity year) and 1987 (low toxicity year).

EXPERIMENTAL

Dinoflagellate abundance and hydrography

Dinoflagellate samples were collected at weekly intervals from January 1986 to July 1987 at the CSIRO wharf station (River Derwent), and at irregular intervals from North West Bay in the d'Entrecasteaux Channel and from Killala Bay in the Huon estuary (Fig.1). Subsurface water samples (2m depth) were collected using a 5 litre Niskin bottle and preserved with Lugol's fixative.Dinoflagellate cells were counted by the Utermöhl inverted microscope technique. Additional qualitative dinoflagellate surveys (present;absent; abundant) were carried out by making vertical hauls (from 0 to 10 m depth) with a 20 µm mesh plankton net. Benthic resting cysts of *G.catenatum* (6,7) were isolated from sediment samples collected with a Craib corer by sonication, size fractionation and density-gradient centrifugation. Temperature and salinity were measured throughout the water column by standard methods.

RED TIDES: BIOLOGY, ENVIRONMENTAL SCIENCE, AND TOXICOLOGY
Okaichi, Anderson, and Nemoto, Editors

Shellfish toxins

The Huon River and d'Entrecasteaux Channel area of southern Tasmania is the site of some twenty mussel (*Mytilus edulis planulatus*) and oyster (*Crassostrea gigas*) farms in various stages of development. No shellfish farms are allowed in the Derwent River because of industrial effluents, and only wild intertidal mussels were sampled in this area. Whenever toxic dinoflagellates were detected in the water column, shellfish farms were sampled at weekly intervals and shellfish stocks tested for PSP toxins at the Institute of Medical and Veterinary Sciences (Adelaide), using the standard mouse assay(8).

RESULTS

Description of the toxic dinoflagellate

G.catenatum is an unarmoured, chainforming dinoflagellate. The individual cells (24 to 35 µm long, 30 to 41 µm wide) vary greatly in shape (subspherical to biconical) and show a surface covering of hexagonal amphiesmal vesicles (Fig.2a).Rapidly growing cells form long chains (mostly 4 to 32 cells long),whereas senescent populations are composed of predominantly single cells.The girdle is deep and shows a left-handed displacement over 1/5 to 1/3 of the cell length, and the sulcus is narrow and deep and extends well into the epicone and hypocone. *G.catenatum* produces a brown, spherical, benthic resting cyst (40 µm diameter) with microreticulate surface markings that reflect the girdle and sulcus (Fig.2b). A detailed taxonomic discussion of this organism and characterization of its unusual PSP toxins will be published elsewhere (6,9). At present, this species is known from Mexico, Argentina, Spain, Portugal, Japan and Tasmania.

Occurrence of *G.catenatum* in Tasmania

In southern Tasmanian waters, *G.catenatum* appears to be confined to the Huon River, d'Entrecasteaux Channel and River Derwent (Fig.1). *G.catenatum* resting cysts capable of seeding new dinoflagellate blooms have been observed in all sediments from this region; there is no evidence for translocation of populations from other areas. The factors that confine *G.catenatum* to the area under direct influence of the plumes of the Huon and Derwent Rivers are not yet understood.

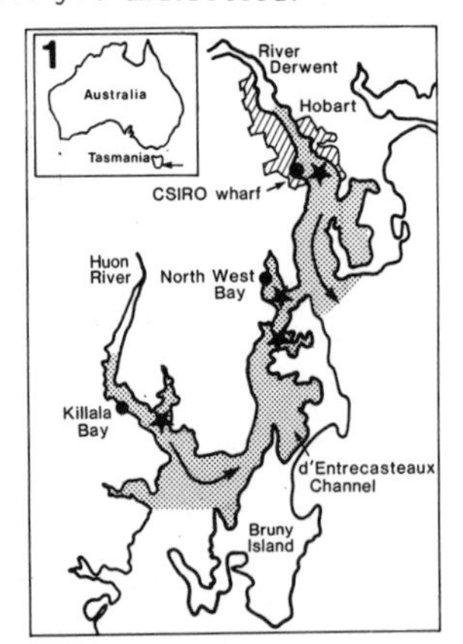

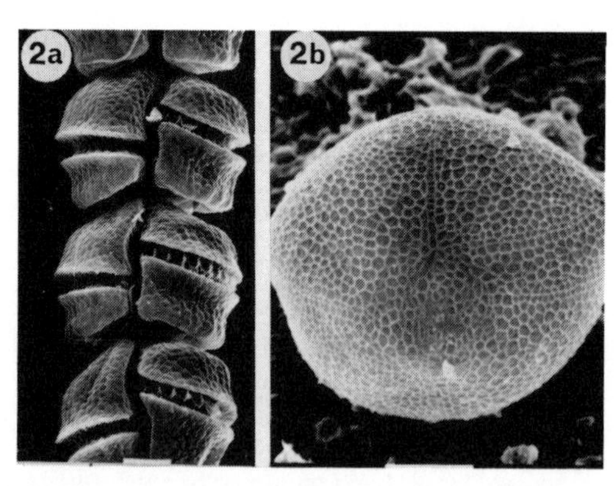

Fig.1. Map of southern Tasmania showing the area affected by *G.catenatum*. The sampling stations (•) and major water movements (arrows) are indicated.Benthic cyst beds of *G.catenatum* are also identified (✱).

Fig.2. Scanning micrographs of (a) a chain of *G.catenatum* and (b) its microreticulate, benthic resting cyst. Scale = 10 µm.

Dinoflagellate abundance and hydrographic environment

In southern Tasmanian waters, *G.catenatum* forms blooms (10^4 to 10^6 cells l^{-1}) from December-January (early summer) until June-July (autumn-early winter). During this period, water temperatures range from 12 to 18°C and salinities range from 28 to 34°/oo (Figs.3a,b). When water temperatures fall below 11 to 12 °C, the dinoflagellate slowly disappears from the water column (Fig.3c).These growth characteristics have been confirmed in laboratory cultures where dinoflagellate growth (one division every 3 to 4 days) occurs at temperatures between 15 and 20°C and salinities between 23 and 34°/oo. Growth is halted at temperatures below 12°C and salinities below 20°/oo (unpublished data).

Dinoflagellate abundance and shellfish toxicity

G.catenatum cell numbers and mussel toxicities in southern Tasmanian waters in 1986 and 1987 are summarized in Table 1. Longline cultured mussels ,on average, became 2 to 3 times more toxic than longline oysters grown at the same localities (data not shown), and longline cultured mussels became 3 to 4 times more toxic than intertidal mussels (compare Figs.3d and 4a,b). Shellfish toxicity problems were more severe in 1986 (up to 14 shellfish farms temporarily closed) than in 1987 (6 farms affected). The highest mussel toxicities, in both 1986 and 1987, were observed in the Huon estuary. Longline mussels from Killala Bay (Fig.4b) in 1986 showed PSP peaks (up to 8350 μg/100g) between March and June, whereas in 1987 dinoflagellate blooms developed only in late May, causing PSP maxima up to 1140 μg/100g. At the CSIRO wharf station, intertidal mussels became toxic only when *G.catenatum* concentrations exceeded 10^4 cells l^{-1} (Figs.3c,d), whereas in the d'Entrecasteaux Channel and Huon River area, dinoflagellate concentrations of only 10^3 cells l^{-1} caused significantly higher toxicity (Table 1).

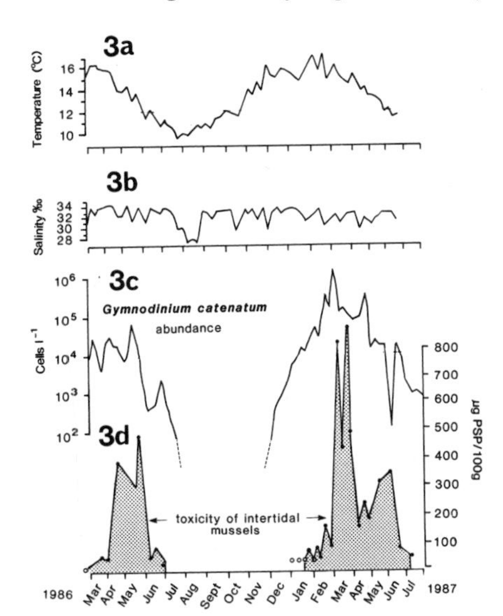

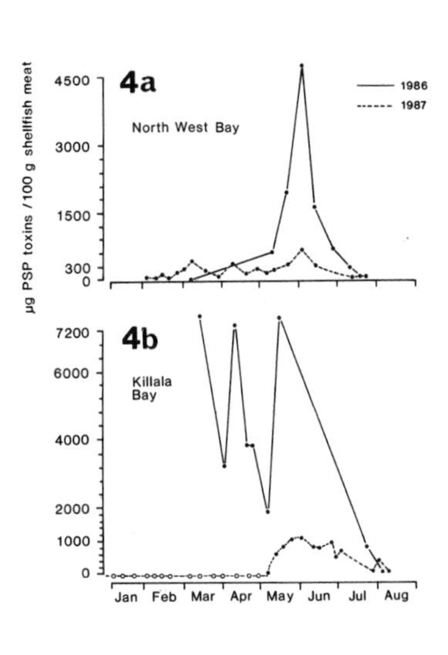

Fig.3.Seasonal variation at the CSIRO wharf station of (a) temperature, (b) salinity and (c) *G.catenatum* abundance at 2m depth, compared with (d) PSP toxicity of intertidal mussels.

Fig.4.Seasonal variation of toxicity of longline cultured mussels, in 1986 and 1987, at (a) North West Bay and (b) Killala Bay.

Table 1. *G.catenatum* abundance and mussel toxicity in southern Tasmanian waters , in 1986 and 1987

Site	Date	maximum *G.catenatum* concentration (cells l^{-1})	maximum PSP concentration (µg/100 g)	number of shellfish farms closed
CSIRO wharf (River Derwent)	March/April 1986	2.10^4	385	-
	June 1986	10^5	480	-
	March/April 1987	2.10^6	940	-
	June 1987	10^4	360	-
North West Bay (d'Entrecasteaux Channel)	March/April 1986	no data	130	0
	June 1986	10^3	4800	9
	March/April 1987	5.10^3	< 50	4
	June 1987	no data	800	1
Killala Bay (Huon River)	March/April 1986	no data	8350	5
	June 1986	10^4	7680	5
	March/April 1987	$<10^2$	< 50	0
	June 1987	2.10^5	1140	2

DISCUSSION

The growth patterns of *G.catenatum* in southern Tasmania are markedly different from the growth patterns of this dinoflagellate in Mexico and Spain (4,5). Shellfish toxicity problems in Tasmanian waters were more severe in 1986 than 1987. In 1986, unusually high early summer and autumn rainfall (more than twice normal) may have stimulated dinoflagellate growth through high levels of inorganic and organic nutrients from land runoff (10). Surface water temperatures in May 1986 were at least 1°C higher than in May 1987 and also may have favoured increased dinoflagellate growth rates. The accumulation of PSP toxins in shellfish strongly depends on the availability of *G. catenatum* as a food source for the filter-feeding shellfish. Relationships between PSP toxicity and dinoflagellate abundance were different for oysters and mussels, for longline and intertidally cultured shellfish, and for shellfish from the Derwent and Huon estuaries. Oysters are known to be more discriminate filter-feeders than mussels, and longline mussels clearly are more exposed to the dinoflagellates than intertidal mussels. Whereas in the River Derwent *G.catenatum* tends to occur in mixed plankton blooms, in the Huon River it often occurred in almost monospecific blooms, thus causing significantly higher shellfish toxicity.

REFERENCES

1. G.M.Hallegraeff, CSIRO Marine Laboratories Report 187 (1987)
2. G.M.Hallegraeff and C.E.Sumner, Australian Fisheries 45, 15 (1986)
3. J.Lüthy, in: D.L.Taylor and H.H.Seliger (eds),Toxic Dinoflagellate Blooms, 15 (1979)
4. L.D.Mee, M.Spinosa and C.Diaz, Mar.Environ.Res. 19, 77 (1986)
5. S.Fraga, D.M.Anderson, I.Bravo, B.Reguera ,K.A.Steidinger and C.M.Yentsch, Est.Coast.Mar.Sci.,submitted
6. G.M.Hallegraeff, D.A.Steffensen and R.Wetherbee, J.Plankton Res.10 (1988)
7. D.M.Anderson, D.Jacobson, I.Bravo and J.H.Wrenn, J.Phycol.(1988)
8. S.Williams (ed), in:Official Methods of Analysis, 344 (1984)
9. Y.Oshima, K.Hasegawa, T.Yasumoto, G.M.Hallegraeff and S.I.Blackburn, Toxicon 25,1105 (1987)
10. A.Prakash, Environ.Lett.9, 121 (1975)

* This study was supported by Fishing Industry Research Trust Account grant 86/84

POPULATION DYNAMICS AND TOXIN COMPOSITION OF PROTOGONYAULAX TAMARENSIS FROM THE ST. LAWRENCE ESTUARY

ALLAN D. CEMBELLA AND J.-C. THERRIAULT
Maurice Lamontagne Institute, Department of Fisheries and Oceans
850 Route de la Mer, Mont-Joli (Québec), Canada G5H 3Z4

ABSTRACT

The population dynamics of *Protogonyaulax tamarensis* was investigated along the north coast of the St. Lawrence estuary, within and adjacent to the region defined by the freshwater plume of the Manicouagan and Aux-Outardes river systems. In spite of the unstable plume boundaries, significant differences between the distribution of *Protogonyaulax* in the core and the marginal areas were observed.

The toxin spectrum and toxicity of *Protogonyaulax* blooms were analyzed by reverse-phase HPLC. Among natural populations from the north shore of the estuary, the toxin composition was virtually identical, although the toxicity per cell varied. The high toxin yield revealed that populations from the St. Lawrence estuary were among the most toxic yet detected.

INTRODUCTION

Previous phytoplankton studies in the lower St. Lawrence estuary indicated that, in general, the proliferation of dinoflagellates was related to prevailing hydrodynamic factors, especially high freshwater run-off and reduced vertical mixing [1]. On a broad spatial scale, the distribution of *P. tamarensis* coincided with the extent of the Manicouagan and Aux-Outardes river plume [2]. Other recent reports indicated the existence of extremely toxic populations, as determined by mouse bioassay, from the Bay of Fundy [3], and of a clinal gradient of toxicity in *Protogonyaulax* isolates, increasing from south to north along the eastern coast of North America [4].

In view of the transient nature of dinoflagellate blooms in the highly dynamic environment of the lower estuary, in the present study, the distributional pattern of *Protogonyaulax* was investigated on a finer spatio-temporal scale. The objective was to achieve a more detailed resolution of potential controlling factors. A second objective was to use a sensitive HPLC method to determine the degree of heterogeneity in toxicity and toxin composition of natural bloom populations from the St. Lawrence estuary.

MATERIALS AND METHODS

The occurrence of *Protogonyaulax*, in relation to the prevailing environmental conditions within and adjacent to the core of the Manicouagan and Aux-Outardes river plume, was followed on a bi-weekly basis, from mid-June to mid-Oct. At each station (Fig. 1) the vertical profile of irradiance, temperature, salinity, dissolved inorganic nutrients, *in vivo* chl *a* fluorescence, and extracted chlorophyll was determined by conventional methods [1]. Phytoplankton samples were collected by pumping from discrete depths and by vertical net (30 µm) hauls. When *Protogonyaulax* was most abundant in the water column, net tows were used to collect material for toxin analysis. The mixed phytoplankton

RED TIDES: BIOLOGY, ENVIRONMENTAL SCIENCE, AND TOXICOLOGY
Okaichi, Anderson, and Nemoto, Editors

sample was size fractionated, and the 20 µm fraction, with Protogonyaulax as the dominant component, was saved for toxin extraction [5] and analysis by reverse-phase HPLC [6].

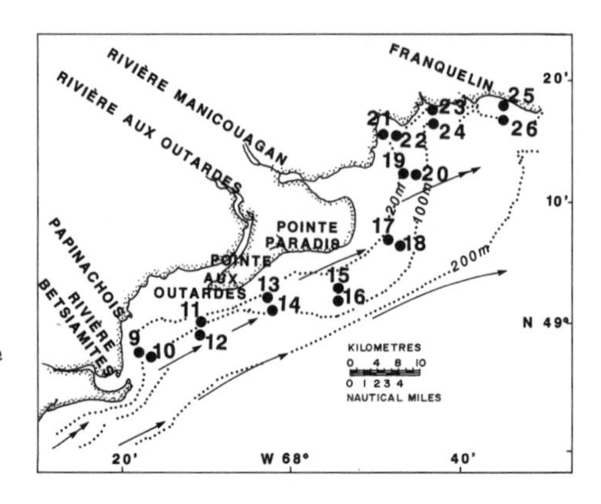

Fig. 1. Sampling stations on the north shore of the St. Lawrence estuary.

RESULTS AND DISCUSSION

The occurrence of Protogonyaulax was highly transient, however, this species dominated the phytoplankton assemblage for a short period during late August, although the blooms never became monospecific. The peak abundance of Protogonyaulax and the chlorophyll a maximum coincided with the core of the plume, which could be identified in the vertical profile (e.g., Fig. 2) as a layer of lower salinity water at the surface. Protogonyaulax aggregations were restricted to a narrow surface layer, typically with a maximum concentration between 2 and 5 m, where nutrient and irradiance levels were apparently sufficient to support high dinoflagellate growth (Fig. 2). Maximal PSP toxin levels in shoreline molluscs (Fig. 3) were not synchronized with the peak in Protogonyaulax abundance, rather, the peak toxicity occurred during the declining phase of the bloom.

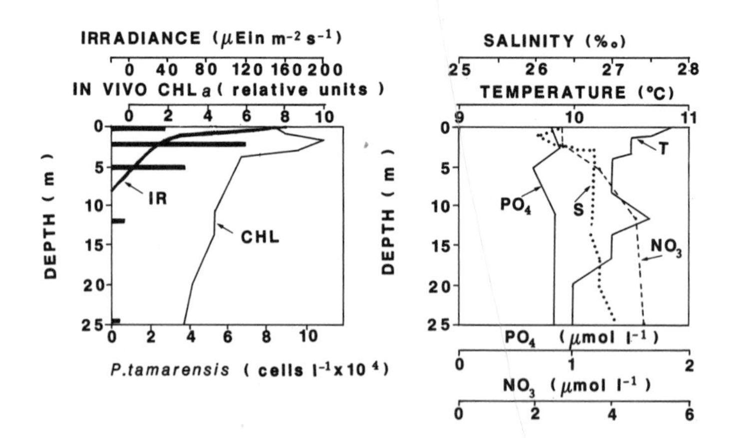

Fig. 2. Vertical profile of P. tamarensis concentration, irradiance, chl a fluorescence, salinity, temperature, and inorganic nutrients at ST 18 during the peak of the bloom.

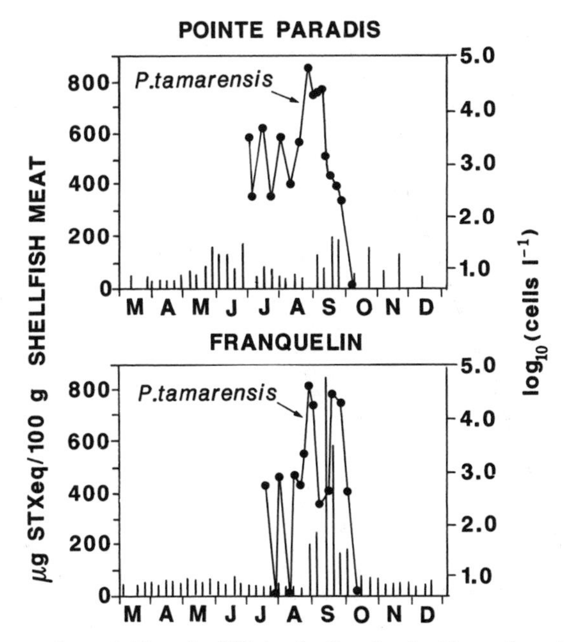

Fig. 3. Seasonal variation in PSP toxin levels in the soft- shelled clam, Mya arenaria, and concentration of P. tamarensis, at a station within the core (Pointe Paradis) and at the margin (Franquelin) of the river plume.

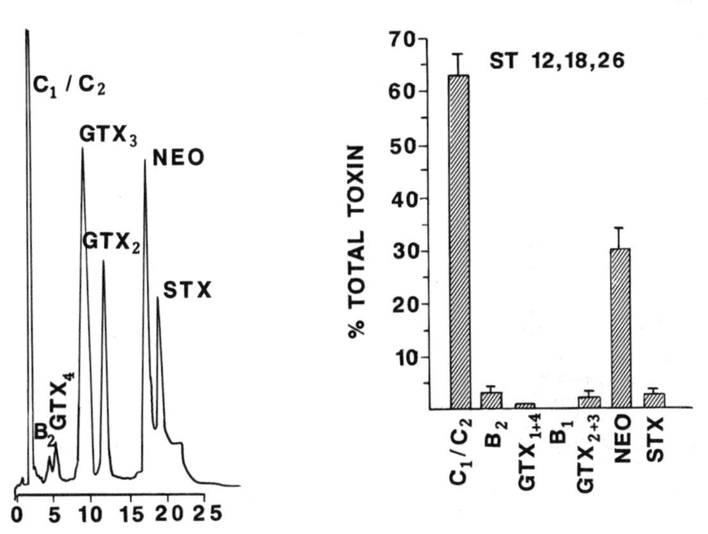

Fig. 4. PSP toxin chromatogram of a natural Protogonyaulax sample collected at ST 12.

Fig. 5. Percent toxin composition ($\overline{X} \pm$ S.E.) of natural Protogonyaulax populations from the St. Lawrence estuary.

The small-scale distributional pattern of Protogonyaulax in the nearshore region of the plume appeared to be strongly dominated by hydrodynamic factors, particularly the degree of water stratification. This is consistent with the previous hypothesis [2] regarding the importance of high freshwater outflow.

The toxin profile (Fig. 4) of natural Protogonyaulax revealed almost no inter-population heterogeneity in toxin ratios (Fig. 5), although there was an approximate three-fold variation among the stations (range: 65-175 fmol per cell) in total toxin content, on a molar basis. Relatively high proportions of the carbamyl N-sulfo compounds C_1/C_2, as well as the more potent carbamate toxin, neosaxitoxin (NEO), were characteristic of St. Lawrence populations.

The extremely high toxin concentration in St. Lawrence Protogonyaulax, as determined by HPLC, is the highest ever reported using this non-degradative method. The conversion of toxins C_1 and C_2 to gonyautoxin 2 (GTX_2) and gonyautoxin 3 (GTX_3), respectively, through acid hydrolysis in the preparation of shellfish and dinoflagellate samples for mouse bioassays, would account for a further enhancement of toxicity, as was measured by the bioassay [7].

The toxin composition of St. Lawrence Protogonyaulax differs substantially from that of isolates from other diverse locations [5]. The profile is distinguishable from that of its closest geographical counterpart, represented by a highly toxic Bay of Fundy isolate [3,5], primarily by the presence of a much higher proportion of toxins C_1/C_2, and a lower relative content of GTX_{1+4}, in St. Lawrence populations. The uniqueness of this toxin spectrum suggests that genetic exchange may be limited, and that populations in the St. Lawrence estuary may be reproductively isolated from other east coast populations.

ACKNOWLEDGMENTS

Reference toxins were provided through the courtesy of J.J. Sullivan (USFDA, Seattle, WA) and G.L. Boyer (SUNY-Syracuse, NY). Lucy Maurice assisted in the preparation and drafting of the manuscript.

REFERENCES

1. J-C. Therriault and M. Levasseur, Natur. can. 112, 77-96 (1985).
2. J-C. Therriault, J. Painchaud and M. Levasseur in: Toxic Dinoflagellates, D.M. Anderson, A.W. White and D.G. Baden, eds., (Elsevier, New York, 1985) pp. 141-146.
3. A.W. White, Toxicon 24, 605-610 (1986).
4. L. Maranda, D.M. Anderson and Y. Shimizu, Est. coast. shelf Sci. 21, 401-410 (1985).
5. A.D. Cembella, J.J. Sullivan, G.L. Boyer, F.J.R. Taylor and R.J. Andersen, Biochem. Syst. Ecol. 15, 171-186 (1987).
6. J.J. Sullivan, J. Jonas-Davies and L.L. Kentala in: Toxic Dinoflagellates, D.M. Anderson, A.W. White and D.G. Baden, eds., (Elsevier, New York, 1985) pp. 275-280.
7. A.D. Cembella, J-C. Therriault and P. Béland, J. Shellfish Res., in press, (1988).

A TOXIC DINOFLAGELLATE FIRST RECORDED IN TAIWAN

SU, HUEI-MEEI*, I-CHIU LIAO* AND YOUNG-MENG CHIANG**
*Tungkang Marine Laboratory, Tungkang, Pingtung, Taiwan 92804;
** Institute of Oceanography, National Taiwan University,
Taipei, Taiwan 10764, ROC

ABSTRACT

An incident of paralytic shellfish poisoning due to ingestion of cultured purple clam (*Soletellina diphos*) occurred in southern Taiwan on January 1, 1986. Monitoring of phytoplankton for the causative organism was undertaken soon after the accident. A species of dinoflagellate - *Protogonyaulax tamarensis* was collected and studied. Acid extract of cultured cells of this alga was found to be toxic with mouse assay.

INTRODUCTION

A food poisoning incident due to ingestion of purple clam (*Soletellina diphos* Linnaeus) occurred in southern Taiwan on January 1, 1986. The clams, which were obtained from a brackish water pond at Tungkang, Pingtung, Taiwan, were found to contain gonyautoxins [1, 2]. However, the source of the toxins was not known. As noted by various authors [3, 4], blooms of some species of dinoflagellates are correlated with reports of paralytic shellfish poisoning (PSP). Therefore, an attempt was made to find out if there were any toxic dinoflagellates growing in the ponds of that area, by collecting phytoplankton right after the accident. Several species of photosynthetic dinoflagellates were collected and one of them was a strain of *Protogonyaulax tamarensis* (Lebour) Taylor which is known to contain PSP [3]. This represents the first toxic dinoflagellate found in Taiwan.

MATERIALS AND METHODS

Phytoplankton collection

The ponds are located in the vicinity of the mouth of Tungkang River, in Tungkang, southern Taiwan. Water samples were taken from the surface of the pond with one liter plastic bottles once or twice a month from January 1986 to October 1987. Five-hundred milliliters were fixed with Lugol's solution [5] and then concentrated to 10 ml for phytoplankton counting and identification. Temperature and salinity of the ponds were also measured. Both living and Lugol-fixed samples were examined for the indentification of dinoflagellates.

Light and scanning electron microscopy

Clonal cultures of *P. tamarensis* were maintained in 15 ppt "K" medium [6], and grown in an incubator at 25°C, 12:12LD and 2,000-4,000 lux. Living cells were observed with DIC (Differential Interference Contrast) and phase contrast microscopy, and their thecae were disassociated by a slight pressure on the cover glass. For scanning electron microscopy, cells were fixed in 0.5% GTA and 0.2% OsO_4 simultaneously, or fixed in Lugol's solution, with the culture medium as the buffer. The cells were transferred to a capsule with an 8 µm Nuclepore filter, and then washed and dehydrated in

RED TIDES: BIOLOGY, ENVIRONMENTAL SCIENCE, AND TOXICOLOGY
Okaichi, Anderson, and Nemoto, Editors

a graded acetone series, critical point dried, shadowed with gold and viewed with a scanning electron microscope Hitachi Model S-520.

Toxicity

Cultured cells of P. tamarensis which were grown in a 15 ppt medium at 20-23°C, 12:12LD, were harvested during the late exponential phase. Acid extracts (pH 4.0) were prepared according to Schmidt and Loeblich III [3] and measured for PSP toxicity by the A.O.A.C. method.

RESULTS

Morphology

Cells of P. tamarensis are globular, slightly longer than broad (Figs. 1, 2). They are 14-34 μm long, and 13-31 μm wide and their L/W ratio is 1.05-1.15. The cingulum is equatorial, descending its own width (Figs. 3, 4). The epitheca is rounded with a slight hump at the apical pore (Fig. 5); the hypotheca is rounded but appears lightly depressed where the sulcus reaches the antapex (Fig. 6). The thecal plates are thin and smooth with pores, and covered with a delicate outer thecal membrane (Figs. 5, 6). The plate formula is: apical pore complex (APC), 4', 6", 6C, 8S, 5"', 2"" (Figs. 7-9). The APC is triangular and has a fishhook-shaped apical pore (Fig. 10). A ventral pore is present (Figs. 3, 5, 8). A large C-shaped nucleus is situated beneath the cingulum, and makes the cell looked evacuated in the center (Fig. 11). Cells are single, and double (Fig. 2) only shortly after cell division. The temporary cyst is round (Fig. 11), while the resting cyst is cylindrical with rounded ends (Fig. 12).

Ecology

In addition to many species of diatoms, blue-green algae and flagelates, five photosynthetic dinoflagellates and several colorless ones were found. P. tamarensis was common in most crab ponds at cell densities of less than 200 cells/ml during winter and spring, but, on December 5, 1986, it was found blooming with cell densities as high as 1600 cells/ml in a crab pond next to the toxic clam pond (which has discontinued culture). During that time, the water temperature of the ponds was around 17.9-30.0°C and the salinity was 10 to 19 ppt.

Toxicity

The toxicity of a two liter culture containing about 2 x 10^7 cells was calculated to be about 450 MU by mouse assay.

DISCUSSION

This alga clearly fits the characteristics of P. tamarensis as described by Taylor [4] and Fukuyo [7]. Some Lugol-fixed specimens sent to Dr. Fukuyo have been confirmed by him to be this species.

Cells of this alga are usually single, and may remain in chains of two only briefly after cell division. Therefore, the attachment pores described by Fukuyo [7] cannot be found. This alga differs from those found in the Pacific Ocean [7] and the Atlantic Ocean [8, 9] in terms of size of theca and growing conditions. The thecae of this alga are usually smaller (15-25 μm) than those of the latter two (30-40 μm), and the temperature for growth is higher (20-30°C) than those reported from other places of

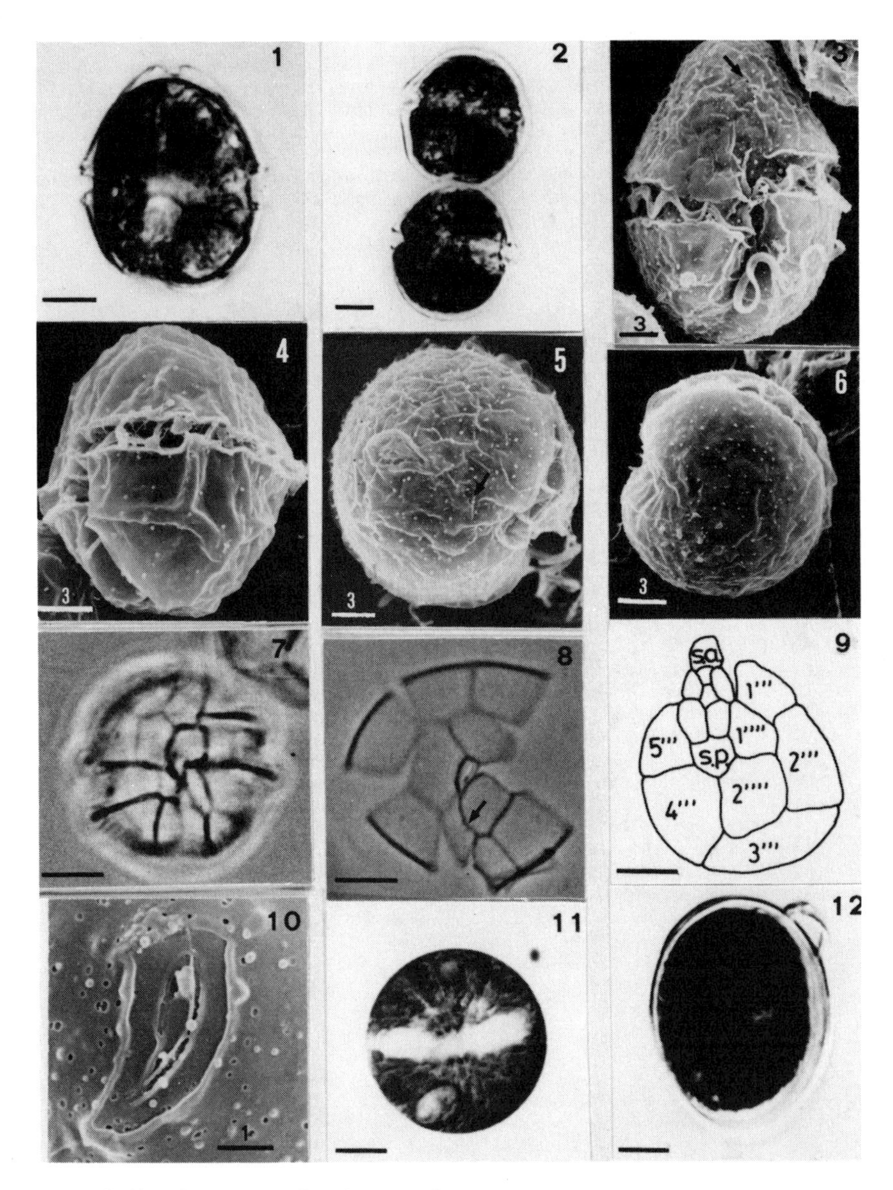

Figs. 1-12. Protogonyaulax tamarensis.
Fig. 1. Ventral view. Fig. 2. Cells in chain. Fig. 3. Ventral view (SEM). Fig. 4. Side view (SEM). Fig. 5. Epitheca (SEM). Fig. 6. Hypotheca (SEM). Fig. 7. Ventral view of theca. Fig. 8. Plate pattern of epitheca. Fig. 9. Plate pattern of hypotheca. Fig. 10. APC (SEM). Fig. 11. Temporal cyst. Fig. 12. Resting cyst.
(Arrows show the ventral pore; all scales = 5 µm except where indicated).

the Northern hemisphere [4, 7, 9]. Therefore, southern Taiwan might be the southernmost limit of this dinoflagellate in its distribution. This alga also grows in less saline water (10 ppt in field, 5 ppt in culture) than those found from other places [4, 9, 10].

This alga is the first record of a toxic dinoflagellate from Taiwan, and also the first report of P. tamarensis from the tropical Pacific region.

ACKNOWLEDGEMENTS

We thank Dr. Y. Fukuyo, University of Tokyo, for kind confirmation of the species name. Our thanks are also due to Dr. I.J. Chen and Mr. J.J. Cheng for conducting mouse assay, Dr. C.W. Li for advice on SEM preparation, Ms. C.Y. Lin and Ms. S.S. Rem for SEM operation and other members of the Tungkang Marine Laboratory for offering various invaluable assistance. This work was supported by N.S.C. and C.O.A., R.O.C. Contribution A No. 83 from the Tungkang Marine Laboratory.

REFERENCES

1. C.C. Chang and S.J. Hong, Toxicon 24, 861-864 (1986).
2. D.F. Hwang, T. Noguchi, Y. Nagashima, I.C. Liao, and K. Hashimoto, Nippon Suisan Gakkaishi 53, 623-626 (1987).
3. R.J. Schmidt and A.R. Loeblich III, J. mar. biol. Ass. U.K. 59, 479-487 (1979).
4. F.J.R. Taylor, in: Seafood Toxins, E. Ragelis ed. ACS Symp. Series 262, (Am. Chem. Soc. 1984) pp. 77-79.
5. T.R. Parsons, Y. Mita, and C.M. Lalli, A Manual of Chemical and Biological Methods for Seawater Analysis (Pergamon Press, Oxford 1984).
6. M.D. Keller and R.R.L. Guillard, in: Toxic Dinoflagellates, D.M. Anderson, A.W. White and D.G. Baden, eds. (Elsevier, New York 1985) pp. 113-116.
7. Y. Fukuyo, Bull. Mar. Science 37, 529-537 (1985).
8. A. Prakash, Fish. Res. Bd. Canada 20, 983-996 (1963).
9. A. Prakash, Fish. Res. Bd. Canada 24, 1589-1606 (1967).
10. A.W. White, J. Phycol. 14, 475-479 (1978).

RED TIDES OFF THE PORTUGUESE COAST

M.A. DE M. SAMPAYO
Instituto Nacional de Investigação das Pescas
Avenida de Brasília, 1400 Lisboa, Portugal

ABSTRACT

The growth and abundance of red tide organisms on the Portuguese coast are looked upon both as a product as well as a process of localized eutrophication in some coastal zones, and mainly related to hydrographic conditions and loading of organic matter from industrial and municipal waste discharge.

The influence of a 1986 red tide formed in Galicia, NW Spain, and the presence of PSP toxins in Portuguese bivalves from the west coast as far as Obidos lagoon are analysed, and the role of coastal geomorphology and hydrographic properties in the control of the problem are emphasized.

Gymnodinium catenatum Graham was the organism responsible for this problem.

INTRODUCTION

Red tides, meaning high concentrations of planktonic organisms that discolor the sea water, are events which on the Portuguese coast occur mainly in restricted coastal zones which we can call prone areas, and are shown in fig. 1. However, the terminology is complicated by the fact that toxic dinoflagellates may not always be of sufficient abundance to discolor the water, but may be numerous enough to toxify shellfish, as was the case in the PSP problem on our coast last year.

Our present work is divided in two parts. In the first we summarize the red tide events with discoloration, and in the second part we will analyse the PSP problem which occurred between October and December 1986 due to the presence in the plankton of the toxic dinoflagellate *Gymnodinium catenatum* G., but not in sufficient concentration as to discolor the water.

RED TIDES WITH DISCOLORATION

The growth and abundance of marine plankton species resulting in red tide conditions is usually confined to coastal waters, or to those regions of the sea where active upwelling takes place. Red tide organisms seem to thrive best under low salinity and high organic enrichment. Both these conditions prevail in coastal waters, particulary in areas of land drainage, mostly in enclosed or semi-enclosed coastal areas.

As red tide blooms correspond to high organic loading in sea water, municipal and industrial waste discharges introducing substantial amounts of terrigenous organic material, trace metals and biologically active substances in coastal waters, create conditions favourable for the development of red tides. Sewage drainage into coastal water contributes for high concentrations of pollutants which tend to be less diluted and dissipated in coastal lagoons, bays or semi-enclosed coastal areas subject to lower circulation and flushing, building up favourable conditions to initiate phytoplakton blooms, and modifying quantitatively and qualitatively the natural population leading to one species dominance.

Studies on unialgal cultures of several species of marine phytoplankton have been useful in elucidating the role of organic enrichment in the initiation of red tide blooms, which is in some way related to the biological conditioning of sea water. The subsquent development and continuation

RED TIDES: BIOLOGY, ENVIRONMENTAL SCIENCE, AND TOXICOLOGY
Okaichi, Anderson, and Nemoto, Editors

of red tides are largely governed by non biological factors such as coastal geomorphology, hydrography and weather conditions, like favourable winds, temperature and light intensity (insolation).

Biotic and abiotic factors are involved in the initiation and development of red tide blooms. Adequate nutrients, temperature, light intensity and a relatively stable water column are some of the environmental factors that stimulate growth and maintenance of red tide species. The primacy of the correct physical (oceanographic/climatological) conditions for bloom development is well established, but biological aspects are of paramount importance as blooms can not occur without seed and each species has its own ecological response.

On the Portuguese coast red tide blooms occur mainly in five restricted areas which are marked in fig. 1. These areas are close to tourist resorts where in summer population can more than double, enlarging municipal sewage discharges. Regions 1 and 3 are also close to the two most industrialized parts of the country, Oporto and Lisbon respectively. We call these areas prone regions as they are bays or embayments, relatively protected, with sewage outlets and freshwater influence either through estuaries and/or coastal lagoons.

O - Prone red water regions
★ - Zone affected with PSP, 1986
➚ - Surface current directions at the time

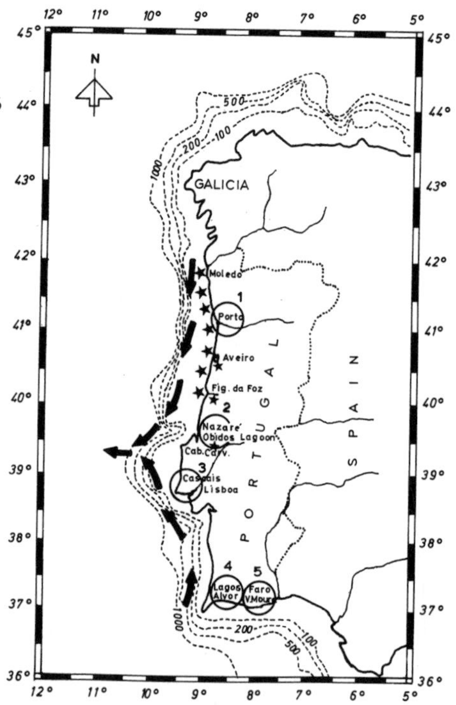

Fig. 1 - The Portuguese coast showing the five prone regions for red tide events and the 1986 PSP affected area. Surface current directions at the time are marked.

As pointed out before, eutrophication of coastal waters due to input of nutrients, particularly from sewage and industrial waste, are favourable conditions for red tide bloom initiation when associated with favourable physical factors (oceanographic/climatological) which account for its development (1,2,3,4). We must also point out that these prone areas are in the Portuguese coastal upwelling system (5,6,7) which can have a seed transport effect as well as a well known fertilizing effect.

Upwelling on the Portuguese coast occurs mainly between July and the middle of September, having a year to year variation related to the North component of the prevailing winds (5,6).

There is little doubt that pollution-induced eutrophication of enclosed or semi-enclosed marine coastal areas provides the potencial for red tide out breaks (2). Most of the red tides that have occurred on our coast (table 1) have had no deleterious effects (4,8,9,10,11,12,13,14,15). Only red waters in side Obidos Lagoon, like those of *Alexandrium lusitanicum*, *Prorocentrum minimum* and *Prymnesium parvum*, have caused PSP, DSP and fish mortality respectively (4,14,15). These have had an enormous socioeconomic impact in the region as the lagoon was closed to shellfish harvesting for 28 years, until 1985.

TABLE 1. - Main Red Tide Events and Responsible Organisms

REGIONS / ORGANISMS	1	2	3	4	5
MESODINIUM RUBRUM	1983,84		1980,81,86	1983,85	1981
HETEROSIGMA AKASHIWO (O.luteus)			1982		
PRYMNESIUM PARVUM		1972*			
A.LUSITANICUM (G.tamarensis)		1958*,59*,62*			
GONYAULAX POLYEDRA		1944,67			
N.SCINTILLANS	1987				1982
PROROCENTRUM MICANS		1955*,87*			
P.MINIMUM (E.baltica)		1958*,61*,62*,87*			
S.TROCHOIDEA		1982			
CHAETOCEROS SPP				1983	
RH.DELICATULA			1984	1983	
SCH.DELICATULA				1983	
SKELETONEMA COSTATUM		1958*,59*,87*			

*Red tides inside Obidos lagoon.

THE 1986 PSP PROBLEM

An outbreak of PSP intoxication due to bivalves occurred betweem 30 September and December 1986 on the NW Portuguese coast from Cabo Carvoeiro 39° 20' N northwards (fig. 1). The problem was due to the presence in the phytoplankton of the chain forming dinoflagellate *Gymnodinium catenatum*, but not in sufficient number to discolor the sea water.

We are almost sure that this outbreak, the first on our coast related to *G.catenatum*, was influenced by a red tide of the species in Galicia (NW Spain). In Galicia, there is the highest production of mussels in the world, and red tide events with discoloration are frequent (16,17). On September 23 we were notified that a bloom of *G.catenatum* was detected inside Vigo Ria, the species being more abundant outside (S.FRAGA, personal communication). On September 30, in clams from Matozinhos (fig. 1), 39.4 ug/100g of PSP toxins were detected in the first positive test for the region on the PSP monitoring program carried out by the National Health Institute (E.SILVA and S.FRANCA, personal communication).

In October all the bivalves between Moledo and Obidos Lagoon were toxic while those farther South were free of PSP toxins. The toxicity increased until November. The highest value detected was 1594 ug/100g in mussels from Moledo on October 27 and began to decline slowly, until December when the shellfish fisheries were reopened.

At the end of summer prevailing surface coastal currents are from the North, moving from Galicia to Cabo Carvoeiro, where they meet surface currents coming from the South (6), and surface sea water is pushed off-shore more or less at that point of the Portuguese coast, fig. 1, (J.PISSARRA, personal communication).

These were the hydrographic properties controlling the problem and they are associated with the coastal geomorphology, a cape close to a submarine canyon with very narrow continental shelf in the vicinity. We can also add that the wind system at the time (Data from National Meteorological Institute) were, at Viana do Castelo (fig. 1), mainly light winds from N and NE. Between

5 and 10 of October the winds were so light that the sea was without waves, which favoured the species transport without damage and its persistance on our coast. The highest mean daily wind speed was 10.7km/h on October 23 until November 10 (5.6km/h) when it began to blow strongly for six days, disrupting the water column stability (17km on 11 November) and disfavouring the species. Waves reached 5m high on the 15 November. On the other hand, winds at Cabo Carvoeiro were blowing more strongly for all the period from September 26 (29.6km/h) with a maximal mean value of 33 km/h on November 18 and a minimum of 4.9km/h on November 8.

SUMMARY

On the Portuguese coast there are prone regions for red tide events with discoloration effect. They are closed, semi-enclosed or more or less protected areas mostly not far from tourist resorts and/or the most industrialized parts of the country. The events are associated with hydrography, coastal geomorphology, climatological-physical processes, and/or municipal, industrial sewage discharges.

A PSP toxicity of bivalves on the NW coast, North from Cabo Carvoeiro, which caused a high negative impact on the shellfish fisheries between October and December 1986 had its primary origin in a G.catenatum red water formed off Galicia (NW Spain). Favourable climatological conditions with light winds and relatively high temperature for the season can be an explanation for the duration of the problem. Affected area was limited by coastal geomorphology.

ACKNOWLEDGEMENTS

Financial support to participate in this Symposium was provided by the Luso-American Foundation for Development and Calouste Gulbenkian Foundation. We also had an accommodation grant from the Organizing Committee. Thanks are due to Mrs Idalina Silva for typing the manuscript and to Mr. João Diogo for the drawing.

REFERENCES

1. B.Dale & C.M.Yentsch, Oceanus 21 : 41-49 (1978)
2. A.Prakash, In: LoCicero (ed.) Proc. 1st Int.Conf. on Toxic Dinoflagellate Blooms. Mass.SciTechnol.Found. : 1-6 (1975)
3. A.Prakash, ICES/S.M. on the Causes, Dynamics and Effects of Except.Mar. Blooms and Related Events, B : 13 (1984)
4. M.T.Moita, M.E.Cunha, M.A.M.Sampayo, ICES/S.M. on the Causes, Dynamics and Effects of Except. Mar. Blooms and Related Events, B : 09 (1984)
5. A.F.G.Fiúza, Proc. of a Seminar. on Actual Problems of Oceanography, JNICT, Lisbon : 46-71 (1982)
6. C.D.Dias, X.Pastor. G.Pestana, C.Porteiro, E.Soareas, F.Alvarez, ICES/C.M., H : 42 (1983)
7. B.Coste, A.F.Fiúza, H.J.Minas, Oceanologica Acta 9 (2) : 149-158 (1986)
8. J.S.Pinto, Bol.Soc.Port.Cienc.Nat. XVII : 94-96 (1949)
9. M.A.M.Sampayo & G.Cabeçadas, Bol.Inst.Invest.Pescas 5 : 63-87 (1981)
10. M.A.M.Sampayo, ICES/S.M. on the Causes, Dynamics and Effects of Excep.Mar. Blooms and Related Events, B : 10 (1984)
11. M.A.M.Sampayo & M.T.Moita, ICES/S.M. on the Causes, Dynamics and Effects of Except.Mar. Blooms and Related Events, B : 12 (1984)
12. M.J.Brogueira & M.A.M.Sampayo, ICES/C.M., E : 35 (1983)
13. G.Cabeçadas, M.E.Cunha, M.T.Moita, J.Pissarra, M.A.Sampayo, Bol.Inst.Nac. Invest.Pescas 10 : 81-123 (1983)
14. E.S.Silva, Proc. 4th Inter.Seaweed Symp. : 265-275 (1963)
15. E.S.Silva, Sep.Arq.Inst.Nac.Saúde, Lisboa, IV : 253-262 (1980)
16. R.Margalef, Inv.Pesq. (Barc.), 5 : 113-134 (1956)
17. S.Fraga, J.Mariño, I Bravo, A.Miranda, M.J.Campos, F.J.Sanchez, E.Costa, J.M.Cabanas, J.Blanco, ICES/S.M. on the Causes, Dynamics and Effects of Except. Mar. Blooms and Related Events, C : 5 (1984).

KEYWORDS: Red tides; Hydrography; Toxicity; Portugal.

PARALYTIC SHELLFISH POISONS IN BIVALVE MOLLUSCS ON THE PORTUGUESE COAST CAUSED BY A BLOOM OF THE DINOFLAGELLATE GYMNODINIUM CATENATUM

SUSANA FRANCA,* AND J. F. ALMEIDA**
*Lab. Microbiologia Exp., Instituto Nacional Saúde (INSA), 1699 Lisboa codex;
**Instituto Nacional Investigação Pescas (INIP), 1400 Lisboa, Portugal

ABSTRACT

Surveillance of PSP in bivalve molluscs from the Portuguese west coast led to detection on high levels of PSP toxins associated with a bloom of Gymnodinium catenatum. PSP was determined by means of a bioassay using mice, and standardization was carried out by the AOAC method using saxitoxin. Phytoplankton samples were collected at the same time as shellfish. The results show the geographic distribution, levels (max.1595 ug/100 g meat) and the duration, October (39 ug/100 g meat) to May (42 ug/100 g meat), of PSP in various species of bivalve (mussels,clams). Different rates of PSP clearance were found for different mollusc species.

Despite information given to the population and prohibition of taking shellfish from affected areas, some illness following consumption of mussels, symptomatic of PSP was reported. The value of surveillance programs in preventing PSP poisoning in humans is so far incontestable.

INTRODUCTION

The first report of toxicity in bivalves on the Portuguese coast associated with a Gonyaulax tamarensis bloom was made by Silva [1] in 1962. Since then, until 1986,the occurrence of toxic dinoflagellates in the Óbidos lagoon was sometimes detected by the same author.Harvesting of shellfish was temporarily stopped, but no incident of poisoning has been reported. In April 1986, a toxicity monitoring program was initiated and subsequent occurrences have been detected. This report concerns the first episode of Paralytic Shellfish Poison detected in bivalves along the NW coast of Portugal associated with a Gymnodinium catenatum Graham bloom.

EXPERIMENTS

The establishment of the sampling map (Fig.1) and the co-ordination of the control program is due to Dr. C. Lima. Shellfish samples in the Aveiro estuary (Fig.2) and the field data in this area were made available by Dr. M. Sobral. The program includes both monitoring of water for dinoflagellates [2] and monitoring of shellfish for toxicity.

Plankton samples related to the episode now reported were examined under light microscope by Dr. E. S. Silva and the dominant organism identified as Gymnodinium catenatum. Silva isolated this species in culture, which has been maintained for further cytological and toxicological studies.

The data on toxin levels presented in this paper were obtained from bivalve molluscs - Mytilus sp., Ruditapes sp. and Cerastoderma sp.- growing in the areas affected by the G. catenatum bloom.

Water-soluble shellfish extracts for PSP determination were obtained by the AOAC method [3]. Mice for bioassays, Charles River CD-1, were furnished by Instituto Gulbenkian de Ciência, Oeiras, Portugal. The Paralytic Shellfish Poison standard was kindly sent to us by J. E. Gilchrist, from the Microbial Biochemistry Branch, Div. of Microbiology, Food & Drug Administration, Dep.

RED TIDES: BIOLOGY, ENVIRONMENTAL SCIENCE, AND TOXICOLOGY
Okaichi, Anderson, and Nemoto, Editors

of Health & Human Branch, Div. of Microbiology, Food & Drug Administration, Dep. of Health and Human Services, U.S.A.. The CF value determined was 0.188. Toxin levels are expressed in micrograms of saxitoxin equivalent in 100 grams of shellfish meat.

Aliquots of toxic crude extracts are kept frozen for further studies.

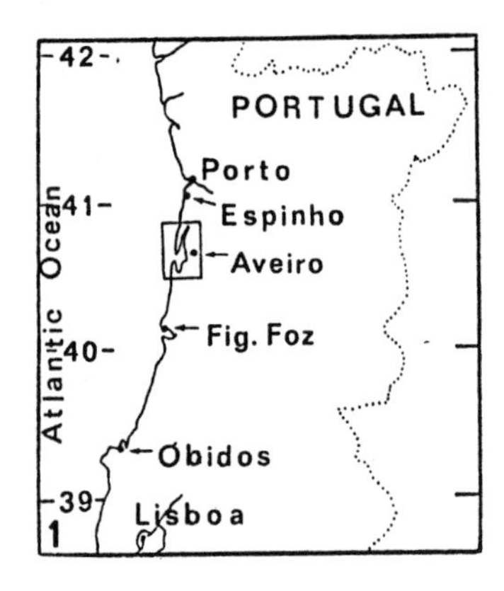

FIG. 1. Distribution of shellfish sampling areas (→), along the Portuguese coast.

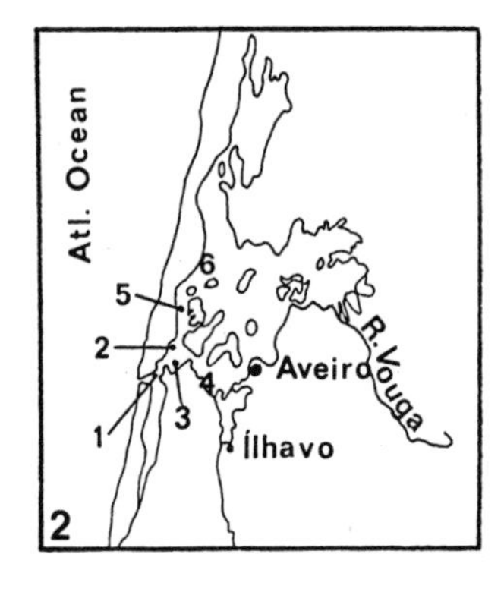

FIG. 2. Aveiro estuary detail (1 - 6, sampling stations).

RESULTS

In October 86 (40th week of 86), the determination of a PSP level of 39.4 ug/100 g on a sample of Ruditapes sp. harvested near Espinho (Fig. 1) led us to request the intensification of shellfish sampling on the NW coast of Portugal.

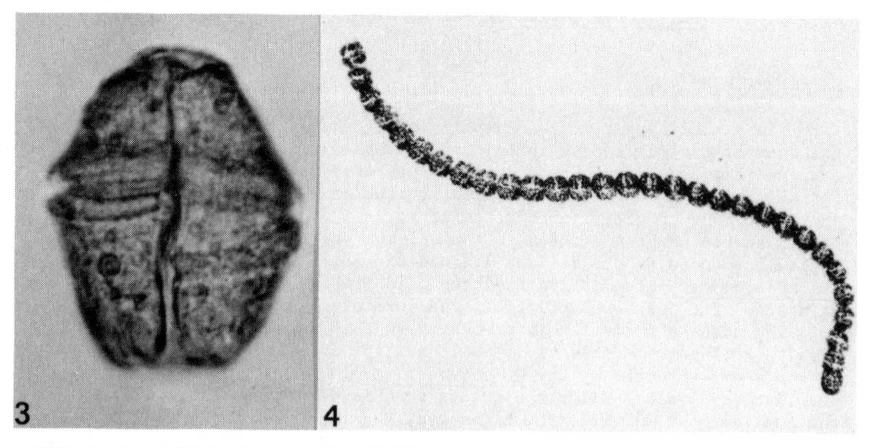

FIG. 3, 4. Light micrographs of live G. catenatum: 3. single cell (x420), 4. typical long chain of 30 cells (x33).

Light microscopic observation of a water sample collected at the time near this area showed the presence of the dinoflagellate G.catenatum (Fig. 3,4). Presence and dominance of G.catenatum was observed in samples of marine phytoplankton collected in the areas of toxic shellfish (Fig. 1; Table 1). This occurrence was not followed by cell counts, but another episode of PSP in Espinho which concerned the same organism was detected in August 87, and the results obtained were: Ruditapes sp.-PSP 95.6 ug/100 g and G.catenatum - 28000 cells/l. (M.A.de M. Sampayo, pers. comm.).

The table summarizes a study of the geographical distribution of toxin in mollusc shellfish samples, and the main results obtained during the period Oct.86 - June 87. Toxicity was detected from Espinho to Obidos lagoon.

TABLE I. Distribution of Paralytic Shellfish Poisons: Toxic levels in Mytilus sp. (M), Ruditapes sp. (R), and Cerastoderma sp. (C).

Portuguese NW Coast - Atlantic Ocean

Lati. 39º 40º 41º 42º

Station	OBIDOS LAG. (com.w.sea)			FIG. FOZ	AVEIRO (estuary) 1		2	3		4	5	6	ESPINHO	
Species	M	C	R	C	M	R	M	M	C	C	C	C	M	R
1986 week														
40														39
41		204	131	445	1023		347		235		s	190		
OCT. 42		236		446					733		722			51
43	779	1096	237	629	1320				615		780		1595	50
(36)* 44	1100	275	55											202
45	468	130	n	181	1436	s		1197	350	388	438	537		
NOV. 46	413	51	s		1030	43	1098	1076	150	391	342	344		
47	174	42	n		756	54	1081	1083	261		257	236		60
(43)* 48	107	n	n	75	358				47		119			
49	47	n				37								
DEC. 50					49						n			
(13)* 51					75	34	65	106	34	45	44	52		
1987														
(10)*JAN.				n	65	40	42	35	45	42	40	54		n
(12)*FEB.		n	n	n	n	36	34	n	49	54	47	56		n
(12)*MAR.	n	n	n		n	n	38	n	43	48	52	44		n
(11)*APR.	n	n	n		n		n	n	n	s	n	39		n
(11)*MAY		n	n		n	s	s	s	n	42	42	n		
(11)*JUN.	n	n	n		n	n	n	n	n	n	n	n		

Toxic levels in ug poison / 100 g of mollusc meat.
s - symptoms and death time longer than 60 min. ; n - no symptoms.
* (number of samples analyzed).

Harvesting of bivalves in the affected areas was stopped when the unsafe level was reached. However two family cases of illness (PSP symptoms) following consumption of mussels (Mytilus sp.) on December were directly reported to us. Dominant symptoms described were: "disturbed equilibrium, paraesthesia mainly circumoral and in limbs, tingling in arms and floating sensation". In one person these neurological symptoms were associated with gastrointestinal distress maintained during three days.

The symptoms observed in mice (with death time 5-7 min.) during the bioassay of both the crude extract and the solution of reference standard were the same. It appears therefore that the toxin extracted from the shellfish is identical to saxitoxin.

No fish kills were reported.

DISCUSSION

Reports on association between paralytic shellfish poisoning and G. catenatum blooms are scarce. Morey-Gaines [4] refers to an outbreak of PSP in Mexico (Pacific coast) in the spring of 1979 and points out this toxicity as an "apparently new characteristic of the species"; Fraga and Sanchez [5] refer to the relation found between G.catenatum and an outbreak of PSP, caused by mussels (Mytilus edulis) cultivated in Galician Rias, that affected more than one hundred people in Western Europe [6]. None of these studies present data about the levels of PSP in shellfish during these occurrences.

The data presented here show the changes of toxin levels related to both the sampling area and the shellfish analyzed. The highest toxin level registered was on Mytilus sp. near Espinho. In this area sampling of this species was not possible again. However, toxin levels found in Ruditapes sp. from this area show a rapid clearance of toxins. In the sampling areas of Aveiro, the data obtained also show that at the oceanic station (Fig. 2: 1) clearance of toxins is faster than at the estuary stations (Fig. 2: 2-6).

We observed that in both Aveiro and in Obidos, the bivalve species, Mytilus sp. and Cerastoderma sp. increase in toxicity faster than Ruditapes sp., although the beds of the three species were close together. In setting up a future control program, these observations will allow the selection of key species for routine surveillance.

In spite of the high levels of toxicity attained during this episode in Portugal, there was no notification of severe poisoning. This confirms the efficacity of this program of public advice and selective banning of shellfish harvesting.

ACKNOWLEDGEMENTS

The technical assistance of Ms. M. L. Pires and M. S-J. Fronteira on the the preparation of shellfish extracts is greatly appreciated.

REFERENCES

1. E.S. Silva, Notas Estud. Inst. Biol. Mar. Lisb. 26, 1-24 (1962).
2. M.A. de M. Sampayo, Red-Tides off the Portuguese Coast (in press).
3. AOAC "Official Methods of Analysis" 63, (14th.,1984), pp. 1336-1343.
4. G. Morey-Gaines, Phycologia 21, 154-163 (1982).
5. S. Fraga and F.J. Sanchez in: Toxic Dinoflagellates, Anderson, White and Baden, eds.(Elsevier, New York 1985) pp. 51-54.
6. J. Luthy in: Toxic Dinoflagellate Blooms, Taylor and Seliger, eds. (Elsevier, New York 1979) pp. 15-22.

PHYTOPLANKTON BLOOMS AND RED TIDES IN THE FAR EAST COASTAL WATERS OF THE USSR

GALINA V. KONOVALOVA
Kamchatka Dept. of Environ. Far East Branch Acad. of Sci.
Partizanskaya St. 6, Petropavlovsk, 683000, USSR

ABSTRACT

Based on long and short-term studies of phytoplankton in the far eastern seas and adjacent waters of the Pacific Ocean, two types of phytoplankton blooms have been established; regular seasonal outbursts of phytoplankton which repeat annually with different intensities, and occasional "red tides". The environmental factors responsible for the appearance of the seasonal phytoplankton "blooms" are analyzed. This paper presents the species causing "blooms" of both types. The "blooms" occur in highly productive eutrophic and hypereutrophic waters and are generally caused by tolerant algal species and by the ciliate *Mesodinium rubrum*. Undoubtedly, the frequency and concentration of "red tides" are directly connected with increased eutrophication of coastal waters under the influence of anthropogenic factors.

INTRODUCTION

Outbursts of phytoplankton growth attaining the strength of "blooms" and red tides have been observed repeatedly in the coastal waters of the far eastern seas and adjacent waters. Long-term, year-round observations of phytoplankton in bays and inlets in the northwestern Sea of Japan were started in 1969 [1-5] and short-term surveys have been conducted in the coastal waters of the Bering and Okhotsk Seas and along the Pacific coast of Kamchatka[6].

MATERIALS AND METHODS

The present study deals with occurrences of seasonal phytoplakton blooms and red tide in the far eastern seas.

Microplankton analysis involved sedimentation of samples preserved in Lugol's or weak (1 to 4 %) formalin solutions and subsequent inverse filtration through Nuclepore filters.

Red tide samples were analyzed *in vivo* or preserved in weak Lugol's solution without concentration.
Cell counts were made using 0.05 to 1 ml chambers, depending on the size of organisms and population density.

RESULTS AND DISCUSSION

Among cyclic seasonal phytoplankton blooms of varying

RED TIDES: BIOLOGY, ENVIRONMENTAL SCIENCE, AND TOXICOLOGY
Okaichi, Anderson, and Nemoto, Editors

strength which recur yearly (one to three blooms a year, depending on region) in the coastal waters of the Soviet Far East, particularly interesting are the winter diatom blooms found under fast ice cover at subzero water temperatures in the northwestern part of the Sea of Japan. An annual algal bloom occurs yearly between December and February in highly productive freezing inlets of Peter the Great Bay, at water temperatures of -0.2 to -1.9 °C and illumination of 2 klx, under a 30 to 60 cm-thick ice cover, and lasts 1 to 1.5 months.

Biomass of microphytic algae was in the range of 10 to 20 g/m^3. An intensive growth of diatoms was associated with strong convective mixing of the water column supplying nutrients from bottom layers. The winter bloom can be explained not only by high concentrations of nutrients [1] but also by low light requirements of the diatom species and their tolerance of subzero temperatures. The boreoarctic diatom _Thalassiosira nordenskioeldii_, accounting sometimes for as much as 40 to 80 % of phytoplankton biomass, and _Chaetoceros debilis_ and _Ch. pseudocrinitus_ should be noted among dominant species of the bloom. _Th. nordenskioeldii_, highly adapted to low temperature and illumination, has no real competitor under the ice cover [7].
Therefore, it can proliferate in the nutrient-rich coastal waters in winter and early spring. In addition, reduced decomposition and grazing by consumers at low and subzero temperatures contribute to an accumulation of phytoplankton biomass.

An autumn bloom in eutrophic bays of the western part of the Sea of Japan, with microalgal biomass at the winter bloom level (8 to 19 g/m^3), occurs between mid-September and mid-November. The bloom is favoured by the disturbance of water column stratification because of convection and strong autumn winds. An upward flow of nutrients produced by the activity of plankters and by bacterial mineralization of summer detritus feeds the bloom. These autumn and winter-spring blooms results from massive growth of the diatoms _Skeletonema costatum_, _Eucampia zoodiacus_, _Ch. compressus_ etc.

A summer bloom lasting about a month between June and August is observed when the water column is stable and organic matter decomposition and nutrient regeneration are maximal due to high water temperatures. The bloom occurs at coastal upwelling areas generated by offshore winds and is supported by increased allochthonous nutrient influx caused by summer rains. Numerous phytoflagellates predominate. The densest populations are produced by the smaller dinoflagellates _Katodinium rotundatum_, _Prorocentrum triestinum_, _P. micans_ and others, by the cryptomonad genus _Plagioselmis_ and the euglenoid _Eutreptia lanowii_. In summer, as well as in autumn, the obligatory eutrophic species _S. costatum_ and the boreotropical _Ch. affinis_ predominate among the diatoms.

Since the early 1980's, _red tides_ associated with the dinoflagellates _Noctiluca miliaris_, _Alexandrium_ (=_Protogonyaulax_)_tamarensis_ etc. and the ciliate _Mesodinium rubrum_ have become more frequent. In 1980-1987, more than ten cases of red tides of different intensities were recorded in the areas of the study, three of them were toxic and resulted in infection of molluscs and in mortality of birds. In April-May 1980 and 1982, strong red tides produced by N. miliaris were observed in inlets of Peter the Great Bay, with population densities reaching 0.5-0.7 x 10^6 cells/l. An analysis of conditions that determined the intensive growth of N. miliaris and development of the red tides showed that along with nutrient abundance and high rates of reproduction in the absence of grazing, the factor of primary importance was a delay in transportation of the population from inshore shallows into the open waters of the bay because of weakening of offshore winds and a change in their paths [8].

Five cases of "red tides" associated with M. rubrum have been recorded since 1983. Two blooms occurred in hypereutrophic Avachinskaya Guba Inlet (Kamchatka Peninsula) in September-October 1983, at water temperature of 9-11 °C and a salinity of 27 ‰ in the surface layer [6] and in late November-early December 1984, at water temperature of about zoro. The second bloom was short-lived and less dense (0.2-0.3 x 10^6 cells/l) compared to the first one (2 x 10^6 cells/l); it was obviously due to decreased reproduction at low water temperature. A weaker red tide associated with M. rubrum was first observed in the Sea of Japan (Vostok Bay) in August 1985, at water temperature of 21-22°C. It occurred after significant desalination of the surface sea water in the zone of the greatest land runoff and anthropogenic impact [5]. Population density in the zone of the bloom was rather high, 1.8 x 10^6 cells/l. Onshore south winds maintained the stability of the water column and contributed to the development of the bloom. Distinct "red tides" caused by M. rubrum were repeatedly observed in August-September, at 8 to 12 °C, in highly productive Kraternaya Bay (Ushishir Island, the Kuriles).
None of these red tides caused any obvious unfavourable effect, though the species diversity of phytoplankton and diatoms in particular was poor in each case.

In July - early August 1984 and 1987, brown colouring of water was observed in Avachinskaya Guba Inlet (Kamchatka) at surface water layer temperature of 14-15°C along with increasing toxictity of mussels. Analysis showed an abundance of dinoflagellates of the Alexandrium (=Protogonyaulax) "tamarensis" group: A. tamarensis and A. acatenella with population densities up to 0.6-7 x 10^6 cells/l. In July-August 1984, the dominant species was S. costatum (8 to 16 x 10^6 cells/l), proliferation of which stimulates growth of most flagellates associated with red tides [9]. This bloom was followed in November-December by a red tide caused by M. rubrum. A brown-red water bloom was also observed in July 1986 near the northeastern coast of Kamchatka (the Bering Sea), at water temperatures of 12 to 14 °C; it caused mortality of marine animals. Here also, water sampling showed high abundance (up to 1-2 x 10^6 cells/l) of A. tamarensis and A. exavata and extremely low diversity of microplankton. It is notable that all of the blooms described above, both seasonal and cyclic in character and "occasional" strongly pronounced red tides, occurred in eutrophic waters rich in biogenic elements. Nobody doubts today that the increased frequency and intensity of "water" blooms are directly related to an anthropogenic impact on neritic zones of the sea. It is established that optimal growth of a species requires a specific combination of biological and physical environmental factors, which seems to develop more readily in nutrient rich waters. This is especially true for eurybionts (and a majority of species mentioned above are eurybionts) which are resistant to changes of environmental physical factors and generate "water" blooms.

ACKNOWLEDGEMENT

The author thanks Dr. A.V. Zhirmunsky, the Director of the Institute of Marine Biology for his initiative and inspiration in these investigaions.

REFERENCES

1. Yu.I. Sorokin and G.V. Konovalova, Limnol. Oceanogr., 18, 6, (1973)
2. L.N. Propp, L.L. Kuznetsov, G.V. Konovalova and Yu.I. Dobryakov, Biologiya Morya, 5, (1982)
3. A.V. Zhirmunsky and G.V. Konovalova, Biologiya Morya, 5 (1982)
4. G.V. Konovalova, Sov. J. Mar. Biol., 10, 1, (1984)
5. G.V. Konovalova and M.S. Selina, Biologiya Morya, 3, (1986)
6. T.Yu. Orlova, G.V. Konovalova and V.V. Oshurkov, Sov. J. Mar. Biol., 12 (1986)
7. J.W. Baars, Mar. Biol., 68, 2, (1982)
8. D.I. Vyshkvartsev, Biologiya Morya, 2, (1985)
9. H. Iwasaki, "Toxic Dinoflagellate Blooms", New York e.a., 1, (1979)

GEOGRAPHIC DISTRIBUTION OF GYMNODINIUM NAGASAKIENSE TAKAYAMA ET ADACHI AROUND WEST JAPAN

MATSUOKA, K.[1], S. IIZUKA[2], H. TAKAYAMA[3], T. HONJO[4], Y. FUKUYO[5] and T. ISHIMARU[6]
[1]: Faculty of Liberal Arts, Nagasaki University, Nagasaki 852, Japan,
[2]: Faculty of Fisheries, Nagasaki University, Nagasaki, 852, Japan,
[3]: Hiroshima Fisheries Experimental Station, Ondo-cho, Hiroshima, 737-12, Japan, [4]: National Research Institute of Aquaculture, Nansei-cho, Mie, 516-01, Japan, [5]: Faculty of Agriculture, University of Tokyo, Tokyo 113, Japan, [6]: Ocean Research Institute, University of Tokyo, 164, Japan.

ABSTRACT

The geographic distribution of *Gymnodinium nagasakiense* was investigated around West Japan in order to clarify its morphological and ecological characteristics. This species has occurred in the following areas; West to North Kyushu, Seto Inland Sea, Coastal areas of the Pacific and the southern part of the Sea of Japan. The morphological features such as cell dorso-ventral compression and the left side position of nucleus are re-confirmed based on materials obtained from various locations, and its physiological characters are also discussed in comparison with *Gyrodinium aureolum* and *Gymnodinium mikimotoi*.

INTRODUCTION

Gymnodinium nagasakiense Takayama et Adachi, 1984, an unarmored autotrophic marine dinoflagellate, was first recorded in Omura Bay of West Kyushu, Japan as a red tide dinoflagellate under an informal name, *Gymnodinium* sp. Type-'65 by Iizuka and Irie[1,2]. After the first appearance, this species has occurred since and much damaged the fisheries culture, mainly in West Japan (Fisheries Agency of Japan[3], Honjo[4]). On the other hand Tangen[5] suggested the possibility that *G.* sp. Type-'65 might be conspecific with another *Gymnodinium*-like red tide species, *Gyrodinium aureolum* Hulbert, 1957 in the North Atlantic. Later, Takayama and Adachi[6] formally described this species as *Gymnodinium nagasakiense* and discussed the morphological similarities to some other species. But recent progress concerning the ecology of this species led to a need for its taxonomical re-evaluation in comparison with similar species, in particular *Gymnodinium mikimotoi* Miyake et Kominami ex Oda, 1935 and *Gyrodinium aureolum*.

METHOD

A questionnaire survey was adopted for obtaining information on the occurrence of *G. nagasakiense* to 29 Fisheries Experimental Stations in West Japan. The main quetions were as follows; Occurrence locations, occurrence date including red tides caused by this species, maximum cell densitv, water temperature, salinity and others.

RED TIDES: BIOLOGY, ENVIRONMENTAL SCIENCE, AND TOXICOLOGY
Okaichi, Anderson, and Nemoto, Editors

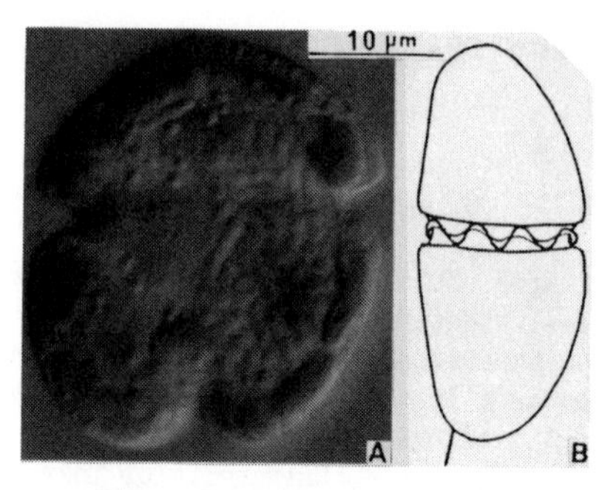

Fig. 1 Gymnodinium nagasakiense Takayama et Adachi A: Optical cross section showing a nucleus located on the left side (Kumano-nada), B: Lateral view showing a flatness of the cell.

For the re-evaluation of morphological features of G. nagasakiense, wild cells were provided from Tanabe Bay, Kumano-Nada and Lake Hamana where this species appeared abundantly this spring. The cultured cells kept in Hiroshima Fisheries Experimental Station and Ocean Research Institute of the University of Tokyo were also observed under optical and scanning electron microscopes.

RESULTS

The re-investigation of morphology of G. nagasakiense is undertaken on the cell shape, cell flatness, and the position of nucleus, features Takayama and Adachi[6] considered to be diagnostic. The outline in

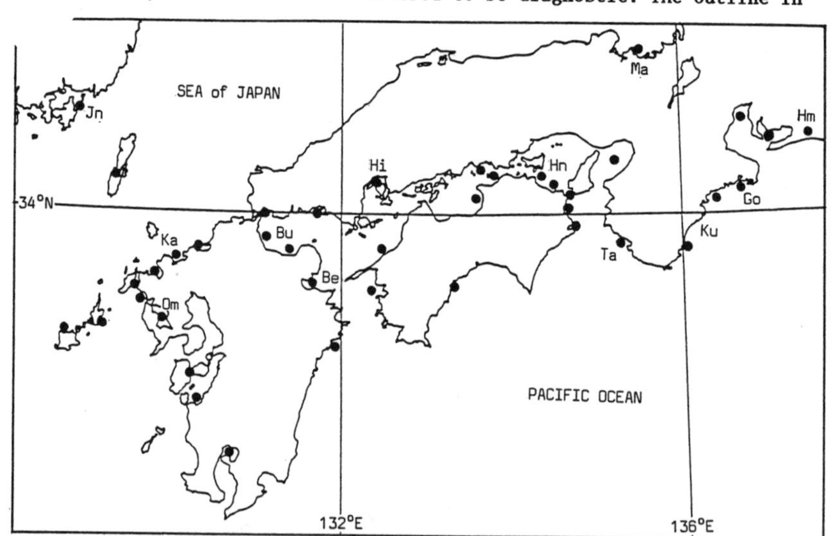

Fig. 2 Geographical distribution of Gymnodinium nagasakiense Takayama et Adachi, Om: Omura Bay, Ka: Karatsu Bay, Bu: Buzen Sea, Be: Beppu Bay, Hi: Hiroshima Bay, Hn Harima-nada, Ta: Tanabe Bay, Ku: Kumano-nada, Go: Gokasho Bay, Hm: Lake Hamana, Ma: Maizuru Bay, Jn: Jinhae Bay.

dorso-ventral view is very variable in cells from different locations. In particular, variation of dorso-ventral cell shape and size in culture is much greater than that in the wild. Both cells in culture and the wild, however, are always flattened dorso-ventrally. This feature may control the swimming of this species which is like a falling leaf, turning over and over through the water. The nucleus is ellipsoidal to reniform, and is always located on the left side.

G. nagasakiense occurs widely around West Japan, mainly from 32° to 36° N; West to North Kyushu (Omura Bay, Karatsu Bay, Hakata Bay, Aso Bay), almost the whole area of Seto Inland Sea (from Buzen Sea to Harima Nada), the coastal area of Kii Suido (Tsubakidomari Bay, Tanabe Bay), the coastal area of the Pacific (Kagoshima Bay, Uranouchi Bay, Kumano Nada, Ise Bay, Chita Bay and Lake Hamana), and the coastal area of the sea of Japan (Maizuru Bay). This species also occurrs in Jinhae Bay, south coast of Korea (Korea Ocean Research and Development Institute and Korea Advanced Institute of Science and Technology[7]).

The most frequent temperatures for G. nagasakiense lie between 24° and 29 ℃, recorded from late spring to early autumn. This supports estimates that approximately 26 ℃ or more is the most favorable temperature for reproduction of this species (Iizuka and Irie[1], Mie Fisheries Techno-logical Centre[8]) and red tides caused by this species frequently occur in summer. However, Honjo[4] reported the winter red tide caused by this species under lower temperature conditions (13° to 21 ℃) in Gokasho Bay.

The salinity range for this species is mainly from 28 ‰ to 33 ‰, except in the following examples; 5 ‰ in Kumano-Nada in 1984, 14.65 ‰ in Hiroshima Bay in 1983 and 14.63 ‰ in Imari Bay in 1986. In culture, Numaguchi and Hirayama[9] reported that the favorable salinity for this species is from 27 ‰ to 32.5 ‰, and the recent study of Mie Fisheries Technological Centre[8] clarified that this species reproduces well between 15 ‰ and 35 ‰ salinity.

The maximum cell densities of G. nagasakiense were 3.0×10^9 cells/ℓ in Hiuchi Nada, 2.4×10^9 cells/ℓ in Beppu Bay, and 1.5×10^9 cells/ℓ in Hiroshima Bay.

DISCUSSION

Since a Gymnodinium-like dinoflagellate, G. aureolum, first caused a red tide around the Norwegian coast, this species has occurred frequently and made red tides with mortality of marine organisms in north European waters. Tangen[5] and Taylor[10] have suggested the possibility that G. aureolum may be conspecific with G. nagasakiense. According to Tangen (pers. com.), the specimens occurring in north European waters have a nucleus located in the left lobe of the hypocone.

Based on the original description of G. aureolum given by Hulbert[11], this species differs from G. nagasakiense and the north European specimens of G. aureolum illustrated by Tangen[5] in having the following features; the cell is only slightly dorso-ventrally flattened, and the nucleus is spherical to subspherical in shape and located near the center of the cell. These morphological differences, therefore, suggest that the specimens which have been referred to G. aureolum in north European waters may be different from G. aureolum and rather similar to G. nagasakiense. From an ecological view point, G. nagasakiense is also similar to the European specimens of G. aureolum in the following way; The former is observed in a wide temperature range of 13 ℃ to 31 ℃ and its favorable temperature is approximately 24 ℃ to 29 ℃. G. nagasakiense made the winter red tide in the low temperature environment of 13 ℃ to 21 ℃ in Gokasho Bay and Buzen Sea. On the other hand, the red tide records caused by the north European specimens of G. aureolum had a temperature range of 8.5 ℃ to 12 ℃ (Tangen [5]), and 11.2 ℃ to 16.2 ℃ (Ballantine and Smith [12]). In salinity, G. nagasakiense has grown in the range of 15 ‰ to 35 ‰ but one very low salinity of 5 ‰ was recorded for this species in Kumano nada. The north European specimens of G. aureolum range from 25 ‰ to 33 ‰ (Tangen[5]).

This evidence indicates that G. nagasakiense is also physiologically similar to the north Europe specimens of G. aureolum, and both organisms are eurythermal and euryhaline dinoflagellates.

Another Japanese Gymnodinium-like species, G. mikimotoi is known to cause a winter red tide with mortality of pearl oysters in Gokasho Bay (Oda[13]). Although Takayama and Adachi[6] distingushed this species from G. nagasakiense in possessing a pentagonal ventral outline, broadly conical epicone and truncated pyramidal hypocone and in being phosphorescent, Taylor[10] considered that both species are con-specific. Accordingly, G. nagasakiense is variable in cell shape and also caused the winter red tide in Gokasho Bay, it is more likely that G. nagasakiense may be conspecific with G. mikimotoi. The evidence that there is no reliable record concerning G. mikimotoi after 1935 probably supports this opinion.

Before making a conclusion regarding the taxonomical problems in G. nagasakiense, G. mikimotoi and G. aureolum, however, we should clarify some of their physiological and biochemical characteristics such as pigment contents and isozyme patterns.

ACKNOWLEDGEMENTS

The authors wish to express their sincere thanks to 29 Fisheries Experimental Stations from Okinawa to Kanagawa and Toyama in West Japan for their kind co-operation with the questionnaire survey.

REFERENCES

1 Iizuka, S. and H. Irie, Bull. Fish. Nagasaki Univ., 21, 61-101,(1966).
2 Iizuka, S. and H. Irie, Bull. Fish. Nagasaki Univ., 27, 19-37, (1969).
3 Fisheries Agency of Japan. Gymnodinium-zoku akashiwo no hassei to higai ni kansuru jyoho no seiri-kaiseki, 9-45, Suisan-cho and Kankyo-cho, (1979).
4 Honjo, T., In Okaichi (ed.) Akashiwo no Kagaku, Kouseisha-Kouseikaku, 228-237, (1987).
5 Tangen, K., Sarsia, 63, 123-133, (1977).
6 Takayama, H. and R. Adachi, Bull. Plank. Soc. Jpn., 7-14, (1984).
7 Korea Ocean Research and Development Institute and Korea Advanced Institute of Sicence and Technology, 242, (1981).
8 Mie Fisheries Technological Centre (1983), Rep. Research Project for Red Tide, 1-37, (1983).
9 Numaguchi, K. and K. Hirayama, Bull. Fish. Nagasaki Univ., 33, 7-10, (1972).
10 Taylor, F.J.R., In Anderson, D.M. et al. (ed.), Toxic Dinoflagellates, 11-26, (1985).
11 Hulbert, E.M., Biol. Bull. Woods Hole Ocean. Inst. 112 (2), 196-219, (1957).
12 Ballantine, D. and F.M. Smith, Brit. phycol. J., 8, 233-238, (1973).
13 Oda, M., Zool. Mag., 47, 35-48, (1935).

THE DOMINANT RED TIDE ORGANISMS IN THE ZHUJIANG ESTUARY, CHINA

YONGSHUI LIN
South China Sea Institute of Oceanology
Academia Sinica, Guangzhou, China

ABSTRACT

This paper reports an investigation of red tide organisms in the Zhujiang Estuary from February to July, 1987. It establishes the species of red tide organisms and their abundance in an important Chinese estuary.

INTRODUCTION

The Zhujiang Estuary, which is adjacent to the South China Sea, is not only an important passage for communication and transportation, but also supports large fisheries. Due to the development of industry and agriculture in the surrounding areas, especially cities such as Guangzhou, Hong Kong, Shenzhun, Macao and Zhuhai, a vast amount of industrial and domestic sewage pours into the Zhuijiang, where red tides frequently occur. In order to determine the organisms responsible for red tides and the important associated environmental factors, a monthly sampling program has been carried out since February, 1987. This paper reports the preliminary results from February to July of this year. This survey is still in progress.

EXPERIMENTAL

The area of study extends from Zhuhai (Xiangzhu) to Guishan Island, where nine stations are located (Fig. 1).

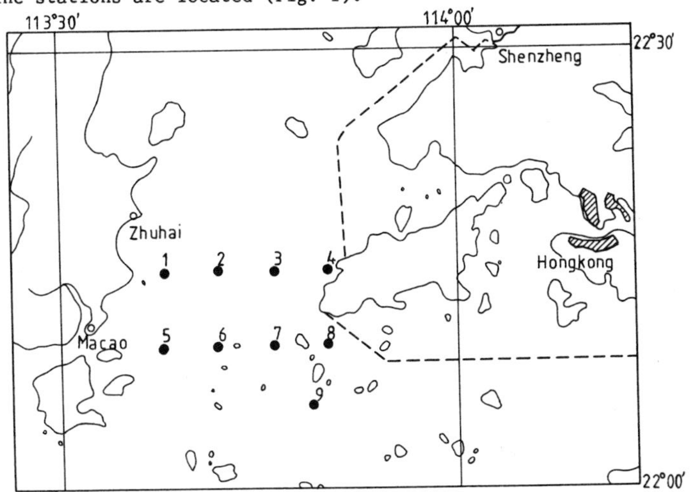

Fig. 1. Map showing the study area and sampling stations

RED TIDES: BIOLOGY, ENVIRONMENTAL SCIENCE, AND TOXICOLOGY
Okaichi, Anderson, and Nemoto, Editors

A small plankton net (37cm diameter, 150cm long, and 68 mesh/cm^2) was used to perform vertical hauls from the bottom to the surface. Samples were fixed with 5% formalin and brought to the laboratory. After settling and concentrating, samples were randomly subsampled and cells counted under the microscope. Species exceeding 5% of the total cell amount were considered dominant.

RESULTS AND DISCUSSION

The phytoplankton population in the Zhujiang Estuary changes dramatically through time. Cell numbers were highest in April (up to 2.3 x 10^8cells/m^3 averaged over all stations), while in May they were lowest (2.5 x 10^5 cells/m^3). In June and July, phytoplankton increased again.
Our results show that many red tide species bloomed in the study area. Major diatom species were Eucampia zoodiacus, Chaetoceros pseudocurvisetus, Skeletonema costatum, Major dinoflagellates species are Noctiluca scintillans, Ceratium breve, C. fusus and C. furca. These were all dominant species at some time during the study.

The Numerical Variation of Dominant Red Tide Organisms:

(1) Eucampia zoodiacus: This diatom's maximum was 3.5 x 10.^{6}cells/m^3 in March, representing 72.2% of the total cell count; the maximum percentage ever achieved was 84.2%. Although concentrations continued to increase into April and reached 1.5 x 10^7 cells/m^3, this species relative dominance decreased to 6.7% of the total cells due to the decreased abundance of other species. In subsequent months, cell numbers were lower as were the proportions of the total cell count.

(2) Chaetoceros pseudocurvisetus: In April and June, concentrations reached 7.5 and 6.2 x 10^7 cells/m^3. respectively. The amount in April accounted for 32.7% of the total cell count, concentrated at Stations 3, 6 and 7 (> 10^8 cells/m^3); The percentage reached 74.2% in June. Red tides of this species are most likely to occur in April and June; there is little possibility of major blooms in other months.

(3) Skeletonema costatum: There is a possibility of red tides of this species in July, because its numbers reached 1.2 x 10^7 cells/m^3 (19.6%). It was abundant at Stations 4 and 8, totalling 6.5 x 10^7 cells/m^3 at Station 4.

(4) Noctiluca scintillans: This was a dominant species at some stations in February, especially at 2 and 3, where it reached 17 and 14 x 10^4 cells/m^3 and represented 42.9% and 10.2% of the total cells respectively. In addition, the proportion was over 7.5% at Stations 1 and 5. This was a dominant species at 2/3 of the stations in May. It is noteworthy that in July, red tides occurred at Station 6 with concentrations of 36.5 x 10^4 cells/m^3 (Figure 2).

(5) Ceratium: The main species were C. fusus, C. furca, and C. breve v. breve. In February, the amount of these three species was high (11 x 10^4 cells/m^3) but their relative proportions were low. Ceratium fusus was most abundant and reached 15 x 10^4 cells/m^3, accounting for 7.4% of the total phytoplankton. In March, these dinoflagellates were also dominant species at some stations. Ceratium breve v. breve was most numerous at Station 5 and reached 51 x 10^4 cells/m^3 (16% of the total cell count). For C. furca, its numbers were also highest at Station 5 and

reached 19.6 x 10^4 cells/m^3 (6.1%). At Station 3, although these three species were not abundant, they accounted for 46% of the total cells. In May, their numbers were also low (11 x 10^4 cells/m^3, but they were again dominant at many stations.

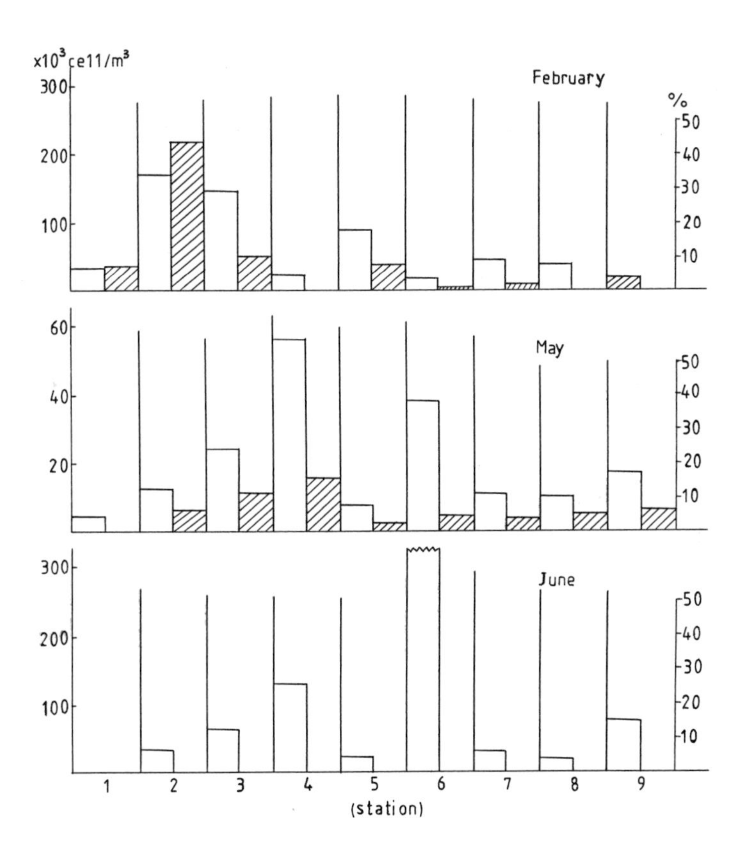

Fig. 2. Concentrations of N. scintillans open bars , cell counts; hatched bars, percent of total phytoplankton.

CONCLUSION

According to those preliminary results, the dominant red tide organisms in the Zhujiang Estuary between February and July may be summarized as follows;

February: N. scintillans.
March: E. zoodiacus is the major species, co-occurring with C. furca, C. breve v. breve and C. fusus.
April: C. pseudocurvisetus, C. compressus, Lauderia borealis, E. zoodiacus.
May: N. scintillans, C. breve v. breve, C. furca.
June: C. pseudocurvisetus.
July: S. costatum, C. affinis.

ACKNOWLEDGEMENTS

The author is grateful for the kind encouragement and advice given by Prof. G. Xue, Directoir of SCSIO, and also expresses his thanks to the members of the research group including the members in EPA of Zhuhai for their kind cooperation. This work was financially supported by a grant from the Scientific Academy of China and by the funds from the Environmental Protection Agency of Zhuhai, China.

REFERENCES

1. C. Y. Xu, Journal of Fisheries of China 6, 174-180 (1982).
2. J. Z. Zou, L. P. Dong and B. P. Qin, Marine Environmental Science 1, 41-55 (1983).
3. Synopsis of Red Tide Organisms, edited by the Working Party on Taxonomy in the AKASHIWO KENKYUKAI c/o Fisheries Agency, Japanese Government (1979-1984).
4. Y. S. Lin, L. M. Huang and W. B. Yuan. The Ecological Characteristics of the Red Tide Occurring in Daya Bay of Guangdong in Spring. Proceedings of the First Symposium on Marine Biology of the South China Sea 18-21, December 1986, Guangzhou.(in press)

EVALUATION OF WATER QUALITY IN THE ZHUJIANG ESTUARY, CHINA

ZHAODING WANG

South China Sea Institute of Oceanology
Academia Sinica, Guangzhou, China

ABSTRACT

This paper presents preliminary results of a monthly comprehensive investigation of eutrophication and red tides in the Zhujiang (Pearl River) estuarine area, beginning in February, 1987. The purpose of this paper is to discuss the trophic level of water quality in the Zhujiang estuary during Spring and Summer. A single parameter as well as a multiparameter Trophic State Index was adopted to assess the trophic status in the estuarine area. Using nutrients (DIN and DIP), COD loading, and changes in phytoplankton (cell counts and chlorophyll a) as the major parameters, a synoptic assessment was made.

INTRODUCTION

Eutrophication is the nutrient over-enrichment of bodies of water. It occurs in inland seas, harbours, estuaries and coastal waters with heavy organic pollution input and poor water exchange. According to many current studies, eutrophication greatly affects marine ecological equilibria, and is an important factor in red tides [1,2,5].

The Zhujiang estuary is subject to the influence of domestic sewage discharged from the cities of Macao, Zhuhai, Guangzhou, Shenzheng, and Hong Kong and by runoff of nitrogen and phosphate fertilizers. In recent years, red tides have occurred several times in this estuary. In order to determine the level of eutrophication in the area and the environmental characteristics conducive to red tides, a comprehensive investigation was initiated in February, 1987. This paper is a preliminary analysis of results obtained during Spring and Summer.

MATERIALS AND METHODS

The study area was delimited by 22°06'30" to 22°15'N and 113°38'05" to 113°49'B and contained 9 sampling stations (Fig. 1) Temperature, transparency, pH, Eh, velocity and direction of wind, velocity and direction of current, salinity, dissolved oxygen (DO), COD, NH_4-N, NO_2-N, NO_3-N, PO_4-P, Fe, Mn, amount and species of phytoplankton and chlorophyll a were measured by standard oceanographic methods.

At present, there are no standard principles, methods or measures for quantifying marine eutrophication. In this paper, the single parameter as well as the multiparameter Trophic State Index (T. Okaichi, 1972) was adopted. Chemical oxygen demand (COD), dissolved inorganic nitrogen (DIN), dissolved inorganic phosphorus (DIP), and phytoplankton (total cell counts and chlorophyll a) were selected as indicators and used to classify waters as oligotrophic, eutrophic or supertrophic. COD=1-3 mg/l, DIN=0.2-0.3 mg/l, DIP=0.04 5 mg/l, and chl. a=1-10 mg/m^3, were the ranges for waters as suggested by Zou [4] (Table I).

RED TIDES: BIOLOGY, ENVIRONMENTAL SCIENCE, AND TOXICOLOGY
Okaichi, Anderson, and Nemoto, Editors

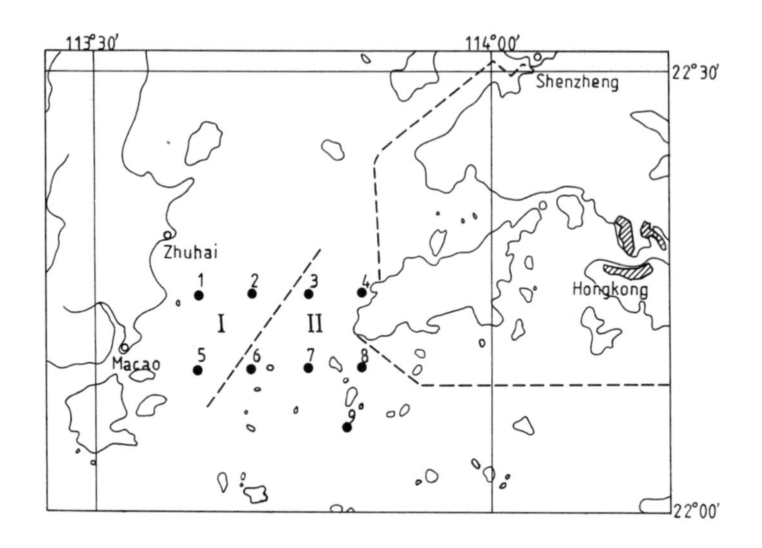

Fig. 1. Map of study area showing the regions and sampling stations used in this study.

RESULTS AND DISCUSSION

Nutrient Concentrations

Mean observed values of DIN, DIP and COD from February to July, 1987 are shown in Table II. DIN was at eutrophication levels except in February whereas DIP just reached mesotrophic levels. COD exceeded the eutrophic threshold, especially in July, when it reached the supertrophic level. Based on Okaichi's Trophic State Index:

$$TSI = \frac{COD \times DIN \times DIP}{4500} \times 10^6$$

If TSI 1, there is eutrophication. The TSI values in Table I show that eutrophication occurred from February to July, particularly in July. This result agrees with COD and DIN values. However, the mean monthly values of DIP are low and not in agreement with DIN, COD and TSI values. Apparently, the TSI values were affected greatly by DIP, so that the TSI was unable to reflect the great changes in phytoplankton standing stock in April 1987. There is a possible reason why DIP is lower. We suppose that the low DIP is related to the low phosphorus content of the estuary which became even lower due to phytoplankton growth [3].

Phytoplankton Standing Stock

An apparent sign of eutrophication is the great abundance of certain dominant species suited to conditions of excess nutrients (N, P) and organic matter. Phytoplankton standing stock is therefore a useful indicator of eutrophication. In the study area, the total phytoplankton cell counts and chlorophyll a are 0.5×10^3 to 2.3×10^5 cells/l and 2.30 to 7.66 mg/m^3, respectively. Based on the eutrophic threshhold of Zou [4], the water in the area became eutrophic in Spring and Summer; it was most serious in April.

TABLE I. Comparison of trophic categories of marine systems and several areas in this study.

	DIN (mg/l)	DIP (mg/l)	COD (mg/l)	Chl.-a (mg/m^3)	Phyto. (cell/l)	TSI*	
	0.2-0.3[a]	0.02		10			b
oligotrophic		0.01[a]		0.3-2.5			c
mesotrophic		0.01-0.03[a]		1-15			
eutrophic		0.03[a]		15-140			
oligotrophic (Kuroshoi)	0.1-0.15	0.01	1	1	10^3		d
eutrophic (threshold)	0.2-0.3	0.045	1-3	1-10	10^3-10^6		
supertrophic			3-10	10-200	10^6-10^9		
Tokyo Bay	0.5-1.0	0.05-0.1	1-3	18-113			
Bohai Bay							
in May	0.24	0.058	2.15	1.69	3.6×10^3	0.33	
in Aug.	0.26	0.231	1.64	3.4	4.6×10^3	67.1	
Zhujiang Estuary							
in Feb.	0.067	0.017	4.80	2.30	4.8×10^3	1.63	e
in Mar.	0.255	0.021	5.33	3.14	4.9×10^3	6.28	
in Apr.	0.317	0.008	6.37	7.66	2.3×10^5	3.73	
in May	0.324	0.013	5.69	3.82	0.5×10^3	5.86	
in Jun.	0.291	0.014	5.21	2.52	8.4×10^4	6.75	
in Jul.	0.506	0.014	21.32	2.78	6.5×10^4	48.66	

*TSI - Trophic State Index (T. Okaichi, 1972)
a. All these data mean total-N or total-P
b. from the Encyclopaedia of China
c. from Vowles and Connell (1980) [6]
d. from Zou, et al (1983) [4]
e. this paper

TABLE II. Mean observed values of several indictors for classifing trophic state.

		DIN (mg/l)	DIP (mg/l)	COD (mg/l)	Phyto. (10^3cell/l)	Chl.-a (mg/m^3)	TSI*	E**
Feb.	I	0.10	0.0225	4.95	1.25	1.63	2.78	6.58(2.78)
	II	0.035	0.0125	4.70	6.00	1.48	0.51	5.50(2.41)
Mar.	I	0.325	0.0175	5.55	2.25	2.92	7.89	8.05(3.05)
	II	0.225	0.0225	5.20	8.50	3.53	6.85	7.45(2.96)
Apr.	I	0.43	0.0125	6.30	195.00	7.71	8.47	9.08(3.33)
	II	0.25	0.0075	6.75	295.00	7.65	3.16	8.30(3.33)
May	I	0.35	0.0225	5.85	1.10	2.71	11.52	8.37(3.31)
	II	0.25	0.0125	5.65	0.45	3.94	4.41	7.53(2.95)
Jun.	I	0.55	0.0150	5.50	40.00	2.74	11.34	9.00(3.09)
	II	0.225	0.010	4.75	240.00	2.62	2.67	6.30(2.48)
Jul.	I	0.70	0.0125	27.50	3.00	1.71	60.16	31.63(13.01)
	II	0.40	0.0150	15.00	155.00	2.51	22.50	17.75(7.29)

*TSI - the calculated values for Trophic State Index
**E - the values calculated by following equation,

$$E = W_i C_i / C_{si} = W_N C_{DIN} / C_{sDIN} + W_P C_{DIP} / C_{sDIP} + W_O C_{COD} / C_{sCOD}$$

when $W_N = W_P = W_O = 1$, and the values in brackets when $W_N : W_P : W_O = 0.10 : 0.45 : 0.45$,
where W_i, C_i, and C_{si} are weighted coefficient, concentration, and standard, respectively, for DIN, DIP, and COD.

From Table II, it is clear that in Spring (February to April) the values of DIN, COD, and Chl. a increase with salinity and correlate with phytoplankton stock in regions I and II. The values of E correlate better than those from the TSI. In Summer (May to July), region I values are similar to those in the Spring; in region II the phytoplankton standing stocks exceed those of region I. Neither TSI nor E reflect this.

CONCLUSION

1. The results reported in this paper show that, based on previously suggested eutrophication thresholds of DIN, COD, Chl. a and phytoplankton, the water in the study area was lightly eutrophic from February to July, 1987. However, the values of DIP are not consistent with this trend. We thus suggest that the eutrophication threshhold for DIP should be 0.020 mg/l, which would then make it consistent with the other indicators.
2. The TSI and E multiparameter indices are not as good as we expected. Therefore, we believe that a task of top priority should be to devise new principles, standards and methods for quantifying eutrophication.

ACKNOWLEDGEMENTS

The author is grateful for the kind encouragement and advice given by Professor G. Xu, Director of SCSIO, and also expresses his thanks to the members of the research group including the members in EPA of Zhuhai, China for their kind cooperation.

This work was financially supported by a grant from the Scientific Academy of China and by funds from the Environmental Protection Agency of Zhuhai, China.

REFERENCES

1. R. J. Reimold, Eutrophication of Estuarine Areas by Rainwater, Chesapeake Science, 8, 132 (1968).
2. C. P. Spencer, Neth. J. Sea Res., 19, (1985).
3. Z. Wang and Y. Lin, Proceeding of the First Symposium on Marine Biology of the South China Sea, 18-21 December 1986, Guangzhou, P. R. of China (in press).
4. J. Z. Zou, L. P. Dong and B. P. Qin, Marine Environmental Sci., 2, 41 (1983).
5. J. Z. Zou, L. P. Dong and B. P. Qin, Hydrobiologia, 127, 27 (1985).
6. P. D. Vowles and D. W. Connell, Experiments in Environmental Chemistry (Pergamon Press, Australia 1980).

TOXIC SHELLFISH POISONING IN GUATEMALA

ROSALES-LOESSENER, F.,* E. DE PORRAS** and M.W. Dix***
*DITEPESCA, Ministry of Agriculture; **LUCAM, Ministry of Health and ***Universidad del Valle de Guatemala, Apartado Postal No. 82, 01901 Guatemala, Guatemala, C. A.

ABSTRACT

On July 30, 1987 a severe outbreak of paralytic shellfish poisoning occurred for the first time near Champerico on the Pacific Coast of Guatemala, Central America. The responsible dinoflagellate was identified as *Pyrodinium bahamense* var *compressa*. Fish, shrimp, blue crabs and bivalve samples from Champerico were bioassayed in mice. Only the small beach clam, *Amphichaena kindermanni*, was toxic, with saxitoxin levels up to 30,000 MU/100g. The clams had been collected in large numbers at low tide and consumed in chowders by the local population. Ingestion was followed directly by severe poisoning of more than 175 persons with 26 deaths.

INTRODUCTION

Paralytic toxic shellfish poisonings, associated with dinoflagellate blooms, have been documented for *Gonyaulax catenella* along the coast of Venezuela, the eastern Pacific from Central California to Alaska [1], and in Chile [2]. Other causative organisms from the Pacific are *G. tamarensis* with a 10-15% mortality rate [1] and *Pyrodinium bahamense* var *compressa* in Papua New Guinea, Brunei and Sabah [3].

In Guatemala, the only previously documented red tide occurred in August 1985 at Puerto Quetzal on the Pacific Coast, but the responsible organism, which resulted in a fish kill, was not identified.

In this study we will report a paralytic shellfish poisoning outbreak in 1987 in Guatemala; results of analysis of seafood and water collected from the beach, ocean and estuary near Champerico and from a commercial shrimp farm to the east of Champerico; identification of the responsible bivalve and dinoflagellate; and the results of interviews with families and medical authorities in the affected areas.

EXPERIMENTAL

When a report was received on July 30, 1987 of a severe outbreak of food poisoning in the Pacific Port of Champerico, Retalhuleu, Guatemala, the following investigations were planned and carried out.

Field Survey

Interviews. Families living in the area, local medical institutions, fishermen, fishing cooperatives and local centers of fishing and mariculture were interviewed.

The following symptoms were considered diagnostic for paralytic shellfish poisoning: severe headache; paresthesia of two of the following, lips, face, ears, fingers, arms, or legs; and at least two of the following:

RED TIDES: BIOLOGY, ENVIRONMENTAL SCIENCE, AND TOXICOLOGY
Okaichi, Anderson, and Nemoto, Editors

dysphagia; shortness of breath; gait abnormality; dysequilibrium; vertigo; motor weakness or paralysis [4].

Water samples were collected with a Van Dorn bottle from the end of a dock about 100m from the shore. Samples were taken from surface water, 2m and 5m depth. Water was also sampled from the surface at about 25m from the shore, from an adjacent estuary, and from a *Penaeus* shrimp farm which uses estuarine water. All samples were adjusted to 5% formalin. In the laboratory they were microscopically examined for dinoflagellates.

Food samples were obtained from an affected family and from a local fishing cooperative. Bivalve, crustacea and fish samples were taken in areas adjacent to the outbreak, a nearby estuary and a *Penaeus* farm.

An aerial overflight was carried out on 4 August, 1987 to try to define the limits of the red tide.

Food analysis

The food samples were assayed at the national laboratory for food and drug analysis, LUCAM (Laboratorio Unificado de Control y Medicamentos). The standard mouse bioassay for paralytic shellfish toxins [5] was used.

RESULTS

In all, 187 cases with 26 deaths (14%) were detected along a 60 km stretch of coast between Semillero, Escuintla and Champerico, Retalhuleu (Fig. 1 and Table I). Twenty-one deaths occurred in Champerico and five at Finca La Verde in Mazatenango (40 km from Champerico). Mortality was highest in children aged 0-6 years (50%) and dropped to 5% in persons above 18 years. Incidence may have been much higher because all families and localities were not visited and persons who did not seek medical aid were not reported.

Several residents, interviewed on 4 August, said that they had noted numbness of the lips after eating clam chowder on 29 July. They also reported that they had been harvesting clams regularly since April.

Table I. Paralytic shellfish poisoning distribution in Guatemala, 1987.

Locality	Cases	Deaths	Date
Champerico	145	21	30/7/87
La Máquina	19	0	30/7/87
Finca La Verde	13	5	30/7/87
Tulate	1	0	1/8/87
Semillero	9	0	3/8/87

In Champerico, severe symptoms appeared on 30 July and the epidemic was first recognized at about 2 pm when patients began arriving at the local clinic. Numbers peaked around 5 pm and intoxication appeared to be related to ingestion of clam chowder or clams at the noon meal. This was confirmed by a case control study which indicated a high correlation ($p < 0.0009$) of poisoning with ingestion of clam chowder [4].

All clams had been collected in the intertidal zone or just below the low tide mark on the ocean beach. The small 2cm clam responsible for the intoxication was identified as *Amphichaena kindermanni*. Laboratory analysis (Table II) indicated that only this species was toxic, with toxicity levels up to 30,000 MU/100gm wet weight, extremely high levels. Biochemical separation of the toxins from a sample sent to the FDA laboratories in Washington DC indicated very high concentrations of 21-sulfosaxitoxin

(Saxitoxin B1), and lower concentrations of saxitoxin and 1-N-hydroxy-saxitoxin (i.e. neosaxitoxin) [6]. Analyses performed by Dr. Kitani on formalin preserved material sent to the University of Tokyo indicated the presence of saxitoxin, neosaxitoxin and gonyautoxins 2, 3 and 4.

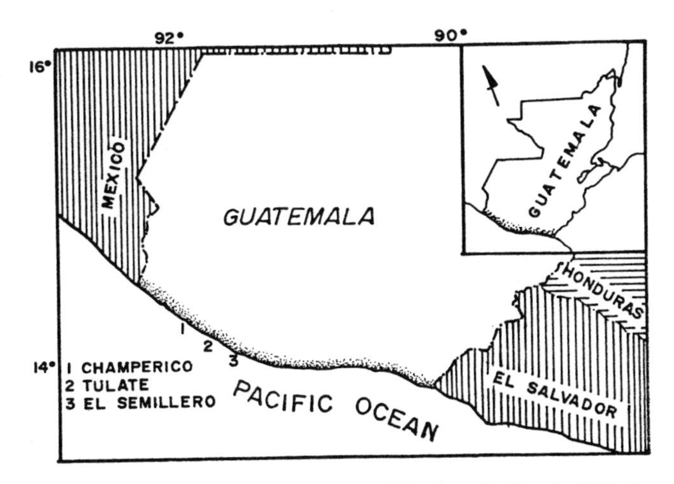

FIG. 1. Guatemalan localities implicated in paralytic shellfish poisoning. August, 1987

TABLE II. Toxicity of samples from Champerico, Guatemala.

	Organism	Date	Toxicity MU/100gm	Collection Site
Bivalves	Amphichaena kindermanni	30/7/87	1,300	Low water
		2/8/87	30,000	Ocean Beach
		7/8/87	11,000	
	Mytella strigata	6/8/87	ND	Shrimp pond
Fresh Fish	Dormitator latifrons	6/8/87	ND	Shrimp pond
	Marine catfish (moribund)	3/8/87	ND	Ocean Beach
	Diapterus sp., Asyanax sp.	6/8/87	ND	Estuary
Dry Salted Fish	Marine catfish, shark,	30/7/87	ND	Ocean
	Scynoscion sp., Diapterus sp.	30/7/87	ND	Ocean
Crustacea	Callinectes sp., Penaeus vannamei	6/8/87	ND	Shrimp pond
	Macrobrachium tenellum	6/8/87	ND	Shrimp pond
	Penaeus spp. (tails)	6/8/87	ND	Ocean
	Clam Chowder	30/7/87	12.7*	Family Dwelling

ND = none detected *MU/ml

Tropical storms delayed aerial recconnaisance until August 4, when we found a red band about 4 km wide and 40 km long parallel to the coast east from Champerico.

As soon as it was clear that toxic dinoflagellates identified as Pyrodinium bahamense [6] were implicated in the poisoning, a moratorium was placed on all seafood harvesting. This was later restricted to molluscs.

Recent samples have contained levels around 390 MU/100g.

DISCUSSION

This is the first time that *Pyrodinium bahamense* has been shown to cause paralytic shellfish poisoning on the Pacific Coast of America. However, *P. bahamense* var *compressa* was reported from the Mexican Pacific in 1942 [3] so the organism is not new to the area. This introduces the possibility that it may have been the causative agent in previous red tides from Central America and Mexico, which have been poorly documented.

Pyrodinium bahamense produces resting cysts which can serve as a source for future blooms when the appropriate environmental conditions appear. At the present time, we must assume that the benthic zone of the continental shelf is well seeded with these cysts. To prevent tragic consequences in the future, a shellfish monitoring program for toxicity has been initiated. The low toxicity of the clams prior to the outbreak suggests that the bloom developed offshore and was brought into contact with the bivalve populations on the beach by currents or mixing associated with a tropical storm. The bloom was maintained or, at least, encouraged, by nutrient run-off in rivers draining agricultural areas, and fed by heavy rains in the preceding two weeks. Moreover, the local estuary is normally separated from the ocean shore by a sand bar which had been breached on July 27. This would have released additional nutrients to the inshore waters.

One might speculate that the combination of tropical storm, increased nutrients, changes in currents because of breaching the sand bar fulfilled a set of conditions necessary for the inshore populations of *Pyrodinium bahamense* to reach sufficient density to cause the clams to become toxic.

The bivalves in the estuary were not toxic on 6 August, although the beach clams still were, suggesting that *Pyrodinium* was not present in significant numbers in the estuary.

We recommend that the monitoring program be coupled with education of rural and marginal populations which are most at risk. Because there was no prior history of shellfish poisoning, the public was unaware of the danger. Indeed, the first cases were treated for pesticide poisoning (a not infrequent occurence in the area) and given atropine, until laboratory tests proved negative for pesticides [6]. Timely intervention by public health authorities in informing the public in the coastal areas prevented further tragedies after July 30.

Investigation is needed to define which bivalve species and which areas should be included in a monitoring program. More information on dinoflagellate populations, and the factors involved in initiating *Pryodinium bahamense* proliferation in our region would also help in predicting outbreaks.

REFERENCES

1. K.A. Steidinger and K. Haddad, Bioscience, 31 (11),814-819. (1981).
2. S.A. Placier, Informe Segundo Taller Progr. de Plancton Pacifico Oriental, Inst. del Mar, Callao, Perú. UNESCO. (1981).
3. K.A. Steidinger, L.S. Tester and F.J.R. Taylor, Phycologia, 19,329-337 (1980).
4. D.C. Rodríguez and R. A. Etzel, Epidemiología de Intoxicación Epidémica por Mariscos, Champerico, Guatemala. Julio 1987. Rep. Min. Salud Pub. (1987).
5. B.W. Halstead and E.J. Schantz, Intoxicación Paralítica por Mariscos. WHO (1984.).
6. S. Hall, Outbreak of Paralytic Shellfish, Poisoning (IPM) in Guatemala. Unpubl. Rep. PAHO, (1987).

RED-TIDES AT THE MOUTH OF SUMIDA RIVER, TOKYO, DURING THE LAST ELEVEN YEARS, 1976 - 1986

HIDEAKI TAKANO
Tokai Regional Fisheries Research Laboratory
Kachidoki 5-5-1, Chuoku, Tokyo 104, Japan

ABSTRACT

The quantities of chlorophyll a, including pheopigments, in the brackish water of Sumida River were measured at a fixed station near the laboratory at least once in ten days from January 1976 to December 1986. Values higher than 30 μg/l were obtained mainly from the middle of April to the end of September, and in February and October in some years. The highest value was 356 μg/l at the end of July 1977. The main growth areas of plankton were in Tokyo Harbour. Populations were swept to the sampling station by tidal currents.

INTRODUCTION

The author's laboratory is located on a reclaimed island in the estuary of the Sumida River. Many people usually think that the river-water is fresh until the mouth, but at the time of high tides seawater enters for several kilometers from the mouth of the river. The author began observations of this river at the beginning of 1976, by dipping a bucketful of water around the time of high tides. At first it was done once a week, but later it was once every ten days. This task was continued for eleven years, and was stopped at the end of 1986. There is a small landing stage floating on the river beside the laboratory. River-water was taken from a short bridge between the stage and the land.

RESULTS

Water Temperature The water temperature was commonly lowest in mid-February and highest in early August. Among the single values, the lowest was 7.0°C obtained in February 1986 and the highest was 30.3°C obtained in July 1978. The lowest values of the eleven years were in the range between 7.0°C and 9.4°C, and the highest values were between 26.3°C and 30.3°C.

Mean annual values show that water temperature was higher than 15°C after mid-April, and lower than 15°C from early December to the next spring.

Salinity Measurement of salinity were carried out for three years, from

RED TIDES: BIOLOGY, ENVIRONMENTAL SCIENCE, AND TOXICOLOGY
Okaichi, Anderson, and Nemoto, Editors

1976 to 1978. Salinity was highest in mid-February (28.66 per mille) and lowest in mid-June (12.86 per mille). The highest value obtained was 30.57 per mille in mid-February 1977, and the lowest value was 1.28 per mille in mid-August 1977. Therefore, it may be roughly seen that the salinity values were smaller than 20 per mille from late April to late September, and they are greater than 20 per mille from October to mid-April.

Chlorophyll a Values For getting the chlorophyll values, microalgae in river-water were collected by filtration with Watman Glassfibre Filter C, and pigments of microalgae were extracted in 90 % acetone. The absorbance of samples was read with a spectrophotometer. For estimating the quantities of pigments, the trichromatic equations described by Jeffrey & Humphrey [1] were used.

The highest value of chlorophyll *a* obtained in the eleven years was 356 µg/l, in late June 1977, when *Heterosigma akashiwo* and *Prorocentrum minimum* were blooming. Chlorophyll *a* values higher than 30 µg/l were mainly seen from mid-April to early September every year. Even in early spring and mid-

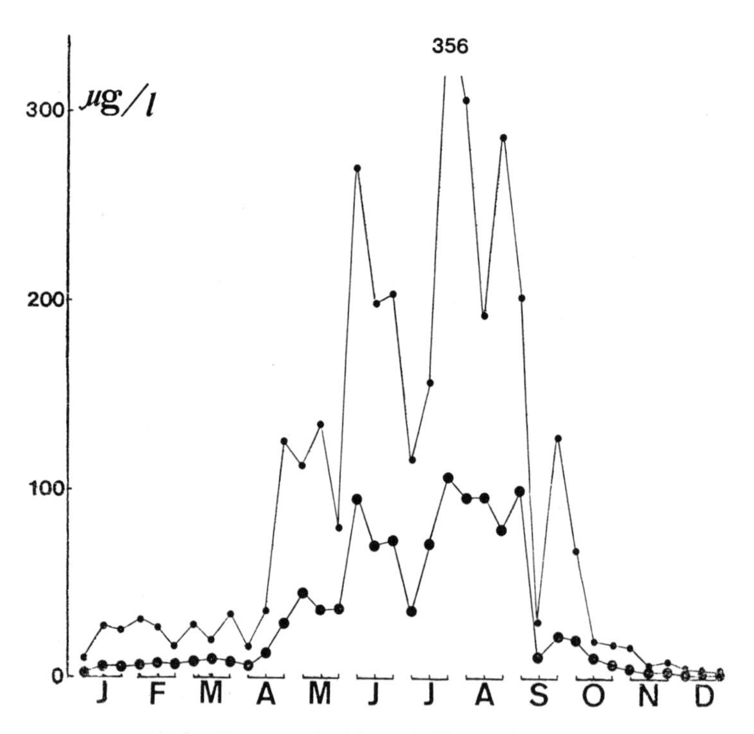

FIG. 1. Changes of chlorophyll *a* values in months.
• - highest values ● - meanvalues of 11 years

TABLE I. Distribution of chlorophyll _a_ values higher than 30 μg/l in the eleven years.

	'76	'77	'78	'79	'80	'81	'82	'83	'84	'85	'86
F											31
M									34		
A				44 77			31		31 125		
M	32	64	67 46 100	61 63	87	39	103 57 61	93		112 134	33
J	30	133	47	50 198 159	89 203	102 98	193 34	214	269 62	84 30	37 118 44
J	114	150 356	82 156 191	115 34	36 65	151	35 35	33 43	76 85	100 35	58
A	31 55 49	154 287	80 181 51	99 115 41	40 136	30 46 44	62 101 57	68 44 38	306 121 53	46 191 192	262 285
S		175		49 44	202	47	57	122	151	48	128
O							30	67	39		

autumn, sporadic blooms of microalgae sometimes occurred.

In this work, the percentages of pheopigments were also calculated by the Lorenzen equations [2]. In the 598 samples obtained in eleven years, samples with pheopigments lower than 10 % were 35 % of the total, with 10-40 % pheopigments were 37 % of the total, and with more than 40 % pheopigments were 28 % of the total.

Main Planktonic Microalgae Although there were many flagellates and diatom species in the river, the following six microplankters frequently caused the red-tides at this station:

1. Euglenophyceae
2. _Heterosigma akashiwo_ (Hada) Hada
3. _Mesodinium rubrum_ (Lohm.) Hamb. & Budd.
   This is a photosynthetic ciliate.
4. _Prorocentrum minimum_ (Pavillard) Schiller
5. _Skeletonema costatum_ (Greville) Cleve
6. _Thalassiosira_ spp.

In addition to these, _Pyramimonas_ aff. _amylifera_ Conrad was once abundant in early June 1984. The diatom _Rhizosolenia setigera_ Brightwell once made a prominent red-tide in winter, but it was before 1975.

Although they were never predominant, the diatoms _Eucampia zodiacus_ Ehr.

and Ditylum brightwellii (West) V.H. were often abundant in spring. Lithodesmium variabile Takano propagated here in autumn.

In other estuaries of southern Japan, Gymnodinium nagasakiense Takayama & Adachi and Prorocentrum micans Ehr. (levantinoides form) sometimes make very heavy blooms. These dinophytes were also found sporadically in the Sumida River, but none of them so far has caused a red-tide here. When they occurred, the author had an impression each time that the cells might have been transported from other waters by the boats carrying rawfish to the Tokyo Wholesale Fish Market. Therefore, they are still thought of as exotic species in this estuary.

DISCUSSION

The salinity of water in the Sumida River estuary during summer is much lower than that in winter. This change may be caused by heavy rainfall and urban drainage during warm seasons. By looking at the topography around the sampling station, it is evident that the marine and brackish water microalgae found in the river-water originally grew in Tokyo Harbour, and entered the river with tidal currents. Therefore, if the samples were taken at the Harumi Futo Wharf, values of chlorophyll a might be higher than those obtained at the river station.

Microflagellates growing in river-water here usually increase after middle April when the temperature becomes higher than 15°C. From early May to early September, blooms of diatoms and microflagellates alternate. In autumn, although the water temperatures are still high, around 20°C, the amount of sunshine is rapidly decreasing, and this causes a striking decrease in the quantity of photoautotrophic microalgae in the river-water.

REFERENCES

1. S.W. Jeffrey and G.F. Humphrey, Biochem. Physiol. Pflanzen 167, 191-194 (1975).
2. C.J. Lorenzen, Limn. Oceanogr. 12(2), 343-346 (1967).

RED TIDES IN CHILEAN FJORDS

CLEMENT, A.* AND L. GUZMAN **
*Instituto Profesional de Osorno, P.O.Box 557, Puerto Montt, Chile; **Universidad de Magallanes, Instituto de la Patagonia, P.O.Box 113-D, Punta Arenas, Chile.

ABSTRACT

This investigation appraises the presence of red tides in Southern Chilean inland waters during the last four years. Scrippsiella trochoidea (Stein) Loeblich has caused nontoxic red tides in three different inland waters; Reloncavi Sound (41°S), Puyuguapi Fjord (44° 30' S) and Union Sound system (52° S). These outbreaks did not occur simultaneously at these locations, but each had planktonic cysts that co-existed with motile cells. Abundant motile cells and spined cysts of this dinoflagellate have been observed in the digestive contents of mussels from Union Sound. No evidence was found of outbreaks associated with PSP and DSP.

A three-year spring survey of hydrography and phytoplankton of Union Sound showed that an index (diatoms to dinoflagellates) decreases toward the summer. Physical features of the inland waters such as the density field and the application of the index are discussed in relation to red tides.

INTRODUCTION

Toxic and nontoxic red tides occur frequently in Chilean fjords, particularly during summer and fall. The available information has focused on descriptive and toxicological aspects, such as paralytic shellfish poisoning (PSP) in the Patagonian fjords (1,2) and diarrhetic shellfish poisoning (DSP) in Reloncavi Fjord (3).

Much work has been done on marine dinoflagellate life history in other regions of the world, with emphasis on laboratory experiments with cysts (4). A recent study of S. trochoidea resting cysts suggests that the relatively short dormancy period combined with the absence of a requirement for a dramatic shift in enviromental conditions could facilitate rapid cycling between resting and vegetative stages in natural populations of S. trochoidea (5). A green-light effect (photomorphogenetic) required to germinate the resting cysts of this species was also found (6).

In the present study we found that resting cysts and motile cells of S.trochoidea co-exist in the euphotic layer and in the digestive contents of mussels in Chilean inland waters.

*Present Address: Oregon State University, College of Oceanography, Corvallis, OR. 97331, U.S.A.

RED TIDES: BIOLOGY, ENVIRONMENTAL SCIENCE, AND TOXICOLOGY
Okaichi, Anderson, and Nemoto, Editors

The primary purpose of this investigation was to assess the presence of red tides outbreaks in Chilean fjords during the last four years. A secondary goal was to understand the S. trochoidea blooms in at least two inland waters, Reloncavi Sound and Union Sound (Fig 1) where more data were available.

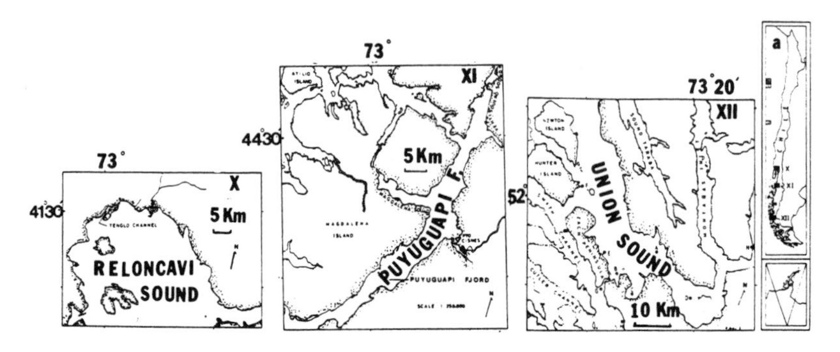

Fig. 1. Maps of study areas. Reloncavi Sound (Tenglo Channel) (X), Puyuguapi Fjord (XI), and Union Sound (XII). Map of Chile (a).

MATERIALS AND METHODS

Water samples were collected from Reloncavi Sound (Tenglo Channel, Fig 1 X) at a fixed station after discolored patches were seen near Puerto Montt, between March 4 and March 8, 1985. Sampling occured daily during spring ebb tides (16:00 -17:00 h local time). Temperature was measured with a reversing thermometer using a Nansen bottle which was also used for salinity samples and for phytoplankton counts.

In Union Sound (Fig 1 XII), the sampling program was part of a planktonic and toxicological project in the Patagonian fjords where extremely toxic events had occurred. This program studied the hydrography and phytoplankton of Union Sound at four stations (A,B,C,and D); during September, October, November and December (austral spring and early summer) for 1982, 1983, and 1984.

Observations of mussel digestive contents and toxicological tests (mice bioassay) were also carried out for a large number of intertidal locations in Union Sound during the same period.

Samples from Puyuguapi Fjord (Fig 1 XI) were collected using a hydroplane, and only phytoplankton net samples taken during March 1984 were available.

Phytoplankton samples were preserved with Lugol's solution and identified to species or genus using a light microscope and counted using the inverted microscope technique.

To synthesize the phytoplankton data from net, water, and digestive contents of Mytilidae (Mytilus chilensis and Aulacomya ater), a simple index was used:

$$G = (A - D)/(A + D) \qquad [-1 \leq G \leq 1]$$

Where A = Diatom abundance, D = Dinoflagellate abundance. Abundance is expressed as cells/ml, or number of species from net samples or from mussel digestive contents.

RESULTS

Outbreaks of S. trochoidea were found in Tenglo Channel, Union Sound and Puyuguapi Fjord.

In Tenglo Channel S. trochoidea was distributed in a thin surface layer and cell density decreased during the sampling period. The most abundant cell density (13368 cells/ml; Secchi depth=0.5m) was in the first meter, on March 4. A very high water temperature (18.5 ° C) and low salinity were also recorded. During the first two days, a weak pycnocline was measured in the upper meter. On the third day the mixed layer increased due to strong winds and mixing from spring tidal currents. The patch moved from Reloncavi Sound to Tenglo Channel due to tidal advection.

The Union Sound system is more complex from an oceanographic point of view. The Patagonian fjords (~50 S) are strongly affected by winds, thermal inversions and lower run-off compared with the Northern fjords (Fig 1). The temperature profiles are highly variable between stations, but salinity shows a more regular pattern. A pycnocline developed between 20-30 m, due mainly to salinity gradients. Yearly surveys were compared within density fields. Different patterns were observed in the same months for 1982, 1983 and 1984. December 1983 was similar to December 1984, but with a shallower mixed layer.

Spined cysts were observed between 5 and 50 m, and the euphotic zone showed an inverse relationship between cyst and motile stages, with the exception of station D, the most inland station at Union Sound. Polykrikos sp. was an important predator of S. trochoidea at the other stations.

Examination of mussel digestive contents showed that Union Sound and Smith Channel were affected by the outbreak in December 1984. Phytoplankton counts at the different stations showed high variability. The lowest value of the index (G $\cong$-1) in the vertical profiles were found in the surface layer for Union Sound and Tenglo Channel.

The bloom in Puyuguapi Fjord (March 1984) was evident from the highly monospecific composition of the net samples.

DISCUSSION

The most frequent red tide outbreaks in Chilean fjords have been caused by the nontoxic dinoflagellate S. trochoidea.

The geographical location of S. trochoidea blooms extended from Valparaiso Bay, (~33 S) [7] to Union Sound (~52 S). In all these outbreaks planktonic resting cysts were found.

At Tenglo Channel and Union Sound the blooms were dissipation events.

These waters have no obvious physical features in common. Tenglo Channel is a shallow canal connected with Reloncavi Sound, and receives large amounts of sewage from Puerto Montt. Puyuguapi Fjord is a typical fjord with a 150 m deep sill at its mouth. Union Sound is a complex system connected with inland waters and coastal waters (Fig 1).

The index G requires more testing and data analysis, but it seems to assess the temporal and spatial variability of phytoplankton composition (diatoms and dinoflagellates).

Mussel digestive contents reflect earlier water column phytoplankton composition. Therefore, the dominance of S. trochoidea in the mussel digestive contents probably indicates when the event was at its maximum.

A large number of resting cysts may be ingested by mussels and other filter feeders in Chilean fjords. It is unknown what percentage of the cysts sink to the sediments; in fact cyst presence in the sediments has never been documented. Since fjords have a significant intertidal and subtidal component due to the large shoreline of this system, ingestion of cysts into mussels may be significant. The anomalous motile stage and cyst in mussel digestive contents after a red tide in Union Sound may be explained by the features of the cysts themselves. Marine dinoflagellate cysts have thick, highly resistant cell walls and thus may be able to survive in the adverse conditions of an invertebrate digestive system. It will be interesting to assess the germination rates of cysts after they go through a digestive system.

ACKNOWLEDGEMENTS

Alejandro Clément thanks Prof. T. Okaichi and the members of the committee of the International Symposium on Red Tides for making his attendance possible. Dr. Patricia A. Wheeler and Lee Tibbitts provided helpful suggestions in the final stages of this work. And finally to Mr. Pablo Muñoz and to all the members of the Hydrobiology Lab. in Punta Arenas. This research was supported partially by SERPLAC, Punta Arenas, Chile.

REFERENCES

1. L. Guzmán, and I. Campodónico, Anales Inst. Patagonia, Monografias 9, (1975).
2. G. Lembeye, Anales Inst. Patagonia. 12, 273 (1981).
3. G. Lembeye, I. Campodónico, L. Guzmán, and C. Kiguel, Abstract. I Jornadas Cs. del Mar. Valparaíso, Chile, 42 (1981).
4. F.J.R. Taylor, The Biology of Dinoflagellates. (Blackwell Scientific Publication, Oxford 1987) pp. 785.
5. B.J. Binder, and D.M. Anderson, J. Phycol. 23, 99 (1987).
6. B.J. Binder, and D.M. Anderson. Nature (London), 322, 659 (1986).
7. P. Muñoz, and S. Avaria, Rev. Biol. Mar., 19(1): 63 (1983).

THE OCCURRENCE AND DISTRIBUTION OF RED TIDES IN HONG KONG - APPLICATIONS IN RED TIDE MANAGEMENT

P.S. WONG
Fisheries Research Station, 100A Shek Pai Wan Road, Aberdeen, Hong Kong.

ABSTRACT

The occurrence of red tides in Hong Kong is reviewed based on reports received by the Agriculture and Fisheries Department. Interesting patterns of distribution of some common red tide species in Hong Kong are obtained. Information regarding toxicity of different species is also deduced. Data on seasonal occurrence and distribution of each species, particularly the toxic ones, would enable better planning of preventing measures or management actions in different coastal regions. Knowledge on toxicity of different species would facilitate the initial assessment of risks involved. Further the consistent weather patterns associated with the blooms of some species suggest the possibility of development of a system for forecasting red tide occurrence.

INTRODUCTION

In recent years red tide has become a growing problem in Hong Kong, particularly as a few red tides have resulted in serious fish kills in some marine fish culture areas. From 1980 to 1986, a total of 12 fish kills was attributed specifically to red tide with a total loss of 89 tonnes of cultured fish valued at HK$4.9 million. Information on red tide occurrence will be useful for management of red tide problems.

DATA ANALYSIS

Records of red tide occurrence in Hong Kong based on reports received by the Agriculture and Fisheries Department from 1975 to 1986 were reviewed. The coastal water around Hong Kong is arbitrarily divided into six regions as shown in Fig. 1A and the seasonal distribution of some major species in the different regions were obtained. Some species are constantly associated with fish kills and by reviewing the development of each fish kill, the toxicities of these species were derived.

RESULTS

A total of 26 species of red tide causative organisms has been identified in Hong Kong. The frequency of occurrence of each species in different regions of Hong Kong and details of associated fish kills are shown in Table 1. In general red tides are more frequent in spring (Fig. 1B) and are mostly found in sheltered regions such as Junk Bay, Port Shelter and Tolo Harbour (Table 1).

The pattern of distribution and toxic nature of some major species are deduced as follows :-

1. *Noctiluca scintillans*

Noctiluca scintillans bloom is ubiquitous in Hong Kong waters. From the seasonal occurrence pattern of *Noctiluca* blooms in different regions of Hong Kong (Fig. 1C), it appears that *Noctiluca* blooms are associated with the mixing of water masses around Hong Kong as has been previously

RED TIDES: BIOLOGY, ENVIRONMENTAL SCIENCE, AND TOXICOLOGY
Okaichi, Anderson, and Nemoto, Editors

TABLE 1. Identified red tide causative species in Hong Kong : their occurrence in different regions of the coastal waters and association with fish kill incidents.

Species	Number of incidents in region							No. with fishkill	Suspected cause	Toxicity test
	1	2	3	4	5	6	Total			
Noctiluca scintillans	2	8	7	4	14	9	44	0	-	-
Prorocentrum triestinum	0	0	0	0	16	0	16	1	Hypoxia	-ve
Prorocentrum micans	0	0	0	0	4	1	5	1	Hypoxia	-
Prorocentrum sigmoides	0	0	0	0	4	0	4	0	-	-
Prorocentrum dentatum	0	0	0	0	3	0	3	1	Hypoxia	-ve
Prorocentrum minimum	0	0	0	0	1	0	1	0	-	-
Gymnodinium nagasakiense	0	0	2	0	4	0	6	4	Toxic	+ve
Gymnodinium simplex	0	0	0	0	4	0	4	0	-	-
Gymnodinium sanguineum	0	0	0	0	1	0	1	0	-	-
Gymnodinium viridescens	0	0	0	1	0	0	1	0	-	-
Gymnodinium sp. X	0	0	1	0	0	0	1	1	Toxic	+ve
Ceratium furca	0	0	4	1	2	0	7	0	-	-
Gyrodinium resplendens	2	1	0	0	0	0	3	0	-	-
Gonyaulax polygramma	0	0	2	0	0	0	2	0	-	-
Peridinium faeroense	0	0	0	0	2	0	2	0	-	-
Peridinium depressum	0	0	0	0	1	0	1	0	-	-
Cochlodinium helicoides	0	0	0	0	1	0	1	0	-	-
Skeletonema costatum	0	0	1	0	6	0	7	0	-	-
Thalassiosira mala	0	0	0	0	2	0	2	0	-	-
Leptocylindrus minimus	0	0	0	0	1	0	1	0	-	-
Leptocylindrus danicus	0	0	0	0	1	0	1	0	-	-
Nitzschia seriata	0	0	1	0	0	0	1	0	-	-
Rhizosolenia delicatula	0	0	0	0	1	0	1	0	-	-
Mesodinium rubrum	2	1	3	0	4	0	10	0	-	-
Cryptomonas sp.	0	0	0	0	6	0	6	0	-	-
Trichodesmium erythraem	0	0	0	2	0	0	2	0	-	-

suggested[1,2]. In early winter (Oct.-Dec.), under the influence of the northeast monsoon, mixing between the South China Sea surface water with part of the Kuroshio takes place along the South China coast west of Hong Kong. *Noctiluca* blooms during this period are therefore located in the western region. In late winter, the territorial waters are influenced by the North China Coast Water whose southerly extension along the coast to the vicinity of Hong Kong is dependent on the strength of the monsoon wind. *Noctiluca* blooms during this season are therefore recorded in the northeastern, eastern or southern waters. In summer (June-August), the predominant factor is the freshwater outflow from the Pearl River and the open sea area bordering the southern boundary of Hong Kong territorial waters consists of an upper layer of relatively low salinity overriding and mixing with the South China Water as it flow eastwards. With southeasterly monsoon, most of the *Noctiluca* blooms are then concentrated in the southern or southeastern bays.

Although *Noctiluca scintillans* has been reported to possess some toxicity[3] and to cause small localised mortality of inshore fish in South Africa[4], in Hong Kong so far it has not caused any fish mortality and is therefore considered as a non-toxic species.

2. *Prorocentrum* species

Prorocentrum blooms are mostly found in Tolo Harbour with only one incident in the northeastern water. Tolo Harbour is semienclosed with high organic loading from human and agricultural waste, hence *Prorocentrum* species can be considered as indicators of polluted waters. *Prorocentrum* blooms are

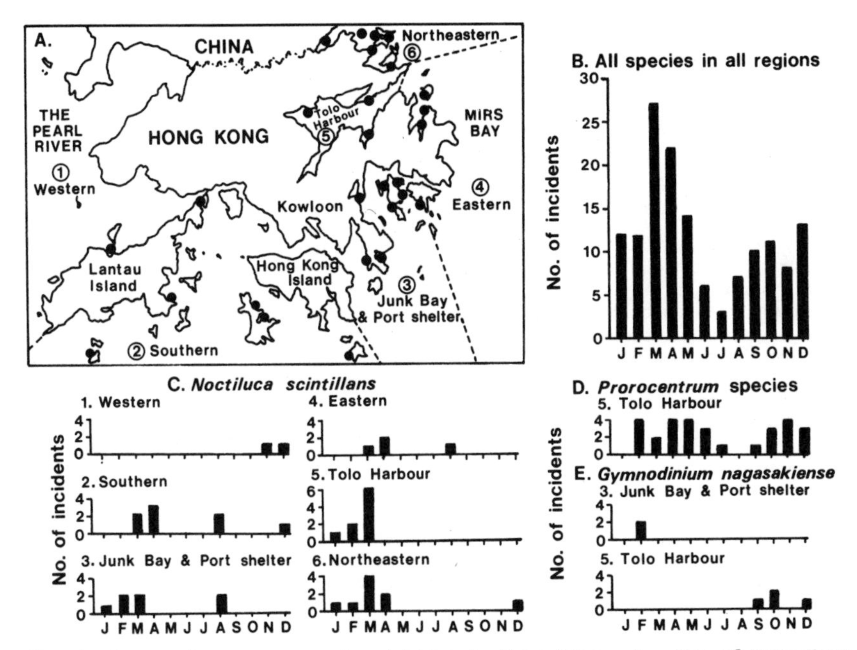

Fig. 1 Seasonal occurrence of red tides in Hong Kong. A - Map of Hong Kong showing the location of 28 fish culture zones (●) and the six coastal regions mentioned in the text, B - all species, C - *Noctiluca scintillans*, D - *Prorocentrum* species, E - *Gymnodinium nagasakiense*.

more common during periods when the weather is unstable, in spring to early summer and autumn to early winter (Fig. 1D).

Some species of *Prorocentrum* were reported to be toxic[4,5]. In Hong Kong, three *Prorocentrum* blooms has resulted in fish kills in fish culture zones but in all these cases oxygen depletion was suspected to be the cause of the fish kills. Toxicity testings of the causative species, *P. dentatum* and *P. triestinum*, confirmed that they are non-toxic.

3. *Gymnodinium* species

Gymnodinium red tides are restricted to the eastern coastal waters of Hong Kong (Regions 3,4 & 5). The most outstanding species is *Gymnodinium nagasakiense*, the bloom of which is often accompanied with fish kills. In one incident, toxicity testings by the standard mouse bioassay have shown that the water and the killed fish contained some toxins. Therefore it is considered to be a potentially toxic species. However, so far no human intoxication has been found to be related to any red tide affected seafood.

Gymnodinium nagasakiense bloomed from September to December in Tolo Harbour and in February in Port Shelter (Fig. 1E). Some interesting weather pattern was found to preceed *G. nagasakiense* bloom. In September, 5-10 days before the bloom there were usually tropical depressions with strong wind and heavy rain followed by warm sunny weather for a few days. In December, the weather pattern usually involved a cold front with strong wind followed by some rain and then a fine sunny period. The fact that *G. nagasakiense* blooms in Hong Kong often occurred around fish culture areas appear to support the suggestion that dissolved organic matter from fish farms may play an important role in the initiation of *G. nagasakiense* bloom[6].

DISCUSSION

In view of the frequent occurrence of red tides and the potential socio-economic and health risks involved, the government of Hong Kong has devised a red tide management strategy which includes an action scheme and a supportive scheme as described by Wong and Wu[7]. The action scheme basically consists of a series of immediate actions to be taken during red tide outbreaks including reporting of red tide sightings, assessment of possible risks involved and appropriate measures to minimise fisheries loss and to protect public health. The supportive scheme caters for problems that cannot be solved immediately and consists of research to supply the information needed, surveillance to monitor the shellfish toxicity and educational programmes to promote public understanding and cooperation to improve the operation of the action scheme. The present review is actually a part of the programme in the supportive scheme. And though the data collected are very limited, the information presented above is very useful for improvement of the action scheme.

Knowledge on toxicity of different species would facilitate the initial assessment of risk involved. For example, previously as *Gymnodinium* blooms are often associated with fish kills and some incidents were found to be mildly toxic, it was deduced that during red tide outbreak whenever *Gymnodinium* is identified to be the causative species, alert emergency action would be taken. Now with more detail on the toxicity of individual species, more accurate assessment could be made without giving any false alarm to the public when non-toxic species such as *Gymnodinium simplex* is involved.

Data on seasonal occurrence and distribution of each species, particularly the toxic ones, would enable better planning of preventive measures or management actions in different coastal regions. The increased frequency of red tide occurrence and the numerous pollution related red tide species found in Tolo have strongly indicated the urgency for strict pollution control in Tolo waters and the need for culturists in fish culture zones within Tolo to equip with aerators for overcoming the problem of oxygen depletion resulting from red tides. Better husbandry techniques to reduce organic loading into the water at the fish culture zones may also help to prevent the blooming of *Gymnodinium nagasakiense*.

As the efficiency of the action scheme relies on prompt reporting of red tide and also prompt actions taken subsequently, forecast of red tide occurrence, if possible, would greatly improve the efficiency of the action scheme as more time would be available for investigation and assessment of the situation. The consistent weather pattern associated with the bloom of some species suggests the possibility of forecasting red tide occurrence from the weather development.

ACKNOWLEDGEMENT

The author wishes to thank the Director of Agriculture & Fisheries, Dr. H.Y. Lee for permission to publish. Thanks are also due to all who have participated in reporting and investiging red tide incidents in Hong Kong.

REFERENCES

1. J.Le Fevre and J.R. Grall, J. Exp. Mar. Biol. Ecol. 4, 287-306 (1970).
2. Y.C. Fung and L.B. Trott, Limnol. Oceanogr. 18, 472-476 (1973).
3. B. Morton and P.R. Twentyman, Environ. Res. 4, 544-557 (1971).
4. D.A. Horstman, Fish. Bull. S. Afr. 15, 71-88 (1981).
5. T. Okaichi and Y. Imatomi, in: Toxic Dinoflagellate Blooms, D.L. Taylor and H.H. Seliger, eds. (Elsevier, Amsterdam 1979) pp. 385-388.
6. A. Nishimura, Bull. Plankton Soc. Japan 29, 1-7 (1982).
7. P.S. Wong and R.S.S. Wu, J. Shoreline Management 3, 1-21 (1986).

AN EXTRAORDINARY, NOXIOUS BROWN-TIDE IN NARRAGANSETT BAY. I. THE ORGANISM AND ITS DYNAMICS

SMAYDA, T. J.,* AND T. A. Villareal*
*Graduate School of Oceanography, University of Rhode Island,
Kingston, RI 02881 USA

ABSTRACT

A brown-tide bloom of a previously unknown chrysophyte Aureococcus anorexefferens (2 μm, autotrophic, non-motile) occurred from May-September 1985 reaching a maximum abundance of 1.2 x 10^9 cells L^{-1}. Mean abundance strongly correlated (r = 0.98) with salinity along the gradient. Its bloom dynamics do not suggest a response to eutrophication. Strong inverse correlations characterized mean abundance and NH_4+NO_3 (r = -0.76) and PO_4 concentrations (r = -0.62). Extensive blooms of diatoms, dinoflagellates, microflagellates and euglenids co-occurred. A highly anomalous sequence of euglenid blooms persisted through November, following termination of the brown-tide. The simultaneous occurrence of a similar brown-tide outbreak in Long Island and New Jersey coastal waters suggests a mesoscale phenomenon associated with complex climatologic and/or hydrographic conditions. Vernal increases in photoperiod and/or irradiance and phagotrophic flagellate abundance are considered to be potential bloom triggering factors. (Proof note: revised appelation Aureococcus anophagefferens now proposed [6]).

INTRODUCTION

The magnitude, duration, causative organism and ecosystem effects of this unusual brown-tide in Narragansett Bay (~41°30'N, 71°15'W) were extraordinary. Weekly phytoplankton analyses carried out since 1959 in lower Narragansett Bay did not previously record Aureococcus, although its small-size and/or limited abundance may have hindered earlier recognition. Maximal 1985 brown-tide concentrations (ca. 2 x 10^9 cells L^{-1}) exceeded by 8-fold previous red-tide bloom concentrations. Local red-tide blooms of dinoflagellates and Olisthodiscus luteus [1,2] usually last several weeks; the 1985 bloom persisted for 5-months. Blooms of the nuisance species Phaeocystis pouchetii [3] and Olisthodiscus luteus [1,2] within Narragansett Bay have not been associated with major deleterious ecosystem impacts, unlike the 1985 bloom, which adversely affected components of the zooplankton, benthos and nekton [4]. This extraordinary bloom occurred as part of a mesoscale phenomenon; similar blooms co-occurred in embayments on Long Island, New York and Barnegat Bay, New Jersey over a distance of ca. 500 km [5]. This implicates a regional climatologic and/or hydrographic event associated with development of the noxious 1985 bloom.

MATERIALS AND METHODS

Seven stations were sampled at three depths along a salinity-nutrient gradient at approximately weekly intervals beginning 25 July 1985 (Fig 1). St 7 is the long-term weekly sampling site. Measurements included: temperature, salinity, NH_4, NO_3, PO_4 and SiO_2; chlorophyll a; phytoplankton and zooplankton numerical abundance and species composition.

RED TIDES: BIOLOGY, ENVIRONMENTAL SCIENCE, AND TOXICOLOGY
Okaichi, Anderson, and Nemoto, Editors

RESULTS AND DISCUSSION

Concurrent, multiple blooms and a species succession characterized the brown-tide. Up to 27 x 10^6 cells L^{-1} of the diatom *Skeletonema costatum*; 14 x 10^6 L^{-1} of a small *Thalassiosira* sp.; mid-August blooms of the dinoflagellates *Prorocentrum redfieldii*, *P. scutellum* and *P. triangulatum* ($>$2 x 10^6 L^{-1}; anomalous euglenid blooms ($>$ 2 x 10^6 L^{-1}), and an extraordinary August pulse (up to 140 x 10^6 L^{-1}) of the small diatom *Minutocellus polymorphus* occurred. The most abundant, persistent organism, however, was *Aureococcus anorexefferens*, described by Sieburth et al. [6] as a new chrysophycean genus and species (Fig 2). (Definitive experimental verification applying Koch's postulates that this organism was responsible for the observed ecosystem disruptions [4] is presently unavailable, although the circumstantial evidence is provocative.) This non-motile organism has resisted isolation into culture, or has appeared in mixed-cultures only [5]. Prominent features of *Aureococcus anorexefferens* are its minute size ($\sim$2 μm), exocellular polysaccharide layer, large, single chloroplast enclosing a distinctive pyrenoid and a voluminous nucleus (Fig 2).

The abundance cycle of *Aureococcus* at St 7 (Fig 3) shows its bloom began precipitously in mid-May ($\sim 10^9$ cells L^{-1}), with a mid-July pulse (1.2 x 10^9 L^{-1}) followed by a prolonged curvilinear decrease (with a brief mid-August resurgence) until early October when it disappeared prior to hurricane GLORIA. *Aureococcus* was most abundant in Greenwich Bay (St 1), with large populations persisting in lower Narragansett Bay (St 6, 7). Mean surface populations at St 1 were twice those in the lower Bay and 3- to 3.5-times upper Bay levels. Regional variations in *Aureococcus* abundance were not correlated with water temperature. Mean abundance was invariant with mean salinity in the upper Bay between 24 and 27.6 ‰ (St 2, 3, 4), but increased from 0.2 to 0.7 billion cells L^{-1} along the salinity gradient (27.6 to 30.5 ‰) from St 1, 4-7 (r^2 = 0.96; Fig 4). Salinity, *per se*, probably was not the causative factor of the regional variations in *Aureococcus* mean abundance, but reflected some factor(s) running in parallel with it. Mean *Aureococcus* abundance at 0 m was inversely and curvilinearly related to NO_3+NH_4 and PO_4 concentrations, progressively decreasing with increasing N and P levels along the gradient from St 1, 7, 6, 5 and 4 (Fig 5). Mean abundance was generally invariant between St 4, 3 and 2. This decrease in mean cellular abundance with increasing NO_3+NH_4 and PO_4 concentrations suggests that the brown-tide development was not fundamentally a response to nutrient enrichment. In fact, high nutrient loadings appeared to suppress *Aureococcus* abundance.

Following maximal abundance on 25 July, *Aureococcus* declined baywide following the pattern depicted for St 7 (Fig 3), and suggestive of a general population control mechanism. Lytic virus infections were commonplace within *Aureococcus* cells during the bloom peak [6], but did not correlate with the bloom demise. A significant surge in phagotrophic flagellates occurred baywide in mid-August and persisted through September, with maximal abundances ranging from ca. 20 to 44 x 10^9 m^{-3}. The coincidence of the decline in *Aureococcus* abundance and increase in phagotrophic flagellates may be causal. A statistically significant direct correlation was found (r = + 0.66).

Aureococcus was last observed on 2 October. Significant decreases in temperature (1.7 to 3.2°C), nutrient concentrations and phytoplankton abundance, and increased zooplankton numerical abundance occurred between then and 9 October. Thereafter, an anomalous phytoplankton community persisted through November characterized by an unique flagellate successional pattern: baywide Euglenid blooms (up to 2.7 x 10^6 L^{-1}) in mid-October causing local green-water displays; October and November blooms of the dinoflagellate *Massartia* rotundatum; September - October blooms of an organism similar to *Fibrocapsa* cf. *japonica* (up to 0.4 x 10^6 L^{-1}); and a brief October bloom of *Olisthodiscus luteus*, the locally common red-tide producer conspicuously absent during the brown-tide.

The winter-spring bloom in early January was typical of Narragansett

Bay. On 14 May 1986, Aureococcus precipitously reappeared in great abundance (53 to 180 x 10^6 cells L^{-1}), but a brown-tide did not develop ; by late June it disappeared. The most significant difference from the 1985 outbreak (comparisons possible only for St 7) was the large population ($\sim$0.5 x 10^6 L^{-1}) of heterotrophic dinoflagellates during 1986. The potential role of grazing as a regulator of Aureococcus blooms is considered by Smayda and Fofonoff [4].

Remarkably, Aureococcus or a similar brown-tide species also reappeared during May/June in Long Island embayments [5]. These regional May bloom-inceptions suggest vernal increases in photoperiod and/or irradiance as potential triggering factors, particularly if an epibenthic stage is present, as found in certain motile chrysophytes [7]. A two-step triggering event would then be required: initially, induction of the morphogenetic transition of the benthic aggregate into its non-motile, planktonic phase, followed by vigorous vegetative growth of the latter regulated by other factors.

ACKNOWLEDGEMENTS

We thank Mr. Paul Fofonoff for his field assistance; Dr. John Sieburth and Mr. Paul W. Johnson for providing the photograph of Aureococcus anorexefferens; Ms. Blanche Coyne for word-processing and drafting services. This work was supported by EPA Cooperative Agreement No. CX812768-01.

REFERENCES

1. D.M. Pratt, Limnol. Oceanogr. 11, 447 (1966).
2. C.R. Tomas, J. Phycol. 16, 157 (1980).
3. P. Verity, T. Villareal and T.J. Smayda, J. Plankton Res. 10, 219 (1988).
4. T.J. Smayda and P. Fofonoff, in: Proceedings of the International Symposium on Red Tides, Takamatsu, Japan, in press.
5. W.M. Wise, Proceedings of the Emergency Conference on Brown Tide and Other Unusual Algal Blooms, Hauppauge, N.Y. (1987).
6. J.M. Sieburth, P.W. Johnson and P.E. Hargraves, J. Phycol., in review.
7. C.R. Tomas, J. Phycol. 14, 314 (1978).

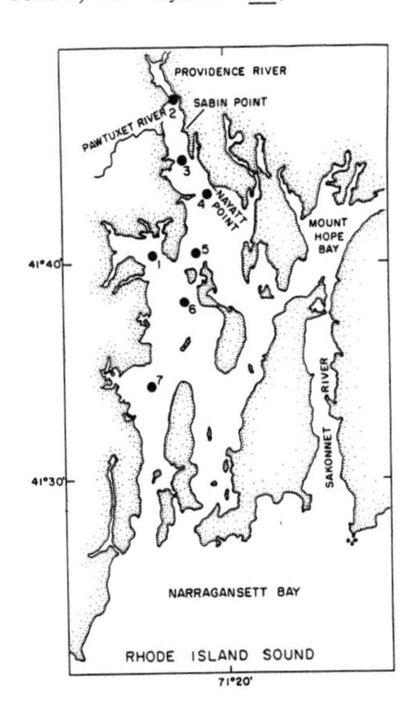

Fig. 1. Station locations in Narragansett Bay sampled during the "brown-tide" study. St 7 site of the long-term station sampled at weekly intervals.

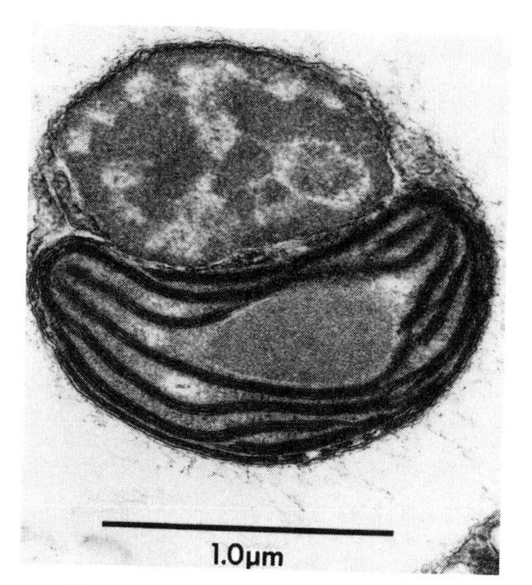

Fig. 2. Ultrastructure of the "brown-tide" organism Aureococcus anorexefferens as shown by transmission electron micrograph of a thin section; magnification 37,000X. (Micrograph courtesy of Prof. J.M. Sieburth and Mr. P.W. Johnson.)

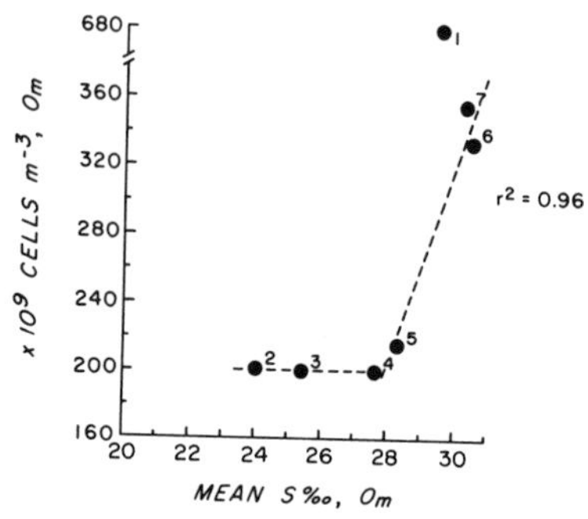

Fig. 4. Mean concentration of Aureococcus anorexefferens at 0 m vs. mean salinity at the seven sampling stations during the "brown-tide" bloom.

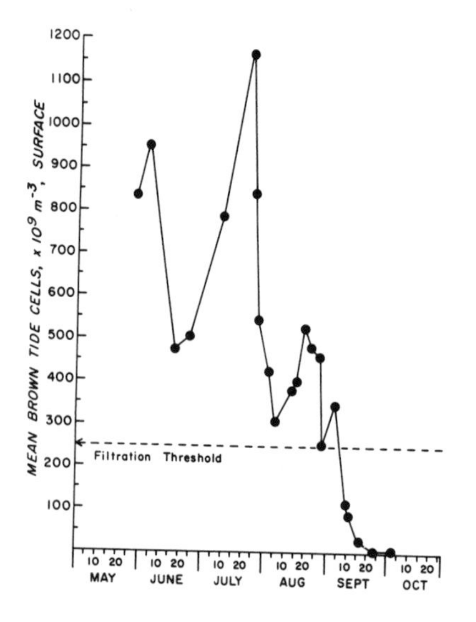

Fig. 3. Abundance of Aureococcus anorexefferens at 0 m at St 7. Filtration threshold designates Aureococcus population density at which filtration by the mussel Mytilus edulis was inhibited.

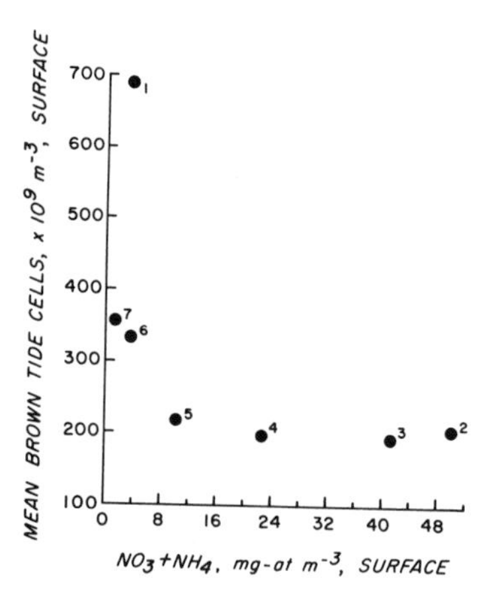

Fig. 5. Mean concentration of Aureococcus anorexefferens at 0 m vs. mean concentrations of NO_3+NH_4 at the seven stations during the "brown-tide" bloom.

AN EXTRAORDINARY, NOXIOUS BROWN-TIDE IN NARRAGANSETT BAY. II. INIMICAL EFFECTS

SMAYDA, T. J.,* AND P. FOFONOFF*
*Graduate School of Oceanography, University of Rhode Island, Kingston, RI 02881 USA

ABSTRACT

A bloom of Aureococcus anorexefferens had significant inimical effects on zooplankton, the edible mussel Mytilus edulis, benthic larval abundance, anchovy fecundity, and on kelp beds. Numerical abundance of the predominant copepod Acartia tonsa inversely correlated (r = -0.91) with Aureococcus abundance. Adult females fed Aureococcus had lower rates of feeding, egg production and reduced body weight. Mean cladoceran (Evadne sp., Podon sp.) abundance was 60-fold lower than during the 1984 and 1986 summers. Natural mussel beds exhibited 30% to 100% mortality, which appeared to reflect starvation. Laboratory-reared mussels ceased to filter when Aureococcus abundance exceeded 500 x 10^6 cells L^{-1}. A kelp die-off occurred when euphotic zone mussels, to which laminarians were attached, died, lost byssal contact and sank into the aphotic zone. Benthic larval numbers inversely correlated (r = -0.58) with Aureococcus abundance. Polychaete and bivalve larvae were 1.5- and 3.6-fold lower, respectively, than previous minima. Egg numbers of the bay anchovy Anchoa mitchilli were about 10-fold lower than in comparison summers. The intensity and prolongation of the Aureococcus bloom seem attributable to a reduction in zooplankton and benthic grazing.

INTRODUCTION

Phagotrophic flagellates increased dramatically during the Aureococcus bloom decline; their mean abundance during the brown-tide correlated directly with Aureococcus [1]. Holozoic dinoflagellates were abundant during the brief resurgence of Aureococcus in May 1986; a brown-tide did not develope subsequently. These heterotrophic dinoflagellates were sparse during the initial stages of the Aureococcus bloom in May 1985; a brown-tide subsequently developed. These associations suggest that phagotrophic, naked flagellates and dinoflagellates actively grazed Aureococcus. This predator-prey relationship contrasts with apparent inimical effects of Aureococcus anorexefferens cells and/or population levels on the feeding, growth, fecundity and viability of certain zooplankters, benthic feeders, and a fish.

RESULTS

There is no field evidence that the Aureococcus bloom was antagonistic to other phytoplankton species. The conspicuous absence of Olisthodiscus luteus may merely reflect inter-annual variability in its red-tide patterns [2]. However, experimental data to evaluate potential antagonistic interactions such as described by Iwasaki [3] are unavailable. The general impression is that the normal summer phytoplankton assemblages thrived in Narragansett Bay during the anomalous and extraordinary brown-tide bloom of Aureococcus anorexefferens.

Total zooplankton community biomass regressed against surface abundance

RED TIDES: BIOLOGY, ENVIRONMENTAL SCIENCE, AND TOXICOLOGY
Okaichi, Anderson, and Nemoto, Editors

of *Aureococcus* revealed two regional subgroupings, both *positively* correlated, in which zooplankton biomass increased with brown-tide abundance (Fig 1). Zooplankton levels were considerably lower (<150 mg dry wt m^{-2}) at lower Bay stations [1] (1, 6 and 7) where mean brown-tide abundance exceeded 300 billion cells m^{-3}. In the upper Bay, where *Aureococcus* levels were <200 billion cells m^{-3}, zooplankton dry weight was considerably greater. For individual zooplankton components, however, a different pattern emerges. *Acartia tonsa* was the dominant copepod during the brown-tide, the result of the normal successional pattern [4]. Strong *inverse* correlations occurred between mean numerical abundances of *Acartia* and *Aureococcus* integrated over the water column (Fig 1). Two distinct regional subgroups are evident: upper Bay St 2, 3 and 4 and lower Bay St 1, 5, 6, 7; $r^2 = 0.83$ for both clusters. *Acartia* was about 2-fold more abundant per unit of brown-tide cellular abundance at the upper Bay stations than in the lower Bay, i.e., the negative affect of *Aureococcus* on *Acartia* numbers was more pronounced in the lower Bay where the brown-tide was most intense. Durbin and Durbin [5] confirmed experimentally that *Aureococcus* was inimical to *Acartia tonsa*. Adult females incubated in the presence of *Aureococcus* had much lower rates of grazing, egg laying, and body carbon development compared to animals fed other phytoplankton species. Daily egg production and body carbon in adult female *Acartia* exposed to *Aureococcus* were about 40% of rates with other phytoplankton diets. Only 44% of the daily body weight was ingested in the presence of *Aureococcus* and 174% with other phytoplankton diets.

Failure of the cladoceran community to develop was the most remarkable zooplankton modification accompanying the brown-tide outbreak (Fig 2). *Evadne nordmanni* and *Podon* sp., which normally exceed 10,000 animals m^{-3} during June - August, failed to appear in 1985. Based on the six-year interannual comparisons at St 7 [1], the mean 1985 May - August abundance of only 80 m^{-3} was 10- to 75-fold lower than the means for the comparison years. The strong recovery during 1986 when a brown-tide did not develope is notable.

The mean depth (9 m) and strong year-round mixing in Narragansett Bay result in a strong benthic-pelagic coupling [6] in which the clam *Mercenaria mercenaria* and mussel *Mytilus edulis* are important benthic filter feeders. Field studies during the brown-tide revealed reproductive failure of gravid mussels and massive Bay-wide mortality approaching 100% by mid-August [7]. In laboratory experiments [7], *Mytilus edulis* actively filtered 2 μm *Synechococcus* in contrast to drastically reduced filtration rates at *Aureococcus* (2 μm) concentrations between 250 to 500 x 10^6 cells L^{-1}, even when provided in combination with *Isochrysis galbana*. Fig 1 in [1] suggests potential feeding inhibition from May - August in the lower Bay. Above 500 x 10^6 cells L^{-1}, marked feeding inhibition occurred, leading Tracy [7] to conclude that starvation stress of the pre-spawning mussels caused the die-off, accelerated by high summer temperatures. However, it is unresolved whether starvation resulted solely or primarily from dense accumulations of unpalatible *Aureococcus* cells. Histopathological examination of field collections during the brown-tide revealed mussel gill filaments were often swollen; the food groove packed with yellow-brown granules; sloughed off mucous; exhibited necrotic foci, and were packed with amoebocytes [8]. The possibility of a supplemental contact-toxin effect of the *Aureococcus* cells, which have a prominent exocellular polysaccharide layer (see Fig 2 in [1]), needs to be evaluated. The mortality of kelp populations (*Laminaria saccharina* and *L. digitata*) was a remarkable side-effect of the mussel die-off [9]. Laminarians attached to mussels sank into the aphotic zone when the dead mussels lost byssal contact and slumped to depth.

Benthic larval abundance and *Aureococcus* abundance were inversely correlated ($r = -0.58$)(Fig3). Mean larval abundance at St 7 during 1985 was the lowest observed during the comparison years (1981-1986). Mean abundances of polychaete larvae and bivalve veligers were 1.5-fold and 3.6-fold lower, respectively, than the 6-year minima. Nonetheless, there is no convincing statistical evidence based on larval abundance that benthic recruitment (excluding *Mytilus edulis*) was impaired by the brown-tide.

An inimical effect at the nekton level was also found. Eggs of the anchovy Anchoa mitchilli were conspicuously sparse during the brown-tide (Fig 4); mean abundance levels were about 10% those during the comparison years. Their strong comeback in 1986 similar to cladocerans (Fig 2) is notable.

DISCUSSION

Contemporaneous blooms of Aureococcus anorexefferens in Long Island coastal embayments during 1985 were also accompanied by inimical shellfish effects. Adult bay scallop, Argopecten irradians, exhibited a considerable weight loss; larval bay scallops 100% mortality; oyster mortality also occurred during Aureococcus blooms [10]. Moreover, light attenuation by the brown-tide reduced the distribution and abundance of eel grass meadows of Zostera marina, a loss which reduced the nursery grounds and habitats for scallops and juvenile fishes [11]. Clearly, Aureococcus anorexefferens is an incomparable, broad-spectrum, nuisance phytoplankter, seemingly without parallel to date. Its blooms appear capable of negatively impacting, directly or indirectly, all major trophic levels, including members of the zooplankton, benthos (bivalves particularly), certain fishes, macroalgae, and even phanerogams. The causes of its extraordinary blooms during the summer of 1985 in Narragansett Bay, the Long Island embayments (in 1986 and 1987 as well) and Barnegat Bay [1] are obscure. The accompanying inimical effects within Narragansett Bay suggest that the intensity and duration of its 1985 bloom were partly attributable to a reduction in grazing by zooplankton and benthos. This grazing impact is in addition to the potential role ascribed [1] to phagotrophic flagellates in influencing the initiation and buildup of Aureococcus blooms. Are such multiple grazing effects a useful general paradigm for evaluating regulation of nuisance algal blooms in the sea?

ACKNOWLEDGEMENTS

We thank Mr. Tracy A. Villareal for assistance during the field studies; Professor Pei Wen Chang for histopathological observations on Mytilus edulis and Ms. Blanche Coyne for word-processing and drafting services. This work was supported by EPA Cooperative Agreement No. CX812768-01.

REFERENCES

1. T.J. Smayda and T. Villareal, in: Proceedings of the Fourth International Symposium on Red Tides, Takamatsu, Japan, in press.
2. D.M. Pratt, Limnol. Oceanogr. 11, 447 (1966).
3. H. Iwasaki, in: M. Levandowsky and S. Hutner (eds.) Biochemistry and Physiology of Protozoa Vol. I, 357 (Academic Press 1979).
4. J.H. Martin, Limnol. Oceanogr. 10, 185 (1965).
5. E. Durbin and A.G. Durbin, Narragansett Bay Symposium Abstracts, R.I. DEM, 97 (1987).
6. C.B. Officer, T.J. Smayda and R. Mann, Mar. Ecol. Prog. Ser. 9, 203 (1982).
7. G.A. Tracy, Narragansett Bay Symposium Abstracts, R.I. DEM, 99 (1987).
8. P.W. Chang, personal communication (1987).
9. S.C. Levings and P.M. Peckol, Narragansett Bay Symposium Abstracts, R.I. DEM, 101 (1987).
10. W.M. Wise, Proc. Emergency Conference on Brown Tide and Other Unusual Algal Blooms, Hauppauge, N.Y., 21 (1987).
11. W.C. Dennison, Proc. Emergency Conference on Brown Tide and Other Unusual Algal Blooms, Hauppauge, N.Y., 14 (1987).

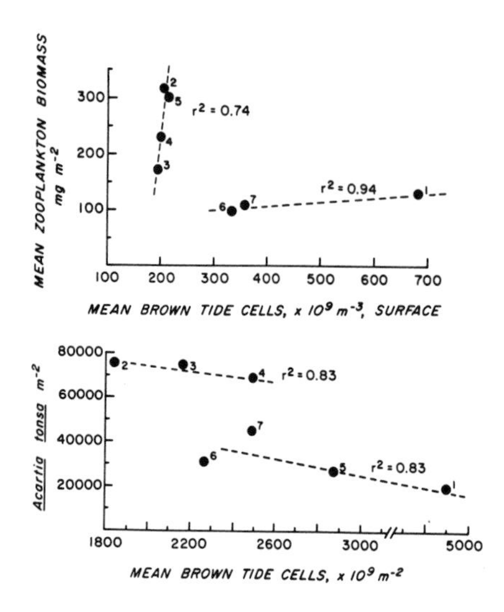

Fig. 1. Mean zooplankton biomass as dry weight during the "brown-tide" bloom vs. mean abundance of Aureococcus anorexefferens at the seven stations (upper panel) and (lower panel) mean numerical abundance of the copepod Acartia tonsa vs. mean abundance of Aureococcus anorexefferens integrated for the water column.

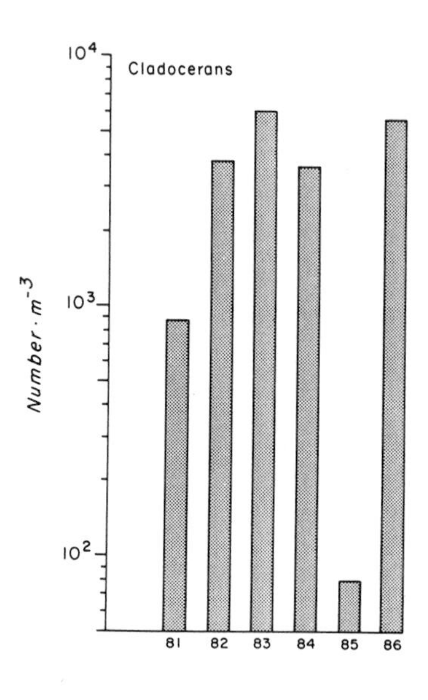

Fig. 2. Mean abundance of the cladocerans Podon sp. and Evadne nordmanni during their summer occurrences at St 7 from 1981 - 1986.

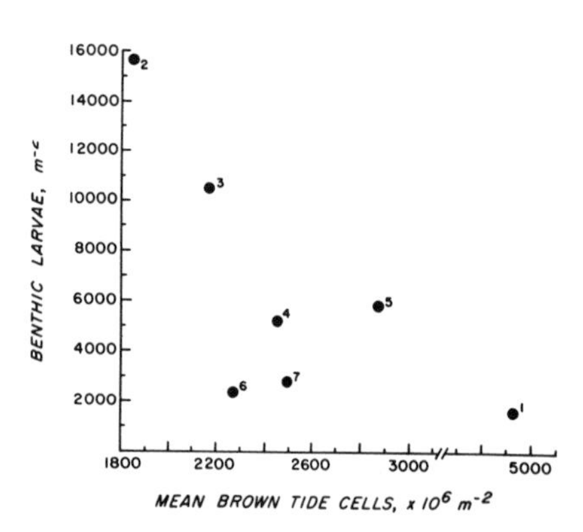

Fig. 3. Mean numbers of benthic larvae vs. mean abundance of Aureococcus anorexefferens integrated for the water column at the seven stations during the "brown-tide" bloom.

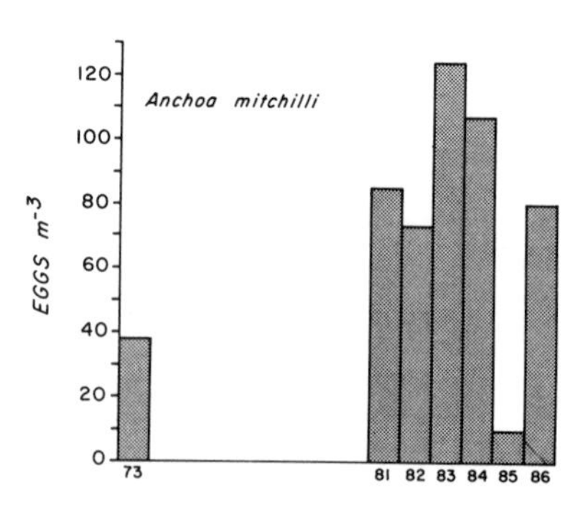

Fig. 4. Mean egg abundance for the anchovy Anchoa mitchilli at St 7 during its summer occurrences in 1973 and from 1981 - 1986.

RED TIDE PROBLEMS IN THE SETO INLAND SEA, JAPAN

TOMOTOSHI OKAICHI
Faculty of Agriculture, Kagawa University, Kagawa, 761-07, JAPAN

ABSTRACT

With the development of industries along the coast of Seto Inland Sea, Japan, aquaculture of yellowtail, oyster and nori (laver) have been suffering from outbreaks of red tides. During these two decades 20 billion yellowtail worth about 20 billion yen were killed. Comprehensive surveys on the cause of outbreaks of red tides are carried out by workers at Universities and Prefectures. Countermeasures to avoid the fishkill are devised and the monitoring systems for red tides, PSP and DSP within Prefectures are organized by the Bureau of Fisheries.

After the first major outbreak of red tide in the Seto Inland Sea (Tokuyama Bay)(Fig. 1) in 1957, red tides increased in proportion with the development of industries in this region. In 1965 a red tide of *Gymnodinium nagasakiense* killed shellfish and other marine organisms in Omura Bay in Nagasaki Prefecture which is located in the western part of Kyushu Island.

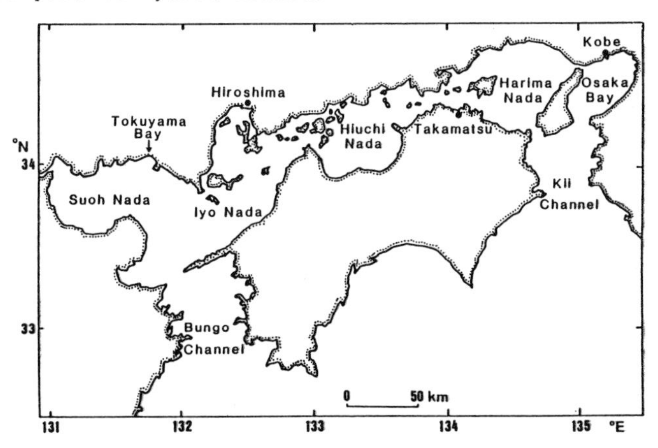

FIG. 1. Seto Inland Sea (area 21200 km^2, mean depth 35 m)

Cooperative studies on the cause of the occurrence of red tide in coastal water area were organized by Prof. T. Hanaoka in 1966, prompted by the *Gymnodinium* red tide which occurred in Omura Bay. Three projects were undertaken by the study group as follows: 1) A study of the relation-

RED TIDES: BIOLOGY, ENVIRONMENTAL SCIENCE, AND TOXICOLOGY
Okaichi, Anderson, and Nemoto, Editors

ship between anoxia in bottom water and outbreaks of red tide (at Omura Bay and Ago Bay, Mie Prefecture), 2) An examination of the effects of organic pollution coming from the land, and 3) Studies of the physiological(nutritional) requirements of red tide organisms. These studies continued for three years.

Red tides in the Seto Inland Sea were exacerbated by the appearance of *Chattonella antiqua* in 1964. Mass mortalities of cultured yellowtail fish occurred in 1970,'72,'77,'78,'79,'84,'85,'86 and '87 mainly in the eastern part of the Inland Sea (Table 1).

Table 1. Fish kills due to Red Tides in the Seto Inland Sea

year	area	plankton	dead fish	No.of fish ($X 10^3$)	lost(¥) million
1972	Harima	Chattonella	yellowtail	14,000	7,100
'77	Harima	Chattonella	yellowtail	3,300	3,000
'78	Harima	Chattonella	yellowtail	2,800	3,300
'79	Harima	Chattonella	yellowtail	1,040	320
	Bungo C.	Gymnodinium	yellowtail	700	500
'80	Bungo C.	Gymnodinium	yellowtail	530	330
'82	Harima	Chattonella	yellowtail	380	770
	Hiuchi	Gymnodinium	red sea bream	290	190
'83	Kii C.	Chattonella	yellowtail	300	300
'85	Suoh & Iyo	Gymnodinium	yellowtail and others	5,600	1.020
'86	Harima	Chattonella	yellowtail	unknown	205
'87	Harima & Kii C.	Chattonella	yellowtail	1,430	2,170

Motivated by the massive outbreak in 1972 in which fishermen lost about 14,000,000 yellowtail worth over 71 billion yen, the Seto Inland Sea Environment Conservation Law (SECL) was passed by the Central Government. The law prohibits large scale reclamation except when permission is obtained from the Bureau of Environment, and it aimed to cut the total chemical oxygen demand (COD) inflow from the land to half of the 1970 level by 1980. According to the Bureau of Fisheries, the COD loading from the land to the Seto Inland Sea was about 2500 tons per day in 1970. Each local government was assigned an amount by which discharged COD was to be reduced. These amounts depended on the inflow from each factory up to 1980.

In accordance with the provisions of the SECL, a survey of the entire Seto Inland Sea was carried out with the cooperation of all of the local governments situated along its coast. Observation stations were set up as square grids with 4 Km between stations. Sea water was sampled at depths of 0, 10, and 20m, and also at 2m above the bottom. Sediments were also collected for measurements of COD. Measurements were taken mainly of organic nitrogen, phosphorus, and COD, along with measurements of phytoplankton and zooplankton composition. There were more than 500 stations. Observations were carried out in each of the four seasons from 1973 to 1976. The data were compiled and analyzed by the Bureau of Environment.

Forty-eight cases of red tide occurred in 1967 in the Seto Inland Sea. Red tides increased in proportion to the increase in industrial production. After the implementation of the SECL, red tides still increased until 1976 (ie. for four years) when they were estimated at 299 cases per year. Then the number of red tides decreased dramatically, and from 1980 onwards about 170 to 200 cases occurred per year(Fig.2).

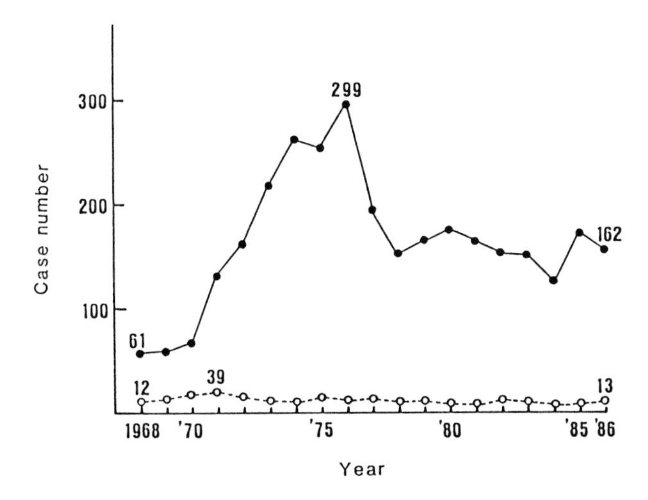

FIG. 2. Red Tides occurred in Seto Inland Sea

●: case number of red tides
○: case number of red tides which caused fish kills.

The regulation of COD forced the construction of secondary treatment facilities by industries and municipal offices along the coast. The facilities are working primaily to reduce COD in sewage, but in the process are also reducing levels of nitrogen and phosphorus. Moreover, phosphorus included in most artificial detergents was removed upon the recommendation of the Government and the request of citizens groups.

During this decade there have been some conflicts between industries and aquaculture. Some of the conflicts became severe enough to require legal settlements, but most were settled by the Central Government's Committee on Public Disorder or by local governments. The Bureau of Fisheries is, in cooperation with local governments, now promoting various countermeasures to avoid the damages resuting from red tides.

In Japan, aquaculture of oysters and " nori " (laver) has taken place since before 1940, and culturing of fish and shrimp developed after 1960.

The total production from aquaculture in 1983 was 1.153 x 10^3 tons, corresponding to 9.6% of the total fish catch of 11.967 x 10^3 tons. The main cultured fishes are yellowtail and red sea bream, and the production in 1983 was 156 x 10^3 tons and 25 x 10^3 tons, respectively. The aquaculture production of yellowtail was as much as 4 times that of natural fisheries, and the aquaculture production of red sea bream was 1.8 times (Table 2).

Table 2. Production of main aquaculture species (1000 tons)

	Yellow tail	Red sea bream	Kuruma shrimp	Oyster	Scallop	Nori
Fisheries	41.8	14.7	3.58	-	128.1	-
Aquaculture(A)	156.2	25.3	1.50	253.2	85.1	360.7
Total(B)	198.0	40.0	5.53	-	213.3	-
A/B X 100	78.9	63.3	35.3	100	39.9	100

Kuruma shrimp culture produced 1,950 tons in 1983. Shellfish culture (oysters and scallops) also is a big industry, together with nori culture. Besides aquaculture controlled by human management, the release of larvae of marine organisms is promoted through assistance from the Government. In 1984, 1.776 x 10^6 scallop larvae and 293 x 10^6 Kuruma shrimp larvae were released. However, after 1970, problems resulting from red tides in the development of aquaculture could no longer be ignored.

Mass mortalities of yellowtail due to Chattonella spp. occurred in 1970 to 1988 successively (Table 1) and 30 million yellowtail were lost which cost about 20 billion yen. Gymnodinium nagasakiense also killed fish in 1984 and 1985. Minimum cell density of Chattonella antiqua lethal to yellowtail was estimated at 110 cells/ml at 29 ℃. These cell densities are lower than those in visible brown patches which are at about 1,000 cells/ml.

The primary causative substances that killed the fish were determined to be highly unsaturated fatty acids(mainly $C_{16:4}$ and $C_{18:4}$)(Okaichi et al. unpublished). Death is caused by the destruction of the surface of the secondary lamella of the gills[1]. This occurs in association with the decrease of pH of the blood which prohibits oxygen transfer by hemoglobin. The special sensitivity of yellowtail compared with other marine fish is explained by the decrease of the Bohr effect in blood with rather high pH values [2].

From 1979 to 1984 about 30 scientists from different Universities were engaged in interdisciplinary studies on red tides. Two reports, " Fundamental Studies on the Effects of the Marine Environment on Outbreaks of Red Tides "(1982)[3] and " Studies on Marine Ecosystem Models on Outbreaks of Red Tides in Neritic Waters "(1985)[4] were prepared.

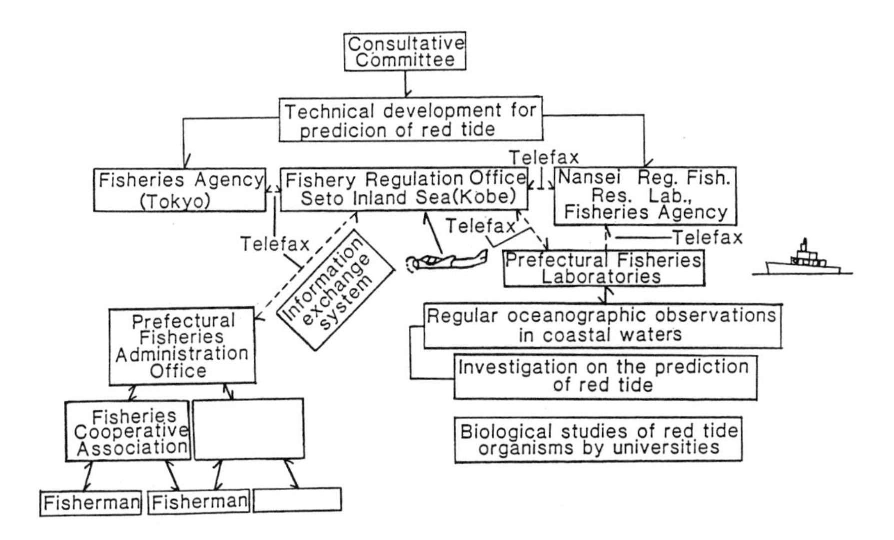

FIG. 3. Information exchange and red tide investigation systems in Seto Inland Sea.

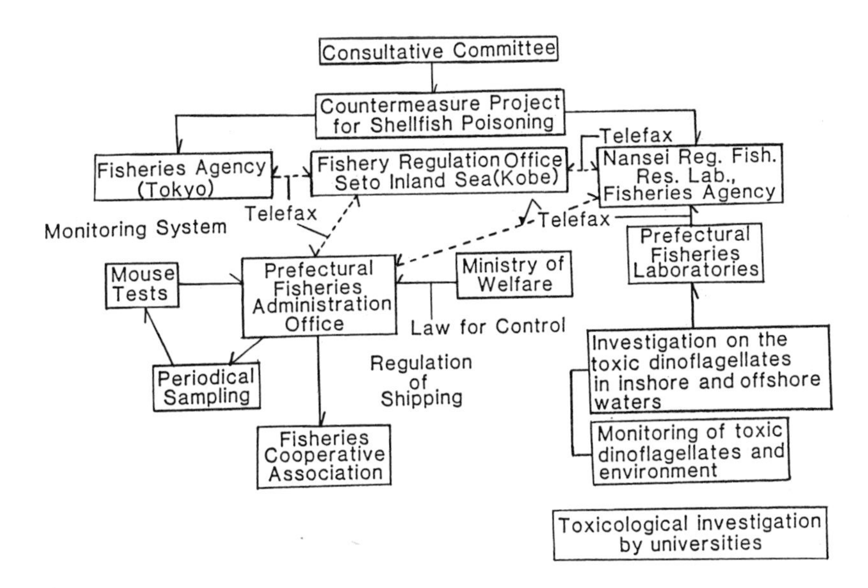

FIG. 4. Shellfish poisoning monitoring and investigation systems in Seto Inland Sea.

Through these studies, oceanographic features of Harima Nada (3430 Km^2, mean depth 26 m) became clear and simulation models on the outbreaks of red tides were also proposed. PSP due to *Protogonyaulax catenella* and *P. tamarensis*, and DSP due to *Dinophysis fortii* are also found in shellfish such as clam, shortneck clam and mussels in early summer.

Monitoring systems connecting Prefectural Fisheries Experimental Stations were organized by the Bureau of Fisheries. The systems which are shown in Figs. 3 and 4 are successfully working to exchange information on the outbreaks of red tides, PSP and DSP [5].

REFERENCES

1. M.Shimada,T.H.Murakami,T.Imahashi, H.S.Ozaki, T.Toyoshima and T.Okaichi, Acta Histo chem. Cytochem. 16, 232-244 (1983)

2. T.Yamaguchi, K.Ogawa, N.Takeda, K.Hashimoto and T.Okaichi. Bull. Japan, Soc. Sci. Fish., 47, 403-409 (1981)

3. T.Okaichi et al, Fundamental Studies on the Effects of Marine Environment on the Outbreaks of Red Tides. (Environmental Science, Monbu Sho). 1-239 (1982)

4. T.Okaichi et al, Studies on Marine Ecosystem Models on Outbreaks of Red Tides in Neritic Waters, (Environmental Science, Monbu Sho). 1-190 (1985)

5. M.Anraku, Toxic Red Tides and Shellfish Toxicity in Southeast Asia, Proceedings of a consultative meeting held in Singapore 11-14 September 1984. 105-109 (1984)

III ENVIRONMENTAL SCIENCE AND BIOLOGY

RELATIONSHIP BETWEEN VERTICAL STABILITY AND THE OCCURRENCE OF CHATTONELLA RED TIDES IN HARIMA NADA, SETONAIKAI

YOSHIYASU IWATA, MASAMI ISHIDA, MASATO UCHIYAMA, ATSUSHI OKUZAWA
Fuyo Data Processing & Systems Development, Ltd.
4-17-14 Akasaka, Minato-ku, Tokyo 107

ABSTRACT

Red tides of C.antiqua and C. marina cause serious damage to cultured fish such as yellowtail and sea bream in Harima Nada. In order to minimize the effects of Chattonella red tides, it is essential to predict the outbreaks in their early stages. The development of techniques useful for the prediction of red tides were investigated in this study. Fluctuations in the vertical difference of density and water temperature from January to August were examined. This study showed that vertical stability was low and bottom water temperature high in the years when severe Chattonella red tides occurred. These parameters appear to have potential in predicting Chattonella red tides in Harima Nada.

INTRODUCTION

It has been reported that vertical stability is often high during Chattonella red tides and low before their occurrence. For this reason, fluctuations of the vertical difference of density from January to August, 1971-1984, were investigated in this study. The fluctuation of bottom water temperature, which affects the vertical stability, also seems to be important in efforts to predict red tides, since suitable water temperatures close to the sea bottom are necessary for the germination of dormant cells.

EXPERIMENTAL

The vertical difference of density, which is calculated from water temperature and salinity, was used as an index. Water temperature and salinity were measured once a month around Harima Nada. The vertical difference of density between the depth of 10m and the bottom was calculated for each area shown in Fig. 1. The difference per meter was also calculated and averaged in each area. The deviation from normal (mean) values of bottom water temperature was also examined for each area.

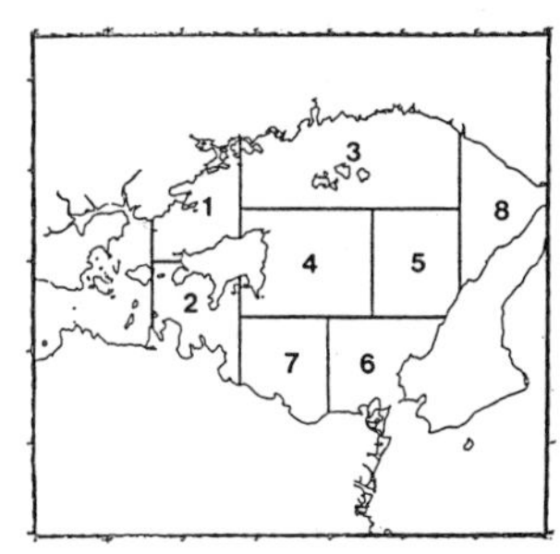

FIG. 1. Eight areal divisions of Harimanada used for calculation.

RED TIDES: BIOLOGY, ENVIRONMENTAL SCIENCE, AND TOXICOLOGY
Okaichi, Anderson, and Nemoto, Editors

RESULTS AND DISCUSSION

Figure 2 shows the variation in the vertical density gradient from January to August during years with and without red tides. The density difference was small from January to April, consistent with unstable conditions in winter. After April, stability increased rapidly. The difference between the years with and without red tides was noticed in May and July (TABLE 1). In May the vertical difference of density tended to be small in red tide years
compared to years when none were observed, with the exception of 1976, 1982 and 1983. In July, the same trends were observed, excluding 1973.

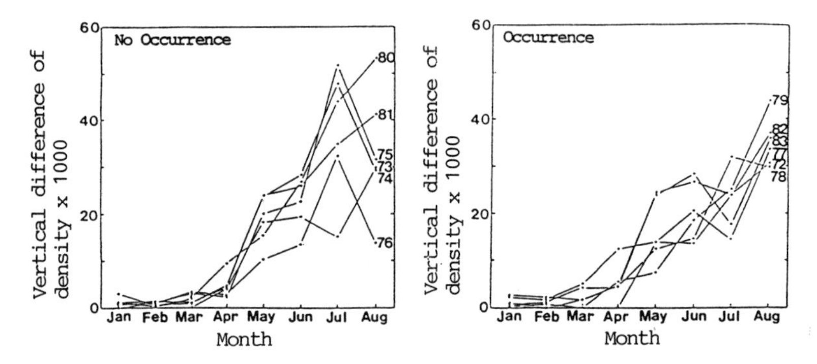

FIG. 2. Fluctuation of vertical difference of density from Jan. to Aug.

TABLE I. The vertical difference of density in May and July in the years with and without the red tides.

Month	Occurrence	No Occurrence
May	Small Value : 1972,77,78,79 Large Value : 1982,83	Small Value : 1976 Large Value : 1973,74,75,80,81
July	Small Value : 1972,77,78,79,82,83	Small Value : 1973 Large Value : 1974,75,76,80,81

Yanagi [1] investigated the vertical difference of water temperature in the central area of Harima Nada. According to his results, the vertical difference in May and June was small in 1972, 1977 and 1978, the years with red tides. It was also noticed that maximum daily wind velocity from May to July was low in 1973 and 1976, the years with no red tide, even though the vertical difference of water temperature was small.

Figure 3 reveals the deviation from the normal or mean bottom water temperature in each grid area. The years when red tides occurred are indicated by a line below the abscissa. The fluctuation of bottom water temperature showed similar trends in each area. The bottom water temperature was typically high in red tide years.

To confirm such a trend, the average deviation from the mean bottom water temperature was then examined (Figure 4). The mean was obtained from February to July. In the years with red tide, high values were observed, excluding 1977.

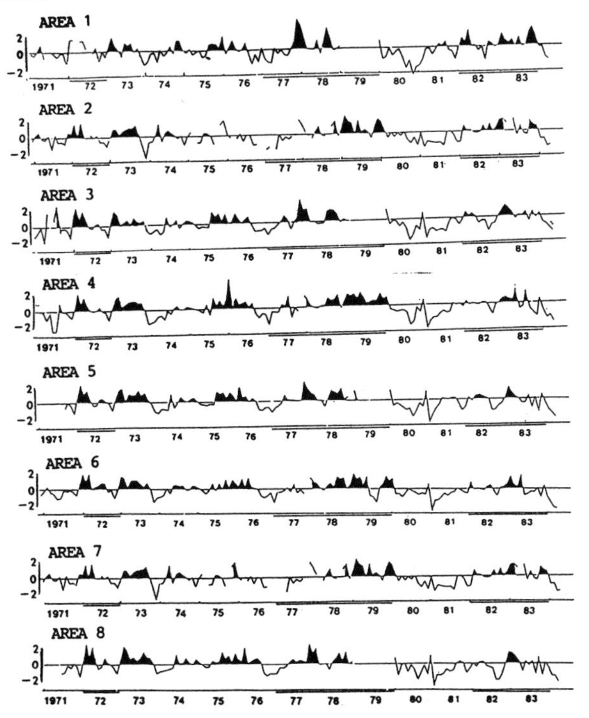

FIG. 3. Deviation of bottom water temperature from normal value in each month from 1971 to 1983.

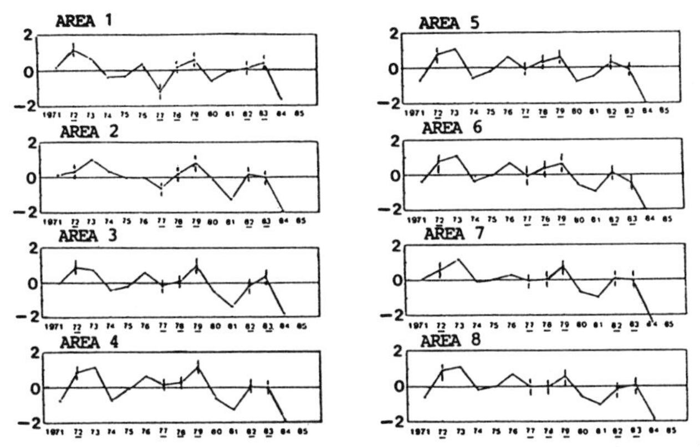

FIG. 4. Averaged deviation from normal value of bottom water temperature from Feb. to Jul. Solid line under the year indicates the occurrence of red tide.

Figure 5 shows the relation between the vertical difference of density in July and the average temperature deviation from the mean between February to July. These values were calculated from data collected by Hyogo Prefectural Fisheries Experimental Station in Area 4 (FIG. 1). The bottom water temperature tended to be high in the years when the vertical difference of density was low. When the opposite occurred, namely when bottom temperature was low and the density gradient high, no Chattonella red tides were observed.

It is presumed that other factors which are necessary for red tide development were not favorable in 1973 and 1976. In 1973, the air temperature was continuously high from the end of the rainy season. In 1976, the summer temperatures were low.

We conclude that vertical stability and bottom water temperature are useful factors in predicting the red tides in Harima Nada prior to their occurrence. Further investigation is being carried out in order to develop a more precise prediction technique.

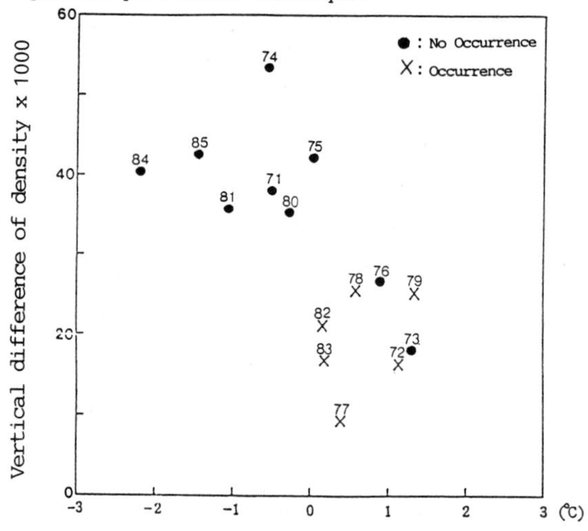

FIG. 5. Relation of vertical difference of density in July and average deviation from normal value of bottom water temperature during a period from February to July.

ACKNOWLEDGEMENT

We would like to express our gratitude to Hyogo, Okayama, Tokushima and Kagawa Prefectural Fisheries Experimental Stations for providing data. We are also indebted to Dr. Masateru Anraku of The Overseas Fishery Cooperation Foundation for his comments.

REFERENCE

1. T. Yanagi, 1984: Marine Science Monthly, No. 199, vol 18:56-59

PHYSICAL PARAMETERS OF FORECASTING RED TIDE IN HARIMA-NADA, JAPAN

TETSUO YANAGI
Department of Ocean Engineering, Ehime University, Matsuyama 790, Japan

ABSTRACT

A scenario for red tide occurrence of Chattonella antiqua in Harima-Nada, Japan is proposed in this paper. The red tide can occur only in the summer when the density stratification is weak at the central part of Harima-Nada, and after a winter in which the westerly monsoon was not strong. Chattonella antiqua can begin to increase at high rate to reach red tide levels following an increased supply of nutrient from the lower layer due to the large vertical mixing component resulting from combined effects of strong winds and tidal currents.

INTRODUCTION

Red tides of Chattonella antiqua have frequently occurred since 1972 in Harima-Nada in the Seto Inland Sea, Japan. The necessary conditions in Harima-Nada of current, water temperature and salinity for a red tide outbreak are fulfilled in summer every year [1]. However, while a large red tide occurred in some years (1972, 77, 78, 79, 82, 83, 87), it did not occur in other years. In this paper, I will investigate, from the viewpoint of physical oceanography, differences between years with and without red tide occurrences in Harima-Nada.

YEAR-TO-YEAR VARIATION OF SEA CONDITION

Relationships between year-to-year variations in physical parameters such as monthly mean air temperature, total monthly precipitation, monthly mean solar radiation, monthly mean wind speed, water temperature, salinity and density at Harima-Nada and the red tide occurrence were investigated, but distinct correlations were not be found [2]. It is reported that the average air pressure difference (ΔP) between Fukuoka and Osaka in December, January and February (winter) was a good parameter for forecasting the outbreak of red tide at Harima-Nada the following summer [3]. That is, ΔP is small in winter just before the summer of red tide occurrence. However, ΔP is not a decisive parameter as shown in Fig.1. On the other hand, it is also reported that the average water temperature difference ΔT between 10 m and 40 m (bottom) below the sea surface at the central part of Harima-Nada in May and June (early summer) was a good parameter for forecasting a red tide outbreak at Harima-Nada in succeeding midsummer [4]. That is, ΔT is small in early summer just before the midsummer red tide occurrence. However, ΔT is also not a decisive parameter, as shown in Fig.1.

Combining these two parameters, ΔP and ΔT, I obtained a good correlation between ΔP, ΔT and red tide occurrence (Fig.2). The red tide of Chattonella antiqua occurred when ΔP and ΔT were both small with the exception of 1973. We can use the parameters ΔP and ΔT for forecasting the red tide occurrence at Harima-Nada one month before it occurs. Red tide outbreaks of Chattonella antiqua occur only in July or August because of the water temperature condition (1).

RED TIDES: BIOLOGY, ENVIRONMENTAL SCIENCE, AND TOXICOLOGY
Okaichi, Anderson, and Nemoto, Editors

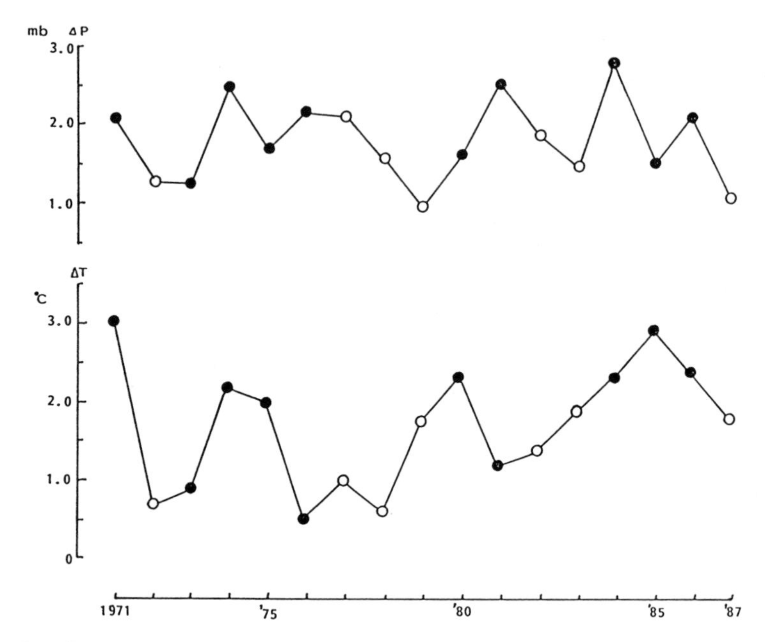

Fig.1 Year-to-year variations in average difference of air pressure (ΔP) in winter between Fukuoka and Osaka and average difference of water temperature (ΔT) in early semmer between 10 m and 40 m below the sea surface in the central part of Harima-Nada. Open circles represent years of red tide occurrence in Harima-Nada; full circles years of no red tide occurrence.

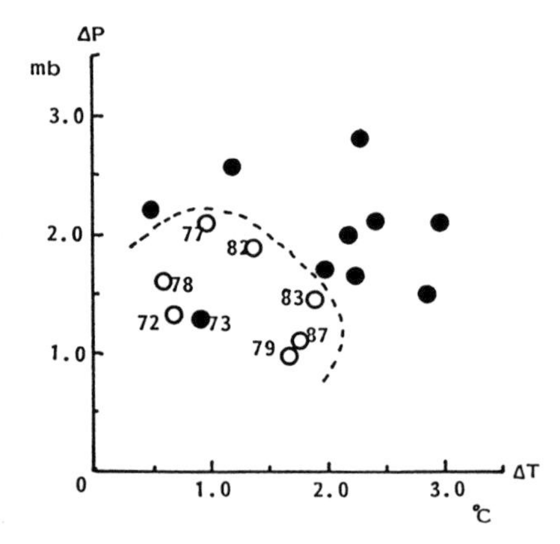

Fig.2 The relations between ΔP, ΔT and red tide occurrence patterns.

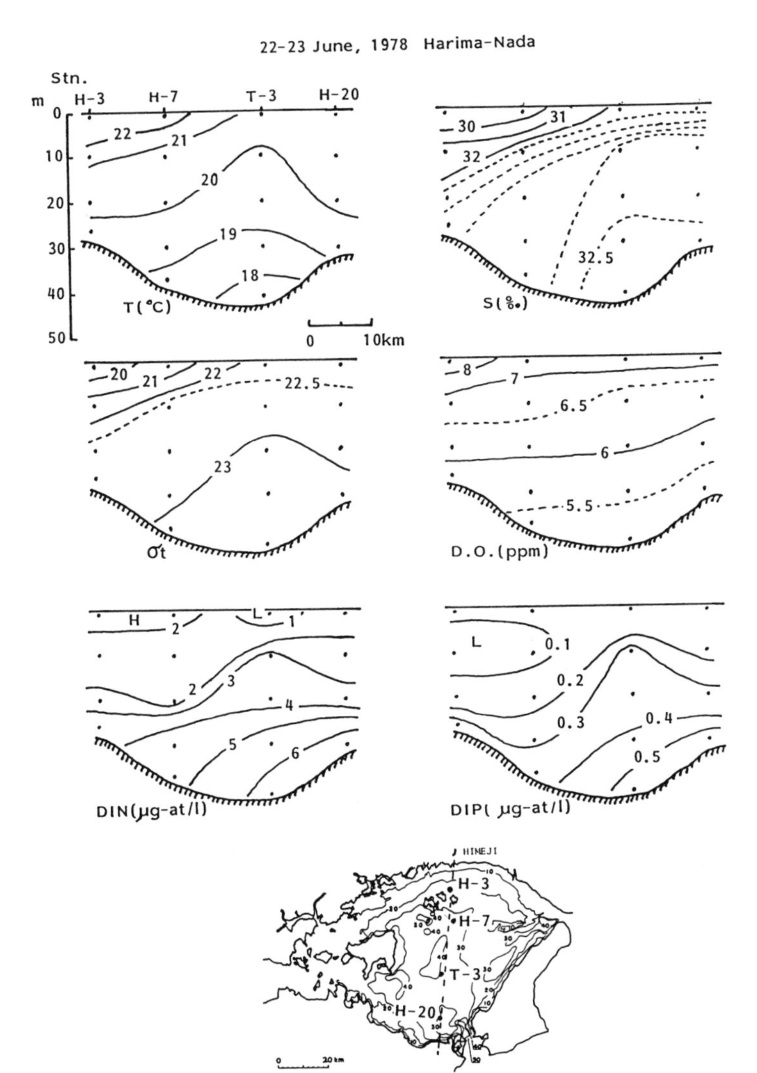

Fig.3 Vertical distributions of water temperature, salinity, density, dissolved oxygen, DIN and DIP along the north-south center line of Harima-Nada on 22-23 June 1978. A large scale red tide of <u>Chattonella antiqua</u> occurred on 28 July 1978.

A small Δ P indicates a weak westerly monsoon in winter and a weak mean current from west to east in the Seto Inland Sea which is driven by this westerly monsoon. I suppose that the cysts of *Chattonella antiqua* are advected eastward by the strong mean current and flow out of Harima-Nada when the westerly monsoon is strong in winter. However, many cysts can stay in Harima-Nada when the westerly monsoon is not so strong i.e. Δ P is small.

A small Δ T reveals a weak density stratification; then salinity is almost uniform vertically in the central part of Harima-Nada. The weak density stratification facilitates vertical mixing due to strong wind blowing at spring tide, which pumps nutrients into the euphotic surface layer from the bottom layer, where much nutrient is stored in summer (Fig.3). Such nutrient supply to the surface layer results in a red tide outbreak of *Chattonella antiqua*.

DISCUSSION

On the basis of above investigations, the following scenario of red tide occurrence of *Chattonella antiqua* in Harima-Nada is proposed. Many cysts of *Chattonella antiqua* can remain at the bottom of Harima-Nada only in winter when the westerly monsoon is weak. Such cysts can trigger a red tide bloom only in summer when the density stratification at the central part of Harima-Nada is weak. The cells of *Chattonella antiqua* begin to increase at high growth rate just after the strong wind blowing at spring tide. The combined effects of wind-driven current due to strong wind and the strong tidal current induce the large vertical mixing then, which supplies much nutrient into the surface layer from the bottom layer. The cells continue to increase resulting in a red tide after about one weak.

To check the above scenario, we have to document the behavior of *Chattonella antiqua* cysts quantitatively and short-term changes in the vertical distribution of nutrients around the time of red tide occurrence.

ACKNOWLEDGEMENTS

The author expresses his sincere thanks to Hyogo Prefectural Fisheries Observatory for providing valuable data.

REFERENCES

1. T.Yanagi, The Science of Red Tide, ed.by T.Okaichi, Koseisha-Koseikaku, Tokyo, 149-159,(1987).
2. T.Yanagi, Report of Environmental Science, B148-R14-8, 121-126, (1982).
3. Fuyo Data Processing and Systems Development, Ltd., Reports of the study on the red tide, 199p., (1984).
4. T.Yanagi, Monthly Marine Science, 16, 56-59, (1984).

FRONTS, UPWELLING AND COASTAL CIRCULATION: SPATIAL HETEROGENEITY OF *CERATIUM* IN THE GULF OF MAINE

FRANKS, P.J.S., D.M. ANDERSON, AND B.A. KEAFER
Biology Department, Woods Hole Oceanographic Institution, Woods Hole, MA, 02543 USA

ABSTRACT

The accumulation of dinoflagellates at tidal fronts has been a popular explanation for the distribution of blooms in coastal waters throughout the world. We sought to test this hypothesis with frequent sampling from March to September, 1987, along a 30km transect from Portsmouth, New Hampshire into the southwestern Gulf of Maine, a region subject to recurrent outbreaks of PSP. No evidence of a tidal front was found in the CTD profiles; however, a dense bloom of *Ceratium longipes* was found offshore in June, closely linked to a wind-driven coastal upwelling. Distributions of *Alexandrium tamarense* (=*Protogonyaulax tamarensis*) suggest that it may respond to the same physical forcings as *C. longipes* even though it remained a small fraction of the total phytoplankton biomass.

INTRODUCTION

The formation of toxic dinoflagellate blooms at tidal fronts has been described for coastal areas the world over [1]. The location of these fronts may be predicted based on the strength of the tidal current, u, and the water depth, h, using the Simpson and Hunter hu^{-3} criterion [2]. The characteristics of such a front are a well-mixed region inshore of the front, and a stratified region offshore. The front itself should be indicated by isopycnals intersecting the surface or the bottom. The dinoflagellate bloom is predicted to follow the pycnocline, with maximal cell densities at the front.

We examined this hypothesis by sampling a transect in the southwestern Gulf of Maine which spanned the depths at which tidal fronts were predicted to occur [3]. With approximately bi-weekly sampling throughout the bloom season we hoped to characterize the temporal changes of the bloom, as aliased by the cruise dates. Measurements were made of conductivity, temperature, *in situ* fluorescence, chlorophyll, dissolved nitrate+nitrite, and cell densities of *Ceratium longipes* and *Alexandrium tamarense (=Protogonyaulax tamarensis)*.

METHODS

The study transect was located in the southwestern Gulf of Maine, extending from Portsmouth, New Hampshire into Jeffreys Basin (Figure 1). The transect was approximately 30km long, with five stations at ~7km intervals. Station 1 was located

RED TIDES: BIOLOGY, ENVIRONMENTAL SCIENCE, AND TOXICOLOGY
Okaichi, Anderson, and Nemoto, Editors

outside the harbour mouth in waters of 10 - 15m depth, while the water column at Station 5 extended to 170m. The two cruises described below took place on May 26 and June 5, 1987. Seven additional cruises were made, but will not be described here.

The ship used was the 45ft vessel, R/V Jere A. Chase, owned by the University of New Hampshire. From the A-frame of this vessel, 40m of 2cm internal-diameter hose was lowered into the water, with a Sea Bird "SeaCat" CTD attached at the hose inlet. A small submersible pump brought water onto deck, where it passed into a bubble trap, through a Turner Designs flow-through fluorometer, and into buckets on deck. The fluorometer was connected to a portable computer which displayed fluorescence in real time as well as storing the data to disk. The hose was raised at two metres per minute, with a flow of two litres per minute. Five litres were collected in each bucket, and subsampled as follows: one litre was filtered through 20μm mesh and the filter backwashed and fixed in 10% formaldehyde for cell counts, while 500ml was filtered through GF/A filters for chlorophyll analysis, and the filtrate frozen for nutrient analyses.

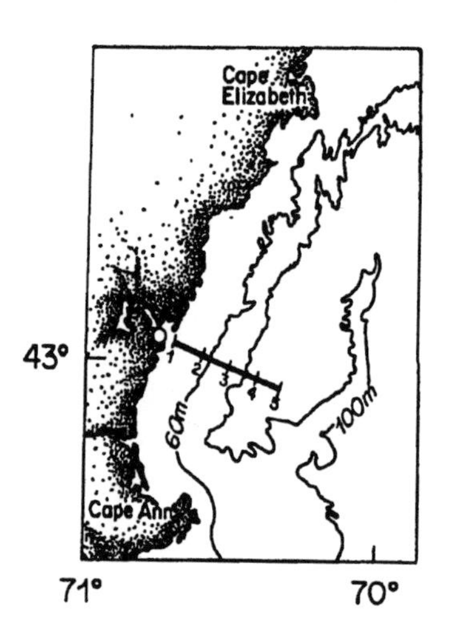

Figure 1. The study transect and the five stations are indicated in this figure. Cape Ann is just north of Boston. The transect begins at Portsmouth, NH on the New Hampshire-Maine border. The transect was approximately 30km long with stations ~7km apart.

RESULTS

One of the objects of this study was to look for a link between dinoflagellate blooms and tidally-generated fronts. No characteristics of a tidal front were seen in any of the CTD transects, although the surface temperature distributions on June 18 and July 2 would resemble a tidal front in satellite imagery. However, analysis of the temperature and salinity distributions indicates that there was an upwelling system present within the transect, centred between Stations 2 and 3. The density profiles of May 26 and June 5 (Figure 2) show this upwelling well. In May, the upwelling occured over a broad horizontal area between Stations 2 and 3, as shown by the upward curve of the isopycnals. By June, the solar heating restricted the upwelling to Station 2, while wind-mixing caused the isopycnals to surface at Station 3. Other more extensive cruises have shown that this upwelling extends 150km northward along the coast up to the Monhegan Island area. The upwelling is likely forced by longshore winds, although buoyancy effects such as freshwater input and solar heating may also play a role. Further data analysis combined with modelling studies are beginning to elucidate the relative contributions of the various forcings.

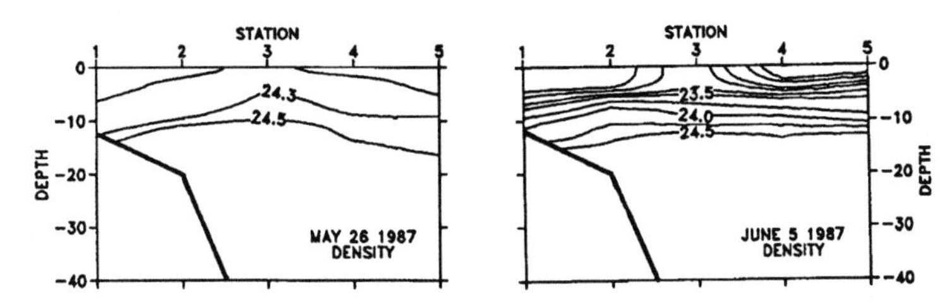

Figure 2. Density (σ_t) contours from May 26 (left panel) and June 5 (right panel). Upwelling is thought to cause the upward curvature at Station 3 in May, and Station 2 in June. The surfacing of the isopycnals in June at Station 3 is caused by surface mixing.

Closely associated with this upwelling was a concentrated band of *Ceratium longipes*, located at Station 3 at 10m depth. This bloom of *C. longipes*, seen in both the fluorescence and cell-density profiles, was developing offshore during May, becoming a band 5m thick and about 15km wide by June 5 (Figure 3). Peak cell denisities reached 5,000 cells l^{-1} in late June, and the bloom had dissipated by early July.

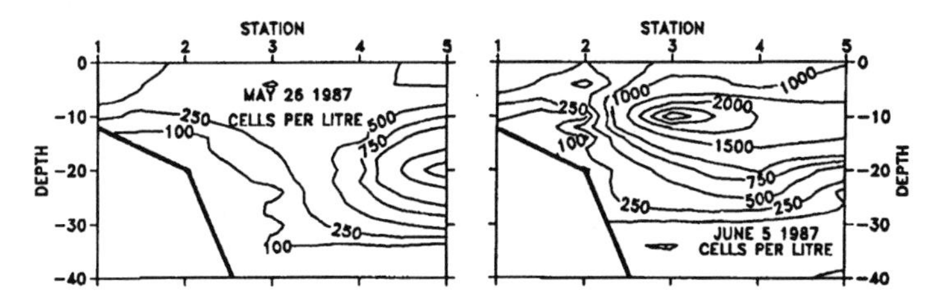

Figure 3. The left panel shows the cell concentrations of *Ceratium longipes* during the May 26 cruise, while the right panel shows the June 5 cruise. Note the high cell concentrations at Station 3 in June, at 10m depth.

Modelling studies combined with data analysis indicate that the strongest upwelling would be at Station 2 (vertical velocities ~0.001 cm s^{-1}), thus the bloom lies offshore of the peak upwelling. The peak horizontal velocities are of the order of 1 cm s^{-1}. Heaney and Eppley [4] have shown *Ceratium furca* to swim vertically at speeds of ~0.03 cm s^{-1}, thus it is probable that the *C. longipes* species which formed this bloom could maintain its depth during upwelling.

DISCUSSION

The close association of the *Ceratium longipes* bloom with the hydrographic features is indicative of the role the water motions play in generating and maintaining a dinoflagellate bloom. Upwelling along the coast can supply nutrients to the cells to maintain *in situ* growth, and can also promote localized accumulation of the cells if the required behavioural response is displayed. In this case, slower swimming in the cooler sub-pycnocline waters [4] would lead to the distribution seen. This scenario remains hypothetical, as vertical migration patterns were not studied during the 1987 field season.

Although the bloom which occurred was a *Ceratium* bloom, cell counts of *Alexandrium tamarense* were also made. Although *Alexandrium* usually blooms in the study region during spring, no such bloom was seen. It is conceivable, though, that such a bloom could occur under the conditions described above. If this were the case, then large subsurface concentrations of toxic cells could build up offshore, to be delivered to shellfish beds when the winds shifted. This mechanism could account for the rapid rise in toxicity seen during PSP outbreaks along the western Gulf of Maine coast. The temporal duration of the bloom, combined with the presence of southward-flowing longshore currents, suggests that the *Ceratium longipes* bloom extended northward in a band about 15km from shore. This spatial pattern is similar to that depicted by Yentsch *et al.* [1] for the area north of our transect. However, the generating mechanism proposed in that study was not upwelling, but tidally-generated fronts. We have shown that wind-driven coastal upwelling will generate biological patterns similar to those proposed for tidal fronts, emphasizing the need for relatively dense, accurate hydrographic measurements coincident with the biological measurements.

ACKNOWLEDGEMENTS

The authors gratefully acknowledge the help and cooperation of Captain Paul Pelletier. This study was supported in part by funding from the Office of Naval Research contract N00014-87-K-0007, and the Office of Sea Grant, National Oceanic and Atmospheric Administration, through a grant to WHOI (NA86AA-D-SG090, R/B-76). WHOI contribution number 6772.

REFERENCES

1. C.M. Yentsch, P.M. Holligan, W.M. Balch and A. Tvirbutas, *Lecture Notes on Coastal and Estuarine Studies 17*, (M.J. Bowman, C.M. Yentsch and W.T. Peterson, eds.), 224, (1986)

2. J.H. Simpson and J.R. Hunter, *Nature*, 250,404, (1974)

3. J.W. Loder and D.A. Greenberg, *Cont. Shelf Res.*, 6, 397, (1986)

4. S.I. Heaney and R.W. Eppley, *J. Plankton Res.* 3,331, (1981)

VOLUMETRIC CONSIDERATION OF SOME SPECIES OF DIATOMS COMPOSING RED TIDES IN THE NORI CULTURE AREA IN ARIAKE BAY

SHIROH UNO* AND KAZUYUKI SASAKI**
* Seikai Reg. Fish. Res. Lab., Kokubu-cho, Nagasaki, 850 JAPAN
**Ariake Fish. Exp. Sta. of Fukuoka Pref., Yanagawa, 832 JAPAN

ABSTRACT

Cultured nori (Porphyra tenera) is often damaged by diatom red tides in Ariake Bay. The authors investigated species succession from the view point of cell volume in the area from September 1985 to March 1986. Settled volume of plankton was low during the period without red tide. From January, the volume showed logarithmic increase up to late February, then quickly decrease to mid March. Inorganic nitrogen in seawater changed conversely. The dominant phytoplankton by cell number during this period was Chaetoceros spp. The five dominant species by population cell volume were Eucampia zoodiacus, Rhizosolenia spp., Chaetoceros spp., Coscinodiscus spp., and Biddulphia sinensis. The amounts of the species with large cells such as Coscinodiscus and Biddulphia, did not fluctuate greatly throughout the observed period. In the laboratory, six diatom species were cultured with Porphyra thalli. E. zoodiacus show the highest growth rate and it proved to be the most important species during the red tide period.

INTRODUCTION

Ariake Bay is the most important area for nori (Porphyra tenera) culture in Japan. The main harvesting season for nori is from November to March. In the winter season (especially from late Jan. to mid Feb.), some species of diatoms form dense blooms. It is a type of spring bloom, but it really ought to be called red tide. The red tide consumes nearly all of the nutrients in the water, resulting in the loss of the dark brown colour of the nori and thus a desrease in its commercial value.

We observed the phytoplankton community, DIN, and nori production in the bay, and addressed the species succession of phytoplankton not only by cell number but also by cell volume.

EXPERIMENTAL

The observation point was situated at the head of Ariake Bay, shown in Fig. 1. Sampling was done at three to ten day intervals. Phytoplankton were collected with a 100µm mesh net. Settled volume was measured first, then determinations of cell number were made. Population cell volume of plankton species was calculated by multiplying mean cell volume (Table I) by cell numbers for the individual species. DIN means the total values of nitrate, nitrite and ammonium nitrogen which were all determined by the method of Strickland and Parsons(1). Nori production and its selling price were obtained from the data of Yamato Fisheries Cooperative Association.

RESULTS

The variation of DIN and settled volume of plankton during

RED TIDES: BIOLOGY, ENVIRONMENTAL SCIENCE, AND TOXICOLOGY
Okaichi, Anderson, and Nemoto, Editors

TABLE I. Mean cell volume of the main diatom species observed in the period.

Species	Mean cell volume
Skeletonema costatum	2.54×10^3
Chaetoceros spp.	6.38×10^3
Nitzschia spp.	5.72×10^2
Asterionella japaonica	1.67×10^2
Thalassionema nitzschioides	1.60×10^3
Thalassiothrix spp.	1.81×10^4
Thlassiosira spp.	2.58×10^4
Rhizosolenia sp. I	3.63×10^5
Rhizosolenia sp. II	1.76×10^4
Rhizosolenia stolterfothii	1.26×10^4
Coscinodiscus spp.	9.05×10^5
Bidduphia sinensis	1.13×10^6
Eucampia zoodiacus	2.40×10^4
Ditylum brightwellii	9.56×10^4
Streptotheca sp.	2.88×10^4
Stephanophyxis sp.	1.57×10^5

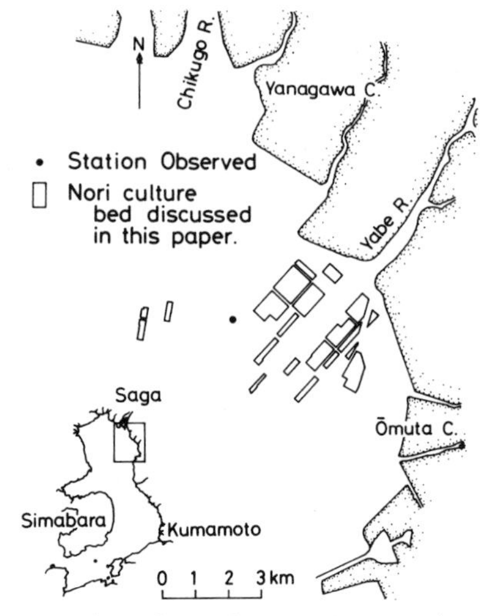

Fig. 1. Observation points and nori culture area of Yamato Fish. Corp. Assoc.

the experimental period are shown in Fig. 2. DIN at the station shows high values except during the diatom red tide. It was greater than 15 µg-at/l from late September to early January. Throuout this period, even with sufficient DIN for phytoplankton growth, no bloom occurred; the settled volume of plankton did not exceed 10 ml/l. In mid-January, plankton volume grew and continued to increase until mid February. The volume reached 531 ml/l in late February, then dropped until mid March. DIN was almost entirely consumed by phytoplankton and nori from mid to late February and it gradually recovered starting in early March. DIN and plankton volume showed reverse patterns.

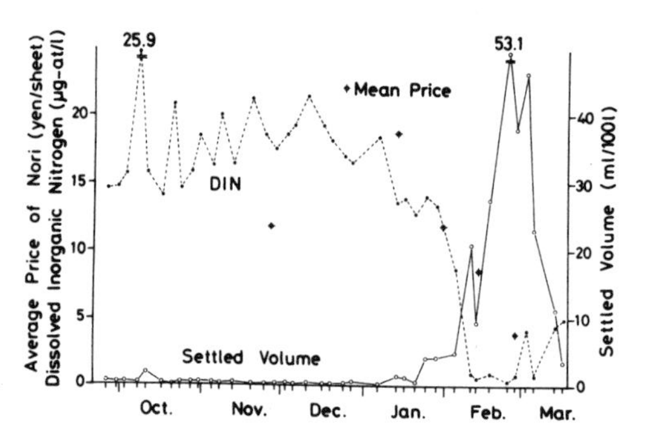

Fig. 2. Variation of DIN and settled volume of plankton and mean price of nori

During this harvesting season, the mean price of nori offered at each tender is also pointed out in Fig. 2. Tenders were held six times; Nov. 26, Dec. 24, Jan. 13, Jan. 30, Feb. 13, Feb. 27. The average price at the first tender (Nov. 26) was rather low, 12.9 yen per sheet. At the second tender (Dec. 24), it jumped over 21 yen, the highest price in the season. Then it declined and never recovered. It finally reached its lowest value, 3.8 yen on February 27. These variations accompanied the change of DIN. Therefore, diatom red tides consume DIN in the field, and depleted DIN lowers the quality of nori. This phenomenon is commonly recognized every year.

Phytoplankton succession was also observed and all members of important species are shown in Fig. 3. Skeletonema was dominant in October, but not during the red tide period. The dominant three during the red tide were Eucampia, Chaetoceros, and Nitzschia. The succession was also examined by the cell volume of each population and important species are shown in Fig. 4. The dominant five species by volume during the period were Eucampia, Rhizosolenia, Chaetoceros, Coscinodiscus, and Biddulphia. The latter two species did not fluctuate so widely throughout the study. Thus the most important were the three former species, especially Eucampia which increased suddenly in late January and occupied the highest position in terms of population cell volume.

DISCUSSION

From the results in Fig. 4, Eucampia appeared to be the most important species among the five during the span of the diatom red tide in Ariake Bay. The other eleven species in Table I were not as ecologically significant. The importance of Eucampia is also discussed in the culture experiments of Sasaki and Uno(2). They compared division rates and maximum cell densities in uni-algal and bi-algal cultures with Porphyra thalli using six species of diatoms; namely Eucampia zoodiacus, Asterionella japonica, Chaetoceros curvisetus, Ditylum brightwellii, Skeletonema costatum, and Streptotheca thamensis. Their results of bi-algal culture are shown in Table II. Using the data in Table II, we tried to make a schematic growth curve of their population cell volume from uni-cell inoculum to the maximum, shown in Fig. 5. The uni-cell volume of the six species varied widely. The volume of E. zoodiacus was about

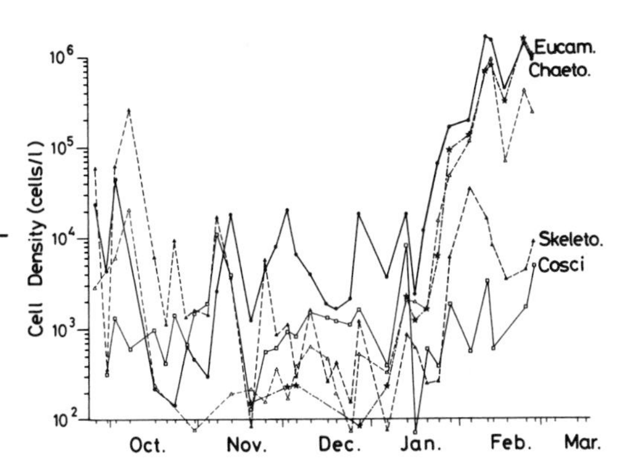

Fig. 3. Phytoplankton succession compared by cell number (density).

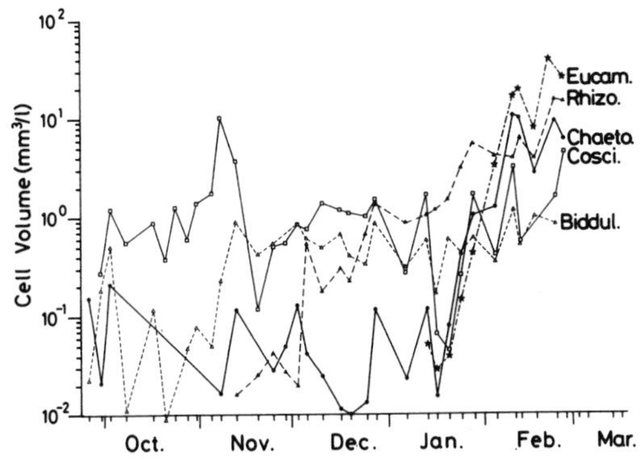

Fig. 4. Phytoplankton succession compared by cell volume.

TABLE II. Comparison of mean cell volume, doubling time, and maximum cell density among diatom species in culture with Porphyra tenera.

Diatom species	Cell Vol. (um^3)	Doubl. T. (days)	Max. Density (cells/ml)
Ditylum brightwellii	9.56×10^4	1.32	3.4×10^3
Streptotheca thamensis	8.63×10^4	2.50	2.4×10^3
Eucampia zoodiacus	2.40×10^4	0.74	1.0×10^4
Chaetoceros curvisetus	1.22×10^4	1.28	3.6×10^4
Skeletonema costatum	2.54×10^3	1.45	4.2×10^5
Asterionella japonica	1.67×10^2	1.11	6.4×10^4

one thousand times larger than that of A. japonica. However, the population cell volumes at the maximum densities of the six species were not so differed; the highest was for S. costatum and the lowest was for A. japonica. The six species were all started from one cell. The most important species by population volume was D. brightwellii for the first three days. Then it was surpassed by the species with the highest growth rate and the first to reach maximum density under these environmental conditions, E. zoodiacus. These bi-algal cultures with Porphyra were conducted using mono-specific phytoplankton cultures. If mixed species were used, the doubling time and maximum cell density results for the six diatoms would be different from Table II. However, our observations show that E. zoodiacus often appears in the early spring in the nori culture area, and forms red tides. The species is considered to be the most important among the many species of diatoms present in nori culture season.

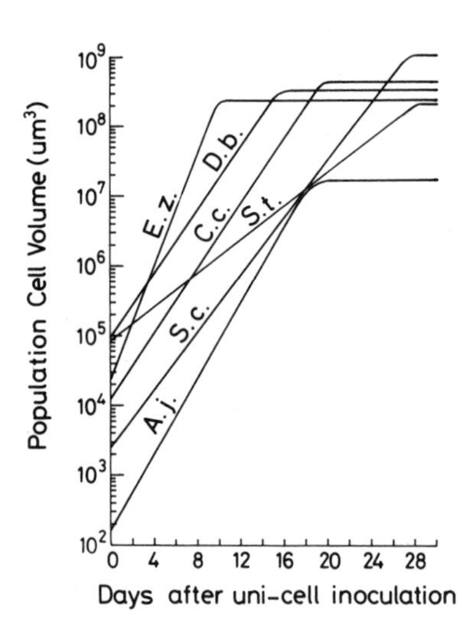

Fig. 5. Schematic diagram of growth curve of six diatom species occuring in the Bay.

Up to now, succession of phytoplankton species has almost always been treated in terms of cell numbers of populations, not by cubic volume. However it should be treated in terms of cubic volume of cells when some aspects of ecological cycling (especially nutrient requirements) are being considered.

REFERENCES

1. J.D.H. Strickland and T.R. Parsons, Fish. Res. Bd. Canada, Bulletin 167 (1968).
2. K. Sasaki and S. Uno, Bull. Plankton Soc. Japan, (in press).

PROMPT INFORMATION ON RED-TIDES IN THE CULTURAL GROUNDS OF NORI (EDIBLE LAVER)

TAMIJI YAMAMOTO AND KOUSUKE FUJISAKI
Aichi Prefectural Fisheries Experimental Station, Aichi 443, Japan

ABSTRACT

Using a cultured diatom, correlations between chl. *a* and DCMU-enhanced fluorescence and between photosynthetic activity and DCMU-induced fluorescence increase were obtained. The DCMU methods were applied to forecast the development of red-tides in Nori cultural grounds. The Ip, an index showing the phytoplankton photosynthetic activity per unit biomass, increased 2-4 weeks before the phytoplankton outbreaks. The methods are simple to operate and considered to be advantageous for their practical values in providing prompt information on red-tide potential in Nori cultural grounds.

INTRODUCTION

The yield of Nori (edible laver) accounts for 37 % of the annual total production of sea fisheries in Aichi Prefecture [1]. The coastal areas around the Chita Peninsula are major cultural grounds of Nori.

An increase of nutrient loading from land in recent years has made the coastal areas prime places for the development of red-tides [2]. Red tides, mostly diatoms, recently have been found even in winter, the major culturing season of Nori, and the occurrence has been increasing year by year [3]. The decrease in nutrient levels in the culturing grounds due to the red-tides damages the growth of the Nori fronds, and makes the color fade.

DCMU (3(3,4-dichlorophenyl-1,1-dimethylurea) is known as one of potent herbicides which inhibits plant photosynthesis [4]. The findings that DCMU-enhanced fluorescence is strongly related to the cellular chl. *a* regardless of the growth conditions and species [5], and that DCMU-induced fluorescence increase is correlated with photosynthetic activity [6], are relevant to the field study.

In the present study, the possibilities of applying the DCMU methods for promptly forecasting the development of red-tides in the cultural grounds of Nori were examined.

METHODS

Laboratory experiments

A diatom ***Chaetoceros decipiens*** was cultured in a 10 *l*-mixture of 1/10 Matsudaira medium [7] and 1/10 Provasoli's No. 8 medium [8]. The incubation was conducted under 20°±1°*C* and *ca.* 2000 *lx* fluorescent lamps and L:D=16: 8 cycle. The culture medium was stirred using a magnetic stirrer during 27-day incubation.

Sample was taken from the vessel at 11:00 every day and used for the following measurements. Chlorophyll *a* was measured fluorometrically [9] using Hitachi 139 UV-VIS spectro-fluorometer. Photosynthesis was measured basically by the ^{14}C method [10] using triplicate 100 *ml* transparent glass bottles, but to one of the three as a control was added $10^{-5}M$ DCMU instead of the conventional dark treatment [11]. After the 2-3 *h* incubation at 20°±1°*C* and *ca.* 10000 *lx*, radiocarbon activities were measured.

Phytoplankton ***in vivo*** fluorescence, both natural (Fn) and DCMU-

RED TIDES: BIOLOGY, ENVIRONMENTAL SCIENCE, AND TOXICOLOGY
Okaichi, Anderson, and Nemoto, Editors

added (Fd) were measured after 30 *min* dark treatment at room temperature [12] using the spectro-fluorometer described above.

Field works

Surface seawater was collected weekly from 18 September 1986 to 4 March 1987 at 3 stations located in Nori cultural grounds (Fig. 1), and *in vivo* fluorescence Fn and Fd measured by the same method described above using a Turner Design 10 fluorometer. Here, Fd was converted to the equivalent chl. *a* amount. Ip=(Fd-Fn)/Fd was regarded as an index corresponding to the photosynthetic activity per unit biomass. Ammonia-N, nitrite-N, nitrate-N and phosphate-P were determined for seawater filtered through 0.45 μm membrane filter [13].

The normal approximated value "Ts", derived from Kendall's rank correlation coefficient "τ" [14], was computed to test the increase and decrease patterns between chl. *a* and nutrient salts.

RESULTS

Correlations between Fd and chl. *a*, and between (Fd-Fn) and ^{14}C-uptake

The cellular chl. *a* of *C. decipiens* increased exponentially from 4.1 to 590.0 μg chl $a \cdot l^{-1}$. Photosynthetic activity also increased from 22.9 to 2372.5 $\mu gC \cdot l^{-1} \cdot h^{-1}$ during the incubation. The Fd was highly correlated with chl. *a* (r=0.999) during the period of log-growth phase (1-21 days), but deviated from the regression line in the senescent phase (22, 24 and 27 days)(Fig. 2). The (Fd-Fn) also showed good correlation with ^{14}C-uptake (r=0.967) during the log-growth phase (Fig. 3).

Variations of phytoplankton and nutrients in Nori cultural grounds

During the observations, three marked chl. *a* peaks (28 October-5 November 1986, 20 January and 24 February 1987) were found at all stations (except for the second peak at St. 1) in synchrony with each other, even though the sampling locations were about 10 *km* apart (Fig. 4). Phytoplankton chl. *a* increased rapidly and often nearly reached maximal red-tide levels within the short period of 1-2 weeks; the outbreaks did not last longer than 3 weeks. The maximum chl. *a* value (30.4 $\mu g \cdot l^{-1}$) was recorded at St. 3 on 24 February 1978; however, the maximum growth rate (0.61 doublings$\cdot d^{-1}$) estimated by the chlorophyll increase was found on 21-28 October 1986 at St. 2.

Concentrations of nutrient salts were generally high at Sts. 1 and 3 compared to St. 2 (Fig. 4). The ranges of dissolved inorganic nitrogen (DIN=NH_4-N +NO_2-N +NO_3-N) and phosphate-P during the period of the observations were 0.31-30.12 μg-at $N \cdot l^{-1}$ and ND (below the limit of detectability)-2.63 μg-at $P \cdot l^{-1}$ at St. 1; 0.76-13.01 μg-at $N \cdot l^{-1}$ and ND-1.18 μg-at $P \cdot l^{-1}$ at St. 2; and 2.60-19.71 μg-at $N \cdot l^{-1}$ and ND-2.88 μg-at $P \cdot l^{-1}$ at St. 3.

Statistical analyses showed negative correlations between chl. *a* and DIN or phosphate-P at all stations within significant concordance ($P \leq 0.1$) at Sts. 2 and 3.

The Ip reached 0.7 or more about 2-4 weeks prior to the phytoplankton outbreaks (except for the chl. *a* peaks on 24 Feburuary 1987 at St. 1 and on 28 October 1986 at St. 2; Fig. 4).

DISCUSSION

The significant negative correlations between chl. *a* and DIN or phosphate-P mean that the nutrient concentrations *in situ* were controlled

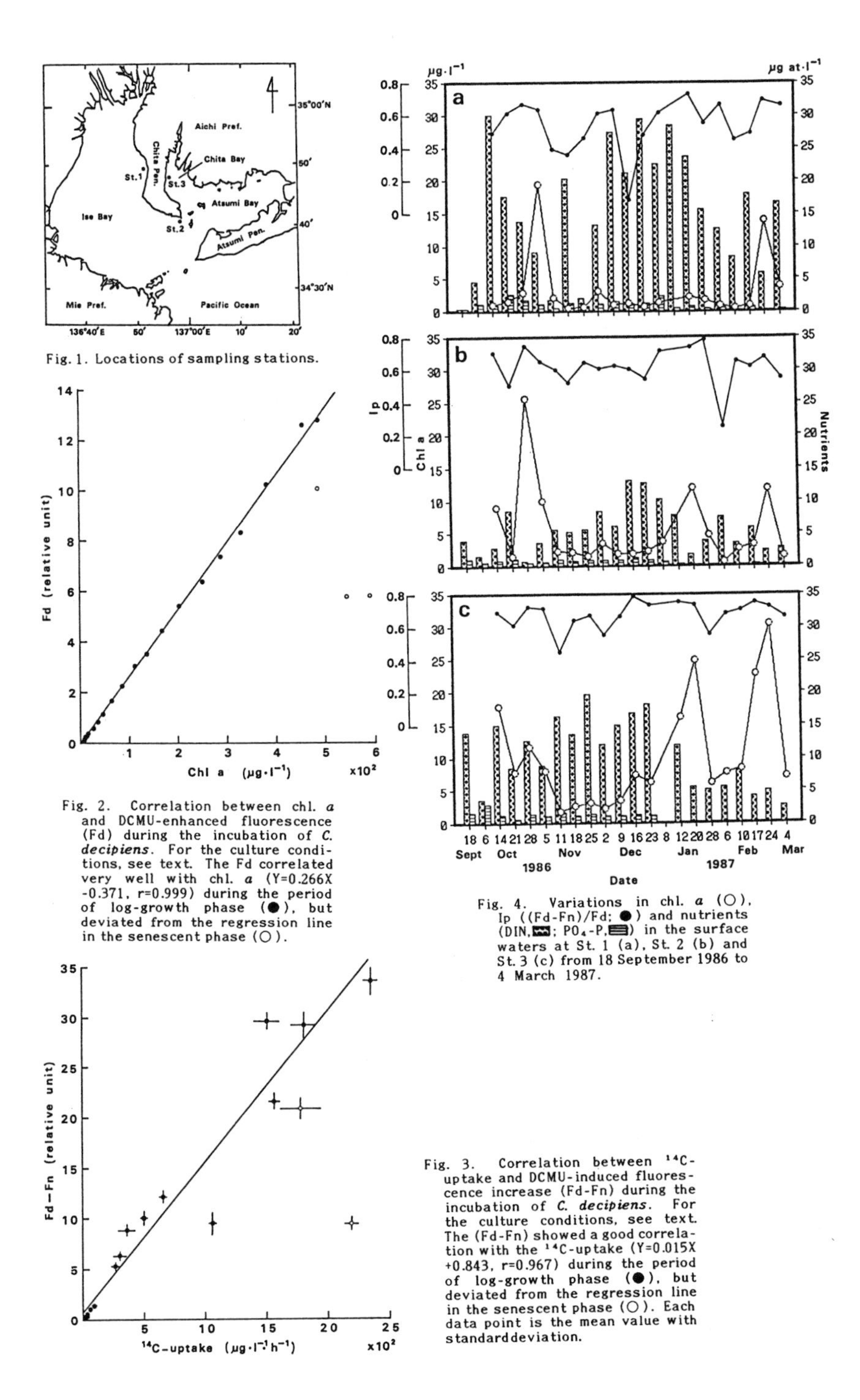

Fig. 1. Locations of sampling stations.

Fig. 2. Correlation between chl. *a* and DCMU-enhanced fluorescence (Fd) during the incubation of *C. decipiens*. For the culture conditions, see text. The Fd correlated very well with chl. *a* (Y=0.266X -0.371, r=0.999) during the period of log-growth phase (●), but deviated from the regression line in the senescent phase (○).

Fig. 3. Correlation between ^{14}C-uptake and DCMU-induced fluorescence increase (Fd-Fn) during the incubation of *C. decipiens*. For the culture conditions, see text. The (Fd-Fn) showed a good correlation with the ^{14}C-uptake (Y=0.015X +0.843, r=0.967) during the period of log-growth phase (●), but deviated from the regression line in the senescent phase (○). Each data point is the mean value with standarddeviation.

Fig. 4. Variations in chl. *a* (○), Ip ((Fd-Fn)/Fd; ●) and nutrients (DIN, ▨; PO_4-P, ▤) in the surface waters at St. 1 (a), St. 2 (b) and St. 3 (c) from 18 September 1986 to 4 March 1987.

mainly by the phytoplankton abundance, not by the cultured Nori. In winter the dominant red tide species typically are ***Skeletonema costatum, Chaetoceros*** spp. and ***Thalassiosira nordenskiöldii*** [3]. The Ks from uptake kinetics of nitrogen by these diatoms are generally smaller than that of Nori. For example, the Ks of *S.* ***costatum*** for ammonia and nitrate are 0.8-3.6 and 0-1.0 μg-at $N \cdot l^{-1}$, respectively [15]. On the other hand, those of Nori are 3.5-13.1 and 0.7-3.9 μg-at $N \cdot l^{-1}$ [16]. Therefore, the areas of relatively low nutrient concentration, such as St. 2, are disadvantageous to the growth of Nori. Actually, Nori of fine quality is harvested at the inner parts of both bays.

The DCMU methods are considered to be useful under the general assumption that natural phytoplankton populations are actively growing, since satisfactory correlations between chl. *a* and Fd and photosynthetic activity and (Fd-Fn) were found during the period of log-growth (Figs. 2 and 3). In the initial stage of natural phytoplankton growth, the physiological state of the population is considered to be in log-growth phase.

It is reported that photosynthetic activity estimated from DCMU-induced fluorescence increased 1-2 days before chl. *a* attained its maximum [17]. Although our sampling interval of once a week was somewhat longer than theirs, similar trends were observed (Fig. 4). The DCMU methods therefore considered to be advantageous for practical use in providing prompt information on red-tide blooms in Nori cultural grounds.

ACKNOWLEDGEMENTS

We thank both the fisheries disseminators of Chita Office, Fishery Division and the fishermen for their cooperation in the field studies.

REFERENCES

1. Aichi-ken, Doukou-chousa Shiryou 78, 76 (1987).
2. K.Sasaki, T.Suzuki, Y.Matsukawa and Y.Satou, Sakana 26, 45-70 (1981).
3. A.Miyamoto, H.Tsuchiya and Suishitsu-chousasen Norikumi-in, Aichi-ken Suisan Kenkyu Gyouseki C70, 58 (1987).
4. J.S.C.Wessels and R.van der Veen, Biochem. biophys. Acta 19, 528-549 (1956).
5. R.E.Slovacek and P.J.Hannan, Limnol. Oceanogr. 22, 919-925 (1977).
6. G.Samuelsson and G.Öquist, Physiol. Plant. 40, 315-319 (1977).
7. C.Matsudaira and R.Kawakami, Suisan Doboku 5, 33-36 (1969).
8. L.Provasoli, J.J.A.McLaughlin and M.R.Droop, Arch. Mikrobiol. 25, 392-428 (1957).
9. C.S.Yentsch and D.W.Menzel, Deep-Sea Res. 10, 221-231 (1963).
10. E.Steemann Nielsen, J. Cons. Int. Explor. Mer 18, 117-140 (1952).
11. L.Legendre, S.Demers, C.M.Yentsch and C.S.Yentsch, Limnol. Oceanogr. 28, 996-1003 (1983).
12. B.B.Prézelin and B.M.Sweeney, Plant Physiol. 60, 338-392 (1977).
13. J.D.H.Strickland and T.R.Parsons, Fish. Res. Bd. Canada 167, 310 (1972).
14. R.R.Sokal and F.J.Rohlf, Introduction to Biostatistics 368 (1969).
15. R.W.Eppley, J.N.Rogers and J.J.McCarthy, Limnol. Oceanogr. 14, 912-920 (1969).
16. T.Yamamoto and M.Takao, Jpn. J. Phycol. 36, 37-42 (1988).
17. N.Fukazawa, T.Ishimaru, M.Takahashi and Y.Fujita, Mar. Ecol. Prog. Ser. 3, 217-222 (1980).

DEVELOPMENT OF CONTINUOUS MONITORING SYSTEMS FOR ENVIRONMENTAL PARAMETERS RELATED TO RED TIDES

KAZUO HIIRO,* JITSUZO NAGAO,** AND TAKASHI KIMOTO**
*Government Industrial Research Institute, Osaka, 8-31, Midorigaoka-1, Ikeda-shi, Osaka-fu 563; **Japan Aqua Technology, Inc., 3-1, Funahashi-cho, Tennoji-ku, Osaka-shi 543, Japan

ABSTRACT

Automatic measuring instruments for water chemistry and physical parameters were developed. Phosphate was measured with the phosphomolybdate method. Total phosphorus was determined by autoclave digestion and the phosphomolybdate method. An ultraviolet spectrophotometric method was applied to the measurement of nitrate and nitrite, membrane enrichment and conductivity measurement was used for ammonia, autoclave digestion and ultraviolet spectrophotometry for total nitrogen. Permanganate oxidation was applied to the measurement of chemical oxigen demand(COD) and the silicomolybdate method to silicate. Temperature, pH, dissolved oxygen(DO), salinity, turbidity and chlorophyll-a were measured with conventional instruments.

An observation station and a barge were constructed. The station and the barge were equipped with all the above mentioned automatic measuring instruments. Electric power generators, an automatic sampling apparatus and a wireless telemetering system were also supplied. The automatic measuring apparatus functioned without major problems. By employing the automatic sampling-analyzing equipment it was found that physical conditions and nutrients showed rapid fluctuations in the water.

INTRODUCTION

It is often believed that one of the main reasons for the occurrence of red tides is eutrophication. Therefore, for the elucidation of the mechanisms behind red tides and for an early warning of a possible red tide outbreak, it is important to establish automatic measuring techniques for plant nutrients in waters.

Several new automatic instruments for substances such as phosphate, total phosphorus, nitrate, ammonia and total nitrogen, were developed. Automatic instruments for COD and silicate were also constructed. An "Aqua Survey System" was established, composed of observation and shore stations. This system can measure the above mentioned substances as well as other water quality parameters, such as temperature, pH, DO, salinity, turbidity and chlorophyll-a, at hourly intervals. A barge equipped with the same apparatus as the observation station was also constructed.

The system was built by Japan Aqua Technology Inc. and Namura Shipbuilding Co. Ltd. under the sponsorship of the Research Development Corporation of Japan. The Government Industrial Research Institute, Osaka provided the fundamental technologies for the design and construction of the equipments.

The technical details of the system and data obtained by the system are reported in this paper.

RED TIDES: BIOLOGY, ENVIRONMENTAL SCIENCE, AND TOXICOLOGY
Okaichi, Anderson, and Nemoto, Editors

DEVELOPMENT OF ANALYTICAL INSTRUMENTS

Phosphate

The automatic instrument for disolved inorganic phosphate was based on the formation of the phosphomolybdate complex, enrichment of the complex on a preconcentration column of Sep-Pak C18, elution with ethanol and measurement with a voltammetric detector (Fig. 1).

Total phosphorus

Total phosphorus was measured by the molybdenum blue spectrophotometric method after digestion in an autoclave with potassium peroxodisulfate under 2kg/cm^2 and at 120°C (Fig. 2).

Nitrate

Nitrate was measured with a specially designed dual wavelength spectrophotometer shown in Fig. 3. The difference in absorbance between 223nm and 232nm showed a good correlation with the concentration of nitrate plus nitrite.

Ammonia

The analytical method for ammonia was based on enrichment of ammonia with a gas-permeable membrane, and conductivity measurement of the ammonia-enriched dilute sulfuric acid solution (Fig. 4).

Total nitrogen

Quantification of total nitrogen was based on digestion under pressure and at an elevated temperature in an autoclave and measurement of the formed nitrate with an ultraviolet spectrophotometric detector (Fig. 5).

Chemical oxygen demand(COD)

Organic compounds were oxidized with permanganage in an alkaline solution and the excess permanganate was converted to iodine by the addition of iodide. The amount of iodine was measured with an amperometric detector (Fig. 6).

Silicate

Silicate was quantified as the silicomolybdate complex using a voltammetric detector (Fig. 7).

Other water quality parameters

Chlorophyll-a was measured with a spectrofluorometric method.

Fig.1 ANALYTICAL PROCEDURE FOR PHOSPHATE

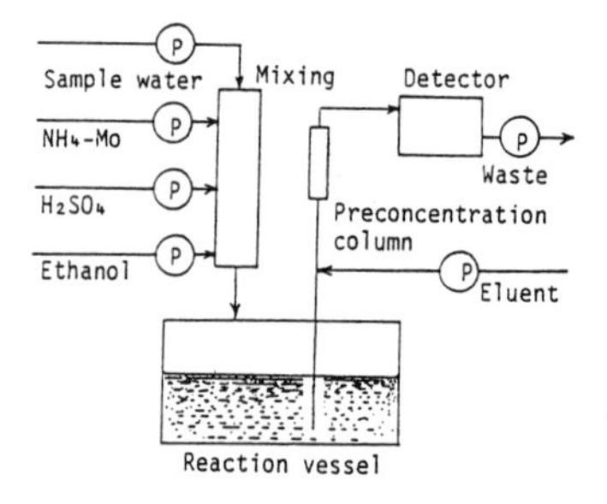

Fig.2 ANALYTICAL PROCEDURE FOR TOTAL PHOSPHORUS

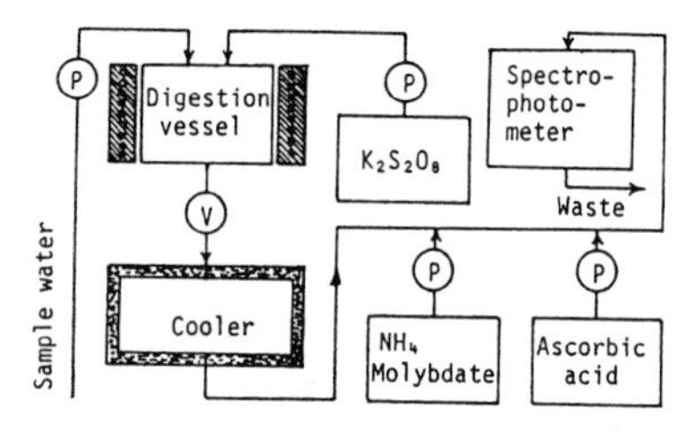

Fig.3 ANALYTICAL PROCEDURE FOR NITRATE PLUS NITRITE

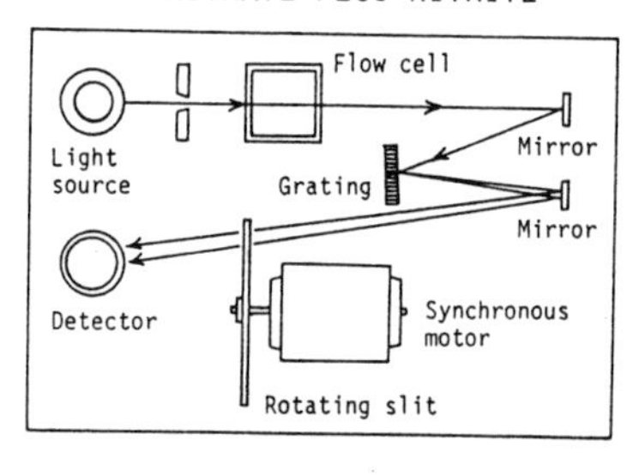

Fig.4 ANALYTICAL PROCEDURE FOR AMMONIA

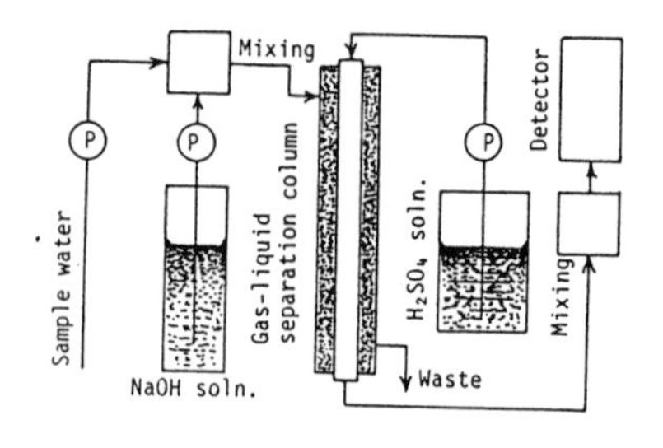

Table 1 SUMMARY OF ANALYTICAL METHODS

Phosphate	Phosphomolybdate method 0 – 200 ppb
Total phosphorus	Persulfate digestion, phosphomolybdate method 0 – 200 ppb
Nitrate	UV dual wavelengths spectrophotometry 0 – 500 ppb
Ammonia	Gas permeable membrane, conductivity measurement 0 – 200 ppb
Total nitrogen	Persulfate digestion, UV spectrophotometry 0 – 500 ppb
COD	Permanganate method 0 – 10 ppm
Silicate	Silicomolybdate method 0 – 2 ppm
Chlorophyll-a	Fluorometric method 0 – 100 ppb
Temperature	-10 – 40°C
pH	Glass electrode method
Dissolved oxygen	0 – 10 ppm
Salinity	0 – 40 $^{o}/oo$
Turbidity	0 – 100 ppm

Fig.5 ANALYTICAL PROCEDURE FOR TOTAL NITROGEN

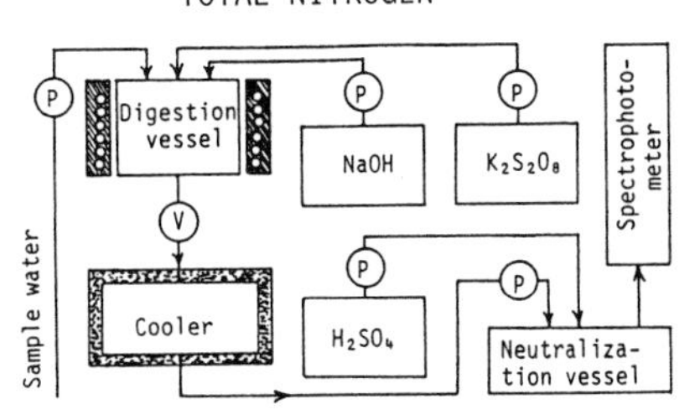

Fig.6 ANALYTICAL PROCEDURE FOR COD

Fig.7 ANALYTICAL PROCEDURE FOR SILICATE

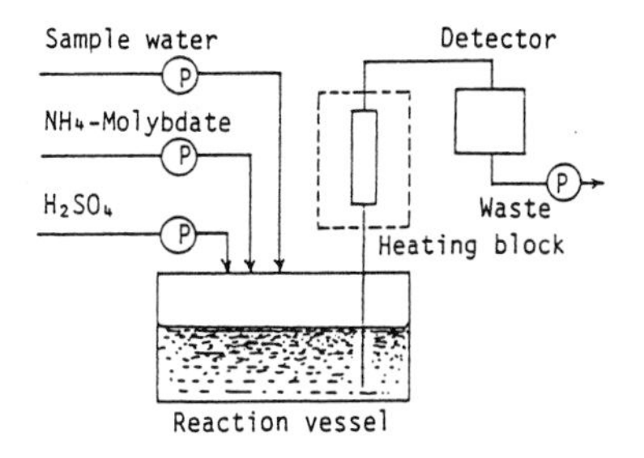

An automatic filtering unit was also constructed. Commercially available instruments were used for the measurement of other water quality parameters (Table 1).

Observation stations

An observation station was constructed as a semi-submerged buoy. The size of the buoy was 6.0x4.5x4.7m (Fig. 8). A transportable barge of 5.4x2.8x2.2m size, was also constructed (Fig. 9). Electric power was supplied by two diesel engine electric power generators. The sations were equipped with telemetering systems.

RESULTS OF MEASUREMENTS

The Aqua Survey System, composed of an observation station and a shore station, was successfully operated during approximately one year(1987). It was found that the concentration of nutrients changed rapidly (on time scales of one hour). These are to our knowledge the first continuous measurements of water quality parameters, including nutrients and chlorophyll-a, reported in the literature.

The concentration of chlorophyll-a changed periodically during the

Fig.8 MOORED SEMI-SUBMERGED BUOY

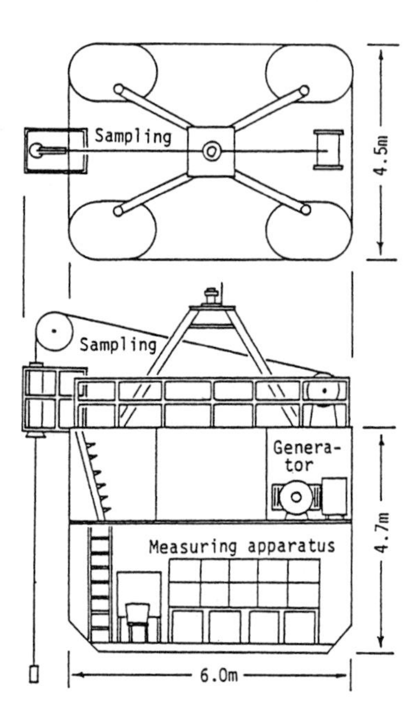

Fig.9 TRANSPORTABLE BARGE

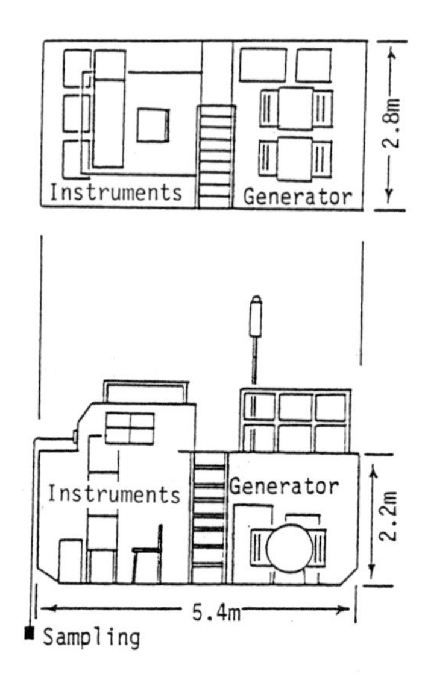

period when red tides were observed with an aircraft (Fig. 10). The concentration of chlorophyll-a was highest during the daytime at the depths of 0.5m and at 5m, and was lowest at 17m. These results obviously show a vertical migration of plankton diurnally.

The maintenance of the automatic instruments was repeated every two weeks. It was found that the precision of the instruments was within 15%. A 1/36 model of the observation station was made and stability of the model was tested. It was found that the station could be used when wave height was less than 4.5m. The cost of an Aqua Survey System is about 80,000,000 yen and the cost of the maintenance is about 10,000,000yen for one year.

Fig.10 PERIODICAL CHANGE OF CHLOROPHYLL-a

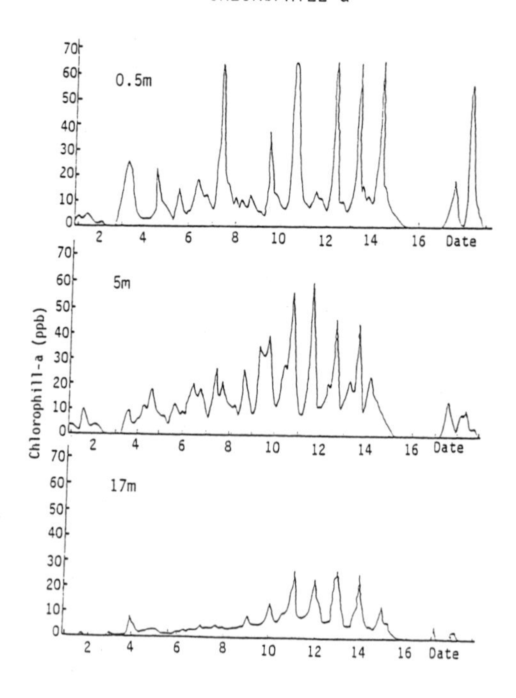

AN IN SITU INSTRUMENT TO MEASURE WATER QUALITY IN THE BENTHIC BOUNDARY LAYER

Tanimoto, T., K. Kawana and A. Hoshika
Government Industrial Research Institute, Chugoku
2-2-2 Hiro-Suehiro Kure Hiroshima, 737-01, Japan

ABSTRACT

An in situ instrument has been developed to make continuous measurements of the vertical distribution of water quality in the bottom water and various parameters of the bottom sediment. The instrument is equipped with specially designed elevators which have in situ sensors for salinity, temperature, dissolved oxygen, turbidity, pH, Eh and water content of sediment. After the instrument is placed on the sea bottom, the sensors are lowered from about 1 m above the bottom to 30 cm below the sediment by moving the elevators vertically. By this procedure, the vertical distribution of some parameters are obtained continuously from the bottom water layer into the sediment.

INTRODUCTION

Oxygen-deficient bottom water is observed in summer in the regions of limited water movement in the Seto Inland Sea, Japan. It contains large amounts of nutrients and some heavy-metals derived from the bottom sediment (Kawana et al., 1980; Shiozawa et al., 1984; Ochi and Takeoka, 1986). Nutrient-rich bottom water is thought to be one of the causes of outbreaks of red tides. In order to clarity the dissolution mechanisms from the bottom sediment, it is important to know vertical gradients of environmental parameters at the sediment-water interface in detail. We have designed and built an in situ instrument to measure water quality in the benthic boundary layer. The vertical distribution of water quality in the bottom water and various parameters of the bottom sediment are obtained continuously by the instrument. We describe the instrument here.

DESCRIPTION OF THE INSTRUMENT

An in situ instrument (Fig. 1 and Photo. 1) has been designed to operate on the sea bottom. It consists of three major components: a cylindrical outer frame and specially designed main and secondary elevators. The outer frame (1 in Fig. 1) is made up of eight legs and the horizontal structural members which join the legs. The elevators are installed in the outer frame with various in situ water quality sensors. The vertical distribution of water quality can be obtained by moving the main elevator (2 in Fig. 1) vertically. The vertical motion of the main elevator is carried out by rotation of the central axis (6 in Fig. 1). Rotation of the axis is achieved by a motor (4 in Fig. 1) mounted at the top of the outer frame. The secondary elevator (3 in Fig. 1) is fitted in the main elevator and is used to measure various parameters in the bottom sediment. Vertical motion of the secondary elevator is achieved by a second motor (5 in Fig.

RED TIDES: BIOLOGY, ENVIRONMENTAL SCIENCE, AND TOXICOLOGY
Okaichi, Anderson, and Nemoto, Editors

1). The main elevator has the in situ sensors for salinity, turbidity and dissolved oxygen. The sensors for Eh, temperature, pH, and water content of sediment are attached to the secondary elevator. The turbidity sensor uses a transmissometer technique of light attenuation (Kawana and Tanimoto, 1981). The water content of sediment is measured by ultrasonic attenuation (Tanimoto et al., 1986). The ultrasonic path between transducers is 3 cm and the wave frequency is 2 MHz.

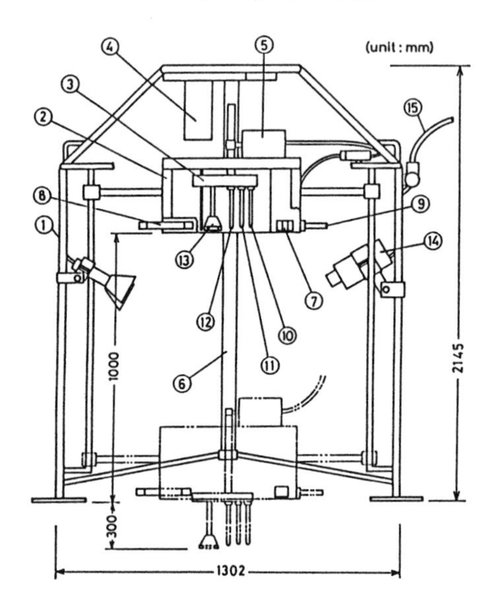

FIG. 1. Schematic view of in situ instrument to measure water quality in the benthic boundary layer.
1: outer frame, 2: main elevator,
3: secondary elevator,
4: main motor, 5: auxiliary motor,
6: central axis,
7: salinity sensor,
8: turbidity sensor,
9: dissolved oxygen sensor,
10: Eh sensor, 11: pH sensor,
12: temperature sensor,
13: water content of sediment sensor, 14: T.V. camera,
15: electrical cable.

PHOTO. 1. In situ instrument to measure water quality in the benthic boundary layer.

FIELD-TEST

The field-test was carried out in summer at Stn. 1 which is located in the eastern region of Hiuchi-Sound (Fig. 2). The horizontal distribution of dissolved oxygen concentration at 0.5 m height above the sea bottom floor is also shown in Fig. 2. Oxygen-deficient bottom water (concentration less than 3 mg/l) is observed in the eastern region.

The water depth of Stn. 1 is about 25 m and the ship was fixed at Stn. 1 by three anchors. The instrument is placed gently on the bottom from the ship. Both elevators are located in the upper part of the frame at the beginning of measurement. All sensors are located about 1 m above the sea floor. After a few minutes to allow disturbance of the bottom to subside, both elevators are slowly lowered to the bottom. Next, the secondary elevator is lowered 30 cm below the sediment surface. By this procedure, the vertical distributions of some parameters are obtained continuously from the bottom water layer into the sediment.

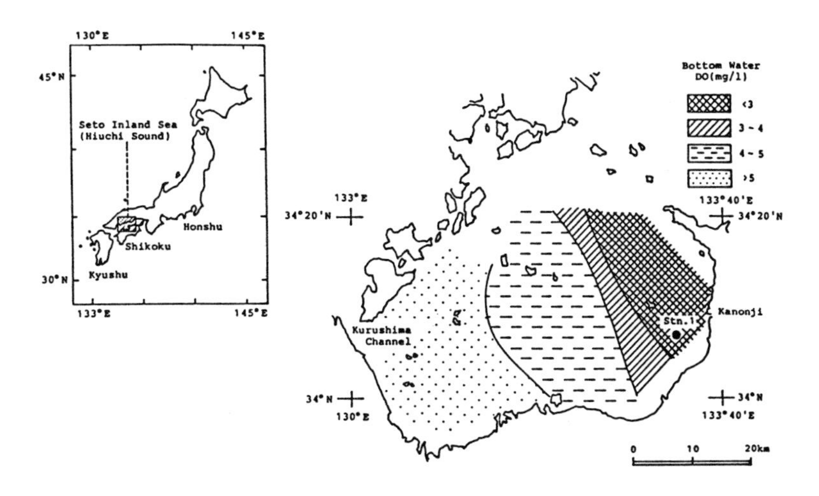

FIG. 2. The location of the field-test and the horizontal distribution of dissolved oxygen concentration at 0.5 m height above the sea bottom in Hiuchi-Sound in summer.

RESULTS AND DISCUSSION

The vertical distribution of environmental parameters from the bottom water layer into the bottom sediment is shown in Fig. 3, measured by the instrument. The dotted line in Fig. 3 shows the sediment surface. The concentration of dissolved oxygen (DO in Fig. 3) is 2 mg/l at about 1 m height above the bottom and decreases to 1 mg/l at the interface. The other environmental parameters are uniform vertically. The temperature (T in Fig. 3) in the bottom sediment decreases gradually with increased distance from the interface. The water content (W.C. in Fig. 3) and Eh decrease suddenly at the interface. We can obtain continuous vertical distributions of some parameters from the bottom water layer into the sediment. It is thought that the instrument introduced here is applicable to events at the sediment-water interface.

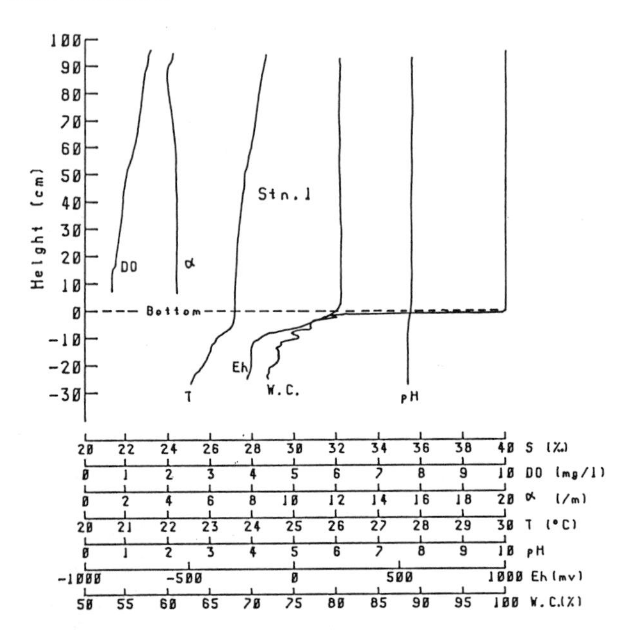

FIG. 3. Vertical distribution of some of environmental parameters from the bottom water layer into the sediment at Stn. 1 in Hiuchi-Sound in summer.
S: salinity, DO: concentration of dissolved oxygen,
α: beam attenuation coefficient, T: temperature,
W.C.: water content of sediment.

REFERENCES

1. K. Kawana, T. Shiozawa, A. Hoshika, T. Tanimoto and O. Takimura, La Mer, 3, 131, (1980)
2. T. Shiozawa, K. Kawana, Y. Yamaoka, A. Hoshika, T. Tanimoto and O. Takimura, Report Gov. Indus. Res. Insti. Chugoku, 21, 13, (1984)
3. T. Ochi and H. Takeoka, J. Oceanogr. Soc. Japan, 42, 1, (1986)
4. K. Kawana and T. Tanimoto, J. Oceanogr. Soc. Japan, 37, 173, (1981)
5. T. Tanimoto, K. Kawana and A. Hoshika, Report Gov. Indus. Res. Insti. Chugoku, 27, 35, (1986)

SIMULATION OF THE AREA OF ACCUMULATION OF CHATTONELLA RED TIDE IN HARIMA NADA, THE SETO INLAND SEA OF JAPAN

M. UCHIYAMA, Y. IWATA, A. OKUZAWA AND M. ISHIDA
Fuyo Data Processing & Systems Development, Ltd., Tokyo, Japan

ABSTRACT

The investigation of the accumulation mechanism of red tide organisms in Harima Nada was carried out by a numerical method. The results indicate that accumulation by flow is important.

INTRODUCTION

In Harima Nada, the eastern part of the Seto Inland Sea, red tides of Chattonella antiqua have often been observed over large expanses. However, areas with dense Chattonella populations are usually limited to particular parts of Harima-Nada. In particular, very dense red tides are commonly observed in the coastal water of Shikoku (Fig. 1).

Concerning the establishment of red tide, the importance of factors such as cell division or local hydrography has not been adequately specified. In order to document the importance of these factors, a simulation of the physical transfer of plankton cells was made using numerical methods.

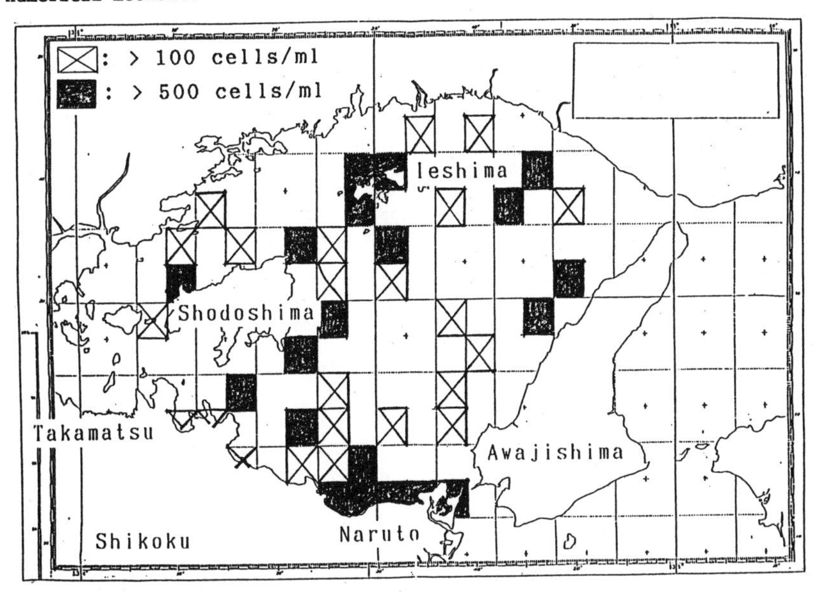

FIG. 1. Occurrence of areas with dense Chattonella populations in Harima Nada, 1983

RED TIDES: BIOLOGY, ENVIRONMENTAL SCIENCE, AND TOXICOLOGY
Okaichi, Anderson, and Nemoto, Editors

METHODS

The transfer or movement of plankton cells can be divided into three different processes:

a. Transfer by regular flow, i.e. tidal and wind currents.
b. Transfer by irregular flow, i.e. diffusion.
c. Locomotion of the plankton themselves, mainly in the vertical direction.

Except for the third process, the increase in the density of Chattonella can be represented by the following equation:

$$\frac{\partial C}{\partial t}=\left\{\frac{\partial}{\partial x}\left(Kx\frac{\partial C}{\partial x}\right)+\frac{\partial}{\partial y}\left(Ky\frac{\partial C}{\partial y}\right)+\frac{\partial}{\partial z}\left(Kz\frac{\partial C}{\partial z}\right)\right\}-\left\{\frac{\partial (uC)}{\partial x}+\frac{\partial (vC)}{\partial y}+\frac{\partial (\omega C)}{\partial z}\right\} \quad (1)$$

In this equation, u, v, w are water velocities in the x, y and z-directions, and Kx, Ky and Kz are diffusion coefficients which indicate the scale of irregular movement of seawater. We modified this equation by the finite difference method, and calculated the increase of cell density. The sea area of Harima-Nada was divided into a grid for these calculations. An assumption was made that seawater velocity and cell density are uniform within one grid cell. In this calculation, only tidal currents were considered, and it has been assumed that the period of the tidal currents is 12 hrs. Water velocity was calculated using another simulation model [1].

The vertical layering of the water column was as follows:

a. Top layer : from 0-5 m.
b. Middle layer : from 5-25 m.
c. Bottom layer : deeper than 25 m.

The vertical migration of Chattonella was not considered in the present analysis. The assumption was made that Chattonella cells are distributed in the top layer all the time and that these cells accumulate where sinking occurs. On the other hand, cells are dispersed in areas with upwelling. For cell division, we applied two cases, i.e. one without division and the other with one division per day.

Conditions at the beginning of calculation were:

a. Chattonella cells present in uniform density, 1 cell/ml in the top layer throughout Harima-Nada.
b. No Chattonella cells found outside Harima-Nada.

RESULTS

Figure 2 shows the distribution of cell density two days after initiation. High density areas were found around the northeast coast of Shodo-shima, the coastal water of Shikoku, the southeast coast of Ieshima island, and the northern area of Naruto Straits. Figure 3 shows the distribution of cells four days after initiation. The high density areas did not change in number relative to the shorter simulation, but did grow larger. In particular, the high density area in the coastal water of Shikoku extended toward the Naruto Straits. These high density areas agree with the areas where red tides have actually been observed.

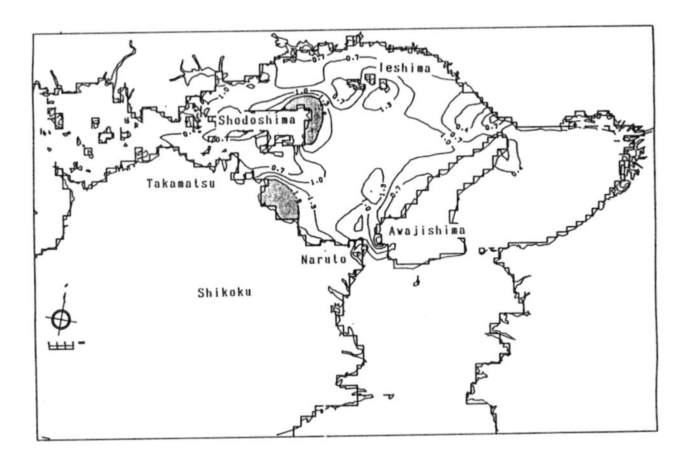

FIG. 2. Distribution of cell density two days after initiation, in the case without cell division.

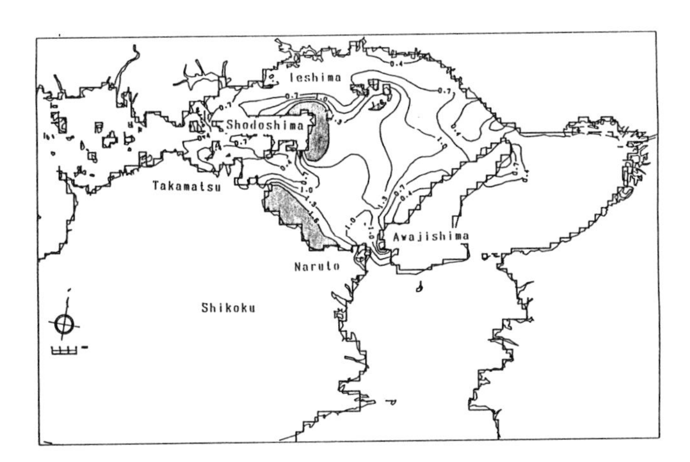

FIG. 3. Distribution of cell density four days after initiation, in the case without cell division.

Figure 4 reveals the distribution of cells two days after initiation with cell division taking place. Compared to the results without cell division, the high density areas are the same. However, cell density in the dense areas was four times higher than in the case without cell division, clearly demonstrating the importance of cell division.

Figure 5 shows the distribution of cell density four days after initiation. The cell density was 16 times higher than in the case without cell division.

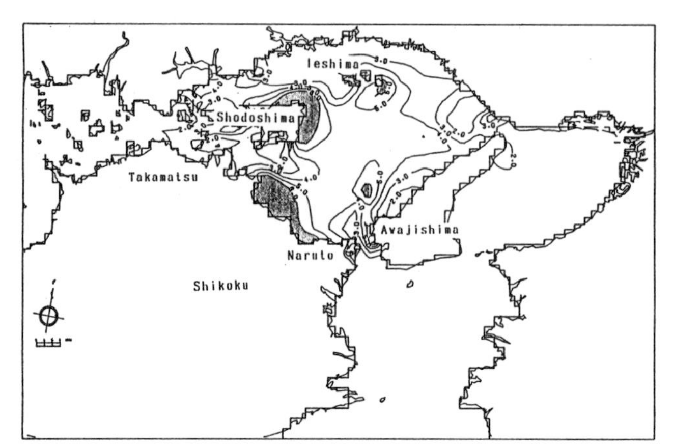

FIG. 4. Distribution of cell density two days after initiation, in the case with one cell division per day.

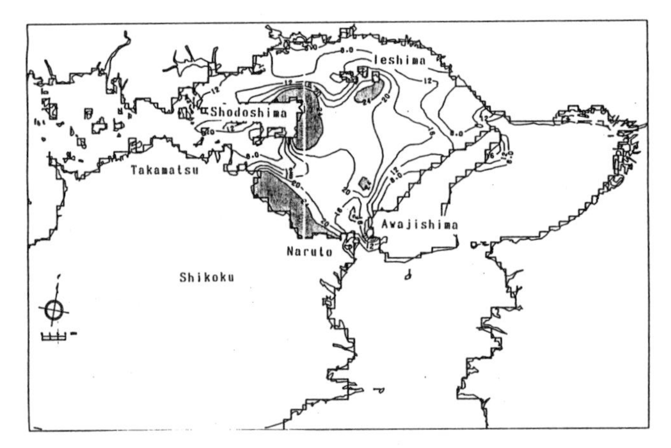

FIG. 5. Distribution of cell density four days after initiation, in the case with one cell division per day.

SUMMARY

Based on the present calculations, it is clear that physical accumulation of cells by local hydrography is important in producing red tides, although the cell density is affected by the rate of cell division. Thus, we presume that both cell division and accumulation by flow must be considered in predicting red tides. In this study, wind effects were not considered. Further investigations are needed with this and other complex environmental factors.

REFERENCES

1. Y. Yoshida and K. Numata, Bull. Japan. Soc. Sci. Fish. 48, 1271-1275 (1982).

NUMERICAL SIMULATION OF RED TIDE AND SENSITIVITY ANALYSIS OF BIOLOGICAL PARAMETERS

Kishi, M. J.* and S. Ikeda**
* Ocean Research Institute, Univ. of Tokyo, Tokyo, 164 Japan
*Institute of Socio-Economic Planning, Univ. of Tsukuba, Ibaraki, 305 Japan

ABSTRACT

This study is concerned with the identification of a model structure of phytoplankton dynamics associated with the outbreak of red tide (*Chattonella antiqua*) in Harima-Nada, Seto Inland Sea, Japan.

A basic one-dimensional model is constructed to demonstrate the importance of the vertical migration of plankton to the outbreak of red tide. The sensitivity of model parameters on the phytoplankton bloom is discussed by means of "statistical principal component analysis". The statistical result also supports the importance of vertical migration and the strong effect of grazing pressure.

A three-dimensional simulation model of two layers in vertical direction is also constructed. The numerical result of the model simulation shows that the tidal and wind-induced currents have a great impact on phytoplankton dynamics.

INTRODUCTION

Recent studies have shown that the oceanographic characteristics of the water have a strong influence on the generation mechanism of red tides. (Iizuka [1], Yanagi [2], and Okaichi [3]).

A simple one-dimensional model is constructed not only to examine the importance of vertical migration of both zooplankton and *Chattonella antiqua* to the formation of a red tides, but also to investigate the role of grazing and/or limiting nutrients on the initial formation of a red tide. As to the vertical migration of *C. antiqua*, Nakamura and Watanabe [4] remarked on its importance to the formation of a "red tide" through their experiment in "micro-cosm" culture. Okaichi *et. al.* [5] pointed out that the existence of inorganic iron in Harima-nada may be essential to the rapid growth of red tide of *C. antiqua*.

Ikeda and Kishi [6] reported that both tidal and wind-driven currents with vertical migration of plankton explain the spatial and horizontal distribution of phytoplankton patchiness. In this paper, we discuss the results of numerical experiments on the outbreak of red tide on the basis of our simulation models [6] calculating the horizontal and vertical profile of *C. antiqua*.

NUMERICAL EXPERIMENTS WITH A ONE-DIMENSIONAL MODEL

Model Structure

In order to simulate the oceanic conditions of Harima-nada in summer, the sea depth was set to 40m (representative depth of Harima-nada). The material cycle in the ecosystem model here is shown in

RED TIDES: BIOLOGY, ENVIRONMENTAL SCIENCE, AND TOXICOLOGY
Okaichi, Anderson, and Nemoto, Editors

Fig. 1. Chattonella antiqua is the red tide species and Skeletonema costatum the competition species (Iizuka [7]. NO_3-N, PO_4-P, and inorganic iron (Fe) are considered limiting nutrients (Okaichi [4]. Paracalanus parvus is a representative zooplankton species in Harima-nada (Uye [8]). The values in the circles in Fig. 1 are the representative observed values in Harima-nada when red tide is not observed.

The photosynthetic and grazing functions are approximated by a Michaelis-Menten type relationship. The values of biological parameters of the equations have been experimentally evaluated in another project (entitled "Studies on Outbreaks of Red Tides in Neritic Waters", principal investigator : Prof. T. Okaichi of Kagawa Univ. funded by special research programs on environmental science, Ministry of Education, Japan 1982-84) as shown in Fig 1.

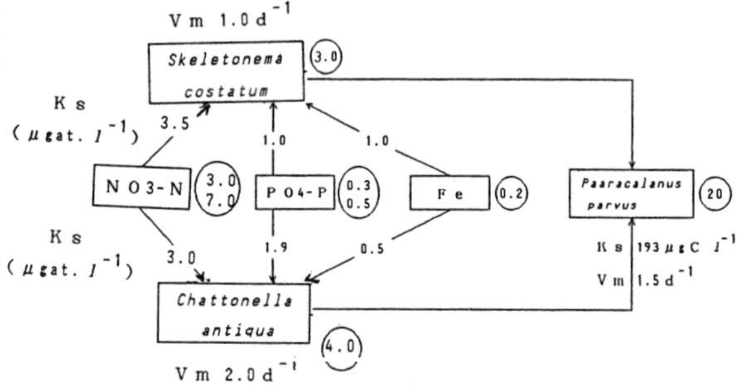

FIG. 1 Model Structure

Case 1 : Without vertical migration of Chattonella antiqua

Figure 2 shows vertical profiles 72 and 78 hours after the initial state without vertical migration of C. antiqua. As grazing pressure is smaller than photosynthesis, neither population of C. antiqua nor S. costatum decrease. In this case red tide cannot be generated.

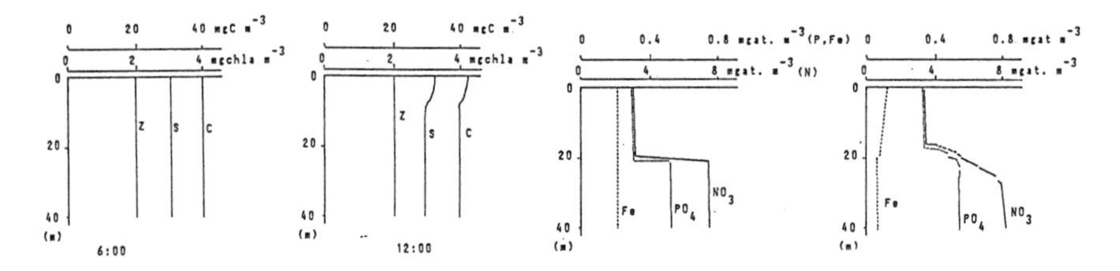

FIG. 2 Vertical profiles in Case 1. (Z: P. parvus, S: S. costatum, C: C. antiqua)

Case 2 : With vertical migration of Chattonella antiqua

Nakamura and Watanabe [5] reported that C. antiqua moves upward in the morning and downward during the night. Okaichi et al [4] observed the maximum level of chlorophyll a of C. antiqua at 20 m depth during the night. The migration velocity of C. antiqua was set at 1 m h^{-1}. Figure 3 shows the time-dependent vertical profiles of each compartment. A high concentration of C. antiqua occurs at the surface due to vertical migration. This plankton patch consumes the inorganic iron in the surface layer, and the photosynthetic capacity is limited by the lack of inorganic iron. Thus the red tide occurs during daytime of the first two days, decreasing gradually due to nutrient limitation, diffusion and the grazing pressure in this ecosystem.

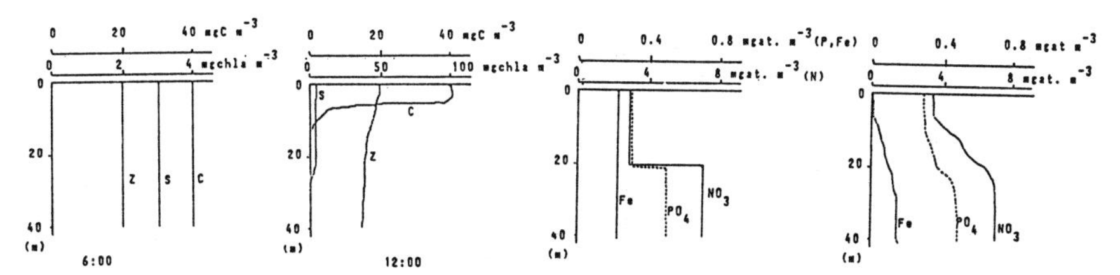

FIG. 3 Vertical profiles in Case 2.

Principal component analysis of biological parameters

A sensitivity analysis was performed on the model using Monte Carlo analysis based on the method proposed by Bax (1982) [9]. After perturbation in biological parameters of the model by means of the Monte Carlo method, we used principal component analysis to reduce a large number of output variables to a more tractable number of principal compononents (P.C.1-P.C.4). Table 1 shows the factor loading and eigenvalues. The first component (P. C. 1) is related to grazing by *Paracalanus parvus*. The second one (P. C. 2) is related with nutrient uptake. This table shows that the grazing parameters play important roles in the perturbation of the ecosystem.

Table 1: Factor pattern 1st principal component to 4th Principal component. (PL, NL, FL, CL, SL, ZL: Concentration at 5m layer at 12:00 on 3rd days from initial state (PO_4-P, NO_3-N, Fe, *C. antiqua*, *S. costatum*, *P. parvus*, respectively). PI-ZI: Initial values. CKSP, CKSN, CSF: Half saturation const. of *C. antiqua* for PO_4-P, NO_3-N, Fe. SKSP, SKSN, SKSF: Same but of *S. costatum*. ZKS: half saturation const. of *P. parvus* for phytoplanktons. ZVM: Maximum grazing ratio. BET: Ratio to biomass of urine by *P. parvus*.)

	P. C. 1	P. C. 2	P. C. 3	P. C. 4
PL	0.108611	0.606114	-0.612914	0.325541
NL	0.051880	-0.705309	0.208747	0.478918
FL	0.145930	0.613271	0.621610	0.095843
CL	-0.975959	0.068962	-0.052543	0.003883
SL	-0.962969	-0.036490	-0.129045	0.028230
ZL	0.973953	-0.055271	0.048014	-0.069114
PI	0.103898	0.588377	-0.624357	0.327123
NI	0.066878	-0.683569	0.226851	0.498292
FI	-0.231279	0.365334	0.347327	0.289331
ZI	0.555990	-0.266905	-0.080084	-0.314775
CKSP	0.252321	-0.124474	-0.233934	0.489967
CKSN	0.080311	-0.036075	0.070857	-0.734222
CSF	0.037285	0.495167	-0.031514	-0.176432
SKSP	0.043466	0.133995	-0.176405	-0.191695
SKSN	0.026639	-0.494392	-0.272147	-0.159785
SKSF	0.135349	0.490076	0.574930	-0.032448
ZKS	-0.162851	-0.002916	0.026391	0.201942
ZVM	0.651232	0.288574	0.119036	0.356719
BET	-0.319539	0.146489	0.514295	0.130847
EIGEN VALUE	3.887208	3.137522	2.171520	1.946974
PROPORTION	20.5%	16.5%	11.4%	10.2%
CUMULATIVE PROPORTION	20.5%	37.0%	48.4%	58.6%

NUMERICAL EXPERIMENTS WITH THE THREE DIMENSIONAL MODEL

Figure 4 shows the calculated tidal plus wind-driven (by westerly wind) currents of the upper layer. The horizontal distributions of biological components are calculated using these currents. Figure 5a shows the horizontal distribution of inorganic nitrogen. A high concentration area can be seen at the mouths of rivers. Figure 5b shows the distribution of S. costatum. The high density also can be seen at the river mouth where the nutrient loading occurs. Figure 5c shows the distribution of C. antiqua as it migrates in the vertical direction. These results coincide with observations of Okaichi [3] .

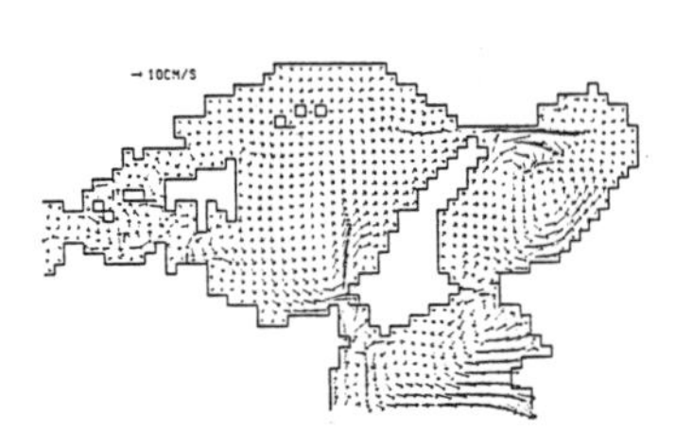

FIG. 4 Calculated tidal residuals.

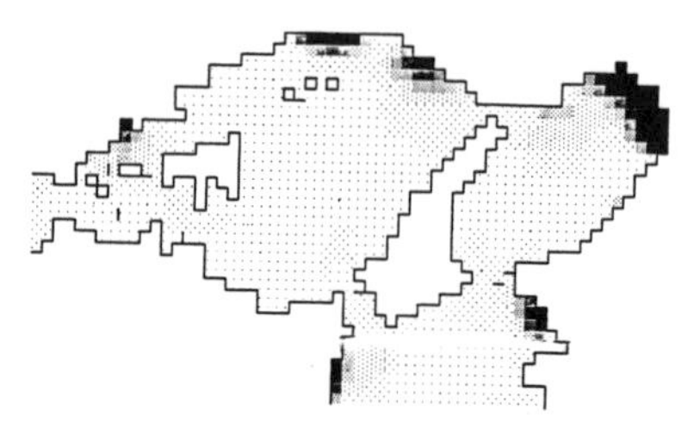

FIG. 5 a) Horizontal distribution of NO_3-N

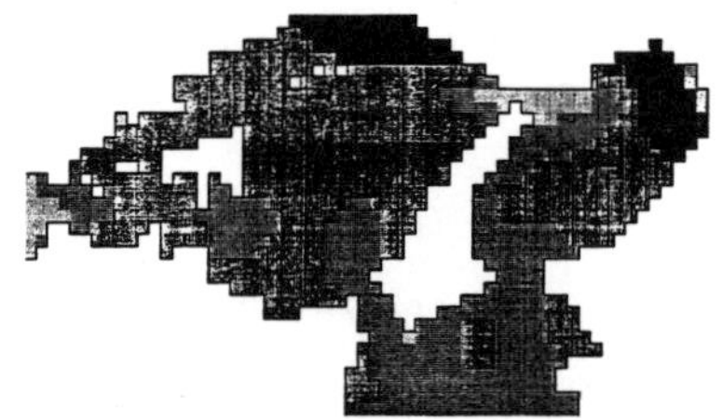

FIG. 5 Horizontal distribution of b) S. costatum, c) C. antiqua.

0 30 36 42 48 54 60 66 72 78 84 90 999 x10⁰ MICRO-G/L

DISCUSSION

A red tide by Chattonella antiqua does not appear at times when there is no vertical migration of C. antiqua because of limitation by inorganic iron in the surface layer and the grazing pressure of P. parvus. By introducing the vertical migration of C. antiqua, a red tide occurs even when there is no surplus nutrient. Competition between S. costatum and C. antiqua does not play an important role in the outbreak of a "red tide" in this model.

ACKNOWLEDGMENTS

The authors thank Dr. S. Ue of Hiroshima Univ. for his helpful discussions, and staff of Fuyo information center (Tokyo).

REFERENCES

1. Iizuka S., "Red tide" Jpn. Soc. Fish. Ser. 34, Koseisha Koseikaku Publ. Japan, (1980).
2. Yanagi T., La Mer., 20, 170-177, (1982).
3. Okaichi T., Jour. Oceanogr. Soc. Japan. 39, 267-275, (1983).
4. Okaichi T., Res. Rep. Environ. Sci., B264-R17-1. Ministry of Education Japan, (1985).
5. Nakamura and Watanabe Jour. Oceanogr. Soc. Japan, 39, 110-114, (1983).
6. Ikeda, S. and M. J. Kishi, Ecol. Mod. 31, 145-174, (1986).
7. Iizuka, S., in Res. Rep. Environ. Sci., B264-R17-1. Ministry of Education Japan, (1985).
8. Uye, S., Mar. Biol. 92, 35-43, (1986).
9. Bax, N., Report for NOAA (USA), NMFS Contract No. 82-ABC-145, (1982).

MONITORING ALGAL BLOOMS, THE USE OF SATELLITES AND OTHER REMOTE SENSING DEVICES

CHARLES S. YENTSCH
Bigelow Laboratory for Ocean Sciences, West Boothbay Harbor, ME 04575

ABSTRACT

Because of the short time scales and large spatial dimensions of many blooms, techniques for remote sensing of blooms have been suggested. Remote sensing as used here refers to the quantitative assessment of the numbers of algae and the dimensions of the patch without directly sampling the water mass. Such a requirement necessitates observation from considerable heights. Instrumentation for satellites and aircraft have been developed which utilize the light absorption and/or light emitted as fluorescence from algae. Similar techniques could be used in conjunction with aquaculture projects. It is important to define what is being sensed remotely, and why and how to do it.

We now know that the appearance of a bloom at some fixed site can result from augmented growth of individual cells in the bloom, and/or a movement of a large patch of the bloom into the site by water movement. Being able to predict allows the observer to understand, to a first approximation, the cause of the bloom and to take actions to prevent destruction of aquaculture products.

INTRODUCTION

Most of us studying the phenomena of toxic algal blooms are asked, can anything be done? The answer is, "Yes" and "No". Once a bloom is established the overall size precludes the use of any destructive agents or mechanisms that would tend to disperse or contain the bloom. However, since many toxic blooms result from interaction of climatological and hydrographic processes, we can <u>predict</u> the occurrence of a toxic bloom. Such predictions are a benefit to commercial fisheries whose major concern is the quality of product.

The ability to predict the appearance of toxic algal blooms from oceanographic variables requires the observer to associate the outbreak with some environmental parameter such as temperature, freshwater runoff, etc. There is no theoretical or empirical means to account for the specific appearances of toxic species of organisms.

If prediction of toxic blooms is to be of use to fish managers then measurements of those algae must be part of an early warning system. Because of the large scales involved in both time and space, remote sensing by satellites, aircraft, or balloons seen to be likely vehicles as part of the early warning system. These vehicles must be equipped with sensors; the design or selection of which requires knowledge of general oceanographic factors known to augment algal blooms in general, the features that characterize the bloom and can be measured remotely.

RED TIDES: BIOLOGY, ENVIRONMENTAL SCIENCE, AND TOXICOLOGY
Okaichi, Anderson, and Nemoto, Editors

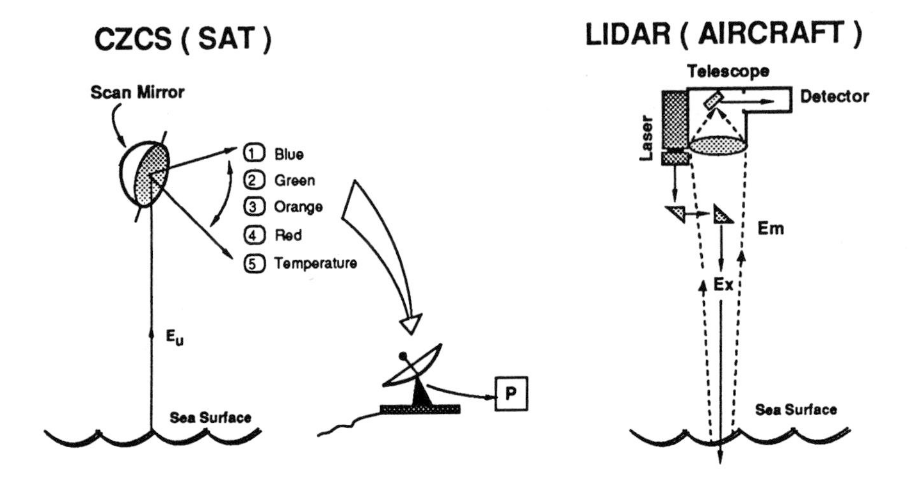

FIG. 1. Remote sensing instruments for ocean color and fluorescence. Satellite (SAT) colorimeter scans upwelled light from the sea. Colors are recorded and transmitted to ground station for processing (P). Lidar (aircraft) fluorometers use lasers to excite (Ex) the fluorescence (Em) from organisms in seawater. The fluorescence is measured by a photodetector.

Enrichment and accumulation produce changes of water color, fluorescence, sea temperature and salinity, and at times sea surface texture. Color, temperature and texture can be measured directly with remote sensing instruments, while salinity can be inferred from certain aspects of seawater color.

With the increase in either algae or dissolved organics, the visible light at short wavelengths is readily absorbed. This causes the color of the water to shift from blue (characteristic of clear open ocean water) to green (characteristic of many coastal regions). For those of us who have worked in the open ocean, where dissolved organic material is sparse, we have found it possible to estimate chlorophyll with a high degree of accuracy by utilizing shifts in water color which result from changes in algal abundance. In coastal areas, where detritus and the yellow substances are abundant, the problem is more complex and involves a competitive interference by these substances. Under these conditions estimates of each color component is difficult.

Measurement of fluorescence (Fig. 1) from the principle photosynthetic pigments of algae and dissolved organic material offer another means of remotely sensing colored substances in water. The primary advantage that fluorescence has over the direct measurement of color is the specificity of the signal. Fluorescent signals arise from the excitation of the molecule by a wavelength of light absorbed by that molecule. In all cases the light emitted as fluorescence is at a longer wavelength from that absorbed. For example, chlorophyll (the universal plant photosynthetic pigment) fluoresces red when exposed to blue light. Phycoerythrin fluorescence is orange when exposed to green light, and yellow substances fluorescence green when exposed to ultraviolet or blue light. This advantage of distinction has applications in separating water color into components, namely detritus versus phytoplankton.

With its inception, high altitude remote sensing was acknowledged as having an important role to the success of commercial fisheries, yet the use of satellite colorimetry for problems associated with red tide has come as an afterthought. The red tide need has arisen primarily as a need to know why certain areas of the world experience problems with toxic algae while other areas are relatively free of the problem. Out of this need to know has come one general principle which is worth repeating: toxic red tides, at least of the Paralytic Shellfish Poisoning variety, generally occur in regions where the overall plankton productivity is high. These are areas where natural nutrient enrichment processes are ongoing. Thus one is led to believe that problems associated with toxic shellfish are part and partial due to oceanographic processes that occur on planetary scales. Although these events occur with a high degree of regularity the growth activities that result from climatology and biology are not easily predicted. Hence there's a great deal of interannual variability which at the moment is not or cannot be reconciled in terms of any predictive scheme.

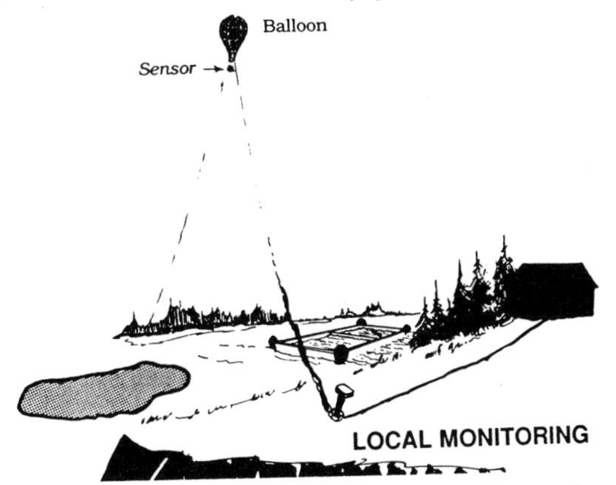

FIG. 2. Schematic balloon/sensor arrangement for early warning of blooms.

It is possible to obtain this type of information remotely and the instruments for that are at hand, but what is not clear are the cost to benefits of such systems: what aspects of the system are necessary? and what are redundant? As a vehicle, aircraft provides an ideal platform for monitoring local algal blooms from which visual and photographic observations can be made. Simple color radiometers can also be employed which are much more quantitive in assessing the magnitude of the bloom concerned. One can envision a number of systems that would become part of an early warning system, perhaps balloons over the entrance of the cove would be useful (Figure 2). National and cooperative local facilities utilizing remote sensing could be employed and the data disseminated in a fashion so that the fish farmer can be alerted to ensuing problems (Figure 3).

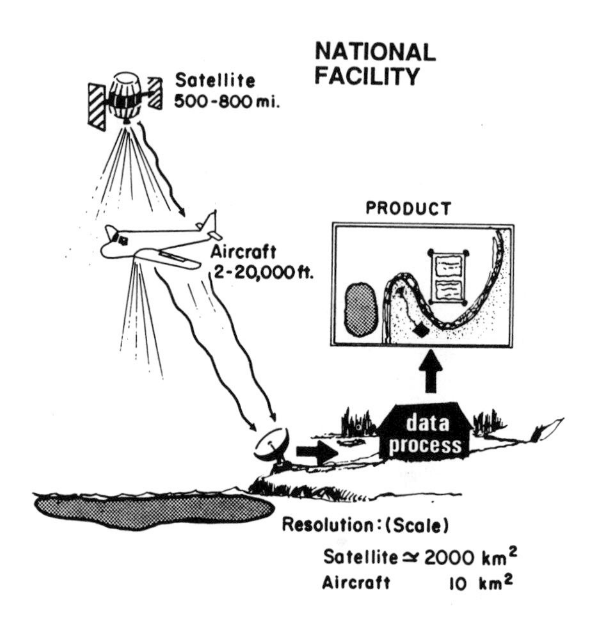

FIG. 3. National facility for remote sensing at toxic algal blooms.

It would appear to this writer that the information that is now needed is whether or not the farmers feel that going to this degree of expense is worth it in terms of protecting their farms. How should the information, if collected, be disseminated and in what political framework should this type of early warning system be placed?

REMOTELY SENSED PHYTOPLANKTON PIGMENT CONCENTRATIONS AROUND JAPAN USING THE COASTAL ZONE COLOR SCANNER

MATSUMURA SATSUKI,* HAJIME FUKUSHIMA,** YASUHIRO SUGIMORI**
*Far Seas Fisheries Research Laboratory, Orido, Shimizu-shi 424 Japan
**Faculty of Oceanography, Tokai University, Orido, Shimizu-shi 424 Japan

ABSTRACT

The phytoplankton pigment concentrations around Japan were mapped using data from the Coastal Zone Color Scanner (CZCS). Detailed mixing patterns could be inferred from the pigment maps. Highly reflective patterns were observed in each channel in an image from late spring. These were caused by coccolithophorids which have calcium carbonate shells.

Although mixing of Oyashio and Kuroshio water can easily be seen in a sea-surface temperature map, pigment and temperature patterns sometimes differ. Transects of both across the front show typical features, suggesting differences in the effect of vertical water motions on surface temperature and pigment concentrations.

INTRODUCTION

In this paper we will discuss the use of CZCS imagery in the study of oceanic pigment distributions. A typical example of the application of the NIMBUS-7 CZCS was the study of warm-core ring structure (Gordon [1]). The CZCS was the best satellite sensor for phytoplankton pigment, however it is no longer in operation. Fundamental to the formulation of a chlorophyll algorithm, Hovis[2] examined the spectral patterns of upward irradiance from waters of various chlorophyll concentrations. In this paper we will use "chlorophyll" or "pigment" to mean chlorophyll plus phaeophytin. Fig.1 clearly shows that the upward irradiance patterns vary strongly with chlorophyll concentration. Thus the concentration may be easily obtained from the data of each channel. The upper boxes in Fig.1

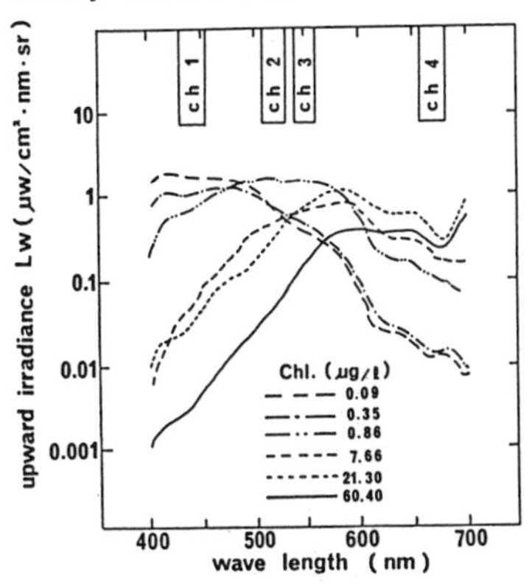

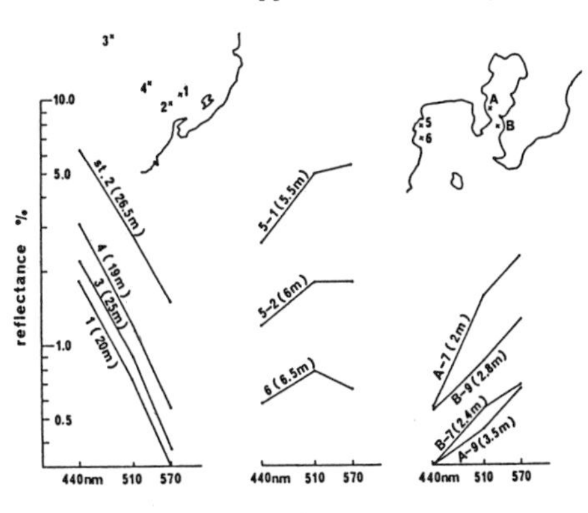

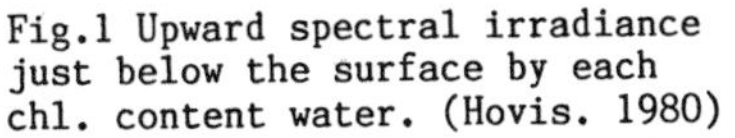

Fig.1 Upward spectral irradiance just below the surface by each chl. content water. (Hovis. 1980)

Fig.2 Surface upwelling irradiance spectra at each water type.

RED TIDES: BIOLOGY, ENVIRONMENTAL SCIENCE, AND TOXICOLOGY
Okaichi, Anderson, and Nemoto, Editors

show the bandwidth of each channel. Okami et al.[3] corrected the spectral irradiance data around Japan, and Matsumura[4] arranged it as shown in Fig.2. At left are spectral irradiance reflectance data from the Japan Sea. Here the water is very clear as shown by the numbers in parentheses (trasparency, m); thus the chlorophyll concentration must be very low. In the center panel are data from turbid coastal waters, and at right are data from the high chlorophyll Tokyo Bay water. The ratio of ch.1, 2 and 3 thus gives an index of chlorophyll content, as inferred from Fig.1 and 2.

CHLOROPHYLL ALGORITHM

Gordon and Morel[5] list the many workers who have tried to determine a chlorophyll algorithm. Fig.3 shows their regression lines(dashed lines). The relationship between chlorophyll content and spectral irradiance varies because of changes with season, region, and species. This is shown in Fig.3 by data from different regions around Japan. M,C,G,S, and S&W refer to Morel, Clark, Gordon, Sugihara and Smith & Wilson respectively.

The solid line in Fig.3 is the regression of our data from around Japan. Thus the chlorophyll content can be found from the spectral irradiance by;

$$\text{Log } C = \text{Log } a + b \text{ Log } R \quad (1)$$

$$R = ch1/ch2 \text{ or } ch2/ch3 \quad (2)$$

Unfortunately, our field data were not coincident with the CZCS images, so we used Gordon's algorithm as the first attempt.

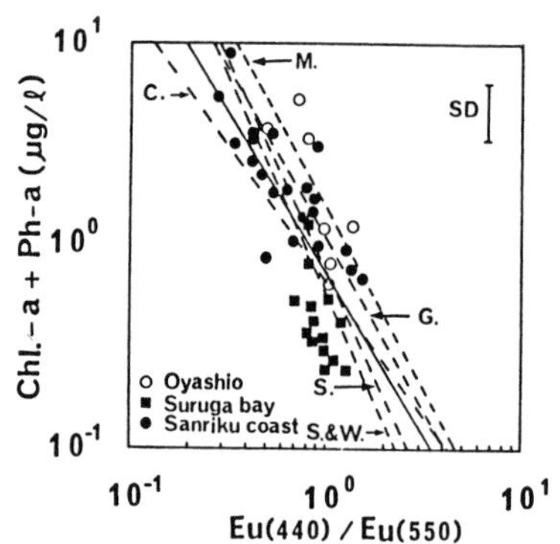

Fig.3 Regresion line from each observers and data around Japan.

CHLOROPHYLL DISTRIBUTION MAPS

Photograph 1 shows the chlorophyll distribution at the southeastern end of Honshu Island. The Kuroshio, which has low chlorophyll concentrations, is flowing from SW to E. The Oyashio and coastal water with high chlorophyll concentrations are trapped by a large frontal eddy. Even south of the Kuroshio, high chlorophyll concentrations can be clearly seen. Thus there must be eddies causing upwelling and vertical mixing to support the primary production.

Photograph 2 shows the pigment distributions off south Hokkaido. Oyashio water flows from the north with high-chlorophyll coastal water. After separating from the Hokkaido coast to the south, this water collides with Tsugaru water. This can be seen in the strong front south of Hokkaido The Oyashio water forms large eddies as it flows toward the south. The temperature map is almost the same as the pigment map in this case.

Photograph 3 shows patterns typical of a perturbed area. The eddies must cause vertical mixing, otherwise the narrow eddies would lose their shape due to lack of nutrient. Although the chlorophyll concentration was calculated using the ratio of chl, 2 and 3 according to eq.(1), the original data of each channel show high values compared to other case.

Photograph 4 shows the upward radiance of ch 1. The data from ch 2 and ch 3 show almost the same tendency. High reflectance at every channel indicates white suspended matter with chlorophyll. Clay is not likely since it is so far from land. Thus these substance must be coccolithophorids: phytoplankton with $CaCO_3$ shells. Holligan[6] compared CZCS data and coccolithophorid distributions along the outer margin of the north west European continental shelf,while Okada and Honjo[7] examined coccolithophorid distributions in the Pacific ocean. Comparing their work to these CZCS data, we determined that some of the eddy patterns must depend on coccolithophorids. Since we do not have ground truth data, we should attempt to create a new algorithm for coccolithophorids in the future.

Photo.1 Chlophyll distribution

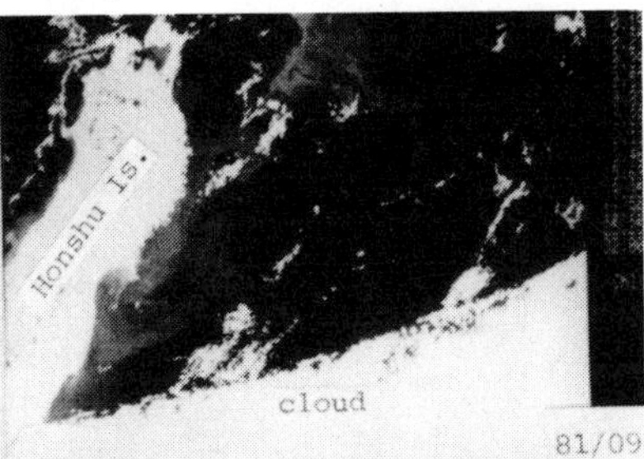

Photo.2 Chlorophyll distribut

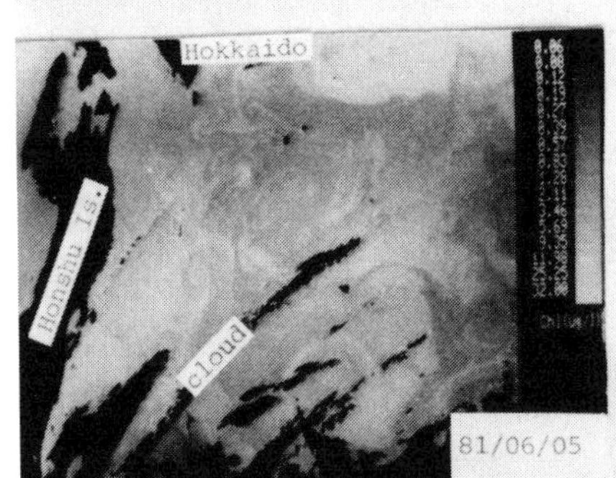

Photo.3 Chlorophyll distribution

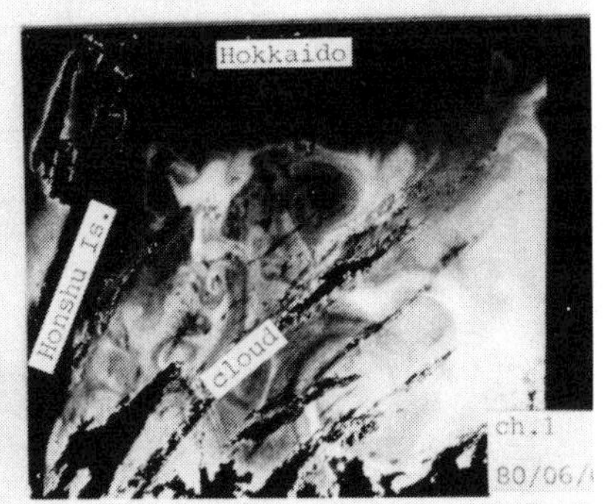

Photo.4 Upward radiance of ch

CHARACTER OF FRONT

The chlorophyll concentration maps show that the shape of the distribution is dependent on the mixing of water masses. Gradual changes in concentration are caused by diffusion; high chlorophyll eddies may be classified as follows:

a : Simple horizontal mixing or collision of two water masses. The mixing of the Oyashio (low temperature, high chlorophyll) with the Kuroshio (high temperature, low chlorophyll) appears the same shape in both temperature and CZCS map (Fig.4-a).

b : Convergence at a front. High chlorophyll results from retention in a convergence as opposed to being supported by nutrient pumping. The high concentrations gradually decrease due to lack of nutrients (Fig.4-b).

c : Eddy-induced upwelling. In this case the eddy dynamics support the increased primary production, and the shape of the eddy will continue as long as the nutrient supply continues (Fig.4-c).

The CZCS could also determine the sea surface temperature from ch 6. Fig.5 shows a W-E transect across the Oyashio branch near Kuroshio frontal eddy (see Photo 1). A high chlorophyll zone, F1, is located on the eastern margin of the warm water region. This suggests that the chlorophyll is a result of upwelling (see Fig. 4-c). F2 is located on the cold side of the front, and so must be the result of convergence (see Fig.4-b). F3 shows the opposite pattern, suggesting that it was formed by simple horizontal mixing(see Fig. 4-a).

Thus consideration of the hydrography is important when estimating primary production from chlorophyll maps. High chlorophyll concentrations do not always indicate high productivity, and coincident analysis of chlorophyll and temperature maps is an important tool for distinguishing mechanisms of chlorophyll production.

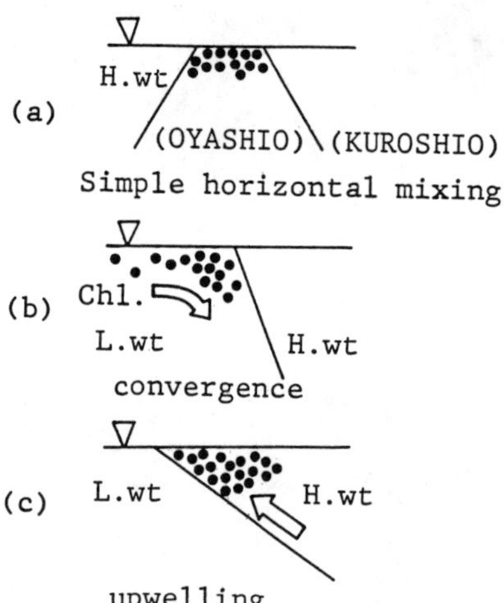

Fig.4 Schematic pattern of chl. front and water temperature front

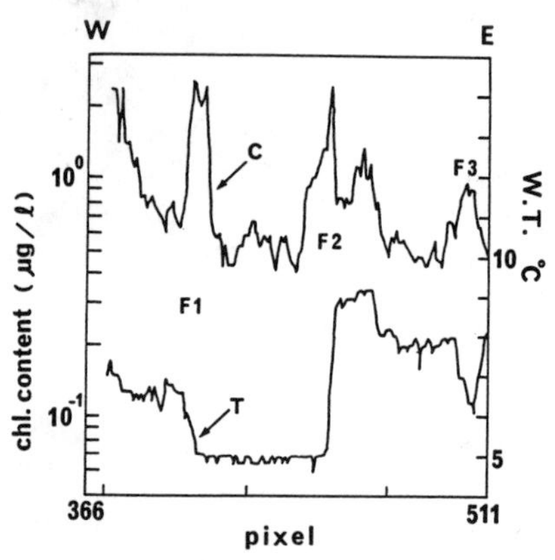

Fig.5 W-E profile of chlorophyll(C) and temperature(T)

ACKNOWLEDGMENTS

This work is partly supported by Research Division of Japan Fisheries Agency. Prof.Natsume, Tokai Univ. and Prof.Yasuda, Chiba Univ. were devoted partners of this work. Prof.Okaichi gave us the opportunity of presentation about this theme to this red tides symposium. The authors thank them all for their assistance.

REFERENCES

1. H.Gordon et al, J.Mar.Res.,40, (1982)
2. W.Hovis,Science, 210, (1980)
3. N.Okami et al, Tec. Report of the Physical Oceanography Laboratory, The Institute of Physical and Chemical Research,2 (1978)
4. S.Matsumura, Sora to Umi, 7 (Japanese with English abstract) (1985)
5. H.Gordon & A.Morel, "Remote Assessment of Ocean Color for Interpretation of Satellite Visible Imagery", Spring-Verlag (1983)
6. P.Holligan et al, Nature 304 (1983)
7. H.Okada & S.Honjo, Deep Sea Res., 20 (1973)

APPLICATIONS OF REMOTE SENSING TECHNIQUES FOR THE MAPPING OF RED TIDE DISTRIBUTION IN COASTAL AREAS

SHIROH UNO* AND MICHIO YOKOTA**
* Seikai Reg. Fish. Res. Lab., Kokubu-cho, Nagasaki, 850 JAPAN
** Asia Air Survey Co. Ltd., Nurumizu 13, Atsugi, 243 JAPAN

ABSTRACT

A remote sensing technique for the mapping of red tides in their early stages has been developed using phytoplankton pigments as an indicator. Pigments in near surface seawater were measured in the Seto Inland Sea by a MSS which was carried on an aircraft. Sea-truth data at 16-27 stations were simultaneously collected by three or four vessels within one hour. Analogue data of the MSS were digitized for image analyses, and the image that indicated the highest correlation with sea-truth data was employed. The most reliable procedure for the pigment distribution was determined to be channel (3 / 2+3+4+5+6+7+8+9). Thus, detailed distribution maps of phytoplankton pigments in near surface water were obtained. This procedure was not used during a period when the water contained a high concentration of inorganic materials. Under such conditions a Multi-Scattering Model was successfully employed.

INTRODUCTION

Since 1960, the frequency of red tides in the coastal areas of Japan have increased. Red tides are usually caused by dinoflagellates and inflict great damages to coastal fisheries, especially to yellowtail cultures. These dinoflagellate red tides do not have a distinct red colour, but are usually dull brown. Sometimes it is difficult to distinguish between a red tide and other types of coloured water. Information on red tide distribution in the early stages of development would be useful in protecting yellowtail cultures.

We have tried to map red tide distributions using airborne remote sensing techniques in coastal or inland sea areas.

EXPERIMENTAL

Remote sensing data were collected by MSS (Multi Spectral Scanner) of Daedulus DS-1250 carried on a Cessna 402B. The MSS has eleven channels; one for infrared, one for near-infrared and the other nine for the visible spectrum (Table I). At each experiment, sea-truth at 16-27 stations were collected by using three or four boats. Sampling was completed within ±30 minutes of the MSS data collection. Seawater samples at each station were

Table I. Range of Wave length at each channels of MSS (Daedulus DS-1250) used in the experiment.

Channel	1	2	3	4	5	6	7	8	9	10	11
Wave length in nm	380 | 420	420 | 450	450 | 500	500 | 550	550 | 600	600 | 650	650 | 690	700 | 790	800 | 890	920 | 1100	8000 | 14000

RED TIDES: BIOLOGY, ENVIRONMENTAL SCIENCE, AND TOXICOLOGY
Okaichi, Anderson, and Nemoto, Editors

taken from the surface and 1/2 Secchi depth. Equal volume of the two samples were mixed and used for the determination of the pigments (chlorophyll _a_ plus phaeopigments), and phytoplankton cell numbers.

RESULTS AND DISCUSSION

Calculated data of the MSS channels at each station were compared to sea-truth data. Different combinations of MSS channels were tested on a computer, and the formula that indicated the highest correlation between MSS data and sea-truth data was adopted. Experiments were done five times in Yashima Bay, Kagawa Pref., and two times in Uwajima Bay, Ehime Pref. (Table II). One experiment in Uwajima was omitted from Table II because of peculiar field conditions described below. Formulas 3/Σ, 7/Σ, 3/5 and 3-5 repeatedly gave higher correlation coeffecients. The best fit between MSS and pigment data was found with the MSS channel combination 3/Σ (Fig. 1). Based on this relationship, a detailed distribution map of phytoplankton pigments was obtained (Fig. 2). This MSS image was measured at 1km height at 11:30 on Aug. 31, 1977. Usually pigment concentrations were higher in the head of the bay, where dinoflagellate red tides have frequently occurred. Concentrations of >100 μg/l were observed in the central part and head of the bay, while low values, <10 μg/l were found around the mouth of the bay.

A high spatial resolution of pigment concentrations is possible using the MSS compared to traditional sampling of sea-water. If a pigment distribution map is constructed using data from stations at transects 500m apart, the pigment distribution pattern is markedly different from that based on MSS (compare Fig 2 and 3-a). When transects at 100m intervals are used, the pigment distribution more closely resembles the MSS image (Fig.3-b). However, it is practically impossible to sample stations at intervals of 100m in the area. Thus, the mapping of pigment

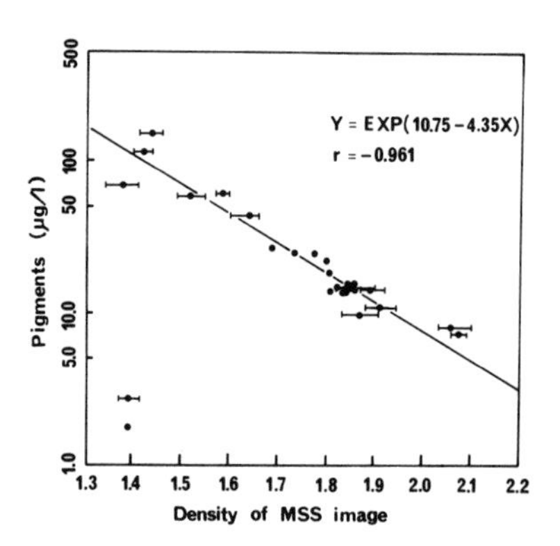

Table II. Correlation coefficients between MSS image density and sea-truth data of phytoplankton pigments by several methods in Yashima and Uwajima Bay.

Formula	3/Σ	7-(8+9)	4/Σ	7/Σ	3/5	3-5	3/7	4/7	3-7	8-5	7-5
Yashima											
1977	-0.961	-0.930	-0.898*	0.805*							
1978	-0.827		-0.756	0.895	-0.720						
1979	-0.779			0.741	-0.816	-0.811					
1980	-0.850			0.854			-0.893	-0.901			
1981	-0.581		-0.696	0.672		-0.592	-0.632		-0.947	0.865	0.845
Uwajima											
1979	-0.902			0.836	-0.877	-0.931					

(Σ : 2+3+4+5+6+7+8+9, * : pigments value was not treated to logarithm)

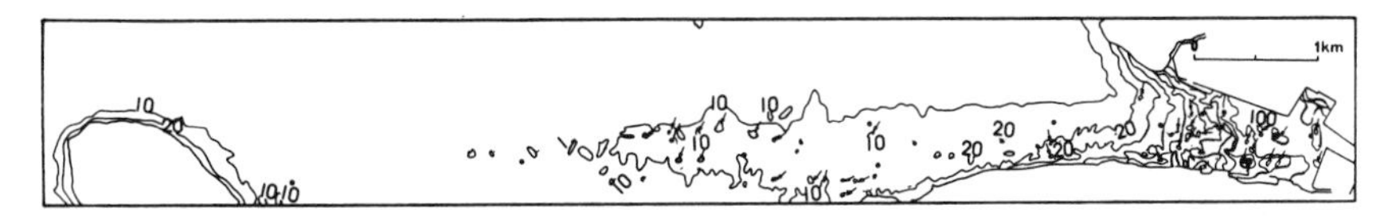

Fig. 2. Distribution map of phytoplankton pigments measured by MSS in Aug. 1977 in Yashima Bay. Concentrations in ug/l.

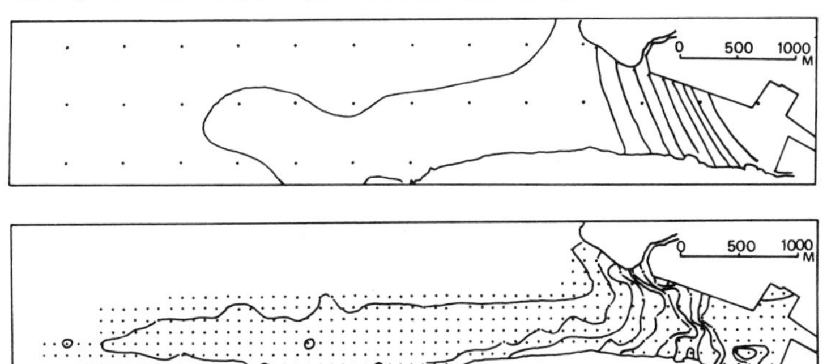

Fig. 3-a, 3-b. Distribution map of phytoplankton pigments when data are plotted for each 500m interval (3-a, upper), and for each 100m intervals (3-b,lower).

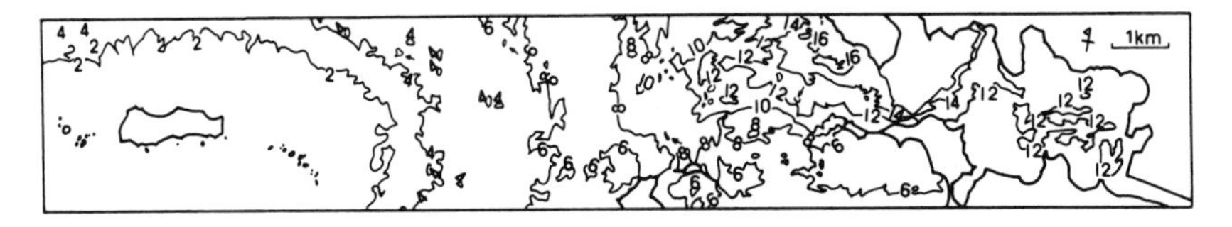

Fig. 4. Distribution map of phytoplankton pigments based on channel 2 of the MSS in Uwajima Bay in Aug. 1980. Numbers are indicated in ug-pigments/l.

Fig. 5. Distribution map of seston based on channel 4 of the MSS at the same time and place as Fig. 4. Numbers are in ug-seston (dry weight)/l.

Fig. 6. Distribution map of phytoplankton pigments estimated from MSS data with the Multi-Scattering Method.

Table III. Correlation coefficients between MSS image density for each channel and sea-truth data of seston and phytoplankton pigments in July 1980 at Yashima Bay, and in Aug. 1980 at Uwajima Bay.

MSS Channel	2	3	4	5	6	7	8	9
Yashima Bay								
Pigments	-0.668	-0.713	-0.697	-0.583	-0.375	0.041	0.699	0.715
Seston	-0.394	-0.429	-0.773	-0.270	-0.099	0.231	0.492	0.452
Uwajima Bay								
Pigments	0.559	0.457	0.369	0.315	0.255	0.179	0.078	0.032
Seston	0.909	0.958	0.965	0.958	0.954	0.917	0.836	0.730

distribution by remote sensing will be more effective than sea-truth data for synoptic studies over a wide area.

The MSS remote sensing method described above occasionally was not useful when the water contained high amount of suspended materials of non-plankton origin. In August 1980, an unusual situation occurred at Uwajima Bay. The day before, there was a heavy rainfall, and the water contained suspended soils discharged into the bay. The correlation coefficient between image density of MSS and phytoplankton pigments or seston dry weight in the bay was compared to that in the Yashima Bay (Table III). In Yashima Bay, under normal conditions, image density of all channels except 6 and 7 indicated higher correlation coefficients with phytoplankton pigments and lower with seston dry weight. In the Uwajima Bay, on the contrary, the coefficient was high with seston and low with pigments on all channels. The map of pigment distribution constructed with data from channel 2 and of seston with channel 4 (each channel gave the highest correlation coefficient) are illustrated in Fig. 4 and 5 respectively. The distribution pattern of pigments in Fig. 4 was not remarkably different from that of seston in Fig. 5.

Under these sea conditions with a high load of suspended solids of terestrial origin, another procedure for mapping the pigment distribution is available using a Multi-Scattering Model based on the theory of Hulbert(1) or Okami(2). Assuming three points; (i) that the spectral characteristics of phytoplankton pigments are not different between the two sea areas, (ii) that scattering light of both seston and phytoplankton are recorded by the MSS, (iii) that scattering light of seston have no apparent spectral characteristics, we constructed a pigment distribution map by the following procedure: Two distribution models of seston were made by using MSS data. One was made from a channel in which light is absorbed by the pigments. The other model was made from channels in which pigment absorption does not occur. Calibration of both models was done by employing sea-truth data at one station. Estimated values were calculated by subtracting the values of the latter model from former model. Then an estimated distribution map of phytoplankton pigments in the bay was constructed (Fig. 7). After these procedures, the correlation coefficient between image density and sea-truth was increased from 0.620 to 0.839.

REFERENCES

1. E. O. Hulburt, J. Opt. Soc. Amer., 33, 42-45 (1943).
2. N. Okami, Bull. Coast. Oceanogr. Japan, 15, 56-66 (1977) (in Japanese).

AIRCRAFT REMOTE SENSING FOR RED TIDE OBSERVATION

SATSUKI MATSUMURA* and MICHIO YOKOTA**
* Far Seas Fisheries Research Laboratory, Shimizu 424 Japan
** Asia Air Survey Co. Ltd., Atsugi 243, Japan

ABSTRACT

Red Tide observations in the Inland Sea were conducted by means of an airborne Multi Spectral Scanner (MSS). Two methods were used for analyzing the MSS data - an empirical method and an analytical method. Although the empirical method gives good results easily and requires no special optical knowledge, it does require a large set of sea-truth data. This need for numerous observations will limit the rapidity of responses in emergencies, and the time lag between remote sensing and marine observations in areas with strong currents may result in large errors. Aircraft observations with minimal need for sea-truth data are thus the preferred method of remote sensing. To this end, an optical model of upward irradiance for phytoplankton pigment concentrations was used. The remotely-sensed chlorophyll density maps which were constructed by this analytical model using only one sea-truth dataset were in good agreement with the data.

INTRODUCTION

Although it is well known that aircraft remote sensing can be a powerful tool for red tide monitoring, a technique for routine use has not yet been developed. As a first step, color photographs have been used from aerial overflights. In cases when a red tide patch has clear fronts, the scale of the red tide can be measured easily from the photograph, although it is difficult to quantify the density of the red tide plankton organisms. MSS is more useful than photography for many reasons. One of them is that MSS data are generated electronically and are, therefore, amenable to calibration. MSS data can be utilized in a numerical form by computer-assisted techniques which tend to be more quantitative than visual interpretations. In this study, chlorophyll pigments (Chlorophyl a and phaeopigments) were used as indices of red tide.

METHODS

The seawater samples were collected in Yashima Bay and Hiroshima Bay in the Seto Inland Sea by research vessels belonging to prefectural fisheries experimental stations. The water quality and optical properties of the sampled water were measured at each station. A MSS (Deadalus DS-1250) was mounted on an Aero-commander-685 aircraft.

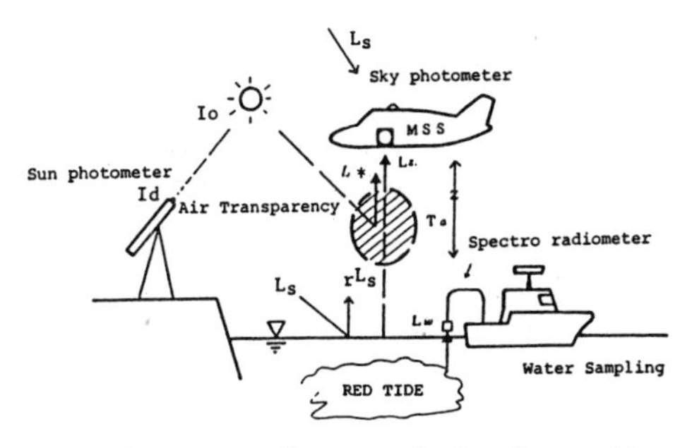

Fig.1 Schematic diagram of the observation system

RED TIDES: BIOLOGY, ENVIRONMENTAL SCIENCE, AND TOXICOLOGY
Okaichi, Anderson, and Nemoto, Editors

Chlorophyll pigments were measured by in situ fluorometry and after extraction in the laboratory. A four channel sun photometer placed on the roof of the experimental station and a sky photometer mounted on the roof of the aircraft were used for air corrections (Fig. 1). Air corrections were made to minimize the effects of pathradiance, optical thickness of the air, and sunlight fluctuations during aircraft observations. Two methods were used for red tide density mapping - an empirical method and an optical method.

1. Empirical Method

After the air corrections, the MSS data for each channel and data calculated from each channel were compared with chlorophyll pigment concentrations. One example of cross correlation coefficients between pigments and MSS data is shown in Table 1. (SUM means sum of values between ch. 2 and ch. 9). The highest linear correlation was found between Ln (chl.) and the difference between ch. 3 and ch. 7. The logarithms of the observed chlorophyll concentrations versus MSS data were plotted (Fig. 2) and chlorophyll concentrations determined from the regression line.

Table 1.
Cross Correlation Coefficient between chl. pigment & MSS data

MSS data	chl.	Ln(chl.)
nm		
ch.2(420-450)	-0.532	-0.684
ch.3(450-500)	-0.665	-0.798
ch.4(500-550)	-0.657	-0.781
ch.5(550-600)	-0.549	-0.672
ch.6(600-650)	-0.332	-0.448
ch.7(650-690)	-0.158	-0.287
ch.8(700-790)	0.303	0.212
ch.9(800-890)	0.396	0.310
ch.11(IR)	0.615	0.599
ch.3/SUM	-0.581	-0.537
ch.5/SUM	-0.749	-0.730
ch.7/SUM	0.672	0.640
ch.9/SUM	0.670	0.613
ch(2-5)	0.485	0.531
ch(4-5)	-0.561	-0.573
ch(7-5)	0.758	0.845
ch(9-5)	0.735	0.830
ch(3/5)	-0.002	0.025
ch(6/5)	0.729	0.755
ch(8/5)	0.768	0.740
ch(3-7)	-0.851	*-0.947
ch(3/7)	-0.667	-0.632

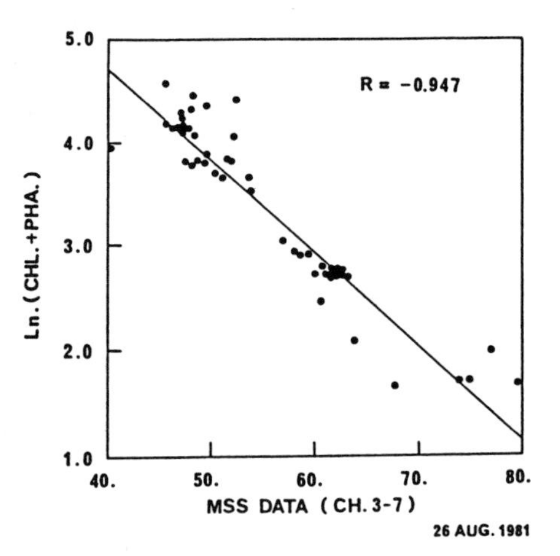

Fig.2 MSS (ch.3-ch.7) versus measured pigment concentration.

2. Optical Model Analysis

Seawater has optical characteristics determined by the water molecules, pigments, suspended substances and dissolved organic material. The sunlight is absorbed and scattered. The ocean color is caused by the back scattered radiation in the visible part of the spectrum. The light absorption coefficient Asw (λ) and scattering coefficient Ssw (λ) of the seawater can be written as follows:

$$Asw(\lambda)=Aw(\lambda)+Achl(\lambda)Dchl+Ap(\lambda)Dss+Ay(\lambda)Dy \quad (1)$$
$$Ssw(\lambda)=Sw(\lambda)+Schl(\lambda)Dchl+Sp(\lambda)Dss \quad (2)$$

Aw,Sw ; coefficients of pure sea water
Achl,Schl ; coefficient per unit chl. concentration
Ap,Sp ; coefficient by unit particle concentration
Ay ; coefficient per unit DOM
Dchl.Dss,Dy; density of chl., SS and DOM, respectively
DOM gives no effect on scattering.

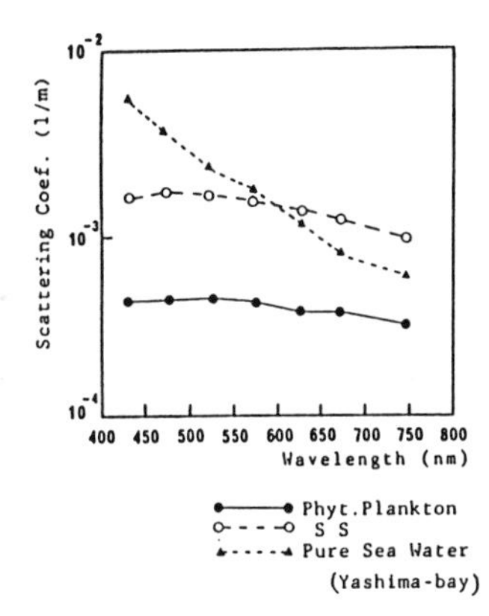

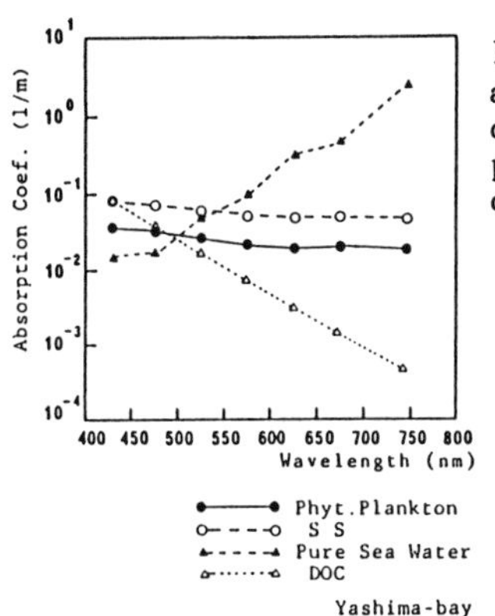

Fig.3 Scattering and absorption coefficient per unit concentration.

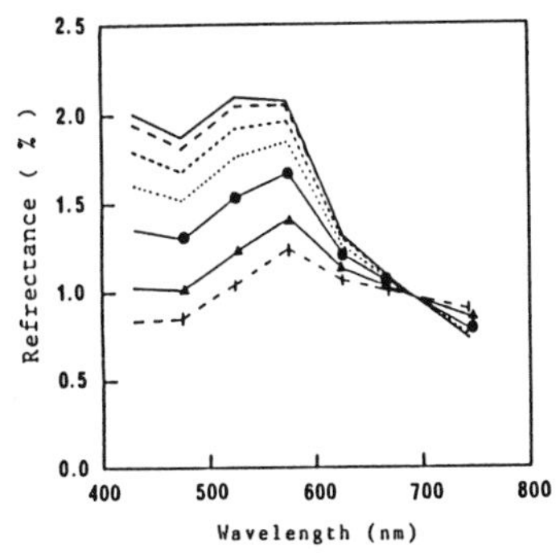

Fig.4 Spectral patterns for different pigment, DOC and SS concentrations.

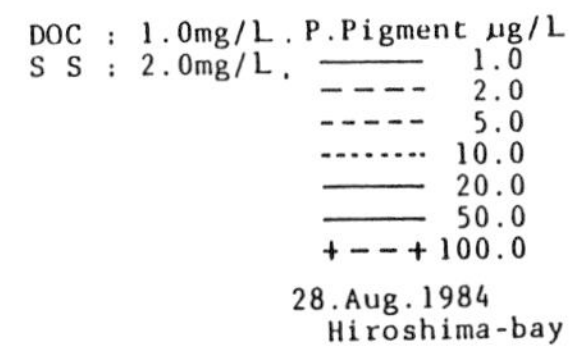

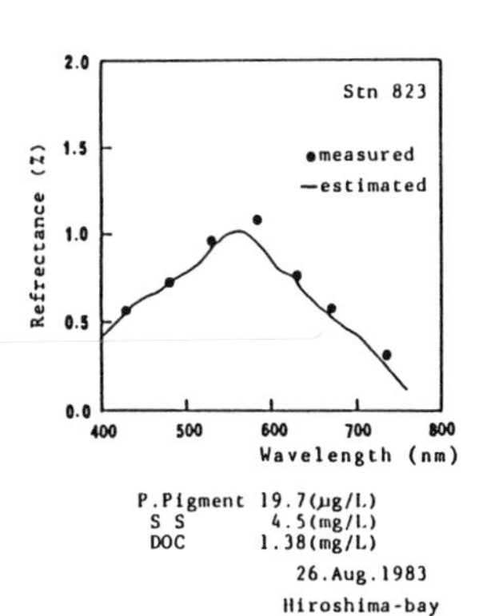

Fig.5 Measured and calculated surface spectral reflectance.

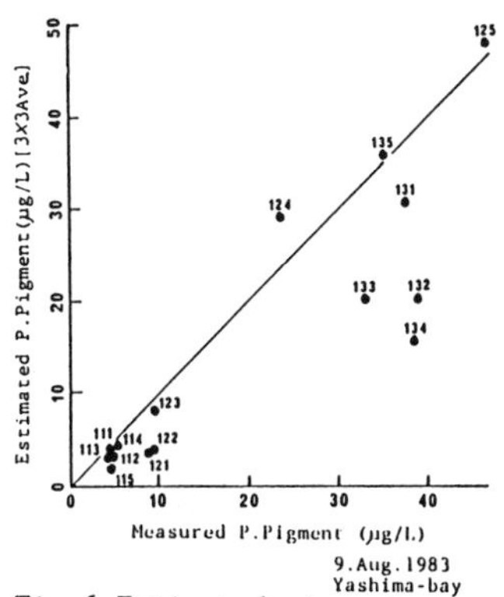

Fig.6 Estimated pigment concentrations using the model analysis versus measured pigment concentations. Number show sampled stations.

Each optical coefficient of the sampled water has a value specific for that water. Measured coefficients are shown in Fig. 3. Assuming a homogeneous surface layer without influence from sea bottom reflection, the light reflection just below the surface, R_∞, can be written as follows:

$$R_\infty(\lambda)=1+Asw(\lambda)/Ssw(\lambda)-\sqrt{(1+Asw(\lambda)/Ssw(\lambda))^2-1} \qquad (3)$$

Using equation (3), several spectral reflectance curves were drawn (Fig. 4). Each curve is characterized by the density of chlorophyll, SS and DOM (i.e . each chlorophyll concentration can be determined from the measured upward irradiance using curve fitting techniques). Fig. 5 shows an example of a calculated curve based on measured water quality and surface reflectance. This example shows that the measured and estimated values correspond reasonably well.

DISCUSSION

The merits of the empirical method are that anyone can use it without optical knowledge or laboratory equipment and that the use of a large field dataset for calibration keeps calculation error small. However, the need for a large sea-truth dataset is a serious limitation because of the time lag between aircraft measurements and boat observations. One of the foremost merits of aircraft remote sensing of red tides is speed. If several boats and men for field and laboratory measurements are needed for a particular method, rapid analysis is not possible. In such a case, remotely-sensed data are used just for interpolation of field data.

For the analytical model, only a few water samples are needed for the determination of optical coefficients. Immediately after the determination of the coefficients, a computer can calculate the density of each factor and draw the chlorophyll concentration map. In this model, a homogeneous composition of water quality over the observing area is assumed. When there are several plankton species with different optical properties, an error could result. Fig. 6 shows the correlation between measured values and estimated values in our study. Samples for Stations 132, 133 and 134 show poor agreement. Those three stations had low salinity due to the influence of river water. For the solution of this problem, a number of algorithms equal to the number of distinct water masses is required. Since we prefer to use remote sensing with minimal sea-truth data, this additional sampling is a serious constraint. In addition, errors may occur at the boundaries between different water masses.

ACKNOWLEDGEMENTS

This work was supported by the Research Division of the Japan Fisheries Agency. Several staff members of the Hiroshima prefectural fisheries station, Akashio research institute of Kagawa prefecture and Asia air survey Co. Ltd. took part in the study under the supervision of Dr. Anraku. Their contribution is greatly acknowledged.

REFERENCES

1. S. Matsumura, Bulletin on Coastal Oceanography, 24 (2) (1987).
2. Ross, W. D., Jour. Paint Teach., 9 (57) (1967).

OCCURRENCE AND BIOMASS ESTIMATION OF CHATTONELLA MARINA RED TIDES IN HARIMA NADA, THE SETO INLAND SEA, JAPAN

SHIGERU MONTANI, MASAHIKO TOKUYASU AND TOMOTOSHI OKAICHI
Faculty of Agriculture, Kagawa University, Kagawa 761-07
Japan

ABSTRACT

The relationship between changes in oceanographic and chemical conditions and the occurrence of red tide outbreaks in northern Harima Nada in the Seto Inland Sea were investigated during the summer of 1983. *Chattonella marina* appeared at a density of 10 cells/ml in several samples throughout most areas of Harima Nada in early July. It then developed into dense patches at the end of July, mostly in the northern area (> 400 km^2). The most dense patch was observed off the coast of Aioi with a cell concentration of 5400 cells/ml on 22 July. The estimated biomass of *C. marina* was 3800, 480, 55 and 14 tons for carbon, nitrogen, phosphorus and iron, respectively. The nitrogen biomass (480 tons) corresponded to about 30% of the total nitrogen in northern Harima-Nada.

MATERIALS AND METHODS

Sampling was carried out at approximately bi-weekly intervals on 7 occasions at 11 stations in northern Harima Nada in the Seto Inland Sea during the summer of 1983. Sampling at each station consisted of measuring water transparency (S.S.) vertical profiles of temperature (W.T.) salinity (S) and dissolved oxygen (D.O.). A cell count of *Chattonella* spp. was also made for each sample. Water samples were collected from various depths using 10L Van Dorn bottles. Suspended matter was collected by filtration through a pre-combusted, pre-weighed GF/C filter for particulate carbon and nitrogen (POC, PON), particulate phosphorus (P-P), and Chl *a* analyses. Other aliquots were collected through a pre-washed Millipore HA filter for iron and nutrient analyses. The filters were rinsed with distilled water, dried and later weighed ashore. POC and PON were determined with a Yanaco MT-2 CHN analyzer. Chl *a*, P-P, nutrients and total iron (T-Fe) were measured by standard methods [1,2].

RESULTS AND DISCUSSIONS

Detailed time series observations were made at St. 64 in northern Harima-Nada (Fig. 1) for S, W.T., D.O., POC, PON, P-P, Chl *a*, (NO_3+NO_2)-N, NH_4-N, PO_4-P, SiO_2-Si and T-Fe. On June 14 and July 2 a density of only a few cells/ml of *C. marina* were observed throughout the sampling area. *C. marina* then appeared at a density of 10 cells/ml in several samples throughout northern Harima-Nada on July 13. Then on July 21 and 22, *C. marina* developed into dense patches covering most of the 400 km^2 northern region (Fig. 1). A low salinity and nutrient enriched water mass appeared on July 21-22. This occurred concurrently with the *C. marina* red tide outbreak (Fig. 2).

RED TIDES: BIOLOGY, ENVIRONMENTAL SCIENCE, AND TOXICOLOGY
Okaichi, Anderson, and Nemoto, Editors

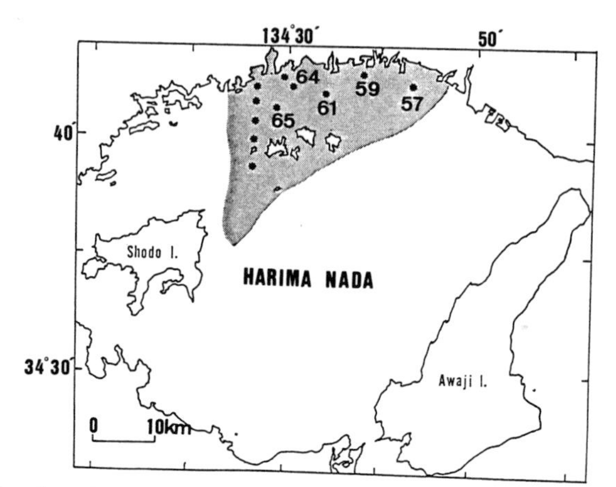

FIG. 1. Sampling stations and red tide area of C. marina.

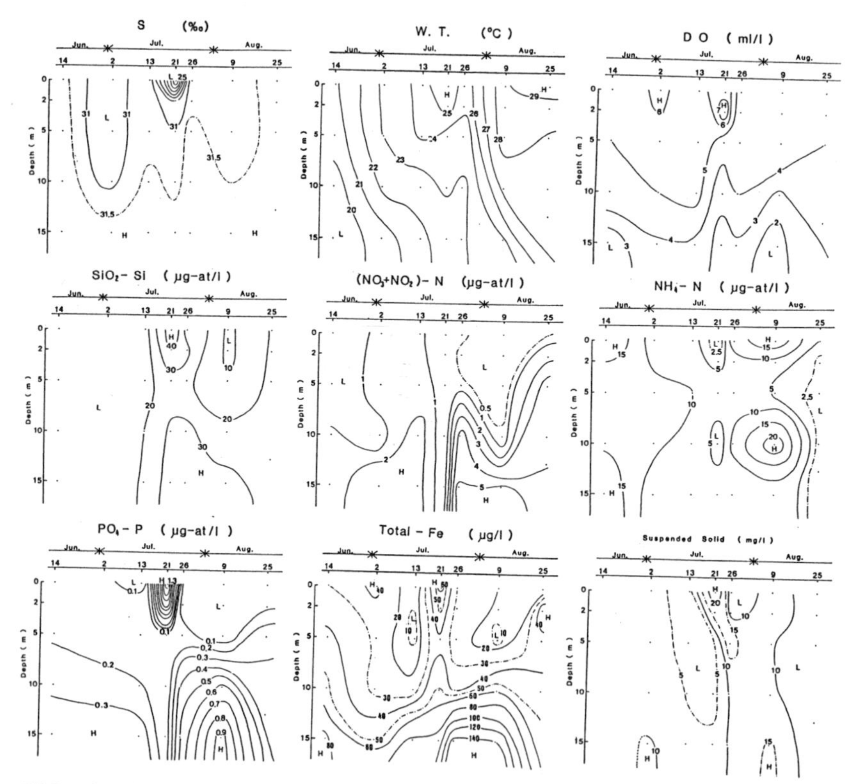

FIG. 2. Variations in the vertical distributions of S, W.T., D.O., SiO_2-Si, (NO_3+NO_2)-N, NH_4-N, PO_4-P, T-Fe and SS at Stn.64 of Harima Nada from June 14 to August 25, 1983.

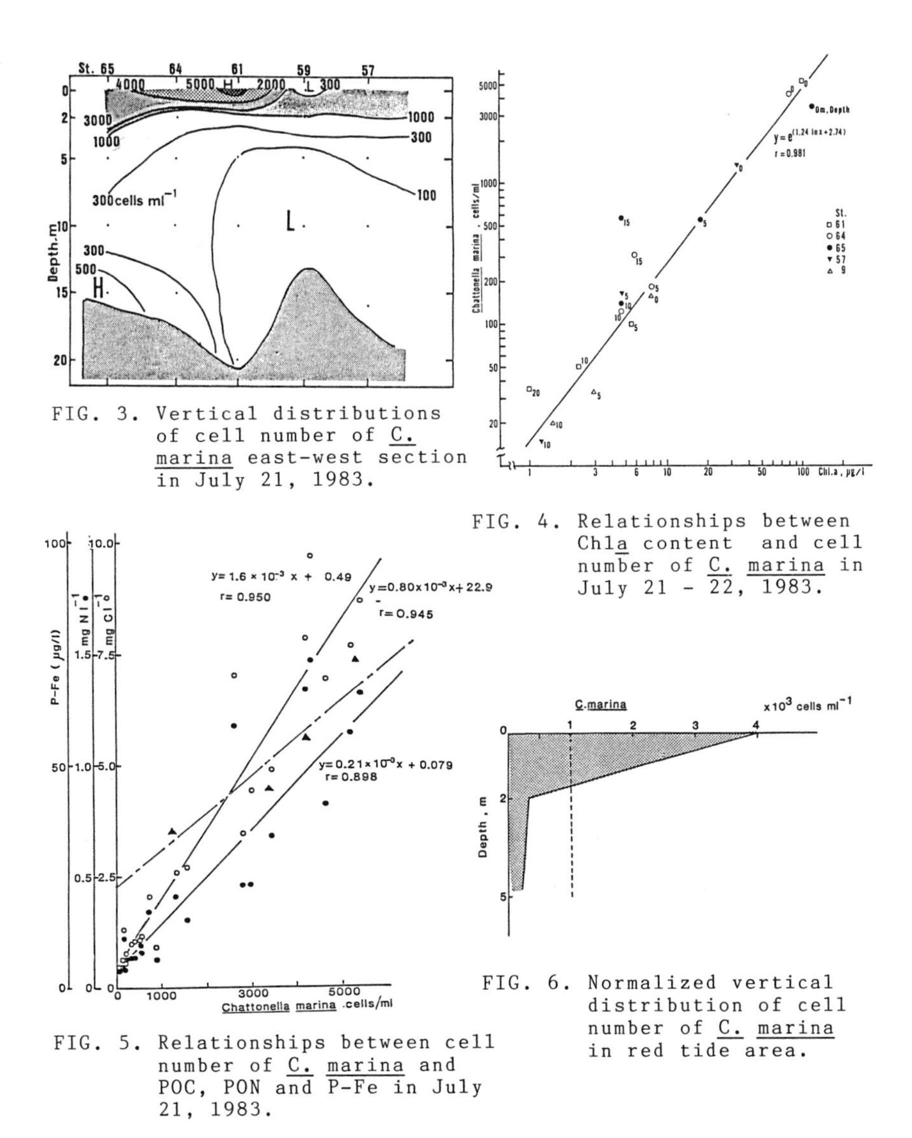

FIG. 3. Vertical distributions of cell number of C. marina east-west section in July 21, 1983.

FIG. 4. Relationships between Chl a content and cell number of C. marina in July 21 - 22, 1983.

FIG. 5. Relationships between cell number of C. marina and POC, PON and P-Fe in July 21, 1983.

FIG. 6. Normalized vertical distribution of cell number of C. marina in red tide area.

The patch with the highest cell number was observed off the coast of Aioi (St.61) with a cell concentration of 5400 cells/ml on July 21 (Fig. 3). Figure 4 shows the relationship between Chl a content and C. marina cell number during red tides. These results indicate that C. marina represents the bulk of the Chl a during the blooms. Increases in PO_4-P and (NO_3+NO_2)-N concentrations coincided with a decrease in dissolved oxygen during the red tide. Silicate concentrations which were relatively stable during the three months of the study (Fig. 2) increased mainly as a result of an increase in the flow of river water coming from an increase in the flow of river water coming from heavy rains between July 15 and 21 (total rainfall, 106mm). Comparisons between C. marina cell number, and POC, PON, P-P and T-Fe during red tide outbreaks at 0, 2 and 5 m depths on July 21 show clear, positive correlations (Fig. 5). The following regression equations were obtained from the data.

$$\mathrm{POC}(\mu g/l) = 1.55 \times \textit{C. marina}\ (\mathrm{cells/ml}) + 486 \quad (r=0.950)$$
$$\mathrm{PON}(\mu g/l) = 0.21 \times \textit{C. marina}\ (\mathrm{cells/ml}) + 79 \quad (r=0.898)$$
$$\mathrm{P.P}(\mu g\text{-}at/l) = 7.2\times10^{-3} \times \textit{C. marina}\ (\mathrm{cells/ml}) + 0.913 \quad (r=0.904)$$
$$\mathrm{T\text{-}Fe}(\mu g/l) = 0.8\times10^{-5} \times \textit{C. marina}\ (\mathrm{cells/ml}) + 22.9 \quad (r=0.945)$$

Fig. 6 shows the normalized vertical distribution of C. marina during the red tide. C. marina was in highest concentrations at depths between 0 and 5m. The average cell number was 1000 cells/ml. The standing stock of carbon, nitrogen, phosphorus and iron in the C. marina red tide were estimated at 3800, 480, 55 and 14 tons, respectively. This large amount of nitrogen biomass (480 tons) corresponded to about 30% of the total nitrogen in northern Harima Nada [3].

REFERENCES

1. T. R. Parsons, Y. Maita and C. M. Lalli, A Manual of Chemical and Biological Methods for Seawater Analysis (Pergamon Press, Oxford 1984).
2. M. Saito, D. Horiguchi and K. Kinou, Bunsekikagaku 30, 635-639 (1981).
3. T. Okaichi and R. Marumo, Studies on Marine Ecosystem Models on Outbreaks of Red Tides in Neritic Waters (Research Report from the Grant in Aid for Scientific Research, Japan, B 264-R14-1, 1985) pp. 1-5.

THE DEVELOPMENT OF ANOXIC WATER AND RED TIDE ASSOCIATED WITH EUTROPHICATION IN HIUCHI NADA, THE SETO INLAND SEA, JAPAN

TADASHI OCHI
Faculty of Agriculture, Kagawa University, Kagawa 761-07
Japan

ABSTRACT

The mechanism of anoxic water mass formation was investigated in the eastern part of Hiuchi-Nada. It was found that the oxygen consumption by the sediment was unexpectedly small and that the oxygen consumption by organic matter freshly deposited on the bottom from phytoplankton blooms, followed by frequent breakdown of the thermocline were important contributing factors in the formation of anoxic water masses in Hiuchi-Nada.

INTRODUCTION

Exchange of sea water in the eastern area of Hiuchi-Nada, which is located in the central part of the Seto Inland Sea, is reported to be relatively limited. In the 1960's, there were serious problems with environmental pollution from organic sewage inputs from pulp-making factories in the southeast coastal region. Occurring simultaneously with the pulp pollution was a progressive eutrophication caused by waste from a chemical fertilizer factory and urban sewage. As eutrophication increased in the 1970's, anoxic water became common in this area and red tides due to *Gymnodinium* spp. and other phytoflagellates were frequently observed. The present study was undertaken to examine the mechanism by which anoxic water masses in the eastern part of Hiuchi-Nada are generated.

METHODS

The main investigation was undertaken in the eastern part of Hiuchi-Nada from 1981 to 1983. Water temperature, salinity and the concentrations of dissolved oxygen, phosphorus, chlorophyll *a*, and the oxygen consumption of the seawater were measured every two weeks. The concentration of organic carbon, phaeophytin and total phosphorus in the sediments were also analyzed. Furthermore, the concentration of dissolved oxygen in the water was continuously measured using a bell jar-type container (diameter 60cm and volume, 43 l) fixed on the sea bed at station T2.

RESULTS AND DISCUSSION

The results clearly indicate that the development of stratification in this basin during the summer season was dependent on the relative difference in water temperature, not salinity. Anoxic water masses were found in that part of the basin in which the water is over 20 meters deep and where organic carbon, phaeophytin and phosphorus have accumulated in the sediments. The upper thermocline was observed at the zone occurring between the sea surface and a depth of 15 meters. A lower thermocline was observed about 5 meters above the sea bottom. The water lying under the lower thermocline had become

RED TIDES: BIOLOGY, ENVIRONMENTAL SCIENCE, AND TOXICOLOGY
Okaichi, Anderson, and Nemoto, Editors

extremely turbid with very low dissolved oxygen (near 0 mg/l). The concentration of phosphate increased to about 2μ-at/l during the summer in the years before 1980, but after 1980 the level of dissolved oxygen has not decreased below 2 mg/l [1, 2]). As shown in Figure 2, although the nutrients in the euphotic zone were very low in the summer, they were high in the seawater below the lower thermocline. In general, the concentration of

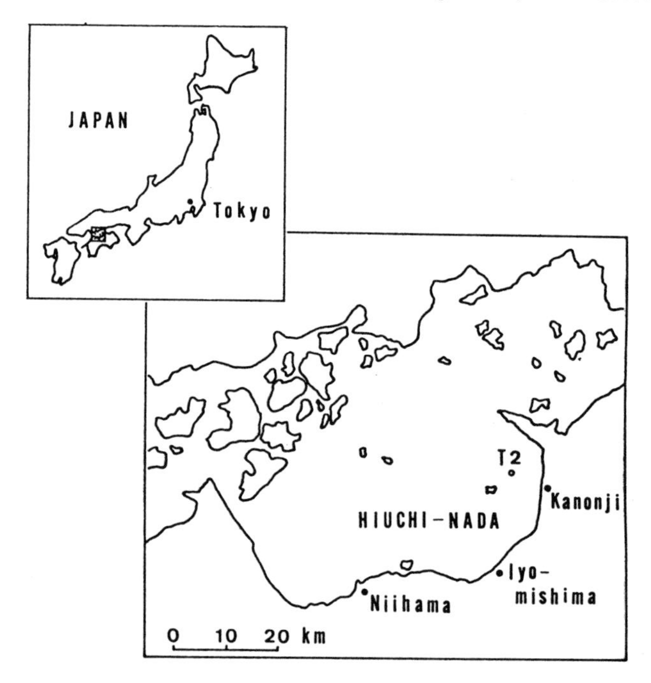

FIG. 1. Location of Hiuchi-Nada and observation station

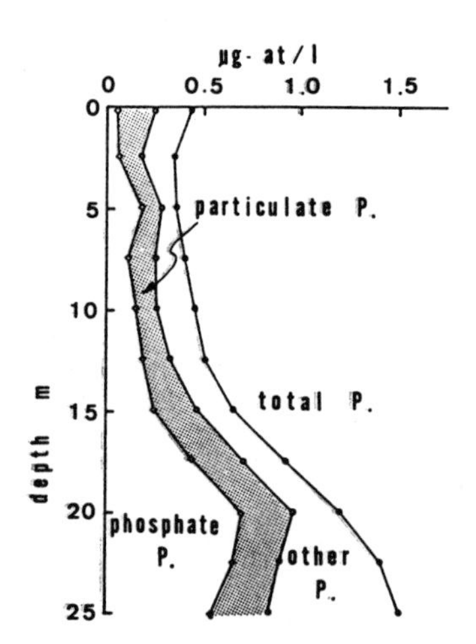

FIG. 2. Virtical distributions of phosphorus at station T2, 10 August, 1982

chlorophyll a was below 1.0 μg/l in the upper 10 meters of the water column, and in the bottom water under 15 meters, it fluctuated in the range from 1.5 to 2.5 μg/l. The maximum concentration of chlorophyll a was typically near the lower thermocline.

The rate of oxygen consumption of the surface sediment was 0.5 $g/m^2/day$. In the water immediately above the sediments, the rate of oxygen consumption and primary production were 1.3 $g/m^2/day$ and 0.6 $g/m^2/day$, respectively. Therefore, oxygen production was insufficient in the bottom water as it was smaller than total consumption by 1.2 $g/m^2/day$. Under these conditions, however, the concentration of dissolved oxygen should not decrease below 2.5 mg/l [3].

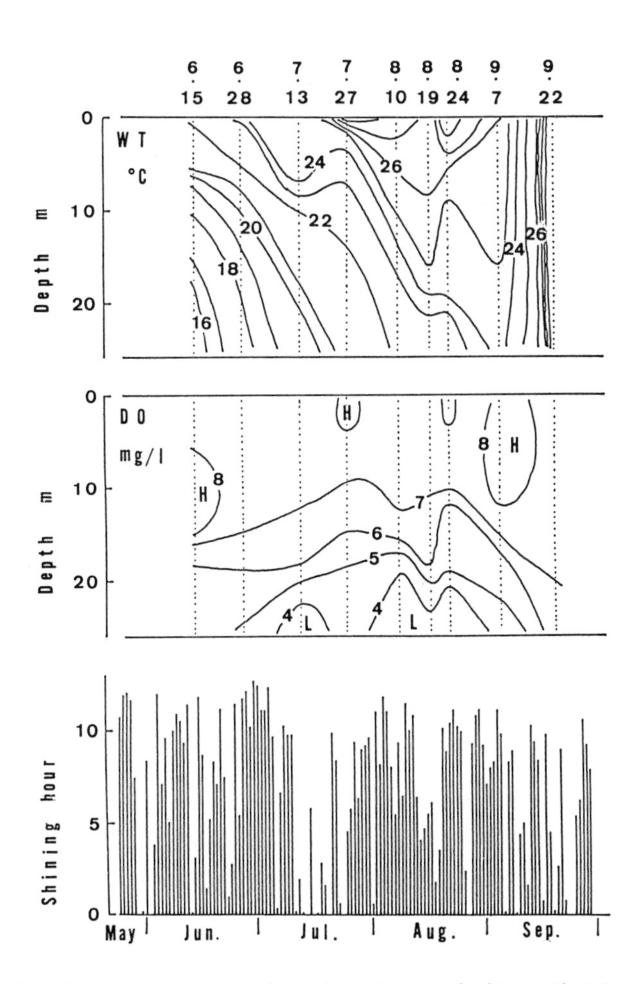

FIG. 3. Seasonal change in the vertical distributions of water temperature, dissolved oxygen and daily hours of sunshine, 1982

From the field results in 1982, (Figure 3) it is clear that a breakdown of the thermocline was brought about by a decrease in daily solar radiation. Subsequently, the zone of maximum chlorophyll a moved up to a depth of about 10-15 meters from its original position near the lower thermocline. Anoxic water masses appeared for about 1-2 weeks after the thermocline disappeared.The concentration of chlorophyll a increased to about 6 μg/l before the appearance of the anoxic water.

CONCLUSION

In the 1970's, it was thought that the formation of anoxic water in the eastern part of Hiuchi-Nada was caused by eutrophication of the water and by the presence of organic pollutants in the sediment. In the 1980's, eutrophication was controlled to some extent. The appearance of anoxic water continued, apparently due to the presence of a lower layer or deep thermocline near the bottom of the water column in which oxygen consumption exceeded supply, including production, and where nutrients were high. When the high nutrients in this lower layer were advected vertically as the thermocline broke down, phytoplankton bloomed and chlorophyll biomass increased substantially. The resulting oxygen demand as the phytoplankton decayed produced the anoxia. Thus, cloudy days led to the erosion of the thermocline and sunny weather then allowed the new nutrients to be converted into phytoplankton biomass, which in turn, led to the anoxia.

REFERENCES

1. T. Ochi, S. Nishio and T. Okaichi, Tech. Bull. Fac. Agr. Kagawa Univ. 29, 297-304 (1978).
2. T. Ochi and H. Takeoka, J. Oceanogr. Soc. Japan, 42, 1-11 (1986).
3. H. Takeoka, T. Ochi and K. Takatani, ibid, 42, 12-21 (1986).

A BACTERIUM HOSTILE TO FLAGELLATES: IDENTIFICATION OF SPECIES AND CHARACTERS

SHINYA ISHIO, R.E.MANGINDAAN, M.KUWAHARA AND H.NAKAGAWA
Faculty of Agriculture, Kyushu University, Fukuoka 812, Japan.

ABSTRACT

A new species of bacterium was isolated from Hakata Bay sediment. This bacterium not only produces a dinoflagellate growth inhibitor (DGI),but attacks and kills the algae of Dinophyceae and Raphydophyceae. When a culture of this bacterium was added to a culture of *Chattonella antiqua*, *Chattonella* migrated upwards and formed a dense layer secreting mucus copiously for self-defense against this bacterium. The cells of this bacterium which attacked *Chattonella* were sent slowly to the bottom with the mucus that *Chattonella* produced. The behavior that *Chattonella* forms a dense layer in seawater seemed to be attributable to their self-defense against this bacterium. One of reasons that various phytoflagellates form monospecific dense layers may also be attributable to self-defense against this hostile species of bacterium.

INTRODUCTION

Cultured yellowtail in Harima Sound were seriously damaged in 1972,'77,'78,'79,'83 and '87 by dense and widespread red tides of *Chattonella antiqua*. The cause of the development of *Chattonella* red tides was ascribed to eutrophication of the Sound water, but the eutrophicated waters of Ariake and Hakata Bays have never affected by red tides of *Chattonella antiqua* . Ishio *et al.*[1] revealed that dinoflagellate growth inhibitor, DGI distributes broadly in the sediments in Ariake and Hakata Bays. Since DGI possessed an antibiotic action to *Micrococcus luteus*, DGI was thought to be produced by a certain species of bacterium. This bacterium has now been isolated from the sediment of Hakata Bay. The method of isolation, the confirmation of DGI production and the characters of this bacterium are described.

MATERIALS AND METHODS

Isolation of DGI Producing Bacterium

A very common saline bouillon (Nutrient Broth of DIFCO 0.5%, Polypeptone of Daigo 1% and filtered seawater) containing crude DGI[2] and DMSO in the concentrations of 300 and 2,500 ppm, respectively was used for the isolation of this bacterium. A diluted suspension of the sediment from the most eutrophic area of Station 18 in Hakata Bay[3] was added to the bouillon and kept at 22±1 °C for 48h.

Confirmation of DGI Productivity of Isolated Bacterium

The isolate was cultured aerobically at 22±1 °C for 48h in 1.25ℓ autoclaved saline bouillon in a turnip form flask. During the culture, the flask was shaken in rotational manner. Since

RED TIDES: BIOLOGY, ENVIRONMENTAL SCIENCE, AND TOXICOLOGY
Okaichi, Anderson, and Nemoto, Editors

the culture did not indicate the presence of DGI through bio-assay, the bacterial cells were centrifuged and freeze-dried. The freeze-dried bacterial cells were extracted with acetone and the acetone extract was subjected to the TLC-bioautography[4], employing *Micrococcus luteus*. Both Rf value and antibiotic action were compared with those of DGIs isolated from Hakata Bay sediment.

Characters of DGI Producing Bacterium

Identification of this bacterium was attempted in accordance with Bergey's Manual of Systematic Bacteriology[5]. The behavior of this bacterium to *Cbattonella antiqua* was examined under a microscope and using five long glass tubes(150 cm long, ϕ 1.2 cm) by adding the culture of this bacterium in the thickness of 1, 2, 4 and 8 mm. Light irradiation was kept constant over the whole length of the five columns of *Chattonella* culture by arranging six 1.2 m long fluorescent lamps.

RESULTS AND DISCUSSION

From the culture with the suspension of sediment and DGI, white colonies with smooth surfaces were obtained on the saline bouillon agar plate treated in the usual ways.

As shown in Fig. 1, this bacterium is a common bacterium with a length of 2.4 x 0.7 μm, a polar flagellum, Gram negative, and facultative anaerobic natures. This bacterium can grow in the range of temperature from 5 to 42 °C and salinity from 5 to 70 ‰, but not more than 80 ‰. This bacterium has the capability of adjusting ambient pH of 6.0-9.0 to the pH of oceanic water of 8.1-8.3.

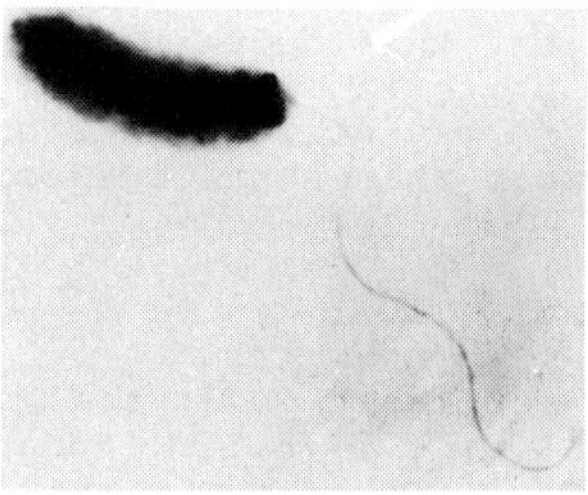

Fig. 1. EM of a bacterium(2.4 x 0.7 μm) hostile to phytoflagellates.

From these characters, this bacterium was regarded to be a typical marine bacterium inhabiting in shallow waters.

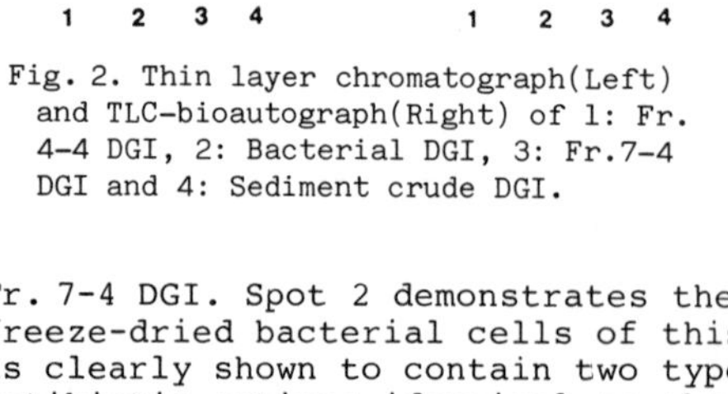

The acetone extracts of freeze-dried cells of this bacterium gave TLC and TLC-bioautographs as indicated in Fig. 2.

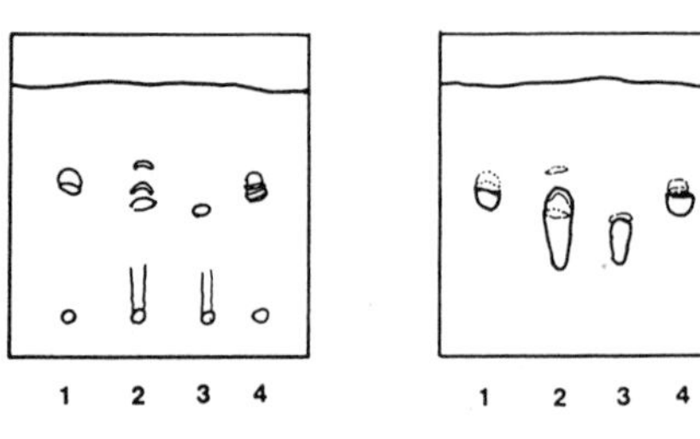

Fig. 2. Thin layer chromatograph(Left) and TLC-bioautograph(Right) of 1: Fr. 4-4 DGI, 2: Bacterial DGI, 3: Fr.7-4 DGI and 4: Sediment crude DGI.

In Fig. 2, spots encircled with a bold line indicate the bacterial free zones of *Micrococcus luteus*.

In these TLC and TLC-bioautographs, spot 1 and 3 indicate two types of DGIs isolated by rechromatography from the acetone extract of Hakata Bay sediment. Spot 4 shows the mud DGI which is shown to consist of Fr. 4-4 DGI; but not Fr. 7-4 DGI. Spot 2 demonstrates the DGI isolated from the freeze-dried bacterial cells of this species. The bacterial DGI is clearly shown to contain two types of DGIs which have Rf and antibiotic actions identical to those of Fr. 4-4 DGI and Fr. 7-4

DGI. Thus, this bacterium was regarded to be a DGI producing bacterium.

A very curious behavior was observed, when the culture of this bacterium was added to the culture of *Chattonella antiqua* under a microscope. Abruptly, the swimming cells of *Chattonella* increased in number in the visual field and their spindle like cells changed into spherical cells and burst as shown in Fig. 3.

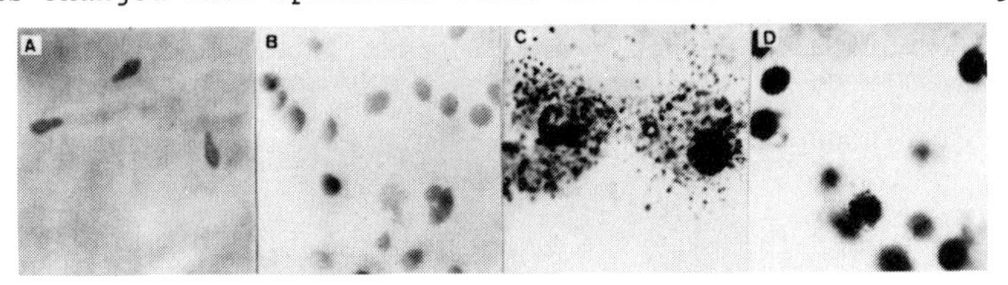

Fig. 3. Normal spindle like cells(A), spherical cells(B) gathered near surface, the burst of spherical cells(C) and remained spherical bodies(D) of *Chattonella antiqua* observed after adding the culture of this bacterium.

When bacteria were added to the column of *Chattonella* culture, *Chattonella* moved upwards and formed a dense layer in each tube, with different thickness depending on the amounts of culture added. Meanwhile, a "string of beads" consisting of the swimming cells of *Chattonella* in strings of mucus was formed in the dense layer and these 'beads" were observed to sink at a rate of 2 cm/min or so. Then the lowermost bead disappeared at a depth of 7∿8 cm. This odd behavior continued for several days. If the disappearance of the lowermost bead occurred by sinking to the bottom, this behavior could not have continued for several days. Therefore, the disappearance must be attributable to their return to the surface. Thus, the distribution of *Chattonella* and this bacterium together with polysaccharides (as glucose) by anthrone method[6] were

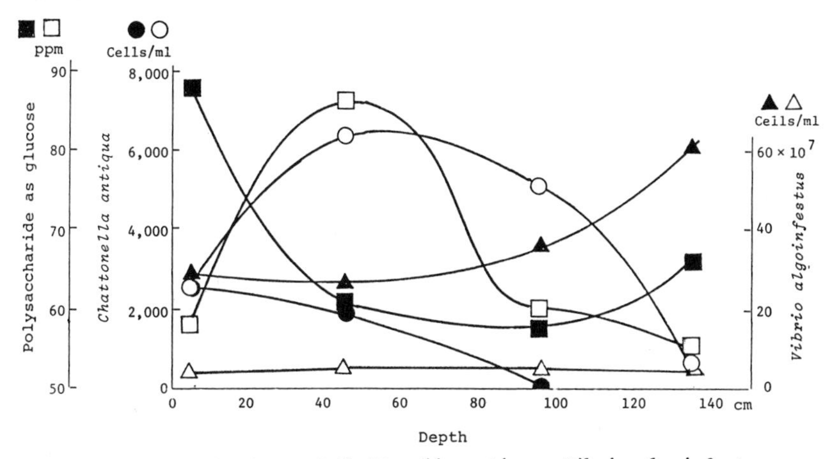

Fig. 4. The distributions of *Chattonella antiqua*, *Vibrio algoinfestus* and polysaccharides in two long tubes, one for test group (solid marks) and another for control group (open marks), respectively. To the control group no *Vibrio algoinfestus* was added. Thus, the bacteria detected in the tube of the control should originate from air.

measured and Fig. 4. was obtained.

Even though this bacterium was added to the top of the column of *Chattonella* culture, they were found most abundantly at the bottom. Polysaccharides were most abundant at the surface but were still present at the bottom.

From these results, it was concluded that the behavior of *Chattonella antiqua* in forming a dense layer of swimming cells near the surface is for self-defense against this enemy bacterium. Formation of dense layers of phytoflagellates at sea surfaces may be associated with self-defense against hostile bacteria of this species, in view of mortal effects on a wide range of Dinophyceae and Raphydophyceae as shown in Table 1.

Table 1. Lowest lethal concentrations (LLC in ppb) of DGI_3 to various species of Dinophyceae and Raphydophyceae.

Dinophyceae	LLC	Dinophyceae	LLC
Fibrocapsa japonica	125.0	*Prorocentrum gracile*	62.5
Heterosigma akashiwo	174.0	*Prorocentrum micans*	62.5
Goniodoma pseudogonyaulax	62.5	*Prorocentrum minimum*	62.5
Gymnodinium breve	62.5	*Prorocentrum triestinum*	91.8
Gymnodinium sp.	62.5	*Protogonyaulax catenella*	43.8
Gyrodinium instriatum	43.8	*Protogonyaulax affinis*	43.8
Gyrodinium sp.	31.2	*Pyrophacus steinii*	21.9
Prorocentrum dentatum	91.8	*Scripsiella trochoidea*	5.5
Raphydophyceae	**LLC**	**Raphydophyceae**	**LLC**
Chattonella antiqua	62.5	*Chattonella marina*	125.0

ACKNOWLEDGMENT

We wish to thank Director O.Goda, Dr. C.Ono and Mr. S.Yoshimatsu of the Akashiwo Research Institute of Kagawa Prefecture for their kind offer of these species of phytoflagellates.

REFERENCES

1. Ishio,S.,T.Nishimoto and H.Nakagawa. Nippon Suisan Gakkaishi, 53(5): 773-787, 1987. Distribution of Dinoflagellate Growth Inhibitor in Sediments in Ariake and Hakata Bays.
2. Ishio,S.,T.Nishimoto and H.Nakagawa. Nippon Suisan Gakkaishi, (in press). Separation of Dinoflagellate Growth Inhibitor from Sea Sediment.
3. Ishio,S.,M.Kuwahara and H.Nakagawa. Nippon Suisan Gakkaishi, 52(5): 901-911,1986. Conversion of $AlPO_4$ to Fe-bound P in Sea Sediments.
4. Kline,R.M. and I.Golab. J. Chromatog., 18: 409-411, 1965. A simple technique in developing thin layer chromatographs.
5. Baumann,P.,A.L.Furniss and J.V.Lee., in "Bergey's Manual of Systematic Bacteriology" Vol. 1. ed. by N.R.Krieg and J.G.Holt. WILLIAMS & Wilkins. Baltimore, 1984. pp. 518-538.
6. Moore,B.G. and R.G.Tischer: Science,N.Y. 145: 586-587, 1984. Extracellular Polysaccharides of Algae: Effects on Life-Support Systems.

IN SITU STUDIES OF THE EFFECTS OF HUMIC ACIDS ON DINOFLAGELLATES AND DIATOMS

Granéli, E., P. Olsson, B. Sundström and L. Edler
Dept. of Marine Ecology, University of Lund
Box 124, S-221 00 LUND, SWEDEN

ABSTRACT

Experiments with 160 l plastic bags were performed in September 1986 *in situ* on the west coast of Sweden. One series of bags was filled with water containing the natural phytoplankton communities dominated by dinoflagellates (>80 % of the total phytoplankton biomass). The other series was filled with 1 µm filtered sea water inocculated with the diatom *Phaeodactylum tricornutum*. Phosphorus or nitrogen was added alone or together with commercially available humic acids to both series. Bags with no additions (C) or with only humic acids added (CH) were used as controls. *P. tricornutum* was stimulated by nitrogen addition, while no positive response was obtained by any nutrient addition to the dinoflagellate bags. Dinoflagellate biomass increased severalfold when humic acids were added together with nitrogen. Also the alkaline phosphatase activity (APA) increased drastically both in the CH and nitrogen + humic acid bags. Tentative conclusions are: 1) The dinoflagellates were inhibited by heavy metals and the humic acids alleviated this inhibition by working as chelating agents. 2) The diatom was not inhibited by heavy metals because it could still grow quite well without humic acids. 3) As APA increased in the dinoflagellate control bag with only humic acids we assume that nitrogen in the humic fraction to some extent was available for the dinoflagellates.

INTRODUCTION

Exceptional plankton blooms and oxygen deficiency have been observed frequently during the last decade in the Kattegat (1, 2). Autumn plankton blooms on the Swedish west coast are often dominated by dinoflagellates, e. g. *Ceratium* species (3). Blooms were heavy in 1980 and 1987 with chlorophyll a values reaching several tens of $mg.m^{-3}$. These years were characterised by exceptionally high precipitation and thus high nutrient transport in rivers.

In areas dominated by forests nitrogen is leached to drainage water mainly in the form of organic compounds, e. g. as humic and fulvic acids. Such areas drain to the Swedish west coast, where rivers are distinctly brown coloured and more than half the nitrogen load is normally in organic form (4).

It has been shown in laboratory studies that both dinoflagellates and diatoms may be stimulated by humic compounds (5, 6), although the mechanisms involved have not been clarified. Humic compounds may act as chelating agents, decreasing the toxicity of metals such as copper (5, 7) or increasing the availability of essential metals such as iron (8). There are also indirect evidence that dinoflagellates may be able to utilize nitrogen in humic compounds (9).

It may be hypothesized that humic compounds in river water are partly responsible for the dinoflagellate blooms observed on the Swedish west coast. The main objective of the present investigation was to test *in situ* in large plastic bags how humic compounds influence phytoplankton assemblages dominated by dinoflagellates or diatoms. The objective was also to test for interactions between humic compounds and nutrients with respect to the influence on phytoplankton growth, and to establish if diatom and dinoflagellate biomass formation is limited by the same nutrient.

MATERIALS AND METHODS

From September 9 to 23 , 1986, 2 m long polyethene plastic bags containing 160 l sea water were exposed *in situ* in the Gullmar fjord, on the west coast of Sweden (Fig. 1). In one series of experiments the bags were filled with 160 µm filtered sea water containing the natural phytoplankton assemblages. In the other experimental series the sea water was filtered through

RED TIDES: BIOLOGY, ENVIRONMENTAL SCIENCE, AND TOXICOLOGY
Okaichi, Anderson, and Nemoto, Editors

cartridge cotton filters (< 1 µm mesh size), and inoculated with $1.5 \cdot 10^6$ cells.ml^{-1} *Phaeodactylum tricornutum*..

Phosphorus (as K_2HPO_4; P bags) or nitrogen (as $NaNO_3$; N bags) was added to the bags to final concentrations of 45 and 80 µM, respectively. P and N were added alone or in combination with 5 mg·l^{-1} humic acids (H bags). The humic compound was a commercially available earth extract (Aldrich). Bags without nutrient additions or with only humic acids (C and CH bags) were used as controls.

Daily measurements were made of *in vivo* chlorophyll a, alkaline phosphatase activity (APA), yellow substances, nitrate, ammonium and phosphate. Samples for cell counting were taken daily, but counting and identification were done only for days considered interesting after inspection of the *in vivo* chlorophyll a and APA curves.

Chlorophyll a was measured with a FM3 filter fluorometer (Umeå Instruments AB); APA was measured in a Turner 101 filter-fluorometer following the method of Petterson (10); yellow substances, nitrate, ammonium and phosphate analyses followed normal procedures for seawater (11). Phytoplankton cells were counted in an inverted microscope.

RESULTS

The phytoplankton assemblages in the dinoflagellate bags were dominated by the following dinoflagellate species: *Ceratium furca, C. fusus, C. lineatum and C. tripos; Prorocentrum micans, P. minimum; Dinophysis norvegica, D. acuta* and the silicoflagellate *Distephanus speculum.* Measured as plasma volume dinoflagellates made up more than 80 % of the phytoplankton biomass in bags with the natural phytoplankton assemblages (Fig. 2).

Fig. 1.Experimental site. Gullmar Fjord.

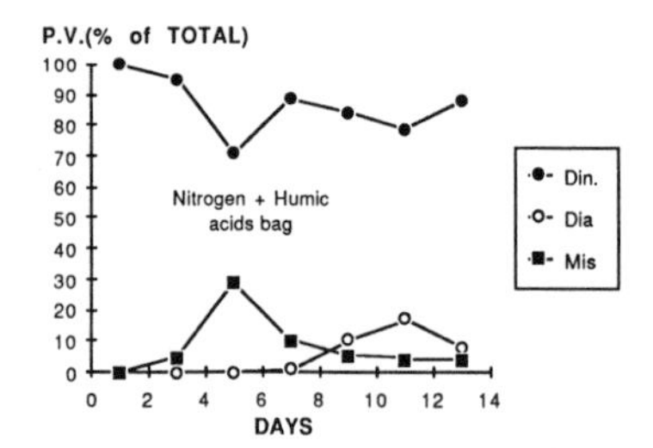

Fig. 2. Percent of the total phytoplankton plasma volume made up by dinoflagellates (Din), diatoms (Dia) and unidentified cells between 1 and 10 µm size (Miscellaneous= Mis)

Nitrogen increased *in vivo* chlorophyll a in the *Phaeodactylum tricornutum* bags when added alone. An even higher increase was seen when nitrogen was added in combination with humic acids (Figs. 3a). Phosphorus addition had no effect compared to control values (Figs.3a)

In the dinoflagellate bags nutrients added alone had no effect on the *in vivo* chlorophyll a whereas a clear increase was seen when nitrogen was added together with humic acids. Phosphorus added together with humic acids increased the *in vivo* chlorophyll somewhat in relation to control bags (Figs. 3 b).

APA was high in the *Phaeodactylum* bags with nitrogen added alone or in combination with humic acids (Fig. 4 a and b), but no APA production was found in the control bags (C and CH). Nitrogen additions also stimulated APA production in the dinoflagellate bags and this effect was also seen in the control bag containing humic acids (CH) but not in the control bag without humic acid additions (C) (Fig.4 c-d).

The added nitrate decreased in all bags but most drastically in the nitrate + humic acids bag with dinoflagellates. Phosphate hardly changed in the *Phaeodactylum* bags but decreased somewhat at the end of the experiments in the dinoflagellate bags. Ammonium increased in all bags but most markedly in the *Phaeodactylum tricornutum* nitrogen + humic acid bag thanks to an extra addition by a seagull which broke through the polyethene roof one night . Yellow substance was appreciably higher in bags containing humic acids than in bags without, and showed no change during the experiment.

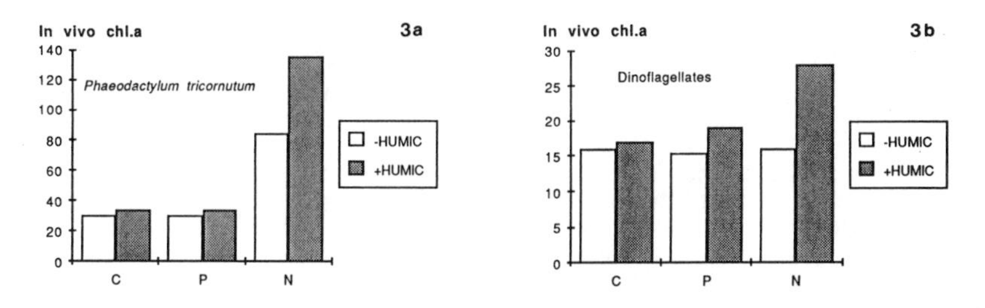

Fig. 3 a-b. Maximal chlorophyll a produced in the bags containing a diatom (3a) and the bags dominated by dinoflagellates (3b). C= control bag, P= phosphorus and N= nitrogen additions.

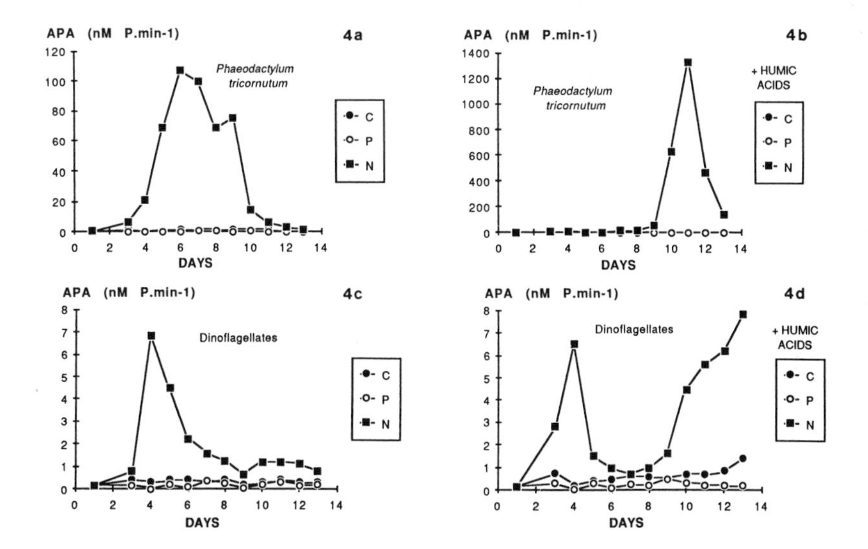

Fig. 4. Alkaline phosphatase Activity (APA) in the diatom (4 a-b) and dinoflagellates (4 c-d). Symbols as in Fig. 3.

DISCUSSION

Competition between different phytoplankton groups or species for nutrients supplied at limiting rates has been proposed as an explanation for phytoplankton dominance and diversity in freshwaters (12, 13, 14, 15). Only recently has experimental research on this subject started in marine waters (16). When comparing diatoms and dinoflagellates it is generally assumed that the former will outcompete the latter if the nutrient supply is in inorganic form, as diatoms normally have higher growth rates than dinoflagellates. It has been shown that humic compounds from soils, rivers, marine sediments and decomposed macroalgae stimulate the growth of dinoflagellates (6,17). Growth enhancement due to humic substances may be one way for dinoflagellates to outcompete the fast growing diatoms.

Granéli et al. (18) working with the diatom *Skeletonema costatum* and the dinoflagellate *Prorocentrum minimum* found that the toxicity of heavy metals was alleviated by chelators (EDTA, NTA), and that the diatom and the dinoflagellate were differently inhibithed depending on the season of the year. The present results show that the diatom *P. tricornutum* was clearly stimulated by nitrogen addition, irrespective of whether humic acids were added together or not. The dinoflagellates could apparently only utilize inorganic nitrogen when it was added in combination with humic acids. This may be an indication that the dinoflagellates were inhibited by heavy metals and that inhibition was alleviated by the chelating action of humic acids.

Addition of humic acids from rivers draining agricultural or forested areas to *Prorocentrum minimum* cultures showed stimulated growth of this dinoflagellate (9). Chemical analysis of the dinoflagellate indicated that it was able to utilize nitrogen from the added humic and fulvic acids. The present results did not show such a stimulation on the growth of a mixed dinoflagellate assemblage. The addition of humic acids did, however, cause an increase of APA in the control (CH) bag, which suggests that organic nitrogen was to some extent available for the dinoflagellates, inducing a phosphorus deficiency. A marked increase in chlorophyll a and APA, as well as a total utilization of all added nitrate in the humus + nitrogen dinoflagellate bag indicate that humic material may be a source of organic phosphorus, or that humic compounds stimulate the production of phosphatases. We conclude that humic material in river water may specifically stimulate the growth of dinoflagellates. However, we do not know which is the underlying mechanism. Most likely both a chelation of toxic metals and a utilization of humic nitrogen and/or phosphorus by the dinoflagellates is involved. More research should be devoted to the possible connection between river transported organic compounds and red tides.

REFERENCES

1. P. Krogh, L. Edler, E. Granéli and U. Nyman, in: Toxic Dinoflagellates, D. M. Anderson, A. W. White and D. G. Baden, eds. (Elsevier Science Publishing Co., Inc.), 501-504, (1985)
2. R. Rosenberg, Mar. Poll. Bull. 16, 227-231, (1985)
3. L. Edler, Limnologica (Berlin) 15, 353-357, (1984)
4. L. Stibe and S. Fleischer, in: Eutrophication situation in the Kattegat, SNV rep. 3272, 26-33, (1986), (in Swedish)
5. A. Prakash and M. A. Rashid, Limnol. Oceanogr. 13, 598-606, (1968)
6. A. Prakash, M. A. Rashid, A. Jensen and D. V. Subba Rao, Limnol. Oceanogr. 18, 516-524 (1973)
7. M.T. Doig, and D. F. Martin, Water Res 8: 601-606, (1974)
8. P. C. De Kock, Science 121, 473-474, (1955)
9. E. Granéli, L. Edler, D. Gedziorowska and U. Nyman, in: Toxic Dinoflagellates, D. M. Anderson, A. W. White and D. G.Baden, eds. (Elsevier Science Publishing Co., Inc.), 201-206 (1985)
10. K. Petterson, Int. Rev. ges. Hydrobiol. 64, 585-607, (1979)
11. R. Carlberg, ICES A, nr. 29, 145 pp, (1972)
12. N. Peterson and D. E. Armstrong, Limnol. Oceanogr. 26, 622-634 (1981)
13. D. Tilman, Science 192, 463-465 (1976)
14. U. Sommer, Arch. Hydrobiol. 96, 399-416, (1983)
15. D. Tilman and R. W. Sterner, Oecologia (Berlin) 61,197-200 (1984)
16. U. Sommer, Limnol. Oceanogr. 30, 335-346, (1986)
17. A. Prakash, in: Fertility of the Sea vol. 2, J. D. Costlow, ed. (Gordon and Breach), New York, pp. 351-368, (1971)
18. E. Granéli, H. Persson and L. Edler, Marine Environ. Res. 18, 61-78 (1986)

SPECIES-SPECIFIC PHOTOSYNTHESIS OF RED TIDE PHYTOPLANKTON IN TOKYO BAY

Myung-Soo Han, Ken Furuya and Takahisa Nemoto
Ocean Research Institute, University of Tokyo, Tokyo 164, Japan

ABSTRACT

Species-specific photosynthesis (SSP) of red tide phytoplankton was investigated in relation to species succession in the inner part of Tokyo Bay, Japan over a one-year period. Carbon uptake rates of individual species were measured by single cell ^{14}C assay for the most frequently occurring five species in surface water. The dense red tides which developed from April through September were formed mainly by *Skeletonema costatum* or *Heterosigma akashiwo*. *S. costatum* occurred rather constantly in the study area throughout a year. In contrast, *H. akashiwo* showed sporadic occurrence with rapid increase and decrease in numerical abundance. SSP tended to be high for both species before blooming but decreased with development of the outburst. Thus, both species had low SSP at their maximum abundance. SSP divided by cell volume was used a as an index of growth rate and defined as volume specific SSP. Volume specific SSP of *S. costatum* tended to increase in less saline waters which were caused by rainfall. It is suggested that elevation of photosynthetic activity is probably a good indication of the initial blooming phase in the inner part of Tokyo Bay.

INTRODUCTION

Tokyo Bay, located in the central part of Japan, is a heavily eutrophicated semi-enclosed body of water. Red tides have been well documented in the inner part. Previous studies revealed that the red tides are characterized by multispecific feature, i.e. several dominant species co-occur [1]. In 1986 and 1987 red tides of *S. costatum* and *H. akashiwo* were frequently observed in the bay. Although species succession between *H. akashiwo* and *S. costatum* has been studied by several workers in different waters [2, 3], succession mechanisms of these populations have remained obscure. Studies on photosynthesis of individual species may explain blooming mechanisms which are a special feature of species succession. This communication reports the species-specific photosynthesis (SSP) of red tide phytoplankton and their succession in Tokyo Bay over a one year period using the single cell ^{14}C assay method [4].

MATERIALS AND METHODS

Field observations were performed at Harumi pier in the inner part of Tokyo Bay, Japan, once or twice a week, from May 1986 through June 1987. Water samples were collected at dawn and prefiltered with 300 µm mesh to remove large zooplankton.

RED TIDES: BIOLOGY, ENVIRONMENTAL SCIENCE, AND TOXICOLOGY
Okaichi, Anderson, and Nemoto, Editors

Water temperature, salinity and nutrients (NO_3, NO_2, NH_4 and PO_4) were measured. Subsamples were fixed immediately with 1 % glutaraldehyde. These were used for species identification, counting, and measuring cell volume.

Other subsamples were prepared by reverse filtration with 20 µm mesh to reduce contamination by nanoplankton (< 20 µm) during the isolation of single cells. The 100 ml subsample was incubated in a teflon-coated bottle with 200 µCi $NaH^{14}CO_3$. Incubation was carried out under 355 $\mu E \cdot m^{-2} \cdot s^{-1}$ at *in situ* temperature. After one to three hours of incubation, individual cells of *Heterosigma akashiwo*, *Skeletonema costatum*, *Thalassiosira binata*, *Eucampia zodiacus* and *Coscinodiscus granii* were isolated under a dissecting microscope at a magnification of x100, and transferred to scintillation vials. Individual cells were acidified directly with 0.5N HCl in the scintillation vials to remove inorganic ^{14}C. Activity was assayed by a liquid scintillation counter using the external standard channel ratio method. Triplicates of labeled medium were taken and acidified for background counting. All ^{14}C measurements were corrected with the background values. ^{14}C uptake for individual cells was calculated by the linear relationship between ^{14}C uptake and number of cells per vial.

RESULTS AND DISCUSSION

S. costatum was the most important species numerically in the bay throughout the year. The main blooming of *H. akashiwo* occurred in late spring, followed by sporadic increases in summer through early autumn.

Cell volume of *S. costatum* fluctuated remarkably over a one year period. Accordingly, volume specific SSP was used to evaluate the activity of individual species. Temporal phase of abundance and volume specific SSP of *S. costatum* (Fig. 1) were not synchronized but rather inversely correlated. Namely, SSP tended to be high during the low cell number period but decreased gradually with the increase of cell numbers.

Spring blooming of *H. akashiwo* continued over about one month in May, 1987 (Fig. 2). In the early stages of blooming, the number of *S. costatum* gradually decreased with the increase of *H. akashiwo* and then increased once again toward the end of blooming. Although its biomass was low, such a trend was also observed for *C. granii*. Thus during the spring blooming period, the volume specific SSPs of *S. costatum*, *H. akashiwo* and *C. granii* were lower than the mean annual value. Volume specific SSP abruptly increased, and then abundance of each species began to rise gradually toward the end of the blooming period. Finally, abundance and volume specific SSP of *H. akashiwo* decreased, while those of *S.*

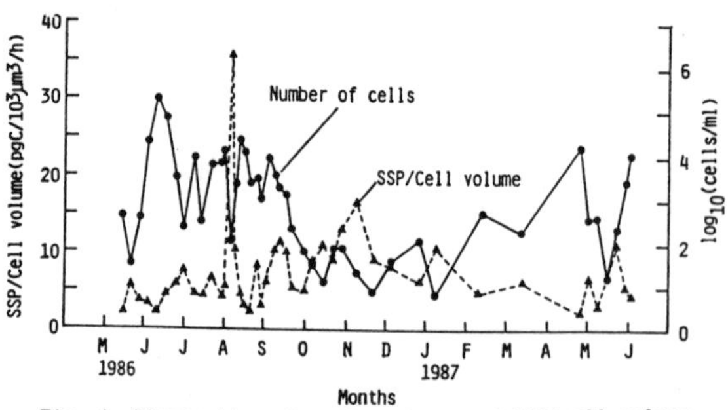

Fig. 1. Fluctuation of cell numbers and SSP/cell volume of *S. costatum*.

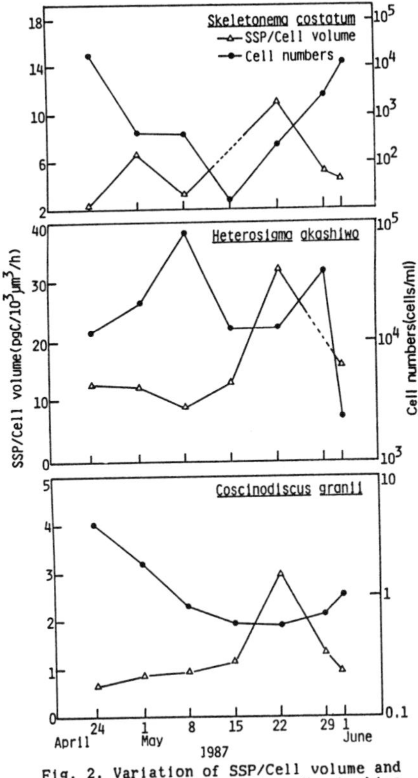

Fig. 2. Variation of SSP/Cell volume and cell number for S. costatum, H. akashiwo and C. granii during the blooming period of H. akashiwo.

costatum increased.

During the this study, salinity changed remarkably (maximum range of variation ± 10) in comparison with water temperature (Fig. 3) within a short time. We examined the effects of short-term variations in salinity on volume specific SSP. Volume specific SSP of S. costatum tended to be high when salinity was greatly depressed due to rainfall. The optimum salinity range of S. costatum, a euryhaline species, has been reported as about 12-29 [5] and the salinity of maximum growth in enriched seawater was found to be 15 [6]. Accordingly, high photosynthetic activity may be attained due to low salinity, which is probably a good indication of the initial blooming phase by different species. Volume specific SSP of S. costatum had no significant relationship with temperature or nutrients (nitrate, nitrite, ammonium and phosphate). Inorganic nutrient salts were high and never limiting to the phytoplankton growth over one year in our study (NO_3: 22.2-118; NO_2:4.4-22.9; NH_4: 88.4-328; PO_4 2.8-12.4 µM).

The relationships between volume specific SSP of S. costatum and temperature, salinity, nitrate, ammonium and cell volume were examined by multiple regression analysis. We found no significant relationship between volume specific SSP and the other variables (5% level, F test). Phytoplankton cells require time to adapt to the changing of environmental conditions, i.e., from a few hours to several days depending upon light intensity [7]. Thus, these statistically insignificant relationships may be a result of time scale discrepancies between the physiological state of individual cells and in situ environmental conditions.

Volume specific SSP were measured for the frequently occurring species in Tokyo Bay. The volume specific SSP on the average throughout the year, of S. costatum (Mean=7.7 pgC/10^3µm^3/h), H. akashiwo (Mean=22.2) and T. binata (Mean=6.7) were high in comparison with large species such as E. zodiacus (Mean=1.7) and C. granii (Mean=1.4). The high volume specific SSP explains the dominance of these relatively small species in the study area. Volume specific SSP of H. akashiwo was, on the average, higher than any other species throughout the investigation, ranging from 8.8 to 4.6 pgC/10^3µm^3/h. This high value seems to be in good accordance with the rapid occurrence

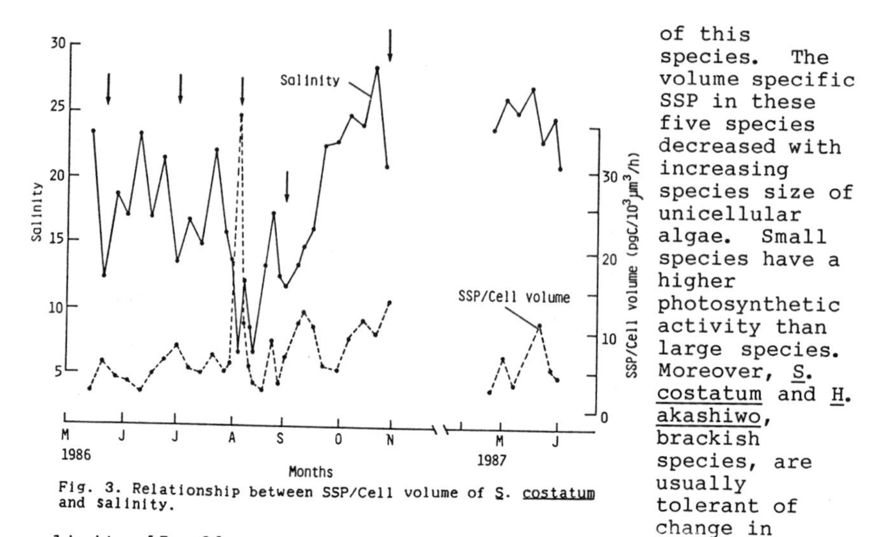

Fig. 3. Relationship between SSP/Cell volume of S. costatum and salinity.

of this species. The volume specific SSP in these five species decreased with increasing species size of unicellular algae. Small species have a higher photosynthetic activity than large species. Moreover, S. costatum and H. akashiwo, brackish species, are usually tolerant of change in salinity [5, 8]. S. costatum has a high relative abundance in all seasons.

We summarize the above discussion as follows: S. costatum and H. akashiwo, eurytolerant species for salinity, have a high photosynthetic activity compared with other species. H. akashiwo showed the highest photosynthetic activity of all species occurring in Tokyo Bay. A high volume specific SSP explains the sporadic occurrence of H. akashiwo in the study area. S. costatum attained an especially high photosynthetic activity in less saline water due to rainfall, which is a good indication of the initial blooming phase. Accordingly, blooms were frequently formed by S. costatum when triggered by low salinity, through spring to autumn in Tokyo Bay.

ACKNOWLEDGMENTS

We wish to thank Drs. E. J. Carpenter and S. Taguchi for their critical comments, Dr. M. Terazaki for cooperation in radioisotope experiments, Drs. S. Nishida and A. Tsuda for their helpful advice and Mr. E.V. Thuesen for correcting the English.

REFERENCES

1. K. Yanami, Bull. Coast. Oceanogr. 16, 112-117 (1979).
2. D. M. Platt, Limnol. Oceanogr. 11, 447-455 (1966).
3. T. Honjo and K. Tabata, Limnol. Oceanogr. 30, 653-664 (1985).
4. R. B. Rivkin and H. H. Seliger, Limnol. Oceanogr. 26, 780-785 (1981).
5. T. Nishizima and Y. Hata, Bull. Japan. Soc. Sci. Fish. 52, 173-179 (1986).
6. H. Curl and G. C. McLeod, J. Mar. Res. 19, 70-88 (1961).
7. A. Richmond, in: Handbook of Microalgal Mass Culture, A. Richmond, eds. (CRC Press, Florida, 1986) pp. 46-69.
8. C. R. Tomas, J. Phycol. 14, 309-313 (1978).

PHOTOPERIODIC REGULATION OF CELL DIVISION AND CHLOROPLAST REPLICATION IN Heterosigma akashiwo

Emi Satoh and Tadashi Fujii
Institute of Biological Sciences, University of Tsukuba, Tsukuba-shi, Ibaraki 305, Japan

ABSTRACT

Cell division and chloroplast replication in Heterosigma akashiwo (Hada) Hada occurred as separate synchronous events during the cell cycle when cells were subjected to light-dark regimes. Under three different photoperiodic cycles of 10L/14D (10-h light/14-h dark), 12L/12D or 16L/8D, cell division began by 19-20th hour and finished at 23-26th hour after the onset of the previous light period, while chloroplast replication began at 20-22nd hour after the onset of the dark period. Almost all the cells divided only once in each 12L/12D cycle. The rate of increase in chloroplast number during one light and dark cycle was always equal to that in cell number in every photoperiod examined.

The minimum light period necessary for both events differed from each other. When the light period was shorter than 6 h, no cell division occurred. When it was shorter than 3 h, no chloroplast replication occurred.

INTRODUCTION

Heterosigma akashiwo (Hada) Hada, which is also known as Olisthodiscus luteus [2], is a marine, raphidophycean, biflagellate, wall-less, unicellular alga. Cell division and chloroplast replication occur as separate synchronous events during the cell cycle under light and dark regimes [1]. However, little is known about the external factors which regulate these divisions. This present study is an attempt to understand the factors regulating cell division and chloroplast replication in H. akashiwo.

EXPERIMENTAL

Growth and maintenance of cells

Heterosigma akashiwo (Hada) Hada was obtained from The Microbial Culture Collection of The National Institute for Environmental Studies, Tsukuba, Japan. A clonal axenic culture strain (NIES-6) was originally isolated in August, 1979 from Tanigawa Fishing Port in Osaka Bay [8]. Cells were cultured axenically in 150 ml of autoclaved modified ASP-7 medium [8] without vitamin mix S_3, in 300-ml Erlenmeyer flasks sealed with silicon caps. All cultures were inoculated with cells at a density of 500-1,000 cells/ml and grown at 20 $\pm$ 1°C under various light-and dark-regimes. Light was applied to the cultures from above using cool day light fluorescent lamps (National; FL 20

RED TIDES: BIOLOGY, ENVIRONMENTAL SCIENCE, AND TOXICOLOGY
Okaichi, Anderson, and Nemoto, Editors

SSD/18) or cool white fluorescent lamps (National; FL 20 SSW/18) and adjusted to an intensity of 0.7 x 10^{16} quanta$\cdot s^{-1} \cdot cm^{-2}$ [7].

Cell counts

Growth was measured by counting the cell number. Approximately 5 µl of 25% glutaraldehyde was added to 1 ml of each culture at various stages of growth, and the fixed cells were counted in a 1-ml counting chamber of the Sedgwick-Rafter type. The relative growth constant (k) during the exponential phase was calculated by the least squares fit :

$$\ln N = \ln N_0 + kt$$

where N_0= initial cell concentration and N = cell concentration at t days after inoculation. The relative cell growth was calculated by the expression: $(N - N_0) / N_0$.

Chloroplast counts

The cultures at various stages of growth were centrifuged at 1,500 rpm for 5 min at 20 °C in a clinical table top centrifuge. The pellet was resuspended in the growth medium to give a final density of approximately 1 x 10^6 cells/ml. Samples of 7.5 µl of the suspension were placed on glass slides, each droplet was covered with a cover glass (18 x 32 x 0.15 mm), and excess fluid was removed with filter paper. The cells were flattened into a single plane of focus, allowing the chloroplasts to be counted easily. Chloroplast counts were carried out under a differential interference contrast microscope. The average number of chloroplasts per cell was calculated from the results of at least 100 cells [7].

RESULTS

Cell growth under different photoperiods

The cell number of each culture was counted daily at hour 2-4 of the light phase. In every culture, the cell number increased exponentially until a cell density of 1-2 x 10^5 cells/ml was limitted by the lack of nitrate. The relative growth constants (k) during the exponential phase were 0.66, 0.44 and 0.41 in the white light, and 0.91, 0.65 and 0.57 in the daylight under 16L/8D, 12L/12D and 10L/14D, respectively. The growth rate increased with the length of the photoperiod and/or the total light energy, and was higher in daylight than in white light.

Synchronous cell division and chloroplast replication

When the cells were cultured under a 12L/12D regime, synchronous cell division and chloroplast replication were observed (Fig. 1). Cell division began by the 7-8th hour of the dark period, and continued until the 0-2nd hour of the next light period. Chloroplast replication began by the 9-10th hour of the light period, and the average number of chloroplasts per cell began to decrease as cells started to divide in the dark period.

Onset of cell division and chloroplast replication

Cells were cultured under three different regimes of

16L/8D, 12L/12D and 10L/14D, and the counting of cell and chloroplast numbers was started on the 5th day after inoculation, when the cell concentration had reached about 1-4 x 10^4 cells/ml. The starting point of each division was determined by examining the point of intersection of a line drawn tangentially to the most steeply increasing part of each curve and the horizontal line representing the lag part of the increase.

Under all regimes, cell division began synchronously by the 4-5th hour before the end of the dark period, i. e., at the 19-20th hour after the onset of the previous light period. Chloroplast replication began by the 2-4th hour before the end of the light period, i. e., at the 20-22nd hour after the termination of the previous irradiation.

Effect of the duration of irradiation for cell division and chloroplast replication

The following experiments were carried out to determine more accurately the requirements of light duration for cell and chloroplast divisidons. Cells were cultured for 5 days under a 12L/12D regime after inoculation, and at the end of the last dark period cells were subjected to light irradiation of various durations. Cell and plastid numbers were counted for 30 h and 18 h, respectively, after the start of irradiation (Fig. 2).

The results showed that (1) no cell division occurred when the duration of light irradiation was shorter than 6 h, (2) the rate of cell division increased with the duration of irradiation when it was longer than 6 h, (3) no chloroplast replication occurred when the duration of light irradiation was shorter than 3 h, and (4) the rate of replication also increased with the duration of irradiation when it was longer than 3 h.

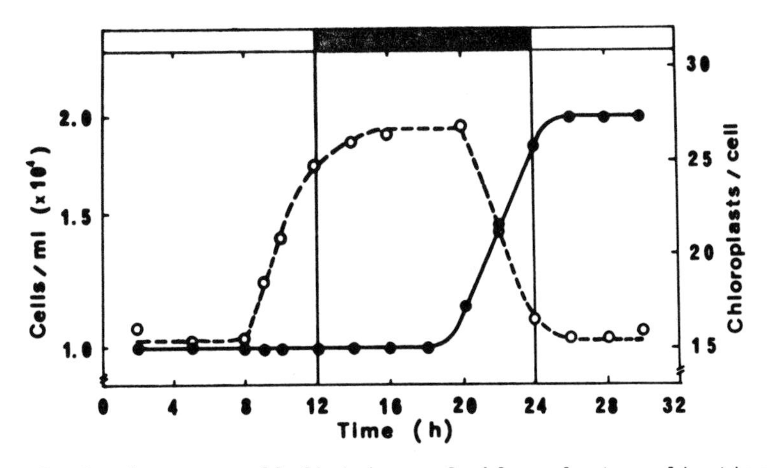

Fig. 1 Synchronous cell division and chloroplast replication in H. akashiwo cultures maintained under an alternating 12L/12D cycle at 20 ± 1℃. During the 12-h light period the culture received illumination at an intensity of 0.7 x 10^{16} quanta•s^{-1}• cm^{-2} using cool daylight fluorescent lamps. Cell number (solid circles) and the average number of chloroplasts per cell (clear circles) were counted every 2 or 3 h.

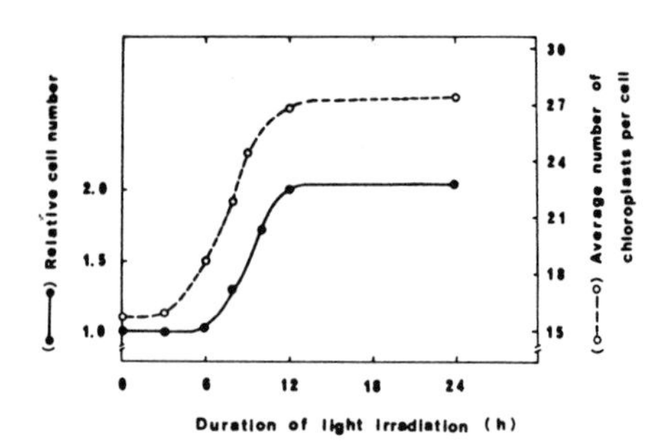

Fig. 2 Effects of light duration on cell division and chloroplast replication. Data showed the relative cell number (solid circles) at hour 30 h and the average number of chloroplasts per cell (clear circles) at hour 18 h after the cells has been subjected to irradiation of different durations.

Discussion

In *Chlorella pyrenoidosa* Emerson strain [5], *Chlamydomous reinhardi* [3] and *Chattonella antiqua* [4], the timing of cell division was shown to be determined by the light-on, while it was determined by the light-off in *Chlorella pyrenoidosa* 211-8b under conditions of high light intensities [6]. In *H. akashiwo* cells the mechanism of the synchronization under light and dark regimes belongs to the former type.

On the other hand, the timing of chloroplast replication in *H. akashiwo* cells appears to be determined by the end of light irradiation under light and dark regimes, although a subsequent irradiation of at least 3 h is required for chloroplast replication (Fig. 2).

These results indicate that the timing of cell divison and that of chloroplast replication are regulated by different signals; cell division by a light-on signal, while chloroplast replication by a light-off signal.

From these findings, *H. akashiwo* cells are regarded as useful for the investigation of biochemical events during cell division and chloroplast replication.

Acknowledgements

We thank Dr. M. M. Watanabe of The National Institute for Environmental Studies, for his help of culture methods.

References

1. Cattolico, R. A., J. C. Boothroyd and S. P. Gibbs, Plant Physiol. 57: 497-503 (1976)
2. Hara, Y., and M. Chihara, Bot. Mag. Tokyo 98: 251-262 (1985)
3. Mihara, S. and E. Hase, Plant Cell Physiol. 12: 225-236 (1971)
4. Nemoto, Y and M. Furuya, Plant Cell Physiol. 26: 669-674 (1985)
5. Pirson, A. and H. Lorenzen, Z. Botan. 46: 53-67 (1958)
6. Pirson., A., H. Lorenzen and H. G. Ruppel, In Studies on Microalgae and Photosynthetic Bacteria (special issue of Plant Cell Physiol.) pp127-139 (1963) The University of Tokyo Press, Tokyo.
7. Satoh, E., M. M. Watanabe and T. Fujii, Plant Cell Physiol. 28: 1093-1099 (1987).
8. Watanabe, M. M., Y. Nakamura, S. Mori and S. Yamochi, Jap. J. Phycol. (sorui) 30: 279-288 (1982)

DETECTION, ENUMERATION AND QUANTIFICATION OF CELL PROPERTIES BY AUTOMATED ANALYSIS

CLARICE M. YENTSCH
Bigelow Laboratory for Ocean Sciences, West Boothbay Harbor, ME 04575

ABSTRACT

Individual particle analysis permits an assessment of variability within a population. Methods are now available which can discriminate various subcomponents of a population as well as size, and simultaneously quantitate various parameters. These methods include flow cytometry/cell sorting and image analysis. Some instruments combine the two basic techniques. Rates of analysis can be up to thousands of particles per minute. Photosynthetic pigments are of primary utility (autofluorescence). Development of various assays permit discrimination of live and dead cells (induced fluorescence via staining) and those cells which are bioluminescent vs. nonbioluminescent (immunofluorescence). Such techniques are useful in laboratory culture experiments as well as permit assessment of properties of toxic dinoflagellates as compared to other subpopulations in the natural environment.

INTRODUCTION

Implicit in particle classification by any automated means is the belief that biological properties are correlated with their volume, composition, form and/or function. To illustrate this point, it is useful to make a list of properties with examples. We compare here particles of 1) sediment, 2) autotroph (microalgae), 3) heterotroph and 4) detritus of equal size/volume. The Basic Biological Index would involve merely measurement of particle volume and plus or minus autofluorescence. The Basic Biological Index is the simplification of biology directed at giving the numbers and sizes of various particles in a given water mass, and characterizing whether or not these are autotrophic cells. With high speed data acquisition and analysis, biological indices can be determined in real time. With additional employment of stains, needed for assessment of live/dead and organic/inorganic, there are unique solutions. This is given in Table 1.

TABLE I. Uniqueness of particle characteristics based on size and fluorescence properties from Yentsch & Spinrad, 1987.

	Basic Biological Index			
	SEDIMENT	AUTOTROPH	HETEROTROPH	DETRITUS
* Size as cell volume or light scatter	10 μm	10 μm	10 μm	10 μm
* Autofluorescence	-	++++	-	±
* Fluorescence live/dead	-	+++	+++	-
* Fluorescence organic/inorganic	-	++	++	++

RED TIDES: BIOLOGY, ENVIRONMENTAL SCIENCE, AND TOXICOLOGY
Okaichi, Anderson, and Nemoto, Editors

The good news:

* Fragile cells, specifically naked dinoflagellates such as *Gyrodinium aureolum* are accurately enumerated and quantified. In that we work with fresh material, the traditional problems encountered with preservation are by-passed.

* Automated analysis can be very rapid, with hundreds to thousands of cells fully analyzed in minutes. Extended data analysis permits useful graphical presentation.

* An *in situ* device is under development by Flow Vision Analyzer, Inc.

* Coupled with a CTD and nutrient analyzer, biological indices of the water column can be assessed simultaneously with physical, chemical properties on shipboard or on buoys.

* Particles in flow can receive simultaneous assessment of size (volume), numbers per unit volume of water analyzed, and plus or minus autofluorescence from chlorophyll.

* Live vs. dead cells can be discriminated by the introduction of an inexpensive viability stain, fluorescein diacetate (FDA) (Sigma Chem. Co.) without detriment to the chlorophyll fluorescence signal (Figure 1).

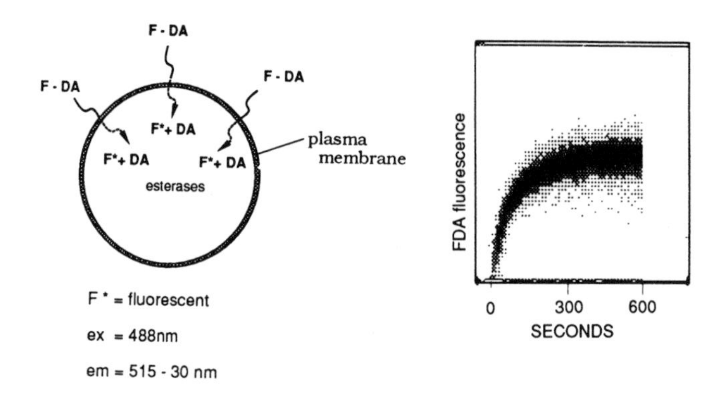

FIG. 1. Fluorescein diacetate (FDA) as a metabolic vigor or live-dead cell stain. Left. Cell schematic: stain FDA is colorless and readily permeates plasma membrane. Inside a living cell esterases cleave the ester linkage. The fluorescein then fluoresces. The fluorescein is not fluorescent in dead cells. Right. Kinetics of FDA stain cleavage in cells of *Exuviella* sp (Exuv) over 600 seconds (10 minutes). Data are for individual cells measured on a Coulter EPICS V Flow Cytometer. For details, see Dorsey *et al.*, in review.

What cannot be accomplished under the current state of development is:

* Simultaneous analysis and processing of small cells (e.g. cyanobacteria ~ 1μm diameter) and large cells (e.g. dinoflagellates ~ 20-50μm diameter). One must focus on one log scale decade. Thus for the purposes here, we focus on cells in the 10-100μm cell diameter size range.

* One cannot discriminate toxic vs. non-toxic species which are similar in size/fluorescence.

* One cannot process more than a few milliliters of water, thus these automated techniques are unsuitable for such organisms as *Dinophysis* which are problematic at only a few hundred cells per liter.

* Species identification is impossible, unless coupled with a whole cell Hiroishi et al, this volume). Such tools are expensive, but when a definitive identification is required, the coupling of techniques is feasible.

Flow cytometry (FCM) is literally the measurement of cells in a flow system. Inert particles as well as cells can be easily measured. Single-celled organisms common in the oceans range in size from <1.0 µm to greater than 100 µm expressed as cell diameter. If expressed as cell volume, this is approximately 0.5 μm^3 to 500,000 μm^3!

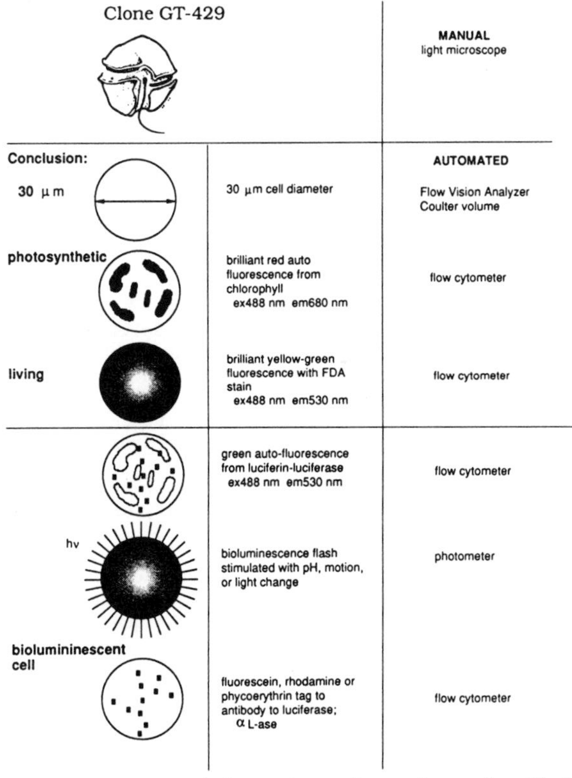

FIG. 2. Manual vs. automated detection. Comparison of cell discrimination based on parameters of cell diameter, chlorophyll autofluorescence, live/dead staining and metabolic vigor with FDA stain, bioluminesced potential based on luciferin-luciferase fluorescence vs. bioluminescence flash vs. immunoreagent, antiluciferase.

Surprisingly, the variability within a clonal population of cells (originating from a single parent cell) can be as great as the variability from one population to another. By defining the limits of this variability, we are able to make some generalizations with confidence, in fact, to compress information. By making observations of a population on a particle-by-particle basis, the variability within a population is easily assessed. In addition, another lever of resolution may be obtained through the simultaneous measurement of two or more parameters for each particle. With flow measurements, statistically significant numbers of particles may be analyzed for a large range of particle concentrations in near real time.

The approach presented here is an extremely simplistic "key" to particles (Figure 2). The approach lends itself to simple, easily operated, automated instrumentation.

Information about the particles *in situ* can be obtained without manipulation, filtration, incubation, extraction and/or preservation, common sources of artifacts. In order to be true partners making real-time shipboard decisions, a "least common denominator" must be universally applied. Cell size, expressed as volume, accounts for greater than two-thirds of the variability found in metabolism in living organisms and has predictable effects on light scatter and absorption. At last, adequate technology is available to test the theoretical basis *in situ* at sea.

Because our objective is to describe, understand and predict the dynamics of optics and biology of a water mass, then classifying the particles into subsets is required. Are the particles: a) a living cell highly interactive with its immediate environment? or b) dead, thus only requiring a suitable environment to degrade and decompose?

REFERENCES

1. Cucci, T.L., S.E. Shumway, R.C. Newell, R. Selvin, R.R.L. Guillard and C.M. Yentsch. 1985. Flow cytometry: a new method of differential ingestion, digestion and egestion by suspension feeders. Mar. Ecol. Prog. Ser., 24: 201-204.
2. Dorsey, J., S. Mayo, C. McKenna and C.M. Yentsch. A rapid analytical technique for assessment of cell viability in marine phytoplankton. In review.
3. Spinrad, R.W. and C.M. Yentsch. 1987. Observations on the intra- and interspecific single cell optical variability of marine phytoplankton. Applied Optics, 26: 357-362.
4. Yentsch, C.M., P.K. Horan, K. Muirhead, Q. Dortch, E. Haugen, L. Legendre, L.S. Murphy, M.J. Perry, D.A. Phinney, S.A. Pomponi, R.W. Spinrad, M. Woods, C.S. Yentsch, and B.J. Zahuranec. 1983. Flow cytometry and cell sorting: a powerful technique for analysis and sorting of aquatic particles. Limnol. Oceanogr., 28(6): 1275-1280.
5. Yentsch, C.M. and S.A. Pomponi. 1986. Automated individual cell analysis in aquatic research. International Review of Cytology, 105: 183-243.
6. Yentsch, C.M. and R.W. Spinrad. 1987. Particles in Flow, Marine Tech. Soc. Journal, 21: 58-68.

DOES *GYRODINIUM AUREOLUM* HULBURT PERFORM DIURNAL VERTICAL MIGRATIONS?

EINAR DAHL* AND UWE H. BROCKMANN**
*Flødevigen Biological Station, N-4800 Arendal, Norway
**Institut für Biochemie und Lebensmittelchemie, Martin-Luther-King-Platz 6, D-2000 Hamburg 13, Federal Republic of Germany

ABSTRACT

A natural water column was enclosed in a transparent plastic bag, 1 m in diameter and 13 m deep. The vertical distribution of selected phytoplankton species, among them *Gyrodinium aureolum* Hulburt, was evaluated by sampling each meter both day and night during four days with clear weather.

The water column was stratified. At the onset of the experiment some measured parameters at the surface and bottom of the bag were: temperature, 8.6 and 11.0^{o}C; salinity, 28.26 and 32.02 o/oo; orthophosphate, 0.19 and 0.42 µM; and nitrate, 0.21 and 4.28 µM.

G. aureolum did not reveal a distinct diurnal vertical migration, although the results may indicate a diurnal positive phototaxis in the scale 1-2 m per 12 h. The alga formed pronounced subsurface maxima in the discontinuity layer both day and night.

INTRODUCTION

Gyrodinium aureolum was recorded for the first time in European waters in 1966 [1]. Blooms of this alga have been widespread [2] ever since, and heavy blooms have affected fish [3] and caused mortality among fish and other organisms [4, 5].

Migration patterns are important factors in the growth strategy of dinoflagellates [6]. In several field studies *G. aureolum* has shown phototaxis [4, 7]. However, in studies of a culture of the alga in a large plastic bag no diurnal migration was revealed [8]. This communication presents results from further studies on vertical migration of *G. aureolum* in a large plastic bag.

EXPERIMENTAL

The experiment was performed at the south coast of Norway (Risør). A natural, 13 m deep, water column was on 28 October 1985 isolated by raising a flat, transparent plastic bag, 1 m in diameter, from 13 m depth to the surface [9].

During four days, 29 October - 1 November, the vertical distribution of selected phytoplankton species was evaluated by sampling 100 ml around noon and midnight, each meter by means of a tube and vacuum. The samples were preserved by iodine and the phytoplankton were identified and quantified with a microscope after concentration by sedimentation and/or centrifugation.

Additional parameters recorded in the bag at the onset and the end of the experiment were: temperature, salinity, oxygen, orthophosphate, nitrite, nitrate, ammonium and chlorophyll *a*.

Light was measured by a Li-Cor quantum sensor mounted on a pier nearby the experimental site.

RED TIDES: BIOLOGY, ENVIRONMENTAL SCIENCE, AND TOXICOLOGY
Okaichi, Anderson, and Nemoto, Editors

RESULTS

The environmental conditions

The natural dark and light cycle was regular during the experiment (Fig. 1) with a light period from about 0800 to 1600.

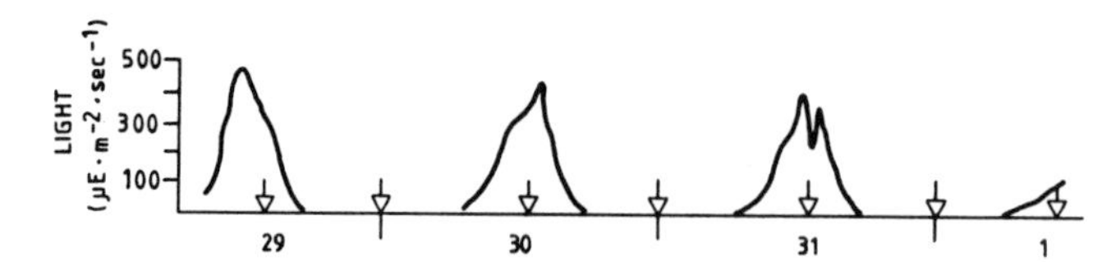

Fig. 1. The light during the experimental period 29 October - 1 November. The arrows show sampling point of time.

The temperature, salinity and nitrate conditions in the bag during the experiment were relatively stable except for a minor temperature increase (Fig. 2). The enclosed water column was stratified with a nitrate poor mixed surface layer of lower salinity and a warmer bottom layer with a higher salinity separated by a discontinuity layer from about 4 to 9 m depth (Fig. 2)

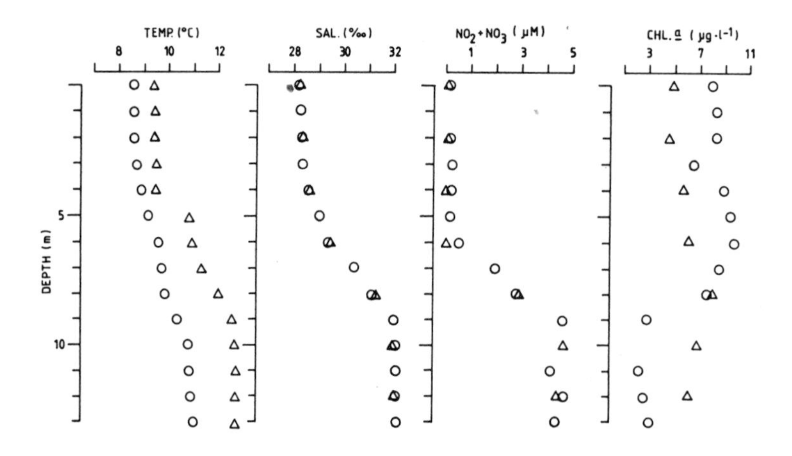

Fig. 2. Temperature, salinity, nitrite + nitrate and chlorophyll a in the bag. o - 29 October, △ - 31 October.

The data on orthophosphate gave a similar picture with 0.2 µM and 0.4 µM as typical concentrations above and below the discontinuity layer respectively. The oxygen saturation above and below the discontinuity layer was 100-110% and 70-80% respectively. Ammonium concentrations were more uniform all through the water column with 0.2-0.4 µM as typical concentrations at the onset of the experiment and 0.5-0.7 µM at the end.

At the onset of the experiment the concentrations of chlorophyll a were about 8 µg per litre above the discontinuity layer and about 3 µg per litre below (Fig. 2). At the end the values were about 5 and 7 respectively due to sinking of diatoms. G. aureolum was the most numerous among the dinoflagellates, but large species as Ceratium spp. contributed more significantly to the total phytoplankton biomass.

Vertical distribution of Gyrodinium aureolum

The first three days of the experimental period G. aureolum formed subsurface maxima in the discontinuity layer during night and day, however, slightly more shallow during day (Fig. 3). The last day and night two subsurface maxima were recorded at night and three small during day. All through the experimental period the concentration of the alga in the surface was always higher during day than during night. Below the discontinuity layer, from 8 m depth and deeper, only low concentrations of G. aureolum were recorded both day and night. The growth rate during the experiment could be calculated to the range 0.4-0.7 divisions per day.

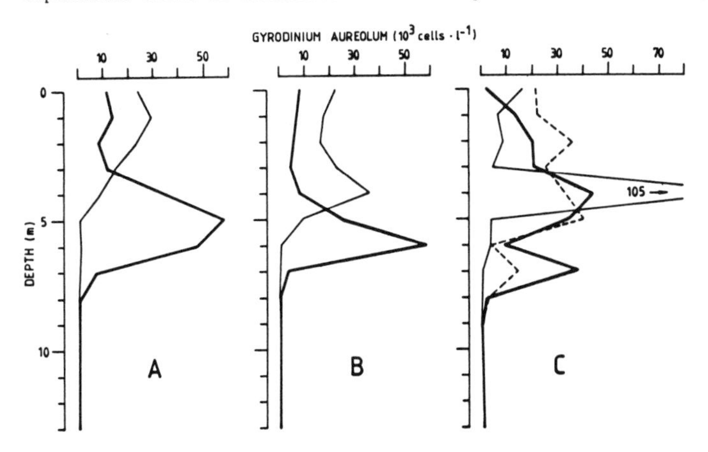

Fig. 3. The vertical distribution of Gyrodinium aureolum in the bag during the experimental period. The heavy lines are night profiles. A - 29 October, B - 30 October, C - 31 October and 1 November (broken lines). See also Fig. 1.

DISCUSSION

The results from the first three days of the experimental period indicate that G. aureolum may perform diurnal vertical migrations in the scale 1-2 m per 12 h which correspond to the swimming velocity recorded in laboratory [10]. Both the changes of concentrations of the alga in the surface and the vertical distribution of the subsurface maximum during day and night indicate a positive phototaxis involved, as have been observed from field studies [4, 7].

Migration was mainly within the discontinuity layer. The migration pattern together with the distinct population maximum formed in the discontinuity layer have implications for the future sampling strategy.

The recording of two and three subsurface maxima of G. aureolum at the end of the experiment complicate the migration pattern. The nutrient status of the cells as well as changes in physical parameters may modulate patterns of the diurnal vertical migration [11, 12]. Studies in a plastic bag of a dense natural bloom with only few other species present would be useful to get more specific information on factors modulating the vertical migration of G. aureolum.

ACKNOWLEDGEMENTS

The authors with to thank V. Fosback, B. Lundin, S.E. Enersen, E.O. Maløen and Ø. Paulsen for technical assistance in the field and in the laboratory, and colleagues for valuable comments on the manuscript.

REFERENCES

1. T. Braarud and B.R. Heimdal, Nytt Mag. Bot. 17, 91-97 (1970).
2. F. Patensky and A. Sournia, Cryptogamie Algologie 7, 251-275 (1986).
3. G.W. Potts and J.M. Edwards, J. mar. biol. Ass. U.K. 67, 293-297 (1987).
4. E. Dahl, D.S. Danielssen, and B. Bøhle, Flødevigen rapportserie 4, 1-15 (1982).
5. K.J. Jones, P. Ayres, A.M. Bullock, R.J. Roberts, and P. Tett, J. mar. biol. Ass. U.K. 62, 771-782 (1982).
6. P.M. Holligan in: Toxic Dinoflagellages, D.M. Anderson, A.W. White and D.G. Baden, eds. (Elsevier, Amsterdam 1985) pp. 133-139.
7. K. Tangen, Sarsia 63, 123-133 (1977).
8. E. Dahl and U.H. Brockmann in: Toxic Dinoflagellates, D.M. Anderson, A.W. White and D.G. Baden, eds. (Elsevier, Amsterdam 1985), pp. 233-238.
9. U.H. Brockmann, E. Dahl, J. Kuiper, and G. Kattner, Mar. Ecol. Prog. Ser. 14, 1-8 (1983).
10. E. Bauerfeind, M. Elbrächter, R. Steiner, and J. Throndsen, Mar. Biol. 93, 323-327 (1986).
11. J.J. Cullen and S.G. Horrigan, Mar. Biol. 62, 81-89 (1981).
12. S.I. Heaney and R.W. Eppley, J. Plankton Res. 3, 331-344 (1981).

EFFECTS OF IODIDE AND IODATE IONS ON MARINE PHYTOPLANKTON

HIROYUKI FUSE, OSAMU TAKIMURA AND YUKIHO YAMAOKA
Government Industrial Research Institute, Chugoku, 2-2-2
Hirosuehiro, Kure, Hiroshima, Japan

ABSTRACT

The effects of iodide and iodate ions on the growth of 5 species of phytoplankton and their accumulation of iodine were investigated. Iodide and iodate ions did not inhibit the growth of *Dunaliella* sp., *Thalassiosira weissflogii*, *Gymnodinium sanguineum*, and *Heterosigma akashiwo* at concentrations of 10^{-4}M or higher. The growth of *Chattonella antiqua* wasn't inhibited by iodate ions up to 10^{-6}M. The concentration of iodide ion which began to inhibit the growth of *C. antiqua* was between 10^{-7} and 10^{-6}M. Iodide was preferentially taken up over iodate. *C. antiqua* and *T. weissflogii* accumulated the most iodine. Most of the iodine in *C. antiqua* was lipid iodine, while that in *T. weissflogii* was water soluble.

INTRODUCTION

Most iodine in seawater is in the form of iodide and iodate. The iodide concentration is about 10^{-7}M and the iodate concentration about 3×10^{-7}M in surface seawater [1]. Some species of algae have been noted for their ability to accumulate iodine [25], often preferentially over iodate [6,10]. Most of the iodine in algae has been reported to be iodide [3,7], but some iodinated organic compounds have been identified [8,9]. The marine brown alga *Ectocarpus siliculosus* required more than 1.5×10^{-9}M iodide for its growth [10]. 4×10^{-6}M iodide favored the growth of the marine diatom *Navicula* sp. whereas the same concentration of iodate prevented its growth [11].

This report describes the effects of iodide and iodate ions on the growth of phytoplankton (including red tide species) and documents their accumulation of iodine.

METHODS

Organisms

Dunaliella sp. (Chlorophyceae), *Thalassiosira weissflogii* (Bacillariophyceae), *Gymnodinium sanguineum* (Dinophyceae), *Chattonella antiqua* (Rhaphidophyceae), and *Heterosigma akashiwo* (Rhaphidophyceae) were used. All but *Dunaliella* sp. were isolated from the Seto Inland Sea. The origin of *Dunaliella* sp. is unknown. The same strains were used in a previous paper on this topic [12].

Growth Experiments and Accumulation of Iodine

Various levels of iodide or iodate were added as sodium salts into AQUIL medium [13] to produce the following concentrations: 0, 10^{-9}, 10^{-8}, 10^{-7}, 10^{-6}, 10^{-5}, 10^{-4}, and 10^{-3}M. 0.2ml aliquots of stock cultures were inoculated into 30ml of medium in glass test tubes.

RED TIDES: BIOLOGY, ENVIRONMENTAL SCIENCE, AND TOXICOLOGY
Okaichi, Anderson, and Nemoto, Editors

Growth conditions and measurements were the same as reported previously [12]. The 30ml cultures were filtered and rinsed once they reached plateau phase growth [12]. The filters were then used for the determination of iodine.

Iodine and Chlorine Content

Each 17,100 strain was grown in Iwasaki medium [14] containing 10^{-3}M iodide (mg Cl/l, 127mg I/l) except for C. antiqua. It was grown on the Iwasaki medium without added iodide and after it grew well, iodide was added to a final concentration of 10^{-3}M and then the culture was allowed to grow for two days. The cultured cells were centrifuged, rinsed with 0.5M $NaNO_3$ and freeze-dried. These were used for the determination of iodine and chlorine content and for the extraction.

Extraction and Assay of Iodine and Chlorine

About 30mg of freeze-dried cells were extracted with 15ml of methanol:toluene (3:1). The extract was separated from the residue by centrifugation. 7.5ml of 1M sodium nitrate solution was added to the extract and the toluene layer was separated by centrifugation [15]. Iodine in the toluene layer and the residue were determined as follows. Samples were first combusted in oxygen flasks [21]. The iodine or chlorine concentration of the absorption solution was then determined as reported by Yonehara [16] and Iwasaki [17].

RESULTS

Effects of Iodide or Iodate on Growth

The growth of five species of phytoplankton was monitored in the media with various concentrations of iodide or iodate. With iodide, the growth of Dunaliella sp., T. weissflogii, and H. akashiwo did not vary between treatments. The growth rate of G. sanguineum decreased at 10^{-3}M.

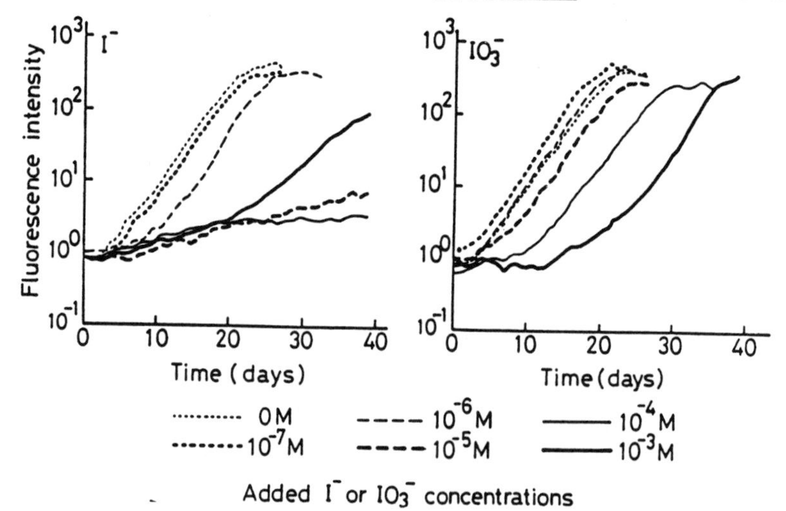

FIG. 1. Effects of iodide or iodate concentrations on the growth of Chattonella antiqua.

C. antiqua had a prolonged lag phase at 10^{-6}M and did not grow at 10^{-5} and 10^{-4}M (Fig. 1). With iodate, the growth of Dunaliella sp. and G. sanguineum was not affected, whereas the growth rate of T. weissflogii and H. akashiwo decreased at 10^{-3}M. C. antiqua had a prolonged lag phase at 10^{-4}M or higher concentrations (Fig. 1).

Effects of Iodide or Iodate Concentrations on Uptake

The amounts of iodine taken up by the algal cells from the culture medium in the uptake experiments are shown in Fig. 2. Of the five species, C. antiqua took up iodine the most effectively at low iodide concentrations; it took up 10% in medium containing 10^{-6}M iodide. This is about the same percentage as reported for Navicula sp. [11]. T. weissflogii took up the largest amount iodide of five species at high iodide concentrations. Iodide was taken up more effectively than iodate when it did not inhibit growth.

Iodine Content

The iodine and chlorine content of the freeze-dried cells of T. weissflogii and C. antiqua and the iodine content of the lipids and the residues were determined. The percent of iodine as a fraction of total iodine in the cells was calculated. Iodine in the extract and in the residue was 65% and 8% in C. antiqua, and 7% and 4% in T. weissflogii, respectively. The remainder of the iodine was water soluble.

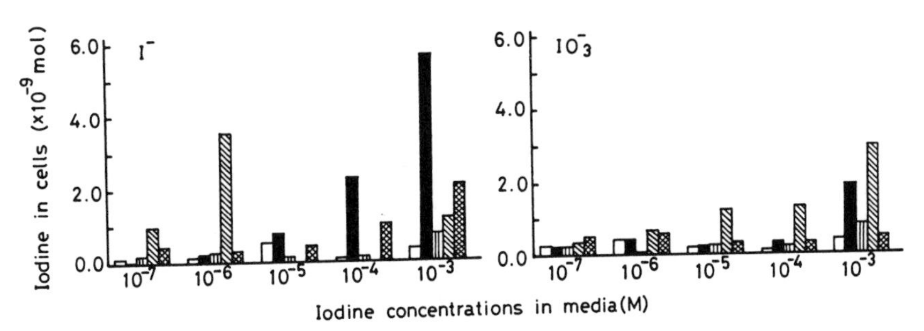

FIG. 2. Uptake of iodide and iodate by phytoplankton

TABLE I. Iodine and chlorine content of algal cells

Species	Cl	I
	(mg / g dry cells)	
Dunaliella sp.	0.6	0.007
Thalassiosira weissflogii	5.0	2.62
Gymnodinium sunguineum	8.8	0.075
Chattonella antiqua	2.6	0.020
Heterosigma akashiwo	22.9	0.853

DISCUSSION

Iodide began to affect the growth of C. antiqua at concentrations between 10^{-7} and 10^{-6}M in this experiment. The iodide concentrations of inshore seawater are reported to be high because of the iodate-reducing substances present in terrestrial run-off [18]. Concentrations of 2.8×10^{-7}M have been measured in the Yarra River estuary, Australia [19]. Iodine can also be oxidized and reduced biologically [20]. Iodide may thus affect the growth of C. antiqua in some areas. The lipid composition of C. antiqua was investigated in detail [22], and fish-toxicity was attributed to free fatty acids [23]. It is noteworthy that this species' organic iodine compounds have not yet been studied in the context of toxicity. Research is now in progress to further investigate iodine in C. antiqua.

ACKNOWLEDGEMENTS

The authors thank Dr. H. Takano for identification of T. weissflogii.

REFERENCES

1. S. Tsunogai and T. Henmi, J. Oceanog. Soc. Japan, 27, 67 (1971).
2. A. Yasui, H. Koizumi, and C. Tsutsumi, Rept. Natl. Food Res. Inst., 37, 163 (1980).
3. M.A. Amat and L.M. Strivastava, J. Phycol., 21, 330 (1985).
4. T.I. Show, Proc. R. Soc. London, Ser. B, 150, 356 (1959).
5. H.G. Klemperer, Biochem. J., 67, 381 (1957).
6. S. Hirano, T. Ishii, R. Nakamura, M. Mitsuba, and T. Koyanagi, Radioisotopes, 32, 319 (1983).
7. H. Megro, T. Abe, T. Ogasawara, and K. Tsuzumura, Agr. Biol. Chem., 31, 999 (1967).
8. S.L. Neidleman and J. Geigert, Biohalogenation (Ellis Horwood, Chichester 1986) pp. 39-45.
9. K.W. Glombitza and G. Gerstberger, Phytochemistry, 24, 543 (1985).
10. M.L. Woolery and R.A. Lewin, Phycologia, 12, 131 (1973).
11. K. Sugawara and K. Terada, Inform. Bull. Planktol. Japan, 213 (1967).
12. H. Fuse, Agr. Biol. Chem., 51, 987 (1987).
13. F.M.M. Morel, J.G. Rueter, D.M. Anderson, and R.R.L. Guillard, J. Phycol., 15, 135 (1979).
14. H. Iwasaki, Biol. Bull., 121, 173 (1961).
15. R.H. White and L.P. Hager, Anal. Biochem., 78, 52 (1977).
16. N. Yonehara, Bull. Chem. Soc. Jpn., 37, 1101 (1964).
17. I. Iwasaki, S. Utsumi, and T. Ozawa, ibid., 25, 226 (1952).
18. V.W. Truesdale, Mar. Chem., 6, 1 (1978).
19. J.D. Smith and E.C.V. Butler, Nature, 277, 468 (1979).
20. R.S. Gozlan and P. Margalith, J. Appl. Bact., 36, 407 (1973).
21. W. Schöniger, Mikrochim. Acta, 869 (1956).
22. Y. Yamaoka and O. Takimura, Nippon Kagaku Kaishi, 1488 (1985).
23. T. Okaichi and S. Nishio, Kagoshimawan Akashio Hassei Gen'in Chosa Kenkyu Houkokusho (1978) p.77.

GROWTH RESPONSES OF NATURAL POPULATIONS OF RED TIDE ORGANISMS TO NUTRIENT ENRICHMENT

Okamoto, K.* and R. Hirano**
* Fisheries Laboratory, Faculty of Agriculture, The University of Tokyo, Bentenjima, Hamana-gun, Shizuoka 431-02, Japan
** Faculty of Agriculture, The University of Tokyo, Tokyo, Japan

ABSTRACT

Nitzschia closterium and *Prorocentrum minimum* are the main organisms causing red tides in a brackish lake, Hamana-ko. Natural populations dominated by one of these species were inoculated into filtered lakewater collected at the same site and enriched with various amounts of nutrients. Inoculations were also made into media prepared using aged sea water which was treated with activated charcoal. *N. closterium* always grew well in nutrient-rich media but lag phase was prolonged when the population was taken from the decline phase of a red tide. *P. minimum* obtained from the initial phase of a red tide grew well in unenriched filtered lakewater, and growth was further enhanced by enrichment. However, *P. minimum* obtained from the decline phase of a red tide declined rapidly even if inoculated into enriched medium.

The maximum growth rates of these species were calculated to be 0.9-1.4 div./day for *P. minimum* and 1.5-3.5 div./day for *N. closterium* .

INTRODUCTION

In natural conditions, the growth of red tide organisms may fluctuate with changes in environmental or biological factors. So far little research has been done on the effects of changes in environmental factors on growth of natural populations in different physiological states. In this study, growth responses of red tide organisms sampled from natural populations were examined following nutrient enrichment.

EXPERIMENTS

Several species of phytoplankton cause red tides frequently in the Hamana-ko, a brackish water lake. In particular, the diatom, *Nitzschia closterium* and the dinoflagellate, *Prorocentrum minimum* cause red tides almost every year. Natural populations dominated by one of these species were collected from the surface layer of the lake and were inoculated within a day into three kinds of culture media which had different nutrient concentrations. These culture media were prepared by using lake water and aged sea water as follows. The first type was prepared by filtering the sample water (millipore HA), and the second medium comprised filtered sample water enriched with nutrients. Nutrient-enriched media corresponded to " F "

RED TIDES: BIOLOGY, ENVIRONMENTAL SCIENCE, AND TOXICOLOGY
Okaichi, Anderson, and Nemoto, Editors

[1] or "SW II " [2]. The third type was prepared using aged sea water, which was diluted with double-distilled water to a salinity equal to that of the sample waters. This aged sea water was treated with activated charcoal, in order to remove vitamins. Red tide water samples were inoculated into these media at the ratio of 1/100 to 1/1000 of total volume and were incubated under the simulated temperature and photoperiod. Non-treated sample water was also incubated at the same time. Cell number of each culture was monitored by microscope .

RESULTS

Growth responses of Nitzschia closterium to nutrient enrichment

Serial culture experiments of Nitzschia closterium were performed on September, 1983. Red tide dominated by N. closterium was monitored every 3 to 4 days and samples from the phase which were close to the maximum stage, declining phase and initial phase of next blooming were used for culture experiments. Nutrient-enriched media corresponded to " F " [1] and inoculum ratio was 1/1000 of total volume.

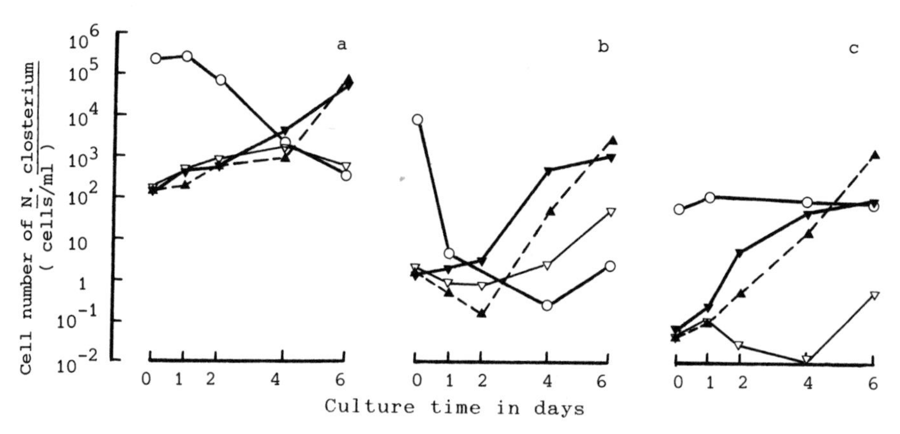

Fig.1 Growth responses of N. closterium to nutrient enrichment. Each symbol shows cell number in the media of ; ▼——▼ : filtered sample water, ▲- - -▲ : nutrient enriched filtered sample water, ▽——▽ : aged sea water treated with activated charcoal, ○——○ : non-treated sample water.

In all three culture experiments (Fig.1 a-c), N. closterium grew well when inoculated into the nutrient-enriched media, however the lag phase was prolonged by about 2 days when the population was obtained from the decline phase of the red tide (Fig.1 b).

The growth rate of N. closterium cells inoculated into the media prepared using aged sea water (Fig.1 c) was faster in the first few days but the cell number after 6 days was lower when compared with the inoculation into nutrient enriched media.

Growth responses of Prorocentrum minimum to nutrient enrichment

Culture experiments of Prorocentrum minimum were not serial, the inoculum having been obtained from three different red tides. The first batch of experiments were done on May 1982 by using two kinds of media (Fig. 2 a). One was filtered sample water and the other was aged sea water treated with activated charcoal. Nutrient enriched media was not prepared. P. minimum grew actively in all the experiments with a mean maximum growth rate of 1 div./day.

The second batch of experiments were carried out on June 1982 with populations in the decline phase of a red tide (Fig. 2b). The media, enriched by adding either N or P to the filtered sample, were incubated at the same time. The cell number of P. minimum declined after 1 or 2 days of plateau phase even though the media were enriched with nutrients.

The third batch of experiments were done in September 1982 with all combinations of media (Fig. 2 c). Nutrient-enriched media corresponded to " SW II " [2]. In all the experiments P. minimum grew well, but growth rate was enhanced by nutrient enrichment, and the cell number after 2 days was higher than in the other treatments.

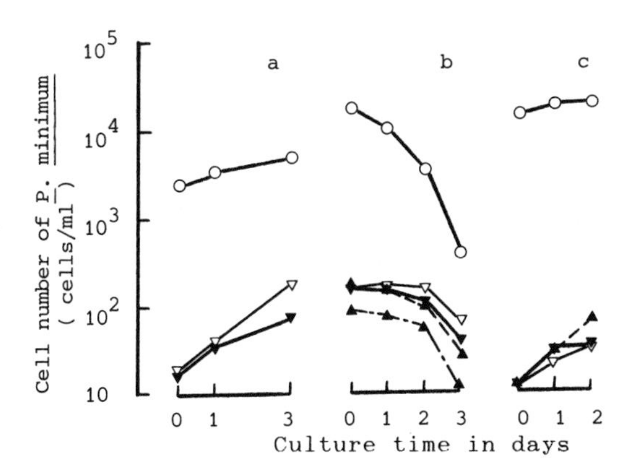

Fig.2 Growth responses of P. minimum to nutrient enrichment. Each symbol shows cell number in the of ; ▼——▼ : filtered sample water, ▼- - - -▼ : nutrient enriched sample water (for Fig.2b, ▼- - - -▼ : N enriched, ▼—-—▼ : P enriched), ▽——▽ : aged sea water treated with activated charcoal, ○——○ : non-treated sample water.

DISCUSSION

Nitzschia closterium grew well in nutrient-enriched media. But there was a delay in the intial phase of growth in nutrient enriched media when compared with medium prepared using the aged sea water. This delay might be due to the lag time needed by cells for adaptation to the environment of nutrient enriched media. Similar observations were also made by Spencer [3] in his culture experiment with Phaeoductylum tricornutum. Prolongation in lag phase with the inocula of populations obtained from the decline phase of red tides also agrees with the results of Spencer [3]. However, N. closterium grew very fast : maximum growth rates were as high as 1.5 - 3.5 div./day.

P. minimum needed longer than N. closterium to recover from decline phase to growth phase, even if enriched with N or P. P. minimum in decline phase might not be able to utilize nutrients for growth as rapidly as a population in the initial phase of a red tide. However, because of the short experimental duration, it was not clear how many hours were needed for the recovery. The maximum growth rates of P. minimum were calculated to be 0.9 - 1.4 div./day. Although these values are somewhat higher than that of natural populations of P. minimum (0.55 div./day, [4]), this result shows that P.minimum has the potential to achieve a maximum growth rate of 1 div./day under natural conditions, similar to the one observed in the case of the naked dinoflagellate, Gymnodinium nagasakiense [5].

One possible reason for the different growth responses of these two species can be due to the difference in storage capacity of nutrients in these two species. Dorch [6] showed that the dinoflagellate, Amphidinium carterae has a comparatively smaller capacity to store nitrate than the diatom, Skeletonema costatum. Hence, it might be possible that P. minimum has a smaller capacity to store nutrients than N. closterium, and has to assimilate nutrients as its internal pools are used. This difference would mean that P. minimum takes a longer time to assimilate nutrients than N. closterium, thus resulting in different growth responses.

ACKNOWLEDGEMENTS

The authors wish to thank Dr. Fukuyo and Dr.Ishimaru for their invaluable advise on preparation of this manuscript. We also thank the staff of The Fisheries Laboratories of The University of Tokyo for their encouragement.

REFERENCES

1. Guillard, R. R. L. and J. H. Ryther, Can. J. Microbiol., 8, 229-239 (1962)
2. Iwasaki, H., Bio. Bull., 121, 173-187 (1961)
3. Spencer, C. P., J. Mar. Biol. Assoc. U. K., 33, 265-290 (1954)
4. Watanabe, K., A. Hino and R. Hirano, Bull. Plankton Soc. Japan, 27, 87-98 (1980)
5. Iizuka, S., in : Toxic Dinoflagellate Blooms, Tayler/Seliger, eds., Elsevier North Holland, 111-114 (1979)
6. Dorch, Q., J. R. Clayton, Jr., S. S. Thoresen, and I. Ahmed, Mar. Biol., 75, 1-14 (1984)

NUTRIENTS AND CHLOROPHYLL *a* VARIATIONS DURING THE RED TIDES IN JINHAE BAY, KOREA

YANG, D. B.
Korea Ocean Research and Development Institute
Ansan, P. O. Box 29, Seoul, 425-600, Korea

ABSTRACT

Tidal cycle time series distributions of nutrients and chlorophyll *a* were measured at a fixed station located at mid-channel of Jinhae Bay, Korea. High nitrate concentrations were observed at the time of low tides whereas high phosphate concentrations occurred occasionally at the time of high tides. Anoxic bottom water appeared to be the major source of high phosphates in the outer bay. Chlorophyll *a* concentrations were positively correlated with nitrate concentrations in April 1981, May 1982 and June, 1983. However, chlorophyll *a* concentrations were also positively correlated with phosphate in June and August, 1981 and September 1982. Despite positive relationships between nutrients and chlorophyll *a*, phytoplankton growth is not likely to be limited by these nutrients in Jinhae Bay. The role of growth stimulators in initiating the extensive blooms of red tide organisms is suggested.

INTRODUCTION

Frequent red tide outbreaks damaging coastal fisheries in Jinhae Bay have been studied by several authors [1, 2, 3]. Masan-Jinhae metropolitan areas have been extensively industrialized in the last two decades, and large amounts of domestic and industrial wastewaters have been discharged into Jinhae Bay. Efforts are underway to study the undesirable phytoplankton blooms in terms of the massive input of pollutants from the adjacent land. Inner Masan Bay waters receiving domestic and industrial wastewaters are connected to the open ocean by a narrow channel. Since the mean tidal excursion in this area is approximately 7km [4], variations of biologically important parameters due to the tidal fluctuations are expected to be largest at mid-channel. This study describes variations of nutrients and chlorophyll *a* concentrations at a fixed station during a tidal cycle.

MATERIALS AND METHODS

Time series observations were made at a fixed station located at mid-channel in April, June and August 1981, May and September 1982, and June, 1983 (Fig. 1). Surface water samples were collected every hour and tidal cycle time series lasted for 7 to 25 hours. Analyses for nitrate and phosphate were carried out with the Technicon Autoanalyzer II using the method of Zimmermann et al. [5]. Chlorophyll *a* was determined on acetone extracts by the SCOR-UNESCO method [6].

RESULTS AND DISCUSSION

Tidal cycle time series distributions of salinity, nitrate, phosphate and chlorophyll *a* in the surface waters are summarized in Table 1. In April, 1981 nitrate concentrations varied from 52-269 μgN/l showing high concentrations at the time of low tide. Chlorophyll *a* concentrations were below 10 μg/l indicating that extensive phytoplankton blooms had not yet begun. In April, high nitrate concentrations were positively correlated with the high chlorophyll *a* (r=0.876).

RED TIDES: BIOLOGY, ENVIRONMENTAL SCIENCE, AND TOXICOLOGY
Okaichi, Anderson, and Nemoto, Editors

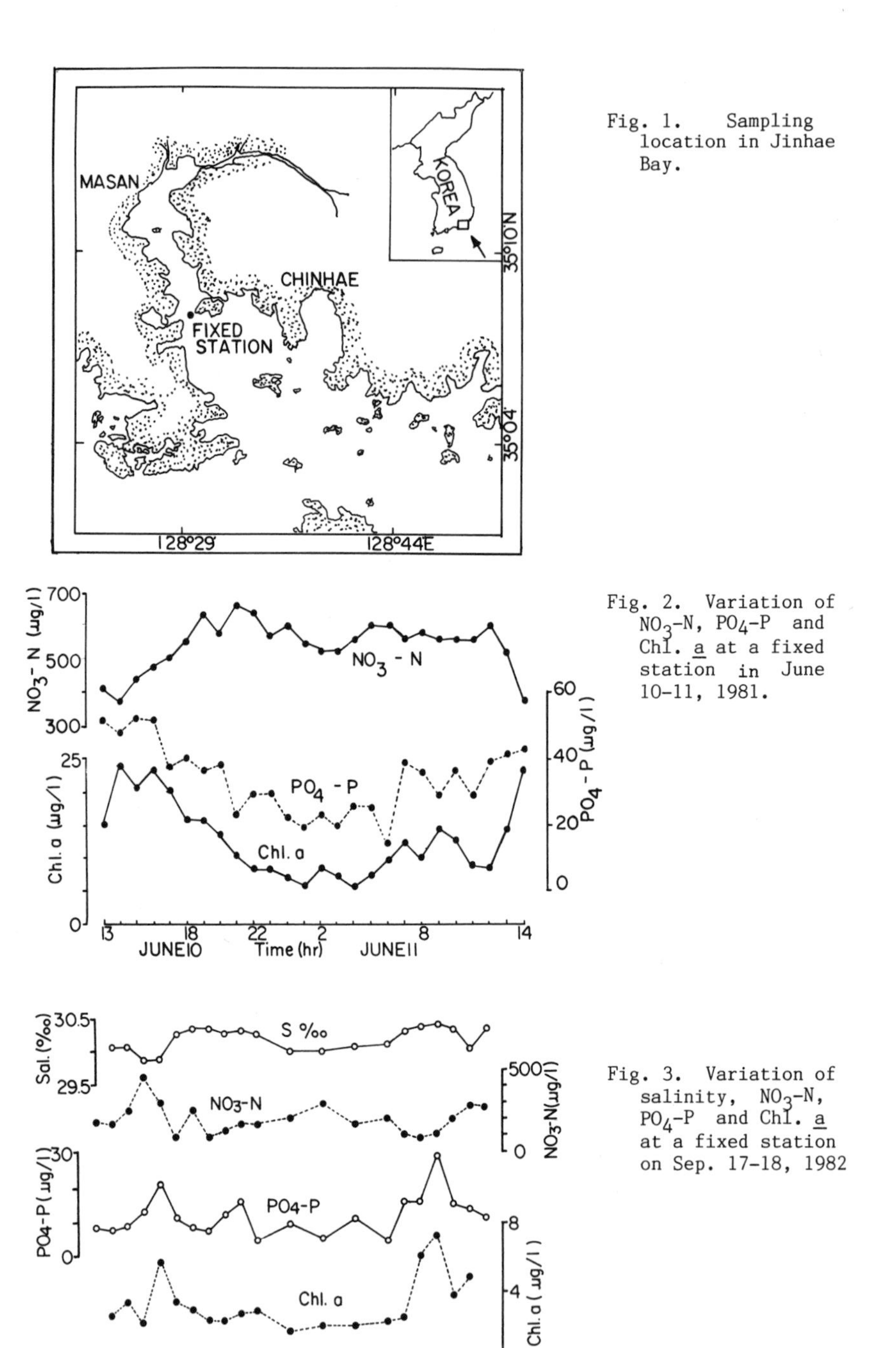

Fig. 1. Sampling location in Jinhae Bay.

Fig. 2. Variation of NO_3-N, PO_4-P and Chl. a at a fixed station in June 10-11, 1981.

Fig. 3. Variation of salinity, NO_3-N, PO_4-P and Chl. a at a fixed station on Sep. 17-18, 1982

In June, 1981 phosphate concentrations between 14.2 and 52.5 μgP/l were maximum at the time of high tide and minimum at the time of low tide. This implies that phosphate was typically higher in the outer bay water than in the inner bay water (Fig. 2). High chlorophyll a concentrations were associated with the high tide. In August, 1981 high nitrate concentrations were found around the time of low tide. Phosphate concentrations were again high at the time of high tide which suggests that phosphate was higher in the outer bay than in the inner bay. Phosphate was positively correlated with chlorophyll a concentrations in August, 1981 (r=0.806).

During the 24 hour observation in May 1982, surface salinity of 29-31 ppt seemed not to be related to the tidal phase due to the wind induced surface water movements. The surface nitrate concentration varyied from below detection to 491 μgN/l, showing a significant negative correlation with salinity (r=0.806). Concentrations of chlorophyll a and nitrate were high in low salinity waters. In September 1982, concentrations of nitrate in the surface waters varied from 88 to 452 μgN/l with a mean of 200 μgN/l (Fig. 3). Phosphate ranged from 5.0 to 29.0 μgP/l, and high phospate concentrations were found along with high salinity. In June 1983, surface seawater samples were taken every 15 min. for 7 hours. Chlorophyll a concentrations (from 10.97 to 28.57 μg/l) were positively correlated with nitrate content (r=0.613).

Tidal cycle time series data obtained in this study demonstrate that surface chlorophyll a concentrations were positively correlated with nitrate in April 1981, May 1982 and June 1983 and with phosphate in June and August 1981 and September, 1982. In Jinhae Bay, nitrate originates from freshwater inputs via inner Masan Bay and are more concentrated in the surface waters. Nitrate content decreases with increasing distance from the shore and it is common to observe high levels of nitrate at the time of low tide. In contrast, high phosphate levels were found in the bottom waters. It was also noted that phosphate levels in the bottom waters increased in summer. In 1981, phosphate levels in the bottom waters of our fixed station in May, June and July were 33.1, 64.1, 110.5 μgP/l respectively. In Masan Bay anoxic conditions often occur in summer due to the seasonal pycnocline and high amounts of discharged wastewater. Fitzgerald [7] has shown that under anaerobic conditions, phosphate in lake muds would diffuse out into the bottom water. Honjo [8] has pointed out that phosphate and ammonia were released into bottom waters from sediments when the concentration of oxygen or pH was low in the bottom layer in Hakata Bay, Japan. Therefore, high phosphate levels which were related to high salinity and high tide in Jinhae Bay probably originate from the anoxic bottom water.

In Jinhae Bay, it is unlikely that nutrients are limiting factors for phytoplankton growth even though correlations could be established between nutrients and chlorophyll a. Nitrate levels measured in this study are too high to limit phytoplankton growth since 100 μgN/l is the minimum level for large scale outbreaks of red tides according to Japanese Association of Fisheries Resources and Protection [9]. Phosphate levels mostly exceeded 20 μgP/l, above which phytoplankton growth is typically not limited according to Kuhl [10]. In Omura Bay Japan, Izuka suggested that red tides of Skeletonema in July were induced by growth stimulators supplied by riverwater runoff and red tides of dinoflagellates in September were driven by growth stimulators formed in the anoxic sediment. Thus in April 1981, May 1982 and June 1983, positive correlations between chlorophyll a and nitrate in Jinhae Bay, Korea suggest that phytoplankton blooms were favored by growth stimulating substances supplied from freshwater, since nitrates mainly originate from freshwater inputs. In Hakata Bay, Honjo [8] suggested that growth stimulators are formed in anoxic sediments and that these materials, upon arriving in surface waters, could enhance blooms of red tide organisms.

In Jinhae Bay, these materials could also be formed in the anoxic sediments which liberate phosphate as well. In this context we have observed high biomass of phytoplankton together with high concentrations of phosphate at the time of high tide. Since accumulation of phosphate in the bottom waters proceeds as seasonal stratification increases, one would expect to observe a positive correlation between phosphate and chlorophyll a levels in summer, and this was what was observed in June and August 1981 and September 1982. However, a more detailed physical oceanographic survey should be performed to make clear whether there is sufficient vertical mixing to bring bottom waters to the surface during the phytoplankton blooming period.

	Salinity (‰)	NO_3-N (μgN/1)	PO_4-P (μgP/1)	Chl. a (μg/1)	correlation with Chl. a
Apr. 1981	—	52-269 (173)	13.0- 27.6 (20.0)	2.31- 11.36 (7.13)	NO_3 (r=0.876)
June 1981	32.18-32.46 (32.31)	375-664 (545)	14.2- 52.5 (31.1)	5.53- 23.97 (12.87)	PO_4 (r=0.870)
Aug. 1981	27.72-30.41 (29.25)	14-432 (148)	50.2-313.5 (122.6)	13.45-181.79 (70.70)	PO_4 (r=0.806)
May 1982	29.1 -31.0 (30.2)	nd-491 (124)	4.0- 14.0 (8.1)	5.48- 38.13 (15.07)	NO_3 (r=0.625)
Sep. 1982	29.89-30.44 (30.21)	88-452 (197)	5.0- 29.4 (12.3)	1.79- 6.07 (3.35)	PO_4 (r=0.767)
June 1983	31.9- 32.0 (32.0)	254-669 (491)	12.2- 36.9 (23.0)	10.97- 28.57 (20.44)	NO_3 (r=0.613)

Table 1. Variation of salinity, NO_3-N, PO_4-P and chlorophyll a at a fixed station in Jinhae Bay (mean values in parentheses).

REFERENCES

1. K. W. Lee, G. H. Hong, D. B. Yank and S. H. Lee, J. Oceanol. Soc. Korea 16, 43-48 (1981).
2. J. S. Park, Bull. Fish. Res. Dev. Agency, 28, 55-88 (1982).
3. K. I. Yoo and J. H. Lee, J. Ocewanol. Soc. Korea, 15, 62-65 (1980).
4. KORDI, KORDI Rep. BSPE 00022-43-7 (1980).
5. C. Zimmermann, M. Price and J. Montgomery, Harbor Branch Foundation Inc. Technical Report, NO. 11 (1977).
6. SCOR-UNESCO, Monogr. Oceanogr. Methodol., 1 (1966)
7. G. P. Fitzgerald, Limnol. Oceanogr. 15, 550-555 (1970).
8. T. Honjo, Bull. Takai Reg. Fish. Res. Lab., 79, 77-121 (1974).
9. Japanese Association of Fisheries Resources and Protection, Water quality standards for fishery environment (1973).
10. A. Kuhl, Bot. Mongor., 10, 636-654 (1974).
11. S. Izuka, Bull. Plankt. Soc. Jap., 19, 22 (1972).

NITROGEN PREFERENCES AND MORPHOLOGICAL VARIATION OF *CHLAMYDOMONAS* SP.

NAKAMURA, Y., H. OGAWA, K. OUCHI AND N. FUJITA
Dept. of Fishery Science, Faculty of Agriculture,
Tohoku University, Sendai 980, Japan

ABSTRACT

When NH_4Cl and KH_2PO_4 were added to seawater in a ratio of N:P=10:1 by weight, *Chlamydomonas* sp. dominated the phytoplankton assemblage and bloomed in the range of 30-120 mg NH_4-N. A marked accumulation of dissolved organic carbon was observed (17-32 mg/l) and the medium was acidic (pH 3.6-5.6). Such phenomena were not observed in the seawater when urea and nitrate were added. The morphological characteristics of this alga were different between the ammonium treatment and the nitrate treatment.

INTRODUCTION

Removal of nitrogen (N) and phosphorus (P) in wastewater is very important for the improvement, stabilization and utiliation of the receiving waters. We have studied the biological removal of N and P from seawater by algae. During the experiments on a variety of algae under high N and P concentrations, we found *Chlamydomonas* sp. grew optimally at high concentrations of ammonium. The growth, physiology, and morphology of this alga are described in this paper.

METHODS

Seawater samples used in the N and P experiments were collected at 4 coastal stations (Onagawa, Soma, Kure and Niigata) and 3 offshore stations (Niigata, Sanriku and Shikoku) around Japan from 1984 to 1987. NH_4Cl, $NaNO_3$ and urea ($(NH_2)_2CO$) were used as N sources at concentrations of 5, 10, 20, 30, 50, 70 and 120 mg N/l of seawater while KH_2PO_4 was used as a P source at 10% of each nitrogen concentration. The appearance and growth of algae were closely observed, and chlorophyll-a, dissolved organic carbon (DOC), pH, and the disappearance of N and P were measured by the methods of Strickland and Parsons (1).

The dominant species observed in this experiment, *Chlamydomonas* sp., grew optionally at high concentrations of ammonium. This alga was isolated and cultured with the Matsudaira medium (TABLE 1) at 20°C and 32 $\mu E/m^2/s$ (L:D=14:10). The nutrient requirements of this alga for ammonium nitrate, vitamins and trace metals were observed for up to 60 days. The growth of this alga was evaluated in 6 qualitative categories. TABLE II. The morphological characteristics of both living and preserved cells were examined under the light and the transmission electron miscroscope.

TABLE I. Composition of Matsudaira medium[a] used in the nutrient experiments with *Chlamydomonas* sp..

$NaNO_3$-N,16.5mg; Na_2HPO_4-P, 1.2 mg[b]; $NaHCO_3$,63mg;
Na_2SiO_3 xH_2O, 10 mg; Fe-EDTA, 200 µg as Fe; Co-EDTA, 1 ug as Co; Cu-EDTA, 1 ug as Cu; Mn-EDTA, 20 ug as Mn; Vitamin mixture 8[c]), 1 ml; Seawater 1,000ml

[a])Matsudaira, unpublished. [b])Stock solution (1.4 g as KH_2PO_4 and 1.8 g as Na_2-EDTA in 100 ml). [c])Provasoli et al. (6).

RED TIDES: BIOLOGY, ENVIRONMENTAL SCIENCE, AND TOXICOLOGY
Okaichi, Anderson, and Nemoto, Editors

yield at 120 mg/l was better than at 20 mg/l. When this alga was cultured in Matsudaira medium (type A) containing 16.5 mg/l of NO_3-N, growth was similar to that obtained when NH_4-N was added (type E) at 50 or 120 mg/l, but less than the optimum growth observed at 70 mg/l. The media in which trace metals and vitamins were added (types C and E) showed better growth compared to the media without trace metals and vitamins (types B and D) (TABLE II). The pH of the medium type E also dropped to the acidic range.

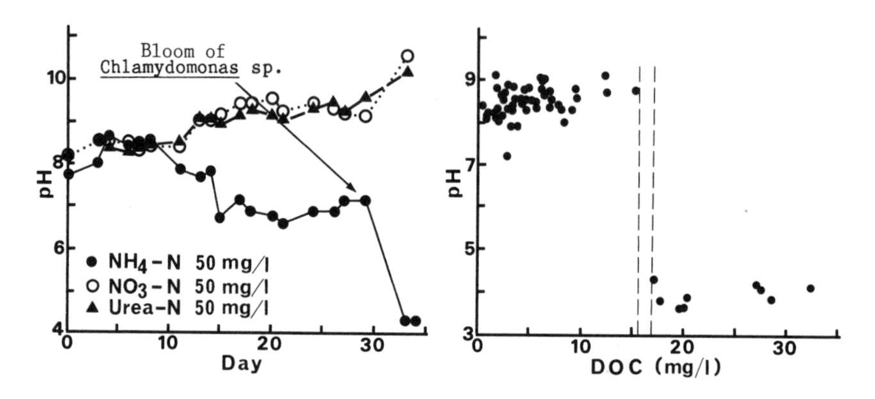

FIG. 1. Variation of pH in seawater loaded with nitrogen and phosphorus.

FIG. 2. Relation of pH and DOC in seawater loaded with nitrogen and phosphorus.

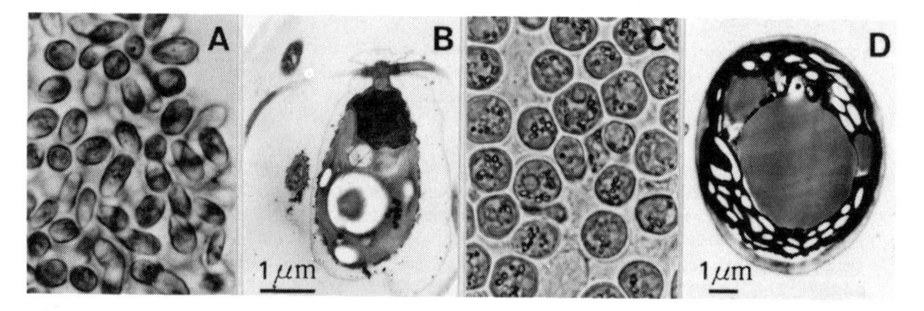

FIG. 3. Chlamydomonas sp.: A) bi-flagellated, swimming cells; B) cross section of the motile cell; C) cells and daughter cells in palmelloid stage; D) cross section of a cell in palmelloid stage.

Morphological Observations

Almost all cells cultured in medium containing 50 mg/l of NH_4-N were oval or ellipsoidal, and biflagellate (FIG. 3-A), but some palmelloid cells which were spherical or broadly ovoid and formed 2 to 4 daughter cells in themselves were also observed. The thickness of the inner layer of the cell wall was 0.02 - 0.12 μm and in some cases separated from the protoplast (FIG. 3-B). In contrast, almost all cells cultured in Matsudaira medium containing NO_3-N as the N source were spherical in shape and lacked flagella (FIG. 3-C). These cells contained 2 to 16

RESULTS

Nitrogen and Phosphorus

The diatoms, Skeletonema costatum, Asterionella japonica, Cylindrotheca closterium and Chaetoceros spp. were dominant in both seawater samples enriched with NO_3-N and urea-N in the range of 5 - 120 mg/l. In the seawater enriched with NH_4-N, the diatoms mentioned above were dominant at concentrations below 20 mg/l, but Chlamydomonas sp., Chlorella spp., pennate diatoms and blue green algae became dominant at higher concentrations. In particular, Chlamydomonas sp. appeared in all seawater samples collected from 7 stations around Japan and always showed abundant growth at ammonium concentratios above 50 mg/l.

An experiment was thus designed to identify the dominant algae that will grow in seawater treated with 50 mg/l of NO_3-N, NH_4-N and urea-N, respectively. The diatoms, S. costatum, C.closterium and Chaetoceros spp., were dominant in $NaNO_3$ and urea-enriched seawater, but Chlamydomonas sp. was always dominant in the NH_4Cl enrichments. At peak concentrations on Day 34, the concentration of chlorophyll-a reached 188.4 ug/l and the pH fell from about 8 to less than 5 (FIG 1). Similar changes did not occur in $NaNO_3$ and urea-treated seawater. The relation between pH and DOC level was then investigated in the NH_4Cl-treated seawater. The initial pH in this experiment was 8-9, while the DOC level was 0.6 mg/l. When DOC was less than 15 mg/l, the pH was in the range of 7.1-9.1. However, when the DOC level exceeded 17 mg/l, the pH suddenly dropped to less than 4 (FIG. 2). When Chlamydomonas sp. was at its highest concentrations, the DOC was 32.3. mg/l and the pH 3.6.

TABLE II. Effects of nitrogen source, trace metal and vitamin on the growth of Chlamydomonas sp.

Medium type	A*	B		C		D	E			
Nitrogen	NO_3-N	NO_3-N		NO_3-N		NH_4-N	NH_4-N			
(mg/l)	16.5	0	20-120	0	20-120	20-120	20	50	70	120
Trace metal and Vitamin	added	not added		added		not added	added			
Growth level	++ ~ ++++	±	± ~ +	±	+ ~ ++	+ ~ ++	+	+++ ~ ++++	++++ ~ +++++	++ ~ ++++

Growth level (±: no or slight (undetectable by eye), +: slight (detectable by eye), ++: well, +++: abundant, ++++: more abundant, +++++: most abundant showing the growth at surface layer and bottom). *Composition of this medium is shown in TABLE I.

When NO_3-N was added as the sole nitrogen source at 20-120 mg/l (medium types B and C), the cell yield of this alga did not differ significantly at any of the concentrations used for each type of medium Between both media, algal growth was better in type C. When NH_4 was the N source (medium types D and E) algal growth in type D was similar to that in type C, but was best in type E with an optimum at 70 mg/l. The cell daughter cells. The inner layer of the cell wall was thick (0.29 - 0.39 μm) and did not separate from the protoplast (FIG. 3-D). The sizes of the motile and palmelloid cells were 1.8-2.9x4.6-6.8μm and 3.8-10.0x3.9-11.2μm respectively, and were not affected by the different N sources.

DISCUSSION

The dominant species in the seawater samples enriched with N and P varied with the different nitrogen sources and their concentrations. Many species of algae can grow when NH_4-N is below 20 mg/l, but their growth is inhibited above that concentration. This has been observed in brown algae (2), phytoflagellates (3) and red algae (4). On the other hand, the micro green alga Chlamydomonas sp. found in all seawater samples collected from 7 stations around Japan, can grow well at concentrations of NH_4-N above 20 mg/l, especially 50-120 mg/l. These results emphasize the differential tolerance of algal species to ammonium. Chlamydomonas sp. is widely distributed in coastal and offshore waters around Japan, but has not been commonly observed in red or green tides because it seems to require such high ammonium concentrations.

The production of extracellular substances by algae is well established (5). The increase of DOC in the seawater (FIG. 2) suggests that Chlamidomonas sp. produced a large amount of organic acids during its bloom. The sudden drop of pH (Fig. 2) could be caused by the following process. When DOC was below 15 mg/l, any decrease in pH due to excreted organic acid was prevented by the buffering capacity of seawater. Then, when DOC increased above 17 mg/l, the buffering capacity was exceeded and the pH fell. Glycolic acid is one of the strongest acids associated with cells of this type and is commonly liberated by algae (5). It is also possible that the change in pH was due to bacterial respiration, which was not monitored.

This alga has two forms, the motile and the palmelloid stage. Both forms are affected by the nature of their nitrogen source. Chlamydomonas sp. is very resistant to the toxicity of ammonium. This alga may thus be useful in the removal of nitrogen and phosphorus from wastewaters.

ACKNOWLEDGEMENTS

We are grateful to Dr. T. Kariya - his advice on this research. Thanks are also due to K. Satake and T. Sato for technical assistance and to all those who assisted in the field and laboratory during this study.

REFERENCES

1. J. D. H. Strickland and T. R. Parsons. A Practical Handbook of Seawater Analysis. Bull. Fish. Res. Bd. Can 167, (1972).
2. H. Ogawa, Hydrobiol. 116/117, 389-392 (1984).
3. H. Iwasaki in: The Cause of Red-tide in neritic Waters, T. Hanaoka et al. (Jap. Fish. Resource Conversation Assn., Tokyo 1972) pp.77-98 (in Japanese).
4. T. Maruyama in: Effects of Environmental Chemicals on Coastal Ecosystem, T. Yoshida, ed. (Koseisha Koseikaku, Tokyo 1986) pp. 109-121 (in Japanese).
5. J. A. Hellebust, Botanical Monographs 10, 838-850 (1974).
6. L. Provasoli, J. J. A. McLaughlin and M. R. Droop, Arch. Mikrobiol 25, 392-428 (1957).

EFFECTS OF *n*-ALKYLAMINES ON MOTILITY AND VIABILITY OF *HETEROSIGMA AKASHIWO* CELLS

Miyagi, N., E. Satoh and T. Fujii
Institute of Biological Sciences, University of Tsukuba,
Tsukuba-shi, Ibaraki 305, Japan

ABSTRACT

Cytotoxic effects of polycationic liposomes and their components on *Heterosigma akashiwo* cells were investigated. Positively-charged stearylamine-liposomes reduced cell motility and eventually caused cytolysis. Negatively- and non-charged liposomes had little effect on cell motility and/or cell viability. Damage was also induced by the single application of stearylamine, one of the *n*-alkylamines. Of the *n*-alkylamines, laurylamine (C_{12}) was the most effective in reducing motility and causing cytolysis of cells. Damage increased with increasing concentration of laurylamine, and with lengthening of the treatment period.

Based on these results, we argue that positively-charged laurylamine could easily fuse with, and incorporate into negatively-charged cell membranes, and eventually cause the breakdown of membranes in *H. akashiwo* cells, which are naturally-occurring cell wall-less organisms.

INTRODUCTION

Most of the red tide organisms, including *H. akashiwo*, are known to be naturally-occurring cell wall-less organisms [1]. The surface of the cell is thus a possible target for attack in exterminating the red tide organisms, because the naked cells can easily fuse with liposomes and/or uptake poisonous agents. *H. akashiwo* cells show diurnal vertical migration and migrate upward during the light period [3]. When the cells migrated upward, liposomes and their components were added to the culture medium. Here, we report the effects of polycationic liposomes and their components upon the motility and viability of *H. akashiwo* cells.

EXPERIMENTAL

Growth and maintenance of cells

Axenic clones of *H. akashiwo* were grown in artificial sea water (modified ASP-7 [4]) and cultured at 20±1°C under a photoperiod of 12:12 LD. Cells were lit from above at an intensity of 6000lux.

Measurement of motility and viability

Culture medium (1ml) containing 1-2 x 10^4 cells/ml was transferred to test tubes (diameter 9mm), mixed with 5ul of liposomes or other chemicals and allowed to stand under the light. After various periods of incubation, total living cell numbers and cell numbers that occurred in the upper 800µl phase of each 1ml culture medium were counted with a microscope (Fig. 1).

RED TIDES: BIOLOGY, ENVIRONMENTAL SCIENCE, AND TOXICOLOGY
Okaichi, Anderson, and Nemoto, Editors

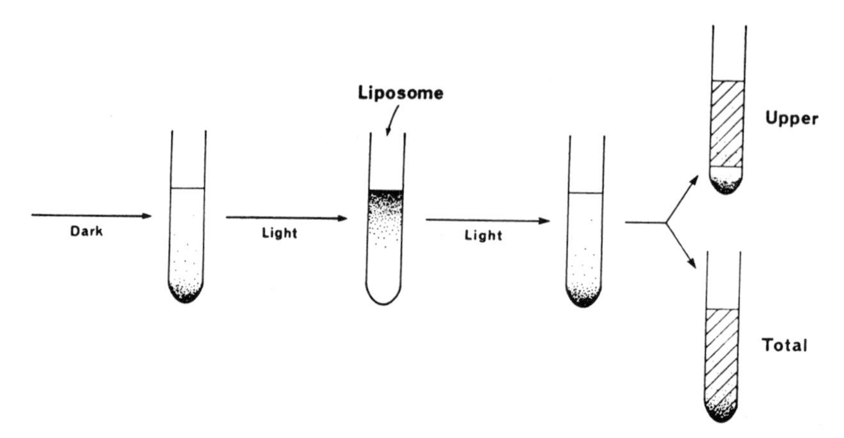

Fig. 1. When the cells migrated upward during the light period, liposomes or their components were added to the culture. Those cells whose mobility was lost sedimented to the bottom of the test tubes and eventually suffered cytolysis. After various periods of incubation, total living cell number (Total) and those in the upper 800 μl of each 1ml culture medium (Upper) were counted.

Preparation of liposomes

This was carried out by the method of Uchimiya [2]. Positively-charged liposomes were prepared from egg-lecithin (50μg), stearylamine (5μg) and modified ASP-7 (5μl). Negatively-charged liposomes were prepared from egg-lecithin (50μg), dicetylphosphate (5μg) and modified ASP-7 (5μl). Non-charged liposomes were prepared from egg-lecithin (50μg) and ASP-7 (5μl).

Addition of various n-alkylamines

n-Alkylamines with hydrocarbon chains of differing lengths were added to 1ml of culture at a final concentration of 0.5μM (from 0.1mM stocks in 100% EtOH). One *n*-alkylamine, laurylamine, was tested at concentrations from 0 to 2.0μM.

RESULTS

Effect of liposomes on motility and viability of cells

After various periods of incubation with positively-charged liposomes, total and upper cell numbers were counted (Fig. 2). Those cells whose motility had been lost sedimented rapidly to the bottom of test tubes and eventually suffered cytolysis. Thus, the total living cell number gradually decreased with lengthening treatment period. Negatively- and/or non-charged liposomes mixed with the culture medium didn't reduce the cell numbers in each fraction.

The culture was mixed with each component, lecithin or stearylamine, of positively-charged liposomes at the same ratio as in the liposome experiment. A single application of stearylamine was effective in causing cell damage. Lecithin caused no damage to cells.

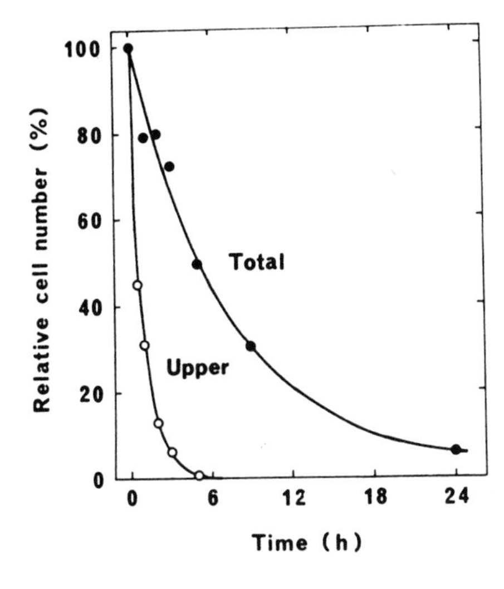

Fig. 2. Effects of positively-chargedliposomes onthemotility and viability of H. akashiwo cells. One ml of culture medium (1.4 x 10^4 cells/ml) was mixed with 5 µl of positively-charged liposome suspension (containing lecithin 50µg and stearylamine 5µg). After various periods of incubation, the total and upper cell numbers were counted.

Table I. Effects of various n-alkylamine on cell viability.

Alkylamine	(n)	Relative cell No.
Control		100
Octylamine	(8)	92.3
Decylamine	(10)	53.8
Laurylamine	(12)	23.1
Mylistylamine	(14)	34.7
Cetylamine	(16)	46.7
Stearylamine	(18)	66.7

(n) = number of carbon atoms

One ml of culture medium (1.6 x 10^4 cells/ml) was mixed with each n-alkylamine (to a final concentration of 0.5µM). After 1hr- incubation, the cell number in the upper phase was counted.

Effect of various n-alkylamines on cell viability

The results are shown in Table I. The most effective n-alkylamine was laurylamine, which reduced the number of living cells in the upper phase of the culture by half within 20min. After 1hr of incubation, 80% of the cells had sedimented to the bottom of the test tubes. The effect of different laurylamine concentrations on cell viability is shown in Fig. 3. About 0.2µM of laurylamine reduced the cell number in the upper phase by half within 1hr.

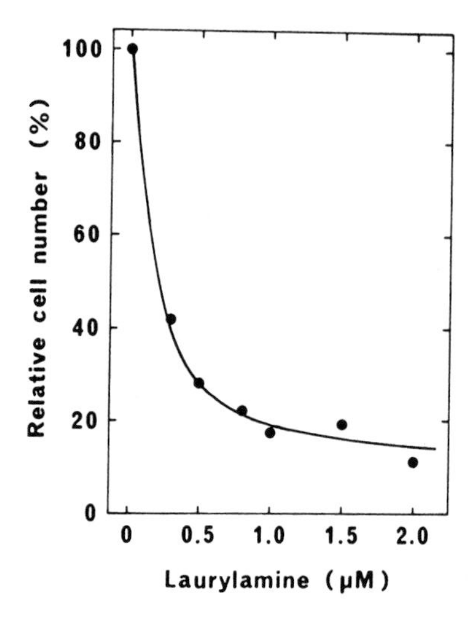

Fig. 3. Effects of laurylamine concentration on cell viability. One ml of culture medium (1.7 x 10^4 cells/ml) was mixed with various concentrations of laurylamine. After 1hr-incubation, the cell number in the upper phase was counted.

DISCUSSION

H. akashiwo cells clearly sustained great damage with positively-charged liposomes, but not with negatively- and non-charged ones. We consider that the former can easily fuse with the negatively-charged cell membranes of wall-less *H. akashiwo* cells, and eventually cause cell damage.

The compound used to make positively-charged liposomes (stearylamine) also caused damage to cells when applied as a solution to the cultures. We tested other *n*-alkylamines and found one, laurylamine, to be particularly effective in causing loss of cell motility and cytolysis. Damage may result from the direct incorporation of these amphipathic *n*-alkylamine, but not from modified ASP-7 which is wrapped inside the liposomes and transfered into the cell when the liposomes fuse the cell membrane. We suppose that the differing effects of the *n*-alkylamines is due to their differing affinities to the plasma membrane.

Although it is not clear how *n*-alkylamines react with membrane components and sustain damage to the cells, these findings may give us a method to exterminate the red tide organisms.

REFERENCES

1. Hara, Y., I. Inoue and M. Chihara, Bot. Mag. Tokyo, 98, 251, (1985).
2. Uchimiya, H., Plant Physiol., 67, 629, (1981).
3. Wada, M., A. Miyazaki and T. Fujii, Plant Cell Physiol, 26, 431, (1985).
4. Watanabe, M. M., Y. Nakamura, S. Mori and S. Yamochi, Jap. J. Phycol. (sorui), 30, 279, (1982).

CHEMICAL ENVIRONMENTS FOR RED TIDES OF CHATTONELLA ANTIQUA

YASUO NAKAMURA,* JUN TAKASHIMA,** AND M. WATANABE*
*Natl. Inst. Environ. Stud., Onogawa, Tsukuba, Ibaraki 305, Japan; **Faculty of Science, Toho Univ., Funabashi, Chiba 274, Japan

ABSTRACT

Environmental parameters that affect the growth of Chattonella antiqua were monitored at Ie-shima Islands, the Seto Inland Sea, in the summer of 1986. Growth bioassay of the seawater was also conducted to estimate the growth rate-limiting factor(s) of C. antiqua. Furthermore, the growth rate of C. antiqua was measured as a function of nutrient concentration. Based on the results obtained from the above observations and experiments, a criterion to assess chemical environments with respect to the outbreaks of red tides of C. antiqua is proposed.

INTRODUCTION

In the last two decades, red tides of Chattonella antiqua (Raphidophyceae) occurred sporadically during the summer in the Seto Inland Sea, Japan. In order to elucidate the mechanisms of red tides of C. antiqua, many field surveys and laboratory culture experiments have been conducted. However, in spite of these efforts, the gap between controlled laboratory experiments and field observations has not completely been bridged and the following questions still remain: (1) What nutrient(s) including vitamins and trace metals control the growth rate of C. antiqua in the Seto Inland Sea in summer? and (2) How fast does C. antiqua grow under a given chemical environment? or in other words, how does the growth rate of C. antiqua relate to the concentration of each nutrient? We have to answer these questions before assessing the role of each nutrient in inducing red tides of C. antiqua. In this context the environmental parameters that affect the growth of C. antiqua were monitored at Ie-shima Islands, the Seto Inland Sea, in the summer of 1986 and growth bioassay of the seawater was conducted. Furthermore, the growth rate of C. antiqua as a function of nutrient concentrations was determined.

The present study represents the results from above observations and experiments and a criterion to assess chemical environments with respect to the outbreaks of red tides of C. antiqua is proposed.

EXPERIMENTAL

Field survey—From July 18 to Aug. 13, 1986, environmental parameters were monitored at Ie-shima Islands, the Seto Inland Sea (134°30'E, 34°40'N). Two sampling stations (A, B) were chosen; the depths of A and B were 30 and 21 m, respectively.

Chemical analysis—Nutrients were analyzed with a Technicon Auto Analyzer for PO_4^{3-}, NH_4^+, NO_2^-, NO_2^- + NO_3^- and DTP. Vitamin B_{12} was analyzed by a microbiological method. An axenic clone of C. antiqua (Ho-1) was used as a test strain [4].

Growth bioassay—Samples were obtained from station A, 0 and 25 m, at an interval of once a week. Growth bioassay was conducted after we returned to our institute. An axenic clone of C. antiqua (Ho-1) was used as a test strain. All incubations were conducted at 25°C, 0.04 ly·min^{-1} and 12:12LD.

RED TIDES: BIOLOGY, ENVIRONMENTAL SCIENCE, AND TOXICOLOGY
Okaichi, Anderson, and Nemoto, Editors

Seawater samples filtered through GF/C filters were enriched with 1.4 $ml \cdot l^{-1}$ of N-metal solution [3] so that metal condition was optimum for the growth of C. antiqua. A control sample was not enriched with NO_3^-, PO_4^{3-} or B_{12}. The other samples were enriched with NO_3^-(10 μM), PO_4^{3-}(1 μM) and B_{12}(2 $ng \cdot l^{-1}$), either singly or in combinations. Then, these samples were filtered through nucleopore filters (pore size 0.4 μm). Five ml of precultured C. antiqua was inoculated to 95 ml of the experimental medium and cultured in a semicontinuous mode (dilution ratio =0.5 d^{-1}). Cell concentrations were monitored daily for 5 days and growth rate (μ) was calculated from the following equation;

$$\ln N_t = \ln N_0 + (\mu - \ln 2) t \quad , \qquad (1)$$

where N_t is the cell concentration at time t.

Growth rate as a function of nutrient concentration---Surface seawater from Kuroshio area, which contained PO_4^{3-}, NO_3^- and NH_4^+ at levels of 0.08, 0.7 and <0.2 μM, respectively, was used for the experiments. Changing the nutrient concentrations systematically (TABLE 1), C. antiqua was cultured in semicontinuous mode. Experimental procedures were essentially the same as those described above. The dilution ratio was 0.5 d^{-1} and cell concentrations were kept below 50 $cells \cdot ml^{-1}$. B_{12}(4 $ng \cdot l^{-1}$) and N-metal solutions (1.4 $ml \cdot l^{-1}$) were added to each sample. The observed growth rates were assessed as a function of initial nutrient concentration since ambient concentrations at the end of the experiment did not change significantly (<25%) from the initial.

TABLE 1. Mode of nutrient enrichment

Experimental mode	PO_4^{3-}(μM)	NO_3^-(μM)	NH_4^+(μM)
A	0.08 - 0.58	5.7	<0.2
B	0.58	0.7 - 5.7	<0.2
C	0.58	0.7	0.7 - 2.1
D	0.08 - 0.58	1.4	<0.2
E	0.18	0.7 - 5.7	<0.2
F	0.13	1.1	<0.2

RESULTS AND DISCUSSION

Field observations---During the survey period red tides of C. antiqua did not occur. Population densities of C. antiqua were always below 10 $cells \cdot ml^{-1}$.

Water was thermally stratified and thermocline was observed at a depth of about 10m. Temperature, salinity and light intensity were optimum for the growth of C. antiqua [3].

Typical vertical profiles of nutrients and vitamin B_{12} are shown in FIG. 1. Nutrient concentrations were very low above the thermocline and high below it. B_{12} concentrations did not change significantly between the surface and near bottom.

Growth bioassay---Experiments were focused on the effects of N, P and B_{12}. Typical results are shown in FIG. 2. The results obtained from other samples were essentially the same as in FIG. 2. At the surface layer, concentrations of N- as well as P- nutrients were too low to support a rapid growth. On the contrary, in the seawater obtained from the depth of 25m, neither N, P nor B_{12} limited the growth.

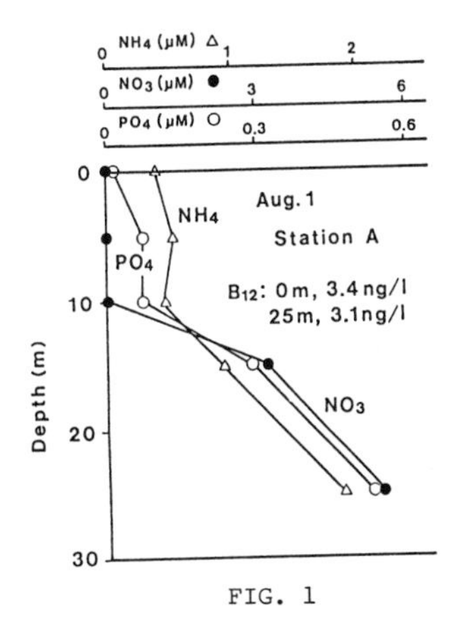

FIG. 1

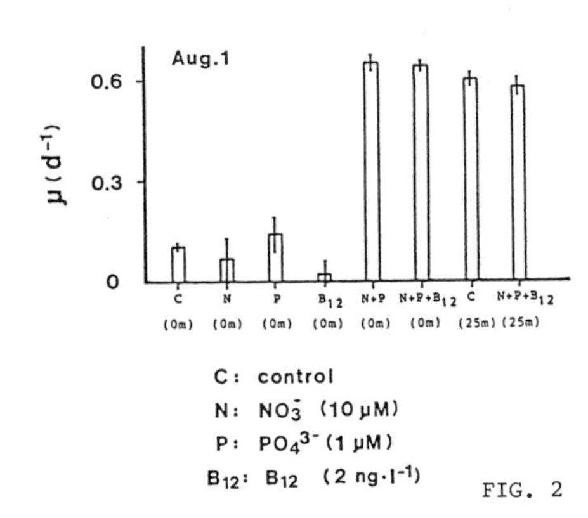

FIG. 2

Relationship between growth rate and nutrient concentrations—In the experimental mode A (TABLE 1), growth rate (μ) as a function of phosphate concentration (S_{PO4}) was well described by the Monod equation:

$$\mu = \mu_{max}^{P} \cdot \frac{S_{PO4}}{K_g^P + S_{PO4}} \quad , (2)$$

where $\mu_{max}{}^{P} = 0.83\ d^{-1}$ and $K_g{}^{P} = 0.11\ \mu M$. In mode B, μ as a function of S_{NO3} was also well described by the Monod equation:

$$\mu = \mu_{max}^{N} \cdot \frac{S_{NO3}}{K_g^N + S_{NO3}} \quad , (3)$$

where $\mu_{max}{}^{N} = 0.81\ d^{-1}$ and $K_g{}^{N} = 1.0\ \mu M$. The replacement of nitrate with ammonium did not change μ significantly (mode C). In mode D and E, observed growth rates seemed to increase hyperbolically with S_{PO4} and S_{NO3}, respectively. In the experimental mode F, μ was measured under the condition that both S_{PO4} and S_{NO3} were fixed at levels close to $K_g{}^{P}$ and $K_g{}^{N}$, respectively (TABLE 2). The observed value of μ was $0.21\ d^{-1}$, significantly lower than that when S_{PO4} (S_{NO3}) was close to $K_g{}^{P}$ ($K_g{}^{N}$) and S_{NO3} (S_{PO4}) was much higher than $K_g{}^{N}$ ($K_g{}^{P}$).

TABLE 2. Growth rate when $S_{NO3} \cong K_g{}^{N}$ and $S_{PO4} \cong K_g{}^{P}$

S_{NO3} (μM)	S_{PO4} (μM)	No. of expts	μ_{obs} (d^{-1})	SD (d^{-1})	μ_{calc} (d^{-1})*
1.1	0.13	8	0.21	0.06	0.27
5.6	0.13	4	0.41	0.06	0.44
1.1	0.57	4	0.43	0.04	0.42

*μ_{calc} was calculated from Eq. 5.

In order to analyze the data obtained from experimental mode D - F, we assumed that μ as a function of S_{NO3} and S_{PO4} can be approximated by a multiplicative function:

$$\mu = f\ (S_{NO3}) \cdot g(S_{PO4}) \quad . \quad (4)$$

Using the data obtained from mode A and B, Eq. 4 changes to the following equation:

$$\mu = \mu_{max} \cdot \frac{S_{PO4}}{K_g^P + S_{PO4}} \cdot \frac{S_{NO3}}{K_g^N + S_{NO3}} \quad , (5)$$

where $\mu_{max} = 0.97\ d^{-1}$, $K_g^N = 1.0\ \mu M$ and $K_g^P = 0.11\ \mu M$. Growth rates calculated from Eq. 5 are shown in TABLE 2 and the observed data (mode D - F) were explained well by this equation.

As shown in experimental mode C, replacement of nitrate with ammonium did not change μ significantly. Thus it is acceptable to replace S_{NO3} with $S_{NO3}+S_{NH4}$ in Eq. 5 and it changes to the following equation:

$$\mu = \mu_{max} \cdot \frac{S_{PO4}}{K_g^P + S_{PO4}} \cdot \frac{S_{NO3} + S_{NH4}}{K_g^N + (S_{NO3} + S_{NH4})} \quad . (5)'$$

Since C. antiqua cannot utilize organic nitrogen and organic phosphorus [4], Eq. 5' means that: if S_{NO3}, S_{NH4} and S_{PO4} in seawater are given, then the rate at which the seawater supports the growth of C. antiqua can be estimated. For simplicity, Eq. 5' was divided by the maximum growth rate observed in laboratory cultures ($0.7d^{-1}$) and a non-dimensional parameter, GP, is defined in Eq. 6;

$$GP = 1.4 \cdot \frac{S_{PO4}}{0.11 + S_{PO4}} \cdot \frac{S_{NO3} + S_{NH4}}{1.0 + (S_{NO3} + S_{NH4})} \quad . (6)$$

GP is considered to be the "growth potential" of seawater with respect to nitrogen and phosphorus.

Ecological considerations——Distributions of GP at station A in the summer of 1986 are shown in FIG. 3. GP was very low above the depth of 10-15m and high below it. Although C. antiqua shows diurnal vertical migrations, the depth where C. antiqua accumulates at night is usually in the range of from 5 to 10m [2]. Thus C. antiqua could not reach the seawater with high GP value and could not grow rapidly. This is considered to be at least one reason that red tides of C. antiqua did not occur in the summer of 1986. On the contrary, GP in the surface layer was usually higher than 0.6 in the red tides of C. antiqua occurring in 1977-79 [1]. A high value of GP in the surface layer and/or shallow "GP-cline" seem to be necessary for the outbreaks of red tides of C. antiqua.

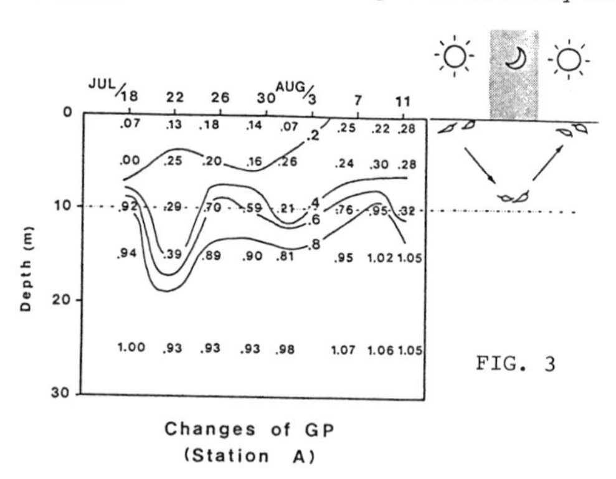

FIG. 3

REFERENCES

1. Fishery Agency, Japan, Data sheets on the red tides occurred in 1977-79.(1982-84).
2. S. Hamamoto, S. Yoshimatsu, and T. Yamada, Reports on red tides by Hornellia occurred in Jun., 1978, Kagawa Prefecture,(1979).
3. Y. Nakamura and M.M. Watanabe, J.Oceanogr. Soc. Japan 39, 110 (1983a).
4. Y. Nakamura and M.M. Watanabe, J.Oceanogr. Soc. Japan 39, 151 (1983b).

MECHANISMS FOR OUTBREAK OF *Heterosigma akashiwo* RED TIDE IN OSAKA BAY, JAPAN

SUSUMU YAMOCHI
Osaka Prefectural Fisheries Experimental Station, 2926-1, Tanigawa
Tanagawa, Misaki-cho, Sennan-gun, Osaka 599-03, Japan

ABSTRACT

Investigations were carried out in Osaka Bay from 1976 to 1985 and the mechanisms involved in outbreaks of *H. akashiwo* red tides were examined. This species showed rapid growth over a wide temperature range of 14-30°C. Growth of *H. akashiwo* was enhanced by addition of chelated iron to seawater samples. *H. akashiwo* could utilize nitrogen and phosphorus in the nutrient-rich bottom layer by downward migration. Incubation of bottom mud, together with field observations of vegetative *Heterosigma* cells, revealed that this species can survive through winter either in vegetative or benthic form in Osaka Bay. This strategy for growth and overwintering is thought to play a major role in the success of *H akashiwo* populations in temperate coastal waters like those of Osaka Bay.

INTRODUCTION

A Raphidophycean flagellate, *H. akashiwo* Hada is a common phytoplankton species that forms red tides in the subartic or temperate coastal waters of the world [1]. In the Seto Inland Sea of Japan this species caused serious damage to the aquacultural fisheries of yellowtail in 1975 and 1981. Therefore, the prediction of *H. akashiwo* red tide is important in the development of coastal fisheries as well as in the preservation of coastal environments in the Inland Sea. The present paper summarizes the results of field surveys and culture experiments on *H. akashiwo* and discusses some of the mechanisms involved in outbreaks of *H. akashiwo* red tides in Osaka Bay.

MATERIALS AND METHODS

Culture experiments

Three axenic clonal cultures of *H. akashiwo* (He-80, He-82 and He-85) were isolated from Osaka Bay in 1980, 1982 and 1985 and their growth responses to water temperature were examined. The cultures were exposed to 22 different temperatures in a Toyokagaku thermogradient incubator (Type TN-3EA) with temperature varying between 5 and 32°C. Growth-limiting nutrients of *H. akashiwo* were determined following the standard algal assay procedure [2]. The seawater samples for the assay were collected from four stations (Off Nishinomiya, off Kishiwada, off Izumisano and Tanigawa Fishing Port) in Osaka Bay in 1980 and 1981 (Fig. 1).

Diel vertical migration

Diel changes in the vertical profiles of *H. akashiwo* cell concentration, chlorophyll-*a* and light intensity were measured at Tanigawa Fishing Port at intervals of 0.5 to 3.0 hr. from August 25 to 26, 1979. At Sano Harbor (Eastern coast of Osaka Bay), vertical distributions of temperature, salinity and *H. akashiwo* were observed at night and the following morning on June 16 and 17, 1980 when the water of the harbor was sharply stratified. Further, red tide waters of *H. akashiwo* were collected at Tanigawa Fishing Port on August 21, 1980 and

RED TIDES: BIOLOGY, ENVIRONMENTAL SCIENCE, AND TOXICOLOGY
Okaichi, Anderson, and Nemoto, Editors

used to fill dialysis bags (Visking Cellulose Tubing) with a capacity of about 1 l). The bags were then incubated for one day at three different layers *in situ*. The first one was exposed at the surface and a second one close to the bottom during the entire period of the experiment. The third bag was placed at the surface in the daytime and on the bottom at night. Immediately after termination of the experiment, the seawater in each bag was filtered through a Whatman type C glass fiber filter and particulate organic carbon and nitrogen on the filter pad were determined.

Liberation of vegetative cells from bottom mud

Bottom muds were taken at three different stations of Osaka Bay (Tanigawa Fishing Port, off Nishinomiya and off Izumisano) in 1982-1985. Mud temperature was measured and the upper 2-cm layer of the mud was sampled. The mud was then transferred to the laboratory and 0.2-0.3 g of the mud was inoculated into 2.5 ml of seawater medium. Forty-eight sets of the mud samples were prepared for incubation at *in situ* mud temperature under illumination of 70 μE $m^{-2}S^{-}$ with a photoperiod of 12:12 LD. Release of vegetative *H. akashiwo* cells from bottom mud was determined by noting the presence of absence of vegetative cells in the medium after incubating for 7 - 14 days. The percentage of release of vegetative *H. akashiwo* cells (R) was calculated by the following equation:

$$R=(Na/Nb)x100$$

where Na=the number of samples in which vegetative cells were observed, and Nb=the total number of samples observed.

RESULTS AND DISCUSSION

Effects of temperature on growth

Cultures of *H. akashiwo* began to grow at 11-14°C and grew steadily at temperatures higher than 14-17°C. This species also showed active growth at a high temperature of 30°C. On the other hand, most of the red tides of *H. akashiwo* appeared in May-July when water temperature in the surface water exceeded 18°C. Such findings indicate that optimum temperature for the growth of laboratory cultures of *H. akashiwo* overlaps with the temperatures prevailing during blooms of this species in Osaka Bay.

Growth-limiting nutrients

The addition of trace metal mixture to the seawater invariably enhanced the growth yield of *H. akashiwo*. Similar enhancement was observed when the seawater was supplemented with chelated iron, but not by the addition of manganese, cobalt or zinc. An exception was found with the seawater collected from Tanigawa Fishing Port. The growth yield of *H. akashiwo* was about 4 times greater in the presence of the trace metal mixture than in the presence of iron alone. In this case, the enrichment of manganese together with iron was effective in accelerating algal growth to the same extent as observed when only the trace metal mixture was added. The single addition of nitrogen, phosphorus or vitamins was completely ineffective in enhancing the growth of this species, although an increase was observed when nitrogen and phosphorus were added together with iron and manganese. These results suggest that iron is a crucial nutrient in the outbreaks of *H. akashiwo* red tide in Osaka Bay.

Diel vertical migration

H. akashiwo migrated toward the sea surface early in the morning at a velocity of 1.0-1.3 m h^{-1}. Downward migration occurred in the afternoon and more than 2.0 x 10^3 cells ml^{-1} aggregated in the bottom layer at night. The upward migration started before sunrise and downward motion occurred prior to sunset. H. akashiwo crossed steep temperature and salinity gradients (6.5°C and 5.7‰) during the diel vertical migration. High values of particulate organic carbon and nitrogen were obtained in the dialysis bags suspended in situ and alternated between the surface and bottom layers, while the values for bags maintained solely in the surface and bottom layers were comparatively low. The results revealed that H. akashiwo migrates toward the sea surface to carry out photosynthesis effectively and then moves to the bottom at night to maximize efficient use of nutrients.

Liberation from benthic cells

Motile cells of H. akashiwo clustered together, lost their motility and eventually formed agglutinated masses of non-motile brown cells showing much variation in size (4-25 µm) and shape. The process of formation of benthic Heterosigma cells observed here is similar to that of Olisthodiscus luteus described by Tomas [3]. The percentage release of vegetative Heterosigma cells from bottom muds tended to decrease with decreasing temperature, but this species also showed high percentages of liberation (40-100%) in March or April when in situ mud temperatures were still low (9.5-15.3°C). On the other hand, field surveys conducted in Osaka Bay during low temperature periods of 1980-1982 showed that H. akashiwo lcan survive through winter as planktonic vegetative cells as well as benthic cysts.

CONCLUSION

From the 10 year data set, the mechanisms involved in outbreaks of Heterosigma red tide in Osaka Bay can be summarized as follows:
The dormancy of this species can be broken by temperatures lower than 10°C in winter. The newly liberated cells, together with vegetative cells which have survived through the winter, begin to grow as the temperature of the seawater increases. The release of vegetative cells from bottom muds becomes maximal at temperatures higher than 15°C. The growth of H. akashiwo is accelerated when the water temperature exceeds 18-20°C. In addition, an increase in river runoff in the rainy season (June-July increases the supply of nitrogen, phosphorus and chelated iron from the land, leading to the success of vegetative H. akashiwo populations in Osaka Bay. Although nutrient concentrations in the surface layer decrease due to the enhanced photosynthetic activity of H. akashiwo populations, the organisms can absorb nitrogen and phosphorus from the nutrient-rich bottom layer by their downward migration. The upward migration of this species produces a dense bloom in the surface layer (Fig. 2).

REFERENCES

1. C.R.Tomas, J. Phycol., 16, 149-156, (1980)
2. U.S. Environmental Protection Agency: Marine Algal Assay Procedure: Bottle test. Corvallis, Oregon, 43pp., (1974)
3. C.R.Tomas, J. Phycol., 14, 314-319, (1978)

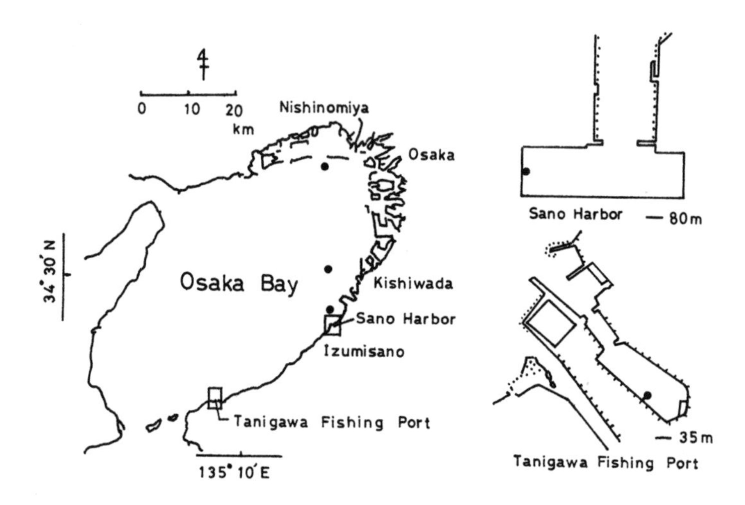

FIG.1. Study area. Symbol: ●, sampling stations of seawater and/or bottom mud.

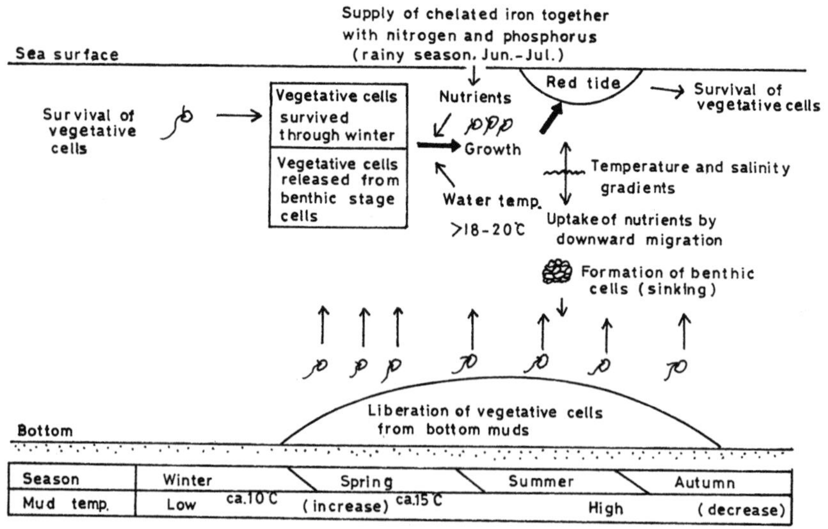

FIG.2. Schematic for the mechanisms of a *Heterosigma akashiwo* red tide in Osaka Bay.

THE DYNAMICS OF VITAMIN B_{12} AND ITS RELATION TO THE OUTBREAK OF *Chattonella* RED TIDES IN HARIMA NADA, THE SETO INLAND SEA

TOSHITAKA NISHIJIMA AND YOSHIHIKO HATA
Kochi University, Faculty of Agriculture, Monobe, Nankoku City, Kochi 783, Japan

ABSTRACT

The vitamin B_{12} requirement of *Chattonella antiqua* and the relationship between the outbreak of *Chattonella* red tides and the dynamics of vitamin B_{12} in Harima Nada were studied.

C. antiqua was found to require vitamin B_{12} for growth, and its vitamin B_{12} half-saturation constant was 0.05-0.35 ng/ℓ. The vitamin B_{12} level which did not depress the growth was estimated at 10-20 ng/ℓ. In Harima Nada, dissolved vitamin B_{12} concentrations ranged from 1.63 to 25.0 ng/ℓ and were generally high in the nearshore area where the red tides have occurred most frequently. According to the budget of dissolved vitamin B_{12} in the nearshore region, the supply almost equally balanced the removal and turnover rates were 1.1-1.3 days. Bacteria and phytoplankton in the water were found to contribute most greatly to both the supply and removal of vitamin B_{12}. The results suggest that the level of dissolved vitamin B_{12} is an important factor controlling the outbreak of *Chattonella* red tides in Harima Nada.

INTRODUCTION

Many marine phytoplankters have been found to be auxotrophic. They require B group vitamins as a growth factor: most commonly vitamin B_{12}, thiamine, and biotin alone or in various combinations [1]. All strains of red tide phytoplankters studied have been reported to require vitamin B_{12}, and some strains have been reported to require thiamine and/or biotin as well [2]. Thus the vitamin supply, especially of vitamin B_{12}, is hypothesized to control not only the growth of red tide organisms but also the selection of the species which develops into red tides.

Chattonella antiqua is a raphydophycean red tide organism which is extremely harmful to marine animals. *Chattonella* red tides have frequently occurred in coastal regions of Japan since 1969 [3]. Along the coast of Harima Nada, in the Seto Inland Sea, approximately 14 million cultivated yellowtail were killed by them in 1972.

The present paper describes the growth characteristics of *C. antiqua* with reference to vitamin requirement and the relationship between the outbreak of *Chattonella* red tide and dynamics of vitamin B_{12} in Harima Nada.

MATERIALS AND METHODS

Vitamin assay

The vitamin B_{12} content of the water was determined by the

RED TIDES: BIOLOGY, ENVIRONMENTAL SCIENCE, AND TOXICOLOGY
Okaichi, Anderson, and Nemoto, Editors

microbioassay method using *Euglena gracilis* strain z [4].

Vitamin requirement of *C. antiqua*

Axenic clonal cultures of two strains of *C. antiqua* were used. The strains Harima 77 and TSU-8011 were isolated from red tides in Harima Nada and in Uranouchi Inlet, respectively. A modified ASP_2NTA medium [5] was used as the basal medium. The B group vitamin requirement and the influence of vitamin B_{12} concentration on the growth, the photosynthesis, and the vitamin B_{12} uptake of *C. antiqua* were examined.

Vitamin B_{12} loads from rivers

The vitamin B_{12} content of the water and the flow rate of the 25 rivers flowing into the northern part of Harima Nada were determined.

Vitamin B_{12} transudation from bottom sediments

The release rates of bottom sediment in Harima Nada were calculated from the amount of vitamin B_{12} released from the sediment into artificial vitamin-free seawater placed on sediment-cores.

Vitamin B_{12} production and consumption by microorganisms

The vitamin B_{12} consumption by bacteria and phytoplankters was computed separately from Co-57 vitamin B_{12} uptake using a photosynthetic inhibitor, DCMU, and using size fractionation with a 5 μm filter. Their production was calculated from the vitamin B_{12} amount released into the sample water during a 24-hour incubation period.

RESULTS AND DISCUSSION

Vitamin requirement of *C. antiqua*

C. antiqua (both the Harima 77 and TSU-8011 strain) was found to require vitamin B_{12} for growth. The growth rate, final growth yields, photosynthetic rate, and Vitamin B_{12} uptake rate of *C. antiqua* depended on the ambient vitamin B_{12} concentration. The kinetic parameters for vitamin B_{12} requirement of *C. antiqua* are summarized in TABLE I. Since the rate is calculated to be saturated at around 10 or 20 times as much as its half-saturation constant, the growth rate of strain Harima 77 is estimated to saturate at about 1 ng/ℓ of vitamin B_{12}. However, the photosynthetic rate and the vitamin B_{12} uptake rate are estimated to saturate at a concentration higher than 20 ng/ℓ. Since cell multiplication is essential for red tide organisms to develop into a red tide, it is estimated that the critical vitamin B_{12} demand for the organism is the level which does not depress its multiplication. Thus, the vitamin B_{12} demand for *C. antiqua* is estimated at 10 or 20 ng/ℓ.

TABLE I. Kinetic parameters for vitamin B_{12} requirements of *C. antiqua*

Parameters	Strains	
	Harima 77	TSU-8011
Vitamin B12 half-saturation constant based on		
growth rate (ng/ℓ)	0.05	0.35
photosynthetic rate (ng/ℓ)	1.72	
vitamin B12 uptake rate (ng/ℓ)	2.74	
Vitamin B12 concentration maximizing growth yields (ng/ℓ)	10	20
Vitamin B12 minimum cell quota (pg/cell)	4.7×10^{-4}	1.6×10^{-3}

Distribution of Dissolved vitamin B_{12} in Harima Nada

Harima Nada is a eutrophic inland area where a large amount of industrial and domestic waste waters

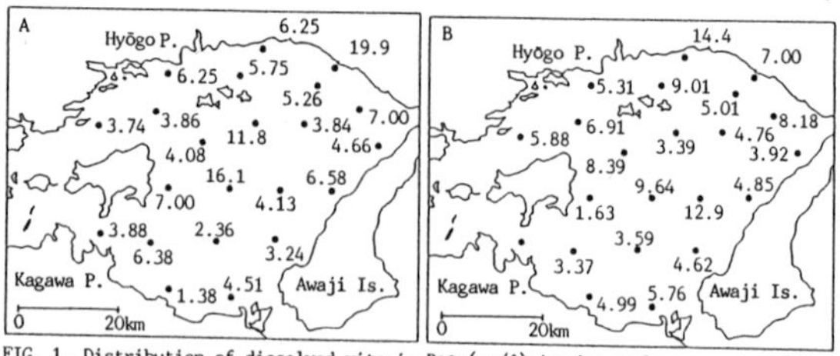

FIG. 1. Distribution of dissolved vitamin B_{12} (ng/ℓ) in the surface seawater of Harima Nada (A:June 23-26,1979; B:November 13-16,1979).

are discharged in the northern part. Red tides have broken out frequently in this area. The contents of dissolved vitamin B_{12} in the water of Harima Nada ranged from 1.63 to 25.0 ng/ℓ. Some of the data [6] are shown in FIGs. 1 and 2. In offshore stations (FIG. 1), the content of dissolved vitamin B_{12} ranged from 1.38 to 19.9 ng/ℓ, generally highest in the northern part. In the northern nearshore and estuarine stations (FIG. 2), the vitamin B_{12} contents were mostly higher than those in the offshore stations. From these results, it is suggested that the water body with a high vitamin B_{12} concentration of more than 10 ng/ℓ appears in the estuaries where organic wastes are being discharged.

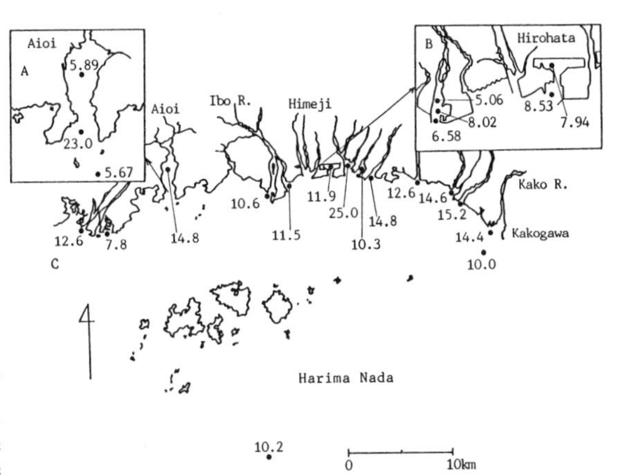

FIG. 2. Distributions of dissolved vitamin B_{12} (ng/ℓ) in the surface water of the nearshore stations in Harima Nada (A,B: June 30,1980;C: May 20,1981).

Consumption and production of vitamin B_{12} by microorganisms in natural seawater

The consumption, the net production, and the gross production by the natural phytoplanktonic and bacterial communities are summarized in TABLE II. The results show that the total rates of vitamin B_{12} consumption by both phytoplankton and bacteria ranged from 5.80 to 6.47 ng/ℓ/day in the light. The gross production rates of vitamin B_{12} by both miroorganisms were considerably high and generally comparable to the consumption rates in the same light conditions. Furthermore, the bacterial community contributed much more to both consumption and gross production of vitamin B_{12} than the phytoplanktonic community did. Though bacteria could produce much more vitamin B_{12} than they consumed, phytoplankters consumed much more vitamin B_{12} than they produced.

TABLE II. Consumption, net production, and gross production of vitamin B_{12} by microorganisms in natural seawater(Uranouchi Inlet, May 13,1983)

Microorganisms	Incubation (lx)	Consumption ng/ℓ/day	Net production ng/ℓ/day	Gross production ng/ℓ/day
Phytoplankton + Bacteria	Dark	4.97	0.47	5.44
	Light(3000)	6.29	0.13	6.42
	Light(9000)	5.80	-0.06	5.74
	Light(18000)	6.47	0.06	6.53
Bacteria + (Ultraplankton)	Dark	3.36	0.62	3.98
	Light(3000)	3.92	0.32	4.24
	Light(9000)	3.45	0.06	3.51
	Light(18000)	3.92	0.10	4.02
Phytoplankton (5.0 μm<)	Dark	1.61	-0.15	1.46
	Light(3000)	2.37	-0.19	2.18
	Light(9000)	2.35	-0.12	2.23
	Light(18000)	2.55	-0.04	2.51

Budget of dissolved vitamin B_{12} in the northern part of Harima Nada

The budget of dissolved vitamin B_{12} was analyzed for the northern part of Harima Nada where *Chattonella* red tides have frequently broken out. The rates were computed from the rates of the production and consumption by microorganisms, the supply from rivers, the transudation from sediments, and the decomposition by daylight. As shown in TABLE III, the results indicate that the total supply of the vitamin was estimated from 34.7 to 41.4 kg/day and that it was almost equally balanced by the removal of the vitamin. Among these agents, both bacterial and phytoplanktonic communities were found to contribute in a major way to the supply as well as the removal of dissolved vitamin B_{12}. According to these rates, turnover rates in this area were estimated to be as short as about one day.

Relationship of dissolved vitamin B_{12} to the outbreak of *Chattonella* red tides

Though not only vitamin B_{12} but also inorganic nutrients are basically necessary for the development of red tide organisms, the contribution of dissolved vitamin B_{12}, apart from the inorganic nutrients, to the outbreak of *Chattonella* red tide is summarized in FIG. 3. *C. antiqua* needs a dissolved vitamin B_{12} concentration around 2 to 5 ng/ℓ to saturate its growth rate. At this concentration *C. antiqua* can maintain the normal growth in competition with other phytoplankters, but cell yields, photosynthesis, and vitamin B_{12} uptake are still depressed. In such eutrophic waters, the supply and the removal of dissolved B_{12} are mainly due to the production and consumption by microorganisms. If a large amount of polluted river water and/or organic waste water is discharged into such a region causing a vitamin B_{12} concentration of more than 10 ng/ℓ, the growth of *C. antiqua* is maximized and its photosynthesis and vitamin B_{12} uptake become more active in that area. As a result of active multiplication the *C. antiqua* community develops into a red tide there. Such high-vitamin B_{12} areas have a tendency to appear in shallow coastal regions into which polluted rivers flow, and after a heavy rain.

TABLE III. Budgets of dissolved vitamin B_{12} in the water of the northern portion of the Harima Nada in summer*

	Item	Rate kg/day	Turnover time day
Supply	Production by bacteria	22.1-26.8	1.73-2.06
	Production by phytoplankton	11.9-13.1	3.48-3.83
	Loads from rivers	0.25-0.77	59.2-182
	Release from bottom sediments	0.45-0.76	60.0-101
	Total	34.7-41.4	1.10-1.31
Removal	Consumption by bacteria	19.8-22.7	2.01-2.30
	Consumption by phytoplankton	13.6-16.0	2.85-3.35
	Abiotic decomposition by light**	2.17-2.89	15.8-21.0
	Total	35.6-41.6	1.10-1.28

* Surface area, 446 km²; mean depth, 14.2 m; volume, 6.31×10^9 m³; standing crop of dissolved vitamin B_{12}, 45.6 kg.
** Specific decomposition rate, 0.099 day^{-1} (by Carlucci, *et al.*) [7]

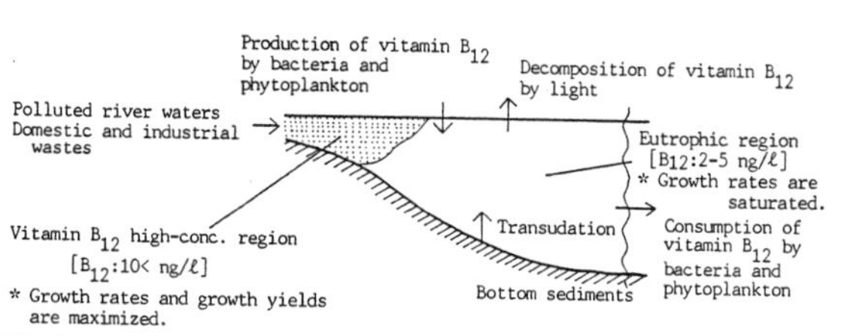

FIG. 3. Relationship of dissolved vitamin B_{12} to the outbreak of *Chattonella* red tides.

REFERENCES

1. L. Provasoli in: "The Sea", M.N. Hill, ed. (Interscience, Vol.1, New York 1963) pp. 165-219.
2. T. Nishijima and Y. Hata in: "Akashio no Kagaku", T. Okaichi, ed. (Koseisya Koseikaku, Tokyo 1987) pp. 181-193.
3. C. Ono and H. Takano, *Bull. Tokai Reg. Fish. Res.* 102, 93-100 (1980).
4. M. Hayashi and T. Kamikubo, *J. Fermen. Technol.* 44, 640-642 (1966).
5. T. Nishijima and Y. Hata, *Nippon Suisan Gakkaishi* 52, 181-186 (1986).
6. T. Nishijima and Y. Hata, *Nippon Suisan Gakkaishi* 52, 1533-1545 (1986).
7. A.F. Carlucci, S.B. Silbernagel, and P.M. McNally, *J. Phycol.* 5, 302-305 (1969).

IS GRAZING OF MARINE CLADOCERANS IMPORTANT FOR THE OCCURRENCE OF RED TIDE?

S. W. KIM, Y. H. YOON, T. ONBÉ AND T. ENDO
Graduate School of Biosphere Sciences, Hiroshima University,
Saijō-cho, Higashi-Hiroshima 724, Japan

ABSTRACT

The grazing of marine cladocerans on red tide organisms was studied using materials collected once or twice a week from April 1986 through May 1987 at Abuto, and monthly during May to September 1987 in Fukuyama Harbor. Both locations are situated in the coastal Inland Sea of Japan.

The gut contents of five species of marine cladocerans which occurred during the present study, *Penilia avirostris*, *Evadne nordmanni*, *E. tergestina*, *Podon leuckarti* and *P. polyphemoides*, were examined with SEM. Phytoplankton species composition was compared with the gut contents.

Skeletonema costatum was found in the gut contents of marine cladocerans throughout the year, although this species showed a great seasonal fluctuation in the abundance (0.87% to 93.47% of total phytoplankton standing crop).

Centric diatoms, *S. costatum*, *Cyclotella meneghiniana*, *Thalassiosira* spp., *Coscinodiscus* spp. and *Chaetoceros* spp., appear to be the main food source for marine cladocerans.

Marine cladocerans grazed also on pennate diatoms, *Navicula* spp. and *Thalassiothrix* sp., and dinoflagellates, *Prorocentrum triestinum*, *P. dentatum* and *Ceratium kofoidii*, albeit in low numbers.

INTRODUCTION

The overall impact of zooplankton grazing on the occurrence of red tide organisms is still obscure. Only a few studies have been conducted [1-4]. For the purpose of understanding red tide phenomena, the ecological role of zooplankton grazing against the outburst of red tide is one of the most important problems to be solved by both zoo- and phytoplanktologists.

Although most of the investigations on zooplankton grazing have focused on the quantitative aspects of feeding, there are some works dealing with the qualitative feeding (*in situ* feeding habits) of copepods [5-7].

The relationship between the abundance of marine cladocerans and phytoplankton pulses has been studied [8-10]. Studies on both quantitative and qualitative feeding of these animals are rare. Bainbridge [8] suggested that peridinians, *Ceratium furca* and tintinnids might be the important food organisms for *Evadne* spp. in the Clyde Sea. Morey-Gaines [9] reported *Podon polyphemoides* grazing on *Ceratium* spp. in the Los Angeles-Long Beach Harbors and White [10] reported *Evadne nordmanni* grazing on the toxic dinoflagellate *Gonyaulax excavata* in the Bay of Fundy.

Gut contents of marine cladocerans were examined to know whether marine cladocerans grazing is important for the occurrence of red tide.

MATERIALS AND METHODS

Quantitative zoo- and phytoplankton samples were collected once or twice a week during the period from April 1986 to May 1987 at Abuto (34°22.2'N, 133°21.0'E) and monthly from May through September 1987 in Fukuyama Harbor (34°24.4'N, 133°24.8'E) in the Inland Sea of Japan.

RED TIDES: BIOLOGY, ENVIRONMENTAL SCIENCE, AND TOXICOLOGY
Okaichi, Anderson, and Nemoto, Editors

The plankton net used for zooplankton collection was of Kitahara type (mouth diameter 22.5cm, mesh aperture 100μm). Samples collected by vertical hauls from near bottom to surface (depth between 6-12m throughout the year) were fixed with 5% formalin.

Two hundred and fifty ml of surface seawater (1m depth) was taken at the same time, concentrated to 5ml with a membrane filter (diameter 47mm, pore size 8.0μm) and examined for phytoplankton species composition and standing crops. Chlorophyll *a* concentration was also measured.

Temperature and salinity of both surface and bottom waters were recorded.

From each zooplankton sample, 1-10 specimens of each species of marine cladocerans were sorted, carefully dissected and their gut contents examined with SEM (JSM-T20).

RESULTS

Water temperature at Abuto fluctuated greatly, with a maximum of 29.8°C in September and a minimum of 9.0°C in February. Salinity ranged between 29.9‰ and 34.8‰ and chlorophyll *a* content was generally high in warmer seasons and low in winter season, ranging between 1.09μg/l and 14.97μg/l.

Five species of marine cladocerans, viz., *Penilia avirostris, Evadne nordmanni, E. tergestina, Podon leuckarti* and *P. popyphemoides*, occurred seasonally (Fig. 1).

Marked high abundance of *P. polyphemoides* occurred during blooms of the dinoflagellates *Scrippsiella trochoidea* and *Prorocentrum dentatum* in June and July 1986 (Fig. 2), whereas the other 4 species did not show this kind of relationship.

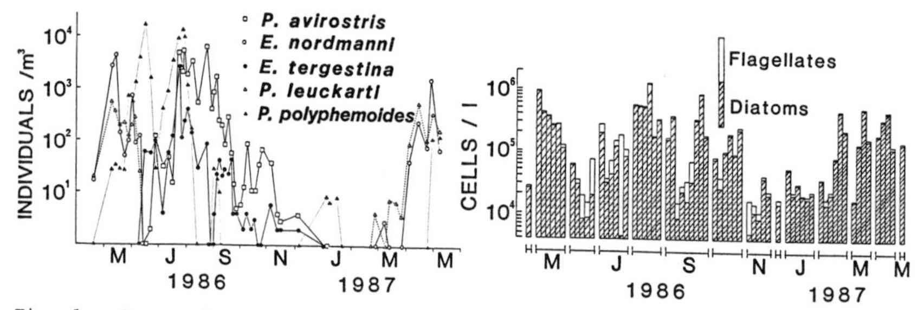

Fig. 1. Seasonal occurrence of five species of marine cladocerans at Abuto in 1986-1987.

Fig. 2. Seasonal change of the standing crops of phytoplankton at Abuto in 1986-1987.

Examination of gut contents of marine cladocerans revealed that *Skeletonema costatum* was grazed by all five species of cladocerans throughout the year although the proportion of *S. costatum* greatly fluctuated with season, ranging between 0.87% and 93.47% of total phytoplankton standing crop . Centric diatoms, *S. costatum* (Fig. 3A), *Cyclotella meneghiniana* (Fig. 3CD), *Thalassiosira* spp., *Coscinodiscus* spp. and *Chaetoceros* spp. (Fig. 3B), were found to be the most important food organisms for this group of animals. Grazing on pennate diatoms and dinoflagellates (Fig. 3E-H) was seldom observed regardless of their abundance in total phytoplankton standing crop . All specimens examined for gut contents had some unidentified materials, which seemed to be derived from digested food.

In Fukuyama Harbor, the red tide samples collected contained *S.costatum, Prorocentrum triestinum* and *Eutreptiella* sp. in May, *P. triestinum* and *Eutreptiella* sp. in July, *P. minimum, P. triestinum* and *Heterosigma akashiwo* in August and *S. costatum* in September. However, gut content examination

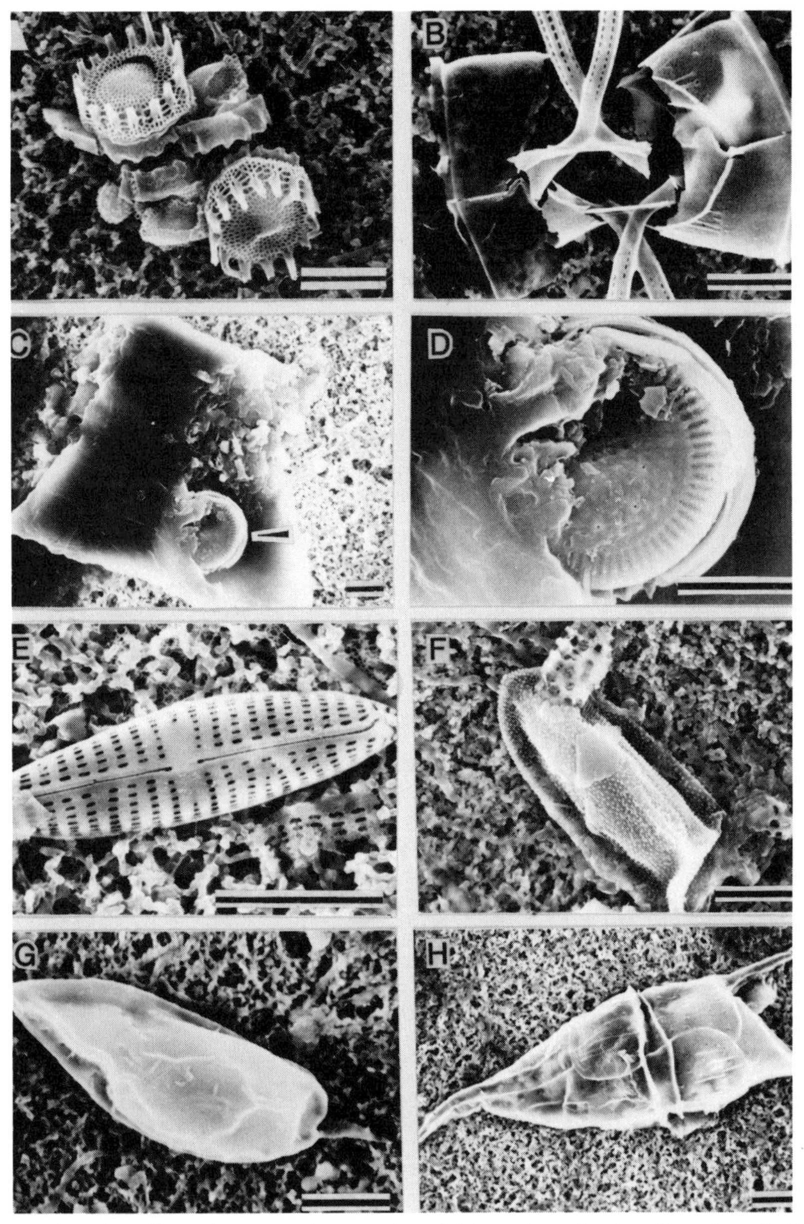

Fig. 3. Scanning electron micrographs of the gut contents of marine cladocerans. (A) *Skeletonema costatum* from *E. nordmanni*, (B) *Chaetoceros* sp. from *P. polyphemoides*, (C) *Cyclotella meneghiniana* from *P. avirostris*, (D) Higher magnification of *C. meneghiniana* indicated by arrow in (C), (E) *Navicula* sp. from *P. avirostris*, (F) *Prorocentrum dentatum* from *P. avirostris*, (G) *P. triestinum* from *P. avirostis*, (H) *Ceratium kofoidii* from *P. leuckarti*. Scale bars indicate 5µm.

revealed that there was no intensive grazing on phytoflagellates by marine cladocerans.

DISCUSSION

Dinoflagellates have been assumed to be important food organisms for marine cladocerans because some species of marine cladocerans occurred in high abundance with dinoflagellate blooms [8-10].

The present study is aimed at understanding (1) What are the natural food items of marine cladocerans? and (2) Is there any selective feeding of marine cladocerans on phytoplankton species? Centric diatoms were found to be the most important food organisms, whereas the grazing on dinoflagellates was seldom observed. The results present a striking contrast to previous reports which mention the grazing of marine cladocerans on dinoflagellates and specific relationship between cladocerans and dinoflagellates.

In the present study, a significant relationship between the abundance of *P. polyphemoides* and two species of dinoflagellates was observed. These dinoflagellates, however, did not occur in the gut of *P. polyphemoides*. This fact suggests that other phytoplankton would be enough for food of *P. polyphemoides*. A recent investigation on food consumption of *P. avirostris* showed that no food deficiency occurs for this species even in the oligotrophic open ocean [11], and the finding of the oceanic distribution of *E. nordmanni* and *E. tergestina* [12-15] further supports that marine cladocerans have no food deficiency in eutrophicated coastal waters even without dinoflagellates, as observed in *P. polyphemoides* during the present study.

To summarize, centric diatoms appear to be the most important food organisms for marine cladocerans and the impact of marine cladocerans grazing on red tide dinoflagellates might be low.

ACKNOWLEDGEMENT

We thank Mr. S. Ohtsuka for his expertise in preparing specimens for SEM.

REFERENCES

1. M.E. Huntley, J. Exp. Mar. Biol. Ecol., 63, 81-91 (1982).
2. K.G. Sellner and M.M. Olson in: Toxic Dinoflagellates, D.M. Anderson, A.W. White and D.G. Baden, eds. (Elsevier, NY 1985) pp. 245-250.
3. S. Uye, Mar. Biol., 92, 35-43 (1986).
4. P.F. Sykes and M.E. Huntley, Mar. Biol., 94, 19-24 (1987).
5. M.V. Lebour, J. Mar. Biol. Ass. U.K., 12, 644-677 (1922).
6. J.T. Turner, NOAA Tech. Rep. NMFS, 7, 1-28 (1984).
7. S. Ohtsuka, A. Fleminger and T. Onbé, J. Crustacean Biol., 7, 554-571 (1987).
8. V. Bainbridge, J. Mar. Biol. Ass. U.K., 37, 349-370 (1958).
9. G. Morey-Gaines in: Toxic Dinoflagellate Blooms, D.L. Taylor and H.H. Seliger, eds. (Elsevier, North Holland, NY 1979) pp. 315-320.
10. A.W. White, Can. J. Fish. Aquat. Sci., 37, 2262-2265 (1980).
11. G.A. Paffenhöfer and J.D. Orcutt, Jr., J. Plankton Res., 5, 741-754 (1986).
12. O.M. Jorgensen, J. Mar. Biol. Ass. U.K., 19, 177-226 (1933).
13. K.F. Wiborg, Rep. Norw. Fish. Mar. Invest., 11, 1-66 (1955).
14. W.W.C. Gieskes, Neth. J. Sea Res., 5, 342-376 (1971).
15. A.R. Longhurst and D.L.R. Seibert, Crustaceana, 22, 239-248 (1972).

CONTROL OF DIEL VERTICAL MIGRATION AND CELL DIVISION RHYTHM OF *HETEROSIGMA AKASHIWO* BY DAY AND NIGHT CYCLES

Masayuki TAKAHASHI* AND Yoshiaki HARA**
*Botany Department, University of Tokyo, Hongo, Tokyo 113, Japan;
**Institute of Biological Sciences, University of Tsukuba, Tsukuba 305, Japan

ABSTRACT

Vertical migratory behavior of *Heterosigma akashiwo* in the laboratory was confirmed to be controlled by the light/dark cycle, not by simple phototaxis, and to be coupled to the diel cell division rhythm. The minimum light intensity initiating vertical migratory behavior was similar to the compensation light intensity for cell division.

INTRODUCTION

It has been suggested that vertical migratory behavior is essential for some flagellates to form red tide by effective absorption of nutrients near the bottom or deeper layer of water column and effective utilization of light energy near the surface [1,2,3,4]. Special concern was given to the evaluation of possible environmental factors controlling vertical migratory behavior and the cell division rhythm in natural waters.

MATERIALS AND METHODS

An axenic culture of *Heterosigma akashiwo* isolated from Tanigawa Harbor was prepared at 20°C under a light:dark (14:10) cycle of daylight-type fluorescent lamps (40 $\mu E\ m^{-2}\ s^{-1}$ on 24 h average) in f/2 medium [5]. At early logarithmic growth phase the culture was transfered to autoclaved glass cylinders (internal diameter, 4 cm; height, 30 cm) for the migration experiments. Vertical migratory behavior was checked by visual observations with minimum light exposure, particularly at the dark period [6]. Cell counts were made using a particle counter (Coulter TA_{II}). Specific growth rate was calculated using the following equation: $\mu_2=(\ln C_2 - \ln C_1)/(T_2 - T_1)\ln 2$ in which C_1 and C_2 are cell numbers at time T_1 and T_2, respectively.

RESULTS

The laboratory culture of *Heterosigma akashiwo* showed clear vertical migratory behavior in a glass cylinder: the cells stayed near the cylinder bottom during the dark period and swam towards the surface during the light period (FIG. 1). Under continuous light, the vertical migratory behavior was maintained during the first 48 hours, followed by homogeneous dispersion within the cylinder. There occurred a transient phase showing a weak vertical migratory behavior between 48 and 62 hours. On the other hand, under continuous darkness the vertical migratory behavior was maintained over a longer period of time: up to 96 hours after turning off the light. An indistinct vertical migratory behavior was also observed between 96 and 120 hours. Changing the light/dark cycle from 24 hours to 12 hours, 6L:6D cycle, the migratory behavior was adjusted to the new cycle after 48 to 62 hours; the same time requirement for the continuous illumination. In the other two cycles of 3L:3D and 1L:1D, migratory behavior was only detected in the 3L:3D cycle although it was not as obvious as in the other longer

RED TIDES: BIOLOGY, ENVIRONMENTAL SCIENCE, AND TOXICOLOGY
Okaichi, Anderson, and Nemoto, Editors

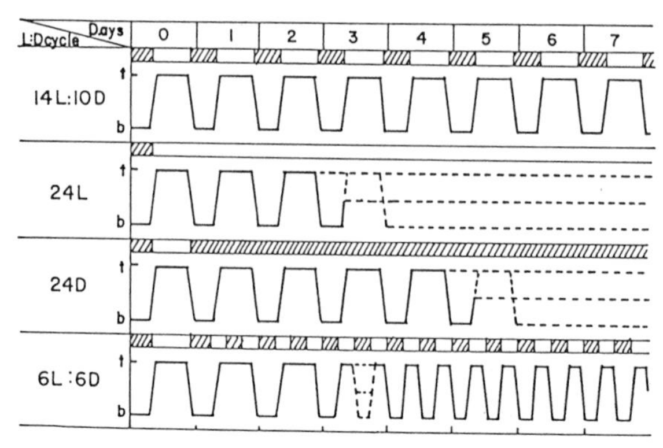

FIG. 1. Vertical migratory behavior of *H. akashiwo* at different lighting regimes. Solid line indicates the densely accumulated position of cells between the top(t) and the bottom(b) of the cylinder, and dotted line shows indistinctive or homogeneous distribution. Shaded area, dark period.

lighting cycles. The experiments mentioned above were all carried out with illumination from the top. When the cylinders were illuminated sideways (in which two cylinders had the top or the bottom half covered by a black sheet), the cells still maintained identical vertical migratory behavior in all the cases [7]. Even during the homogeneous dispersion under continuous light or darkness, each cell of *Heterosigma akashiwo* showed active movement under the microscope which was no different from the cells showing active migratory behavior.

Vertical migratory behavior was tested under various light intensities of daylight type fluorescent lamp ranging from 0.1 $\mu E\ m^{-2}\ s^{-1}$ to 200 $\mu E\ m^{-2}\ s^{-1}$ (TABLE I). Clear vertical migratory behavior was noticed at light intensities over 0.3 $\mu E\ m^{-2}\ s^{-1}$. At lower light intensities, the migratory behavior became unclear with time, the cells losing their migratory behavior faster with time at the lower the light intensities.

TABLE I. Time changes of vertical migratory behavior of *H. akashiwo* under different light intensities of 14L:10D lighting cycle. Migratory behavior; +, clear; ±, indistinctive; -, no.

Light intensity		Days									
$\mu E\ m^{-2}\ s^{-1}$	$E\ m^{-2}\ d^{-1}$	1	2	3	4	5	6	7	8	9	10
200	10.1	+	+	+	+	+	+	+	+	+	+
100	5.04	+	+	+	+	+	+	+	+	+	+
60	3.02	+	+	+	+	+	+	+	+	+	+
30	1.51	+	+	+	+	+	+	+	+	+	+
15	0.76	+	+	+	+	+	+	+	+	+	+
6	0.302	+	+	+	+	+	+	+	+	+	+
3	0.151	+	+	+	+	+	+	+	+	+	+
0.6	0.030	+	+	+	+	+	+	+	+	+	+
0.3	0.015	+	+	+	+	+	+	+	+	+	+
0.2	0.010	+	+	+	±	±	±				
0.15	0.008	+	+	±	±	±	−				
0.1	0.005	+	+	±	−	−	−				
0.0	0.000	+	+	±	−	−	−				

Using a culture maintained under the 14L:10D lighting cycle for many generations, the increase of cell numbers was followed over 48 hours at 2 to 4 hour time intervals under the same lighting cycle for pre-culture (FIG. 2). Cell numbers hardly changed during the light period but increased obviously during the night period, which was also confirmed by several other similar sets of experiments. Total cell volume, on the other hand, showed an opposite trend indicating a logarithmic increase during the light period and no change or even slight decrease during the dark period, although there was some increase in the first dark period (FIG. 2). A similar trend in volume change was repeatedly observed in several other separate experiments including 6L:6D cycle. Since there was no change in the cell numbers during the light period, it became obvious that each cell increased the size during the light period.

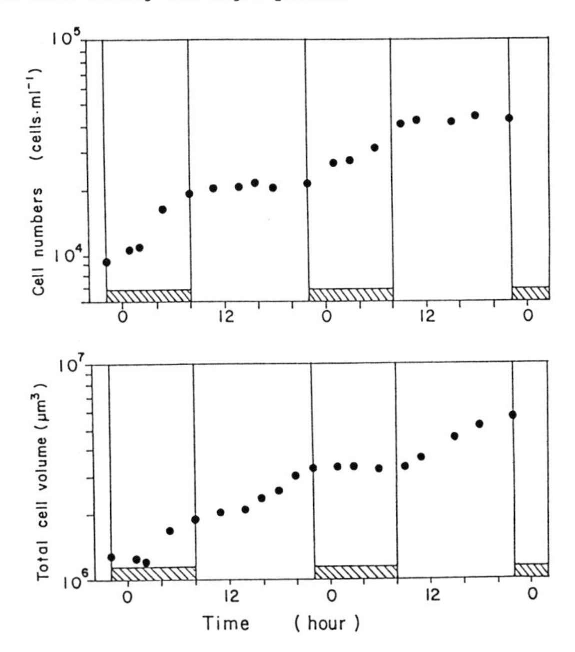

FIG. 2. Changes in cell numbers and total cell volume of H. akashiwo under the 14L:10D lighting cycle at 40 $\mu E\ m^{-2}\ s^{-1}$. Hatched ares, dark period.

Since it is expected that cell division depends strongly on light intensity, effects of light intensity on cell division was examined over 10 days under various light intensities (average of light and dark periods) ranging from 0.1 to 200 $\mu E\ m^{-2}\ s^{-1}$ with the 14L:10D lighting cycle. Cells were inoculated at a density of 10^3 cells ml^{-1} and put into glass cylinders. Cell counts were performed at 8 hours after the light was on with mixing by gentle swirling each time. The cell density increased more than two orders of magnitude during the experiment at high light intensities. The average specific growth rate showed no positive value below 0.6 $\mu E\ m^{-2}\ s^{-1}$ and showed a rapid increase with an increase in light intensity over 1 $\mu E\ m^{-2}\ s^{-1}$, and some light saturation around 60 $\mu E\ m^{-2}\ s^{-1}$ (FIG. 3). The maximum rate attained in this experimental setting was 1 doubling per day.

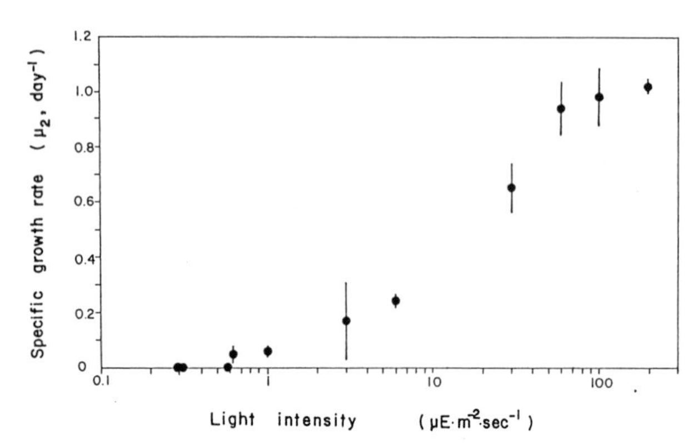

FIG. 3. Relation between the specific growth rate of H. akashiwo and culture light intensity under the 14L:10D lighting cycle.

DISCUSSION

The present study revealed that the vertical migratory behavior of H. akashiwo was controlled by the light/dark cycle. Active cell mobility observed during homogeneous distribution without showing any consistent vertical migratory behavior in the continuous light or dark strongly suggested that the lighting cycle regulated the swimming direction, towards the surface or away from the bottom at light and the opposite direction at night, but not the cell mobility itself. Insensitivity to lighting direction, and carryover behavior of vertical cell migration by switching to a new lighting cycle, both indicated that the migratory behavior was not due to simple phototaxis but was controlled by possible internal clock mechanism in the cell. The timing of cell division was strongly centered in the dark period under a given lighting cycle. However cell division itself even occurred under continuous lighting with no division rhythm [7], which indicated that cell division itself was not as strongly connected to the lighting cycle as division timing. The light energy requirement for initiating the migratory behavior was similar to that for cell division, which suggested that the principal mechanism involved in the migratory behavior could be coupled with that for cell division such as photosynthesis. There is no doubt that the day/night rhythm occurring under natural conditions causes rhythmic behaviors such as the vertical migration and cell division of H. akashiwo.

REFERENCES

1. R. W. Eppley, O. Holm-Hansen and J. D. H. Strickland, J. Phycol., 4, 333-340 (1968).
2. S. I. Heany and R. W. Eppley, J. Plankton Res., 3, 331-344 (1981).
3. N. Fukazawa, T. Ishimaru, M. Takahashi and Y. Fujita, Mar. Ecol. Prog. Ser., 3, 217-222 (1980).
4. M. Takahashi and N. Fukazawa, Mar. Biol., 70, 267-273 (1982).
5. R. R. L. Guillard, In Symposium on Marine Microbiology (C. H. Oppenheimer, ed.), Charles C. Thomas, Springfield, Ill. 1963, pp.93-104.
6. T. Hatano, Y. Hara and M. Takahashi, Jap. J. Phycol., 31, 263-269 (1983).
7. K. Itagaki, Rhythmicity of vertical migration and cell division of a red tide flagellate, Heterosigma akashiwo, MS thesis, Environmental Science, University of Tsukuba, pp.56 (1984).

POPULATION GROWTH OF *Gymnodinium nagasakiense* RED TIDE IN OMURA BAY

SHOJI IIZUKA, HIROSHI SUGIYAMA, AND KAZUTSUGU HIRAYAMA
Nagasaki University, Faculty of Fisheries, Bunkyo Machi 1-14, Nagasaki 852, Japan

ABSTRACT

Natural population density of *Gymnodinium nagasakiense*, a causative organism of common red tides in Western Japan, was surveyed at Omura Bay, Nagasaki Prefecture. Integrated water samplings were done twice a week with a silicon hose which was 15 m long and 18 mm in inner diameter, at every 0.5 km^2 interval. Samplings covered an area where *Gymnodinium* red tides had frequently occurred. A maximum weekly growth rate was observed in the surface layer (0-5m) population as 0.301 (k'/day), while growth rates in the middle layer (5-10m) and deep layer (10-15m) populations were as low as 0.165 and 0.148, respectively. The high growth rate in the surface population lasted for 11 days and led to an occurrence of the red tide which finally reached a dispersed cell density of about $2.0x10^6$ cells/l. The capacity to support growth estimated from nutrient levels in the bay-water was consistent with the actual maximum cell density. Natural population of *G.nagasakiense* was assumed to grow exponentially to the maximum cell density at the maximum growth rate. The results suggest that little or no grazing may have affected the surface population.

INTRODUCTION

It has been estimated that the maximum growth rate of a natural population of *Gymnodinium nagasakiense* Takayama & Adachi might be one cell division per day (k'/day= 0.301) on the basis of preliminary field work [1] and laboratory culture work [2]. This study treats growth processes of a natural population of the *G.nagasakiense* red tide in Omura Bay in 1980 and also the above-mentioned problem.

Omura Bay, about 320 km^2, located in the central part of Nagasaki Prefecture, Western Japan is connected with Sasebo Bay by a narrow strait, being 200 m in width and 50 m in depth. Water volume passing through the strait in a quarter day is 5.5% in a spring tide and 2.8% in a neap tide of the total bay-water volume. Movement of the bay-water is stagnant. Therefore, strong stratification, and anoxic and nutrient rich bottom waters are formed in the summer season.

STUDY AREA AND METHOD

The presumable initial occurring areas of *G.nagasakiense* red tides are distributed throughout the central bay. Among them, an area of 20 km^2 with the depth of 17 to 19 m was selected in the south waters of the Nagasaki Airport Is. as a study area and 12 stations of diamond shape were sampled in the area. We used a non-toxic silicon hose for *G.nagasakiense*, which is 15 m long with two simple ways of separating a 15 m water column into three layers: 0-5m, 5-10m and 10-15m. Thirty-six integrated samples were collected from the three layers at all the stations. As microscopic observations and cell counting of all samples were impossible due to time restriction and cell fragility, all samples were collected into three large vessels by lay-

RED TIDES: BIOLOGY, ENVIRONMENTAL SCIENCE, AND TOXICOLOGY
Okaichi, Anderson, and Nemoto, Editors

er. Then, a subsample from the vessels was brought back to the laboratory, while the remaining water in the vessels was allocated to the in situ experiment for determining the daily growth rate of *G.nagasakiense* .
Each sample of 200 ml water, filtered and unfiltered with a 40 µm mesh net, was enclosed separately in dialysis tubing (Visking Company size no. 20/32) and two pairs of tubes were returned to and suspended at the median depth (2.5, 7.5 and 12.5m) of each water column layer for one or two days. The regular water samplings were carried out twice a week and the in situ suspending experiments were done once a week. The growth rate was expressed by the relative growth rate (k'/day) in both natural and dialysis tube populations:

$$k' = \frac{1}{t} \log (N_t / N_0)$$

where N_0 and N_t are cell densities on the 0 and t day, respectively. Note that this is a growth rate using logarithims to the base 10.

RESULTS

Weekly growth rate

The surface layer population (0-5m) developed to visible red tide cell density on September 1. The weekly mean highest growth rate was 0.304. Population growth rate was highest in both the periods immediately before the red tide, that is August 28 to September 1, and 1.5 to 0.5 week before, that is Auguat 21 to 28. The highest growth rate was nearly the same as the maximum value presupposed in the previous studies [1, 2]. The highest growth rate of the middle layer population (5-10m) was 0.165 in the week from August 14 to 21, one week earlier than the growth rate of the surface layer population reached its maximum. The growth rate of the deep layer population (10-15m) was the lowest of all, with the maximum of only 0.148 in the week from August 21 to 28.

Daily growth rate

The growth rates in the surface layer population suspended at 2.5 m in depth were 0.305 (averaging two filtered and two unfiltered tubes) 0.5 week before the red tide, that is August 28, 0.270 at 1.5 week before, that is August 21, and 0.140 at 2.5 weeks before, that is August 14. The first value was as high as the weekly mean growth rate (0.304) of the in situ population of the same layer and same period. It is apparent that the surface population must have maintained the high growth rate from August 21 to the outburst of the red tide.

Summary of the results

Growth phases of the present red tide population were divided into three parts: (a)exponential phase, (b)red tide phase (= stationary phase) and (c)death phase(Fig. 1).

Exponential phase : The exponential growth phase with a relative growth rate of 0.301 continued for 11 days until just before the occurrence of the red tide.

Red tide phase : The population cell density increased by two thousand times compared with the beginning of the exponential phase and finally became about 2.0×10^6 cells/l, which seems to be the maximum cell density under the present nutritional conditions of Omura Bay. After that, the cell density was maintained at almost the same level for 10 days of the red tide phase. A mean cell density in the period is 1.7×10^6 cells/l.

Death phase	: The red tide population decayed abruptly due to a typhoon on September 11.
Carrying capacity	: The capacity of Omura Bay to carry cell density is about 2.5x10^{6} cells/l in terms of the cell density of *G.nagasakiense* (by unpublished data). The maximum cell density in the red tide phase did not reach the carrying capacity. "Why didn't the maximum cell density in the red tide phase reach the carrying capacity ?"
Evenly dispersed cell density	: The above-mentioned cell densities were based on an assumption that cells were evenly dispersed in each water column. The cell density of red tide patches is expected to reach a level of 10^{7} cells/l by accumulation. Therefore, the present red tide is not a small red tide but moderate scale red tide in terms of the cell density.

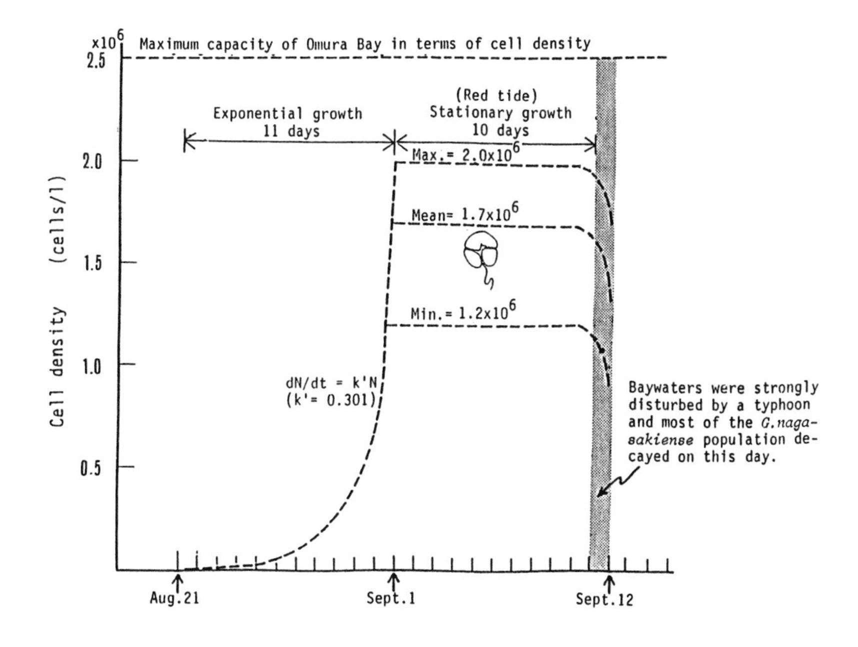

Fig. 1. Diagramatical expression on the growth processes of a *G.nagasakiense* red tide surface population in Omura Bay in 1980.

DISCUSSION

In order to estimate correctly the cell densities of unevenly distributed *G.nagasakiense* populations in time and space, we have used the "Boring method" to collect integrated water samples at as many points as possible with a silicon hose of 15 m long. We call this the "Boring method" as it is used for research of petroleum and underground water.

Consequently, we obtained a value of k'= 0.30 as the mean population growth rate during the exponential phase and this value was precisely identical with the maximum value expected from laboratory culture work. The dispersed maximum cell density was about 2.0×10^6 cells/1 and the mean cell density of 1.7×10^6 cells/1 was maintained during the red tide phase. It seems that the apparent cell density in a red tide patch should reach a level of 10^7 cells/1 by accumulation. By the way, carrying capacity of cell density is about 2.5×10^6 cells/1, and the mean cell density of *G.nagasakiense* in the red tide phase was 1.7×10^6 cells/1. The difference of 0.8×10^6 cells/1 was assumed to be occupied by other phytoplankters such as diatoms and other naked flagellates except *G. nagasakiense*. Actually, *G. nagasakiense* was 68% of total even in the red tide period. The remaining 32% was composed of general phytoplankton. Perhaps, the predators had no difficulty in catching sufficient prey as many phytoplankton other than *G. nagasakiense* were available. The mean growth rate in the exponential phase was identical with the maximum value expected from laboratory work. These facts suggest that little or no grazing affected the accumulation of the surface population of *G. nagasakiense*.

ACKNOWLEDGEMENTS

This study was a work in Special Research Project on Environmental Science supported by Grants in Aid for Scientific Research of the Ministry of Education, Culture and Science, Japanese Government granted to Dr.T.Okaichi, Faculty of Agriculture, Kagawa University (Project No. 56030066). The authors wish to acknowledge the continuous encouragement of him throughout this work.

REFERENCES

1. S.Iizuka, Maximum growth rate of natural population of a *Gymnodinium* red tide. in: Toxic Dinoflagellate Blooms, D.L.Taylor and H.H.Seliger, eds. (Elsevier/North-Holland,N.Y.1979) pp.111-114.
2. S.Iizuka and K.Mine, Maximum growth rate of *Gymnodinium* sp.(type-'65) expected under culture conditions. *Bull.Plankton Soc.Jap.*,30,139-146 (1983).

CYCLONIC DISTURBANCE AND A PHYTOPLANKTON BLOOM IN A TROPICAL SHELF ECOSYSTEM

MILES J. FURNAS
Australian Institute of Marine Science, Townsville 4810, AUSTRALIA

ABSTRACT

Following cyclonic disturbance of tropical shelf waters of the Australian Great Barrier Reef, a widespread (ca. 10^4 km^2) diatom-dominated phytoplankton bloom developed within days. Dinoflagellate and flagellate populations in the post-cyclone bloom were little different from normal summer levels. The observed dominance of the bloom by diatoms is consistent with patterns of rapid *in situ* growth rates measured in Great Barrier Reef waters prior to the cyclone.

INTRODUCTION

On February 1, 1986, tropical cyclone Winifred struck the North Queensland (Australia) coast near 17°30'S after passing over the Australian Great Barrier Reef (hereafter GBR). Winifred strongly affected shelf waters over an area on the order of 10^4 km^2. Figure 1 shows representative hydrographic, chlorophyll and dissolved nutrient profiles from the cyclone affected area as compared to profiles normally observed in the central GBR during the summer. Dissolved nutrient levels, particularly of dissolved inorganic nitrogen (DIN) species increased sharply (Fig. 1E-I). Preliminary nutrient budgets indicate coastal runoff, rainfall and porewaters from disturbed shelf sediments accounted for most of the additional phosphate and silicate in shelf waters and phytoplankton standing crop. In contrast, most of the added nitrogen apparently came from microbial mineralization of organic nitrogen released from resuspended shelf sediments.

Within 1-3 days, a pronounced phytoplankton bloom (Fig. 1D), with chlorophyll concentrations 3-10 times levels normally observed in GBR waters had developed throughout the disrupted area. This bloom was dominated by diatoms, including the genera *Chaetoceros*, *Rhizosolenia*, *Thalassionema*, *Thalassiothrix*, *Leptocylindrus*, *Skeletonema* and *Nitzschia*. *Chaetoceros curvisetum*, several *Nitzschia* spp. and other small pennate diatoms were particularly important. In contrast, populations of microflagellates and dinoflagellates did not increase appreciably over concentrations normally observed during the summer in GBR waters.

GROWTH EXPERIMENTS

During 1983-85, a series of *in situ* diffusion culture (1,2) experiments were carried out in GBR waters with natural populations to quantify phytoplankton growth rates. Inoculum assemblages, usually pre-screened (10 or 35 µm) to remove grazers, were incubated in a range of natural environmental conditions. Figure 2 shows mean growth responses over the 2-3 day incubation periods for aggregate assemblages of *Chaetoceros* spp., small pennate diatoms, small gymnodiniaceae and microflagellates in relation to mean *in situ* irradiance levels and DIN concentrations. Growth responses of individual species were similar to the trends shown for the aggregate groups, but occurred or could be quantified in fewer experiments. In a considerable number of experiments, vigorous growth of a particular taxon occurred, but growth rates could not be calculated as the taxon was not observed in the inoculum. Table 1 summarizes maximum growth rates measured for species and groups important in post-cyclone phytoplankton populations. Where available, maximum division rates measured using similiar methods in a more eutrophied temperate estuary are given for comparison (2).

RED TIDES: BIOLOGY, ENVIRONMENTAL SCIENCE, AND TOXICOLOGY
Okaichi, Anderson, and Nemoto, Editors

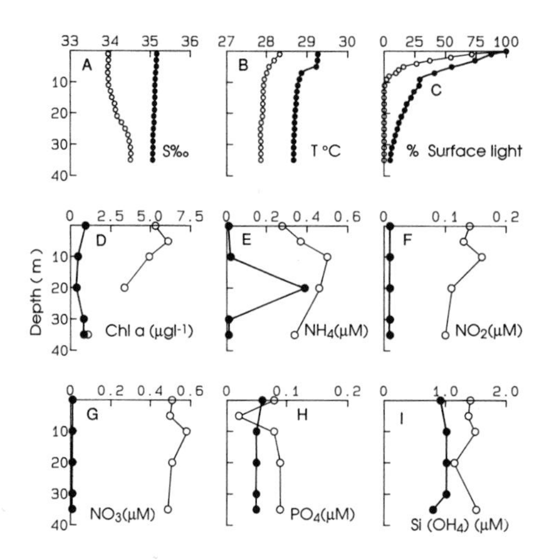

Figure 1.

Vertical profiles of water column parameters at a mid-shelf site four days after cyclone Winifred (o) and one year later (10/2/87) under normal summer conditions (●).

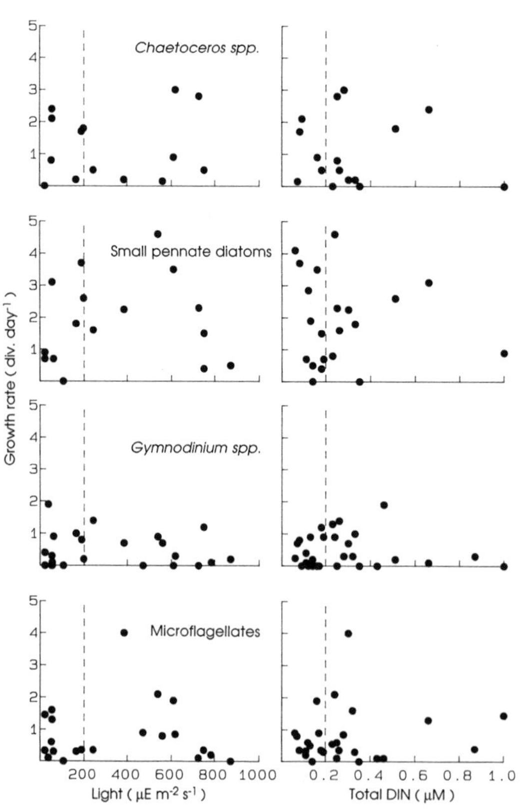

Figure 2.

Mean growth rates in diffusion chambers (doublings day^{-1}) of A) Chaetoceros spp., B) small pennate diatoms, C) small gymnodineaceae and D) microflagellates in relation to *in situ* irradiance levels and dissolved inorganic nitrogen concentrations.

Growth rates of Chaetoceros and small pennate diatom (chiefly Nitzschia spp.) populations in diffusion chambers were frequently greater than 2 doublings day^{-1}. Aggregate microflagellate population growth rates exceeded 4 day^{-1} on one occasion, but otherwise, were less than 2 day^{-1}. The consistency observed in the remaining, lower, flagellate growth rates suggests the highest value might be spurious. Growth rates of small gymnodineaceae, the most rapidly growing category of dinoflagellate, were consistently less than 2 day^{-1}. Significant growth of individual dinoflagellate species was not observed. Near-maximal growth rates in all four groups were recorded at mean daytime irradiance levels of less than 200 $\mu E\ m^{-2} s^{-1}$, approximately 10% of midday surface irradiance, and at total DIN concentrations below 0.2 µM. These patterns suggest that phytoplankton populations present in shelf waters after the cyclone encountered nutrient and light conditions capable of supporting maximal or near-maximal growth rates in the upper half of the water column.

DISCUSSION

Despite apparent ideal conditions for rapid growth, flagellate and dinoflagellate species were not dominant components of the post-cyclone bloom. Growth rates estimated from diffusion culture experiments indicate that diatom species, apparently regardless of cell size, appeared to have higher potential growth rates than did populations or recognizable species of flagellates and dinoflagellates. If the single high growth rate observed for one flagellate population is indeed spurious, the difference between maximal growth rates of a number of diatom species and flagellate populations was on the order of 1-3 doublings day^{-1}. These data, together with data from other temperate systems (2,4) indicate that diatom assemblages should consistently dominate phytoplankton blooms developing after short-term nutrient injection events on the basis of higher intrinsic growth potentials.

A survey of the literature for directly determined in situ growth rates of dinoflagellates and other flagellate species forming red tides (Table 2) suggests that some of these flagellates have maximum growth rates on the order of 2 day^{-1}. While high, these maximum growth rates are still not as rapid as rates recorded for a number of coastal diatom species. Further, maximum growth rates of a number of toxic dinoflagellate species appear to be 1 day^{-1} or less (3,5).

Collectively, these data reinforce the idea that exceptional or accelerated flagellate growth rates after nutrient enrichment events will not by themselves lead to red tide bloom formation, as diatom populations should consistently grow faster than known red tide flagellates. Rather, second order processes such as vertical migration, behavioral responses to hydrographic structural features, life cycle strategies and selective grazing (or discrimination against red tide forming species) by herbivores are likely more important determinants of red tide bloom formation.

ACKNOWLEDGEMENTS

Alan Mitchell assisted ably throughout. This is AIMS contribution no. 395.

REFERENCES

1. Furnas, M.J. 1982a. Mar. Biol. 70: 63-72.
2. Furnas, M.J. 1982b. Mar. Biol. 70: 105-111.
3. Iizuka, S. and K. Mine 1983. Bull. Plank. Soc. Japan. 30: 139-146.

4. Ishizaka, J., M. Takahashi and S. Ichimura 1983. Mar. Biol. 76: 271-278.
5. Karentz, D. 1983. J. Protozool. 30: 581-588.
6. Owen, O.V.H., P. Dresler, C.C. Crawford, M.A. Tyler and H.H. Seliger 1977. Chesapeake Sci. 18: 325-333.
7. Rivkin, R.B., E. Swift, W.H. Biggley and M.A. Voytek 1984. Deep-Sea Res. 31: 353-367.
8. Vargo, G.A. 1984. pp. 113-128 in Lecture Notes on Coastal and Estuarine Studies No. 8. Springer-Verlag, Berlin.
9. Weiler, C.S. 1980. Limnol. Oceanogr. 25: 610-619.

Table 1. Maximum *in situ* growth rates (doublings day^{-1}) of species and groups important in the post-cyclone bloom (GBR). Comparable maximum growth rates from a temperate estuary (NB) are also given.

	GBR	NB
Centric diatoms		
Cerataulina pelagica	1.2	2.6
Chaetoceros compressum	2.1	
C. curvisetum	3.2	
C. peruvianum	1.1	
C. spp.	3.0	
Leptocylindrus danicus	3.2	3.3
Rhizosolenia alata	1.5	
R. delicatula	2.7	
R. fragilissima	2.3	3.4
R. setigera	2.3	
R. stolterfothii	1.7	
Skeletonema costatum	3.2	5.9
Pennate diatoms		
Nitzchia closterium	4.7	3.0
N. pungens/pacifica	1.9	2.2
N. delicatissima	3.8	
Thalassionema nitzschioides	3.1	2.9
Thalassiothrix spp.	2.1	
small pennate diatoms	5.1	
Dinoflagellates		
Gymnodiniaceae "small"	0.5	
Gymnodineaceae "medium"	0.7	
Oxytoxum scolopax	0.7	
Peridinium sp.	1.0	
Prorocentrum micans	1.5	
P. marina	1.4	
Flagellates and picoplankters		
microflagellates	4.1	1.5
non-motile ultraplankton	1.6	1.4
Synechococcus spp.	1.7	
eukaryotic picoplankton	2.0	
autotrophic flagellates	1.3	
Heterotrophs		
ciliates	1.1	1.5

Table 2. Maximum growth rates (doublings day^{-1}) of dinoflagellates and a red tide forming flagellate species measured using direct *in situ* methods.

Taxon	Growth rate (div. day)	Source
Katodinium rotundatum	1.9	6
Gymnodinium simplex	1.7	6
Gryodinium dominans	1.6	6
Proroc. mariae lebouriae	1.2	6
Ceratium spp.	0.11-0.27	9
Pyrocystis noctiluca	0.14	7
P. fusiformis	0.16	7
Peridinium quinquicorne	1.6	8
Prorocentrum minimum	0.9	8
P. minimum	2.0	2
P. triestinum	1.5	2
Olisthodiscus luteus	2.0	2

A COMPARISON OF THE ENVIRONMENTALLY MODULATED SWIMMING BEHAVIOR OF SEVERAL PHOTOSYNTHETIC MARINE DINOFLAGELLATES

Kamykowski, D., S. A. McCollum and G. Kirkpatrick
MEAS, Box 8208, NCSU, Raleigh, NC 27695, U.S.A.

ABSTRACT

The temperature-dependent swimming speeds of 11 species of photosynthetic marine dinoflagellates determined using video techniques were described by previously published five parameter equations. These equations were assigned direction to yield velocities and were combined with species-specific equations describing an assumed photokinetic velocity increment extrapolated from data for Gyrodinium dorsum and a sinking velocity based on Stokes Law. The resulting equation set for each species was applied to an imposed diel vertical migration in a standard neritic water column to examine the relative capabilities of the different species. Inter-specific cell size differences contribute to the wide range of observed capabilities distinguished by the ratio of the magnitudes of temperature-dependent swimming velocity and sinking velocity. These motility attributes contribute to a species vertical progress and, therefore, to its light exposure. Light, in turn, induces a photokinetic response that may counterbalance the effect of cell sinking in certain species. Red tide species occur at both extremes of the observed motility range.

INTRODUCTION

Photosynthetic marine dinoflagellates possess a flagellar apparatus that provides the potential for effective cell movement in the vertical water column. Some species use this capability to position and maintain themselves within a stratum that changes little over a 24 hour period. Other species undergo diel vertical migrations that allow the cells to select various depth locations within the upper ocean over a 24 hour period (1). The choice between these different strategies often depends on the existing environmental conditions and on the physiological state of the species (2). More basically, dinoflagellate species vary in swimming ability. A published survey for 11 different dinoflagellate species suggests that the differences are related to cell morphology, flagellar characteristics and cell size (3).

The present paper extends an instantaneous translational velocity model developed for Gyrodinium dorsum based on laboratory measurements of temperature response, photokinesis and buoyancy (4) to 10 other species, some represented by multiple clones. This comparison examines how different species ranging in cell length from 16 to 58 µm progress under an imposed diel vertical migration while exposed to the same environmental conditions.

METHODS

Published equations describing the temperature-dependent swimming speed of 11 species of photosynthetic marine dinoflagellates (3) formed the foundation of this interspecies comparison. The vector form of the general equation for temperature-dependent velocity (V_T) is

$$V_T = V_A[1-e^{-a(T-T_L)}][1-e^{-b(T_H-T)}]$$

where V_A is the asymptotic rate for the two expontential curves, a and b are the initial and final slopes, and T, T_L and T_H represent the ambient, low and high temperatures. These equations for each species were adjusted to represent swimming in the dark and were applied to ascents or descents in order to provide a baseline swimming velocity. A photokinetic velocity increment (V_L) was derived by extrapolating the measured relationship for Gyrodinium dorsum.

RED TIDES: BIOLOGY, ENVIRONMENTAL SCIENCE, AND TOXICOLOGY
Okaichi, Anderson, and Nemoto, Editors

$$V_L = V_M[\mathrm{TANH}(0.55I/V_M)], \qquad 1$$

where V_M is the maximum photokinetic velocity increment for the species and I is the light intensity. The magnitudes of the temperature-dependent swimming velocities for the other species were multiplied by 0.26, the measured magnitude of the photokinetic effect under saturating photosynthetically active radiation (PAR) for Gyrodinium dorsum, and substituted for V_M. The effect of cell sinking (V_D) was calculated through a species-specific application of Stokes Law for spherical bodies,

$$V_D = [2GR^2(1.07-D)/9N)], \qquad 2$$

where G is the acceleration of gravity, D is the water density at a salinity of 30‰ such that

$$D = 1.024 - 6.0X10^{-5}\,T - 4.9X10^{-6}\,T^2, \qquad 3$$

and N is the water viscosity such that

$$N = 1.8X10^{-2} - 5.5X10^{-4}\,T - 7.3X10^{-6}\,T^2 \qquad 4$$

where T is water temperature. The main biological variable was the species-specific cell radius (R) estimated as one-half the cell length. Cell density was maintained at 1.07 g cm^{-3} for all species based on iso-osmotic measurements using Percoll, a density gradient medium (5).

The equations were combined to yield the translational velocity (V) for vertically ($V = V_T + V_L + V_D$) or horizontally ($V = V_T + V_L$) swimming cells. Sinking velocity opposed the swimming velocity of ascending cells and complemented the swimming velocity of descending cells. The Statistical Analysis System (6) was used to manipulate the data sets and to generate the plots. The standard environmental condition considered in this paper was a 25°C water column with light attenuation coefficient of K = .085 m^{-1} under a half-sinusoidal light cycle (no negative values) with a maximum PAR intensity of 1500 µEinst m^{-2} s^{-1} and with a 12-12 hour day-night cycle.

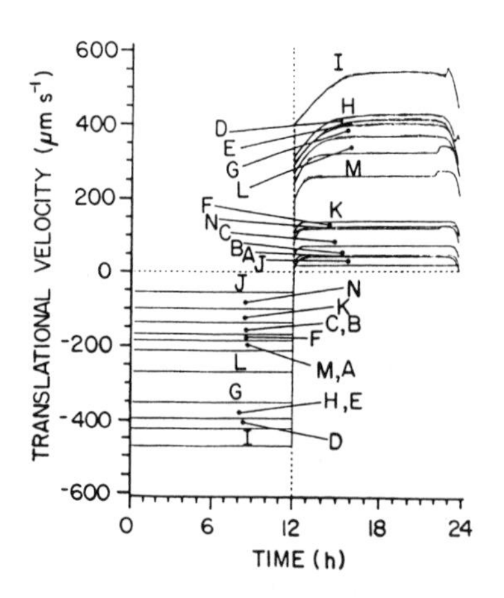

Fig. 1. Comparison of species-specific velocities during imposed migration

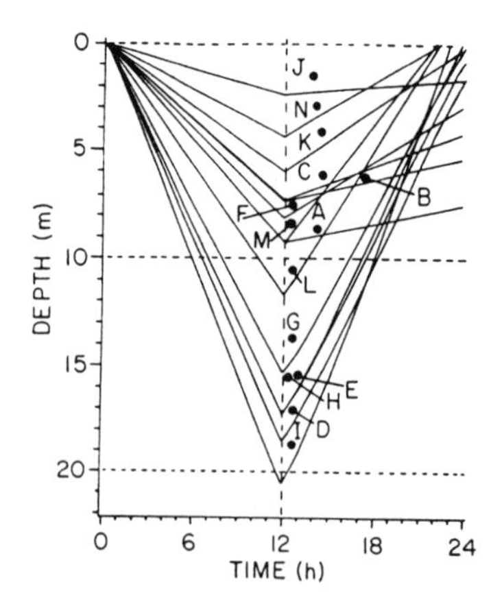

Fig. 2. The depth distributions that correspond to Fig. 1

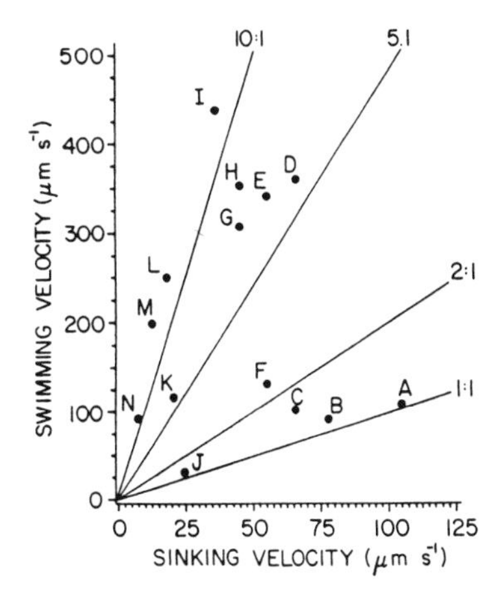

Fig. 3. The species-specific swimming and sinking velocities

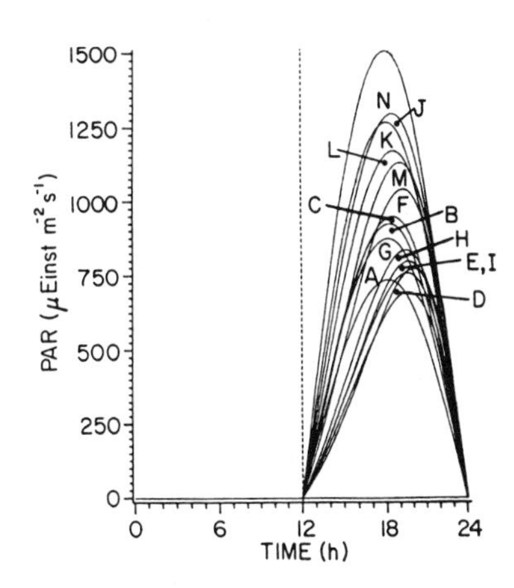

Fig. 4. The species-specific PAR of the trajectories in Fig. 2

RESULTS

The species considered are Gymnodinium splendens (A), Prorocentrum micans (B,C,F) Proteroceratium reticulatum(D), Gonyaulax polyedra (E,G), Gyrodinium dorsum (H), Peridinium foliaceum (I), Gymnodinium nelsonii (J), Amphidinium carterae(K), Heterocapsa illdefina (L), Heterocapsa niei (M) and Glenodinium sp. (N).

Figure 1 emphasizes the wide range of capabilities among the species examined and also demonstrates the expected translational velocity changes between light and dark due to photokinesis and between ascent and descent due to cell sinking. In Figure 2, the descent characteristics are due to the additive effects of the temperature-dependent swimming velocities and the sinking velocities. The mid-sized species (34-46 m) descend the furthest over the 12 hour period. The ascent characteristics suggest at least two groups of species. One group ascends as readily as it descends (e.g. species I); the other group descends much more effectively than it ascends (e.g. species A).

Figure 3 demonstrates that these two groups are distinguished by the ratio of the magnitudes (lower case letters) of the temperature-dependent swimming velocity (v_t) and the sinking velocity (v_d). The species that ascend or descend with equal ease swim over 5 times faster than they sink. The species that exhibit reduced ascents swim 2 or less times as fast as they sink. All species examined exhibit at least a minimal ability to swim faster than they sink (7) at 25°C.

Figure 4 displays the PAR experienced by the various species that migrate according to the trajectories in Figure 2. If light saturation of photokinesis (I_P) is represented by $I_P = v_m/0.55$ (8), then the various species maximize photokinesis at PAR intensities between 20 and 255 μEinst m^{-2} s^{-1} and experience these intensities between 8-12 hours per day. Assuming that maximum photokinetic enhancement of velocity acts over 12 hours while sinking velocity acts over 24 hours, then the modeled photokinesis can compensate for sinking for all species except those that exhibit reduced ascents. This conclusion is tentative until the actual photokinetic responses of more species are determined.

DISCUSSION

Although the imposed diel vertical migration is simplistic (2), the previous comparison of transitional velocities during an imposed diel vertical migration clearly demonstrates significant differences among photosynthetic marine dinoflagellate species. Acknowledged red tide species (e.g., Gymnodinium splendens. Prorocentrum micans, Gonyaulax polyedra Amphidinium carterae and Heterocapsa illdefina) occur throughout the sampled size range and in both species groups based on ascent capabilities (9, 10). The clear differences in capability among these red tide species suggest that alternate motility strategies can yield large dinoflagellate blooms in response to a variety of environmental conditions. For instance, the better swimming species may be able to succeed in more energetic environments than the less capable swimmers. Creative field studies that include the elusive development period preceding red tides will be required to clarify these relationships.

ACKNOWLEDGEMENTS

Research supported by NSF grants 82-00159 and 85-00589, ONR grant N0014-85-C-3095 and a NCSU ICRAC grant.

REFERENCES

1. D. Kamykowski, Northeast Gulf Science 4: 39-43, (1980).

2. J. J. Cullen, M. Zhu, R. F. Davis and D. C. Pierson, In D. M. Anderson, A. W. White and D. G. Baden (eds), Toxic Dinoflagellates, Elsevier, (N.Y.), p. 189-194 (1985).

3. D. Kamykowski and S. A. McCollum, Journal of Plankton Research 8: 275-287 (1986).

4. D. Kamykowski, S. A. McCollum and G. Kirkpatrick, Limnology and Oceanography 33: 66-78, (1988).

5. R. L. Oliver, A. J. Kinnear and G. G. Ganf, Limnology and Oceanography 26: 285-294, (1981).

6. SAS Institute Inc., Box 8000, Cary, NC 27511, (1985).

7. T. Smayda and P. K. Bienfang, Marine Ecology 4: 289-300, (1983).

8. T. R. Parsons, M. Takahashi and B. Hargrave, Biological Oceanographic Processes, 3rd ed, Pergamon Press, 330 p.

9. E. M. Herman and B. M. Sweeney, Journal of Phycology 12: 198-205, (1976).

10. M. A. de M. Sampayo. In D. M. Anderson, A. W. White and D. G. Baden (eds), Toxic Dinoflagellates, Elsevier, (N.Y.), p. 125-132, (1985).

CHAIN-FORMING DINOFLAGELLATES: AN ADAPTATION TO RED TIDES

SANTIAGO FRAGA,* SCOTT M. GALLAGER** AND DONALD M. ANDERSON**
*Instituto Español de Oceanografia. Apdo. 1552, 36280 Vigo, Spain; **Woods Hole Oceanographic Institution. Woods Hole, MA 02543, USA.

ABSTRACT

Swimming speeds of two chain-forming dinoflagellates, the toxic *Gymnodinium catenatum* and the non-toxic *Alexandrium affine*, were measured as a function of chain lengths. Long chains swam faster than short chains. The increase in speed from a single cell to a chain of four cells for both species was about a factor of 1.5-1.6. Populations of both dinoflagellate species were coincident with red tides in areas of coastal upwelling relaxation and downwelling in the Ria de Vigo, northwest Spain. The higher swimming speeds of long chains may allow more cells to remain in the photic zone during downwellings or convergences. This may be a mechanism for local concentration of cells leading to a red tide.

INTRODUCTION

Two chain-forming dinoflagellates, *Gymnodinium catenatum* Graham and *Alexandrium affine* (Inoue and Fukuyo) Balech, bloomed simultaneously in the Ria de Vigo, NW of Spain, in 1985 [1] and again in 1986 (unpublished data) coincident with a relaxation in the persistent coastal upwelling that typifies that region. The upwelling relaxation was associated with areas of downwelling within the ria into which the dinoflagellate species were advected and concentrated in visible red tides. Chain-forming dinoflagellates are often responsible for red tide outbreaks throughout the world. This generalization, plus the simultaneous dominance of *G. catenatum* and *A. affine* in the downwelling regime suggests that chain formation may provide an evolutionary advantage leading to red tide formation in certain hydrographic environments.

The hypothesis being tested in this paper is that the drag force on cells in a long chain is less than that on those same cells if they were unattached. The chain should thus travel faster and remain in the photic zone when downwelling or other advective processes would remove non-motile cells or less-vigorous swimmers.

MATERIALS AND METHODS

Model

If we assume the shape of a single cell to approximate a sphere, Stoke's Law may be used to calculate the force required to propel a cell at a given swimming velocity (v). The total force applied by a chain during swimming (F_s) will be proportional to the number of cells in the chain (n):

$$F_s = n\ (6\pi\mu av)$$

RED TIDES: BIOLOGY, ENVIRONMENTAL SCIENCE, AND TOXICOLOGY
Okaichi, Anderson, and Nemoto, Editors

where a is the radius of a single cell and μ is the dynamic viscosity of the liquid. To propel a chain of cells at a given velocity, F_s must equal the total drag force:

$$F_d = 6\pi\mu A_s v$$

where A_s is the Stoke's radius of a sphere with the drag equivalent to a prolate spheroid oriented parallel to flow as given by Happel and Brenner [2]. We have interpolated values for F_s between those given by the latter authors assuming a linear relationship to exist.

Since the dynamic viscosity varies with temperature, we can construct a matrix of calculated swimming speeds as a function of chain length and temperature using the measured swimming speeds of single cells at given temperatures.

Measurements

Cultures of *Gymnodinium catenatum* (GC1V) and *Alexandrium affine* (PA5V) isolated in the Ria de Vigo during the 1985 bloom, were grown in "K/2" medium [3] at 20°C with a 14:10 L:D period. The last transfer of the cultures just before the experiments was to "K/2" medium without nitrate, with only ammonia as a nitrogen source.

Swimming speeds were recorded with the aid of a Zeiss light microscope fitted with a low-light video camera conected to a VCR. Swimming chambers were constructed with a coverslip separated from a microscope slide by four small pieces of modelling clay. Slides were held on an aluminum microscope stage that was cooled with circulating water from temperature controlled bath. Swimming velocities of chains of *G. catenatum* and *A. affine* were measured by frame analysis on a monitor. A more accurate method of measuring absolute swimming velocities of dinoflagellates was described by Kamykowsky [4], but in the present study we were interested in relative values as a function of the chain length more than in absolute values of swimming velocities *per se*.

RESULTS

Figure 1 shows the calculated swimming speeds of *G. catenatum* and *A. affine* as a function of chain length and seawater viscosity, using a single velocity of 247 μm/sec at 23 °C (average of 36 values) for *G. catenatum* and 410 μm/sec at 23 °C (average of 29 values) for *A. affine*. Greatest increase in swimming velocity as chain length extended was observed for relatively short chains of two to four individuals. As the length of the chain increased, the incremental advantage decreased. The advantage of swimming in a chain was greater at high rather than low temperatures.

Comparing curves of calculated and observed velocities (Fig.2), it was clear that the measured increase in swimming velocity was less than our simple model predicted. This may have been due to the effect of the flow field created by the anterior-most cells in the chain that may have forced trailing cell to swim less effectively.

The greater increase in velocity associated with longer chains *A. affine* compared to *G. catenatum* agree well with the idea that fast swimmers benefit more from long chain formation than do slow swimmers.

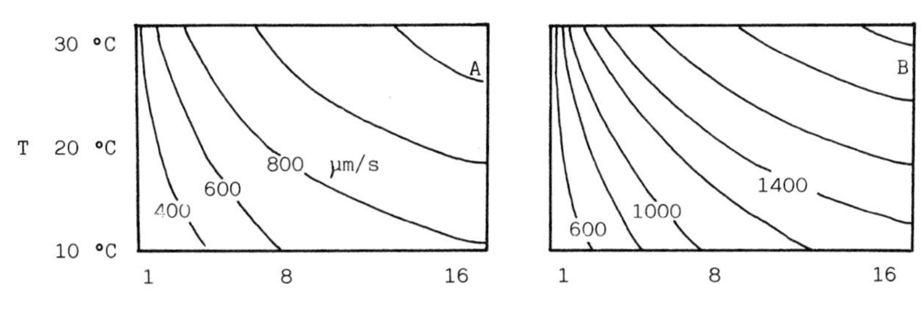

FIG. 1. Calculated swimming velocities of G. catenatum (A), and A. affine (B), as a function only of chain length and temperature.

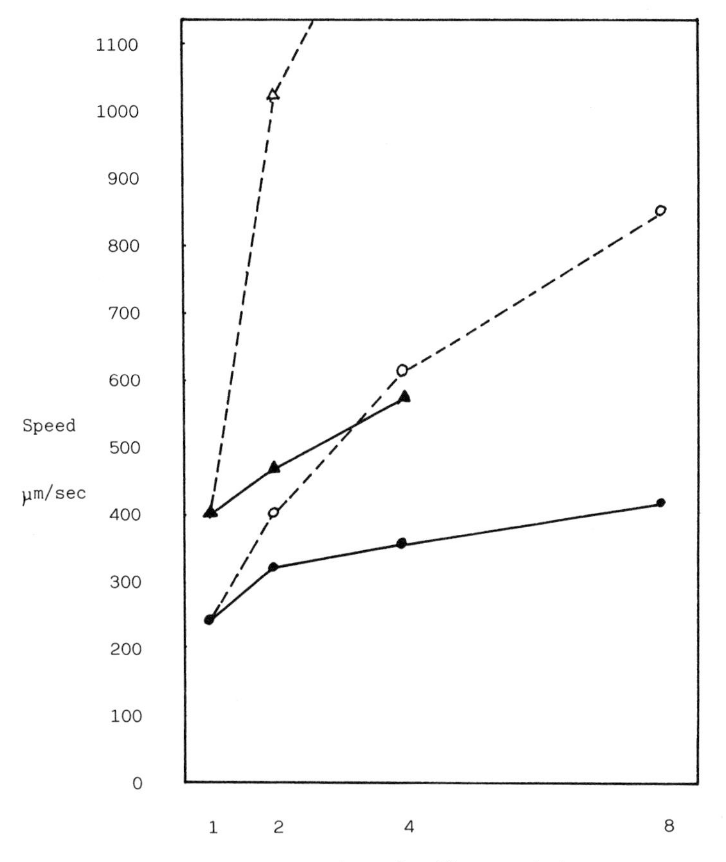

FIG. 2. Calculated (------) and observed (————) swimming velocities of G. catenatum (○ ●) and A. affine (△ ▲).

DISCUSSION

The fact that chains of cells swim faster than single cell is an important concept in the development of red tides in areas of convergences or downwelling. When chain forming species have to compete with non-chain formers, an advantage accrues to those requiring the least amount of energy to stay within the photic zone. This may be true even when the energy needed for motility is a small fraction of total metabolic requirements [5]. Another important fact is that variation in swimming speed may be due to changes in seawater viscosity as a function of temperature. The differences in swimming speeds between chains of different lengths increases with temperature. As cells swim towards the surface, they encounter warmer temperatures which may favor the presence of long chains over individual cells. Non-motile species, like diatoms, will sink from the water column faster at warmer temperatures. This may explain why rapid changes from diatom-dominated populations to single-cell dinoflagellates and to chain-forming dinoflagellates have been observed in the Ria de Vigo when a change of wind direction has caused a relaxation in coastal upwelling ([1] and unpublished data). We have not considered in our simple model physiological factors that clearly play a decisive role in determining the shape of the curves of speed vs temperature shown in the literature [4,6].

ACKNOWLEDGEMENTS

The help of Dr. K. Stolzenbach in clarifying ideas at the begining of this research is gratefully acknowledged. Dave Kullis provided technical assistance with the cultures. This paper was supported by the USA-Spain Joint Committee for the Scientific and Tecnological Cooperation, grant No. CCA-8411089 and a grant from the National Science Foundatiom No. OCE-8711386, and by NOAA National Sea Grant Program Office, Dept. of Commerce under Grant No. NA86-AA-D-SG090, WHOI Project R/B 76. Contribution No.6810 from the Woods Hole Oceanographic Institution.

REFERENCES

1. S. Fraga, D. M. Anderson, I. Bravo, B. Reguera, K. A. Steidinger and C. M. Yentsch, Est. Coast. Shelf. Sci. (in press).
2. J. Happel and H. Brenner, Low Reynolds Number Hydrodynamics. Martinis Nijhof, Publisher, (1983).
3. M. D. Keller and R. R. L. Guillard, in: Toxic Dinoflagellates, Anderson, White C. Baden, eds. Elsevier, p. 113-116 (1985).
4. D. Kamykowsky and S. A. McCollum, J. Plank. Res., 8 (2), 275-287 (1986).
5. J. A. Raven and K. Richardson, New Phytol., 98, 259-276 (1984).
6. W. G. Hand, P. A. Collard and D. Davenport, Biol. Bull., 128, 90-101 (1965).

CONSERVATIVE TAXONOMIC CHARACTERS IN TOXIC DINOFLAGELLATE SPECIES IDENTIFICATION

Steidinger, K., C. Babcock, B. Mahmoudi, C. Tomas and E. Truby
Bureau of Marine Research, Florida Department of Natural Resources, St. Petersburg, FL 33701 USA

ABSTRACT

Several species of Gymnodinium, Gyrodinium, and Ptychodiscus have similar morphological features and pigment composition; however, they differ by a prominent overhanging carina, cingular thecal ridges, position and length of apical groove, shape and length of sulcal intrusion onto the epitheca, and possibly toxins. Sulcal intrusion and apical groove characters may be useful conservative characters for these gymnodinioids. Prorocentrum species also have conservative morphological characters such as pore patterns and possibly shape/size descriptors. The use of optical pattern recognition systems to quantify shape is suggested as a possible tool for numerical taxonomy.

INTRODUCTION

Proper identification of toxic dinoflagellate species is critical, particularly to areas which lack a historical record of the species occurrence, its life history, biochemistry, and growth characters. Species identification allows scientists and officials to project potential impacts to public health, aquaculture, and coastal community economy, and to establish monitoring programs. Although tremendous advances in biological research, e.g., genetics and biochemistry, have occurred, we are still at the morphospecies level in taxonomy of marine flagellates. Identifying meaningful conservative morphological characters or descriptors of low variability to separate closely related species, whether described or new, would be a marked improvement insuring uniform, objective criteria for species identification.

EXPERIMENTAL

Light and scanning electron microscopy.

Cultures of Ptychodiscus brevis, Prorocentrum lima, and P. concavum were grown in either NH-15 [1] medium adjusted to 34 $^{o}/oo$, or modified f/2-Si medium [2]. Ptychodiscus brevis was fixed hypertonic to the medium. Unarmored cells were simultaneously fixed in GTA/OsO_4 in the cold, rinsed, dehydrated, critical point dried, coated with gold, and observed in a Hitachi HHS-2R. Prorocentrum were fixed in 0.5% paraformaldehyde, some were post fixed in 2% OsO_4 then rinsed, dehydrated, and air dried in freon TF. Phase-contrast and differential interference contrast (DIC) optics were used to observe living and preserved cells at the light microscopy level.

Optical pattern recognition and statistics.

SEM micrographs, and silhouettes drawn from them, were analyzed with a BioSonics® Optical Pattern Recognition System (OPRS) using video images. Each image can be enhanced, measured, and analyzed; the digital data are saved to computer files for subsequent analysis. OPRS morphometry software (MOR) allows extraction of spatial or shape-related data via single

RED TIDES: BIOLOGY, ENVIRONMENTAL SCIENCE, AND TOXICOLOGY
Okaichi, Anderson, and Nemoto, Editors

point, line and curves, shape boundary point coordinates, Fourier descriptor 1, and Fourier descriptor 2. In this study, the cell boundary in a micrograph was enhanced with a marking pen, or silhouettes were used. The image was digitized and pixels on image boundaries were identified. Coordinates of the pixels were passed to the Fourier analysis programs (PDI and PD2) which then generated a harmonic series for each image. OPRS was used for circularity (perimeter^2/area occupied) and rectangularity (shape's area/area of its minimum enclosing rectangle) (fig. 1). Although the Fourier-based descriptors of shape have not been analyzed as yet, a Statgraphics® program using direct discriminant analysis whereby all equations are solved simultaneously (equal weight to descriptors) is planned once the program is modified to compile a complete array of harmonic series. Statistical analyses were done with SAS [3] and statgraphics [4].

RESULTS AND DISCUSSION

Prorocentrum, circularity, and rectangularity.

The surface pore pattern of *P. lima* is conservative in that the center of both valves lacks pores. The number of pores statistically covaries with surface area of the valve (minus growth margin) and there is no significant difference between left and right valves (F_{res} (48,48)=1.06; F_{slope} (48,48)=0.307). In comparing within *P. lima* and *P. concavum* variation and between *P. lima* and *P. concavum* variation, using OPRS circularity and rectangularity, *P. lima* and *P. concavum* could be separated by an ANOVA with H_0:diff=0 (circularity: $F_{0.05}(1,38)$=31.34**; rectangularity: $F_{0.05}(1,38)$=28.64**). Although the age of cells can cause wide variation in shape/size due to the growth margin of older cells, quantitative differences between species were still resolved. Such analyses should be attempted with other dinoflagellates. Silhouettes present an averaged two dimensional image (fig. 1) that can be used for circularity, rectangularity, and Fourier analyses.

Gymnodinioids, conservative characters, and Fourier analyses.

The species *Gymnodinium nagasakiense* (fish kills), European *Gyrodinium aureolum* (fish kills), central western Atlantic *Ptychodiscus brevis* (fish kills, shellfish poisoning, and human respiratory disorder), and the Japanese *Gymnodinium breve*-like species (apparently nontoxic) appear to be related morphologically and those that have been analyzed lack peridinin but have fucoxanthin derivatives (K. Tangen, personal communication). Clearly, *P. brevis* is distinct by its toxins and impacts. Tangen (this symposium; abstract) concluded that *Gymnodinium nagasakiense* was conspecific with the European *Gyrodinium aureolum* based on morphology and biochemistry of pigments. Both the Japanese *Gymnodinium nagasakiense* (fig. 2) and the European *Gyrodinium aureolum* have identical apical groove/sulcal intrusion juncture characters. The extent of sulcal intrusion and the length of the apical groove are similar in *P. brevis*, but *P. brevis* is concave/convex, and has a prominent overhanging apical carina, cingular thecal ridges, and a more convoluted right margin to the sulcal intrusion (fig. 3). The Japanese *G. breve*-like isolates (Hirayama and Yoshimatsu isolates, personal observation, this symposium) appear under DIC to have a short apical groove on the ventral surface and a longer sulcal intrusion onto the epitheca than *P. brevis* (figs. 2d and e). The Japanese *G. breve*-like isolates had an upturned proximal cingular curvature. Davis' original illustration of *G. breve* [5] did not portray the cingulum and sulcus accurately (compare figs. 2b and 2c). More work has to be done on these species before a conclusion is reached regarding generic and specific affinities. The two Japanese isolates appear to be different from *P. brevis* and may represent new species, but both isolates should be studied for within and between isolate morphological variation, e.g., shape, apical groove, sulcal intrusion, right sulcal intrusion margin, as well as pigments and potential toxins. Shape alone, based on an averaged or standard image, may separate all these species when analyses are eventually run on Fourier descriptors.

Potential for pattern recognition and numerical taxonomy.

Size and shape, although assumed to be the most variable morphological characters, are conservative to the point that taxonomists after years of experience can identify individuals of many species based on image recognition without looking at small scale detail such as plates or surface markings, e.g., within Dinophysis, Prorocentrum, Protoperidinium, Ornithocercus, Ceratium, Polykrikos, etc. Size and shape, therefore, are meaningful characters. The potential for image or pattern recognition that can be quantified without bias has its place in identification of toxic dinoflagellate groups. With OPRS that use Fourier analyses, within species shape/size variation can be quantified and a standard or average image can be created. Standard images can then, theoretically, be compared for similarity by cluster analyses or by discriminant function analysis to group or separate morphospecies.

ACKNOWLEDGEMENTS

The authors thank the symposium organizing committee, Drs. Ono, Yoshimatsu, Takayama, Fukuyo, and Iizuka, M. Tringali, and H. Neal for their assistance.

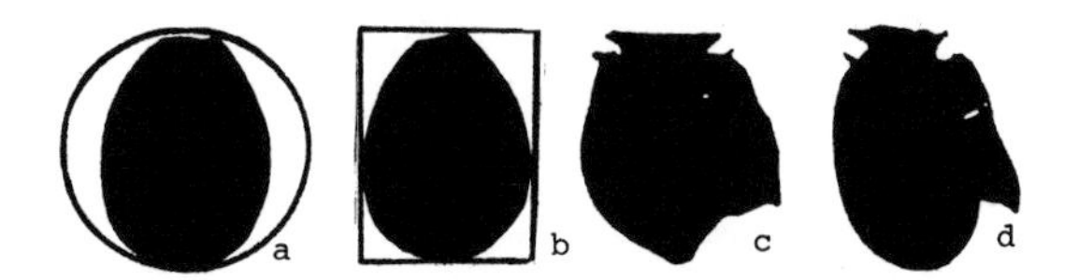

Figure 1. Diagrammatic illustration of silhouette images and use of circularity and rectangularity. a and b. Prorocentrum lima, c. Dinophysis acuta redrawn from Dodge [6]. d. D. dens redrawn from Dodge [6].

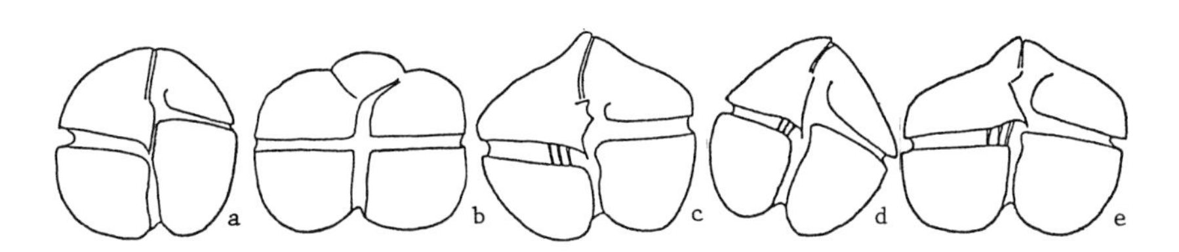

Figures 2. Schematic illustrations. a) Gymnodinium nagasakiense redrawn from Takayama [7]. b) G. breve as originally illustrated by Davis [5]. c) P. brevis from Florida isolates. d) G. breve-like species redrawn from Takayama [8]. e) G. breve-like species redrawn from personal observations of a Japanese strain and micrographs courtesy of Dr. Takayama.

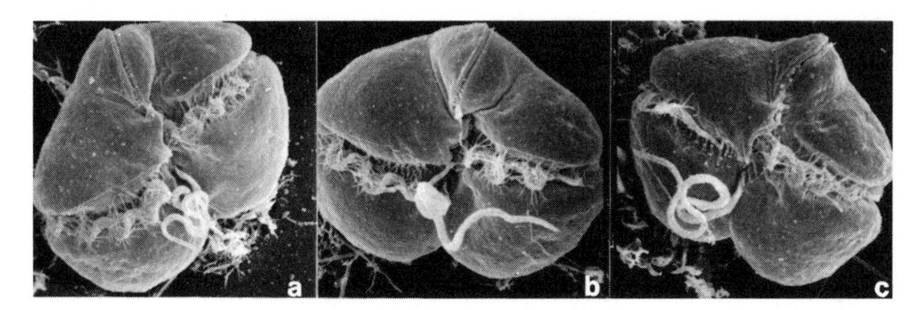

Figures 3. P. brevis a) X 1780; b) X 2115; c) X 2020.

REFERENCES

1. J.A. Gates and W.B. Wilson, Limnol. Oceanogr., 5, 171, (1960).
2. R.R.L. Guillard, In Culture of Marine Invertebrate Animals, W.L. Smith and M.H. Chanley, eds., Plenum Press, 29, (1975).
3. SAS User's Guide: STATISTICS, SAS Institute, (1985).
4. Statgraphics: statistical graphic system, STSC. Inc., (1986).
5. C.C. Davis, Bot. Gaz., 109, 358, (1948).
6. J.D. Dodge, Atlas of Dinoflagellates, Farrand Press, (1985).
7. H. Takayama, Bull. Plankton Soc. Japan, 32, 129, (1985).
8. H. Takayama, Bull. Plankton Soc. Japan, 28, 121, (1981).

DORMANCY AND MATURATION IN THE CYSTS OF CHATTONELLA SPP. (RAPHIDOPHYCEAE), RED TIDE FLAGELLATES IN THE INLAND SEA OF JAPAN

ICHIRO IMAI[1], KATSUHIKO ITOH[1] AND MASATERU ANRAKU[2]
[1]Nansei Regional Fisheries Research Laboratory, Ohno-cho, Saeki-gun, Hiroshima-ken 739-04, and [2]Overseas Fishery Cooperation Foundation, Akasaka Twin Tower 17-22, 2-chome Akasaka, Minato-ku, Tokyo 107, Japan

ABSTRACT

Effects of temperature on dormancy and maturation (acquisition of germinability) in the cysts of *Chattonella antiqua* and *C. marina* in sediments were studied. For maturation of the cysts in the state of spontaneous dormancy, low storage temperatures of 11°C or below were needed, whereas no maturation was observed at temperatures higher than 20°C. Temperatures of 15° and 18°C were critical for maturation. Germinable cysts in sediment maintained germinability at the storage temperatures of 11°C or below. They lost germinability gradually at 15° and 18°C during storage, and rapidly at 20°C or more. The cysts which lost germinability at 22°C recovered the ability after the lowering of storage temperature to 11°C. These results suggest that the cysts which missed a chance of germination during summer can be carried over to the next year through secondary dormancy. Overwintering cyst populations, both new and old (carried over) groups, obtain germinability during winter, and play an important role in initiating the summer red tides as seed populations in the Inland Sea of Japan.

INTRODUCTION

Chattonella antiqua (Hada) Ono and *Chattonella marina* (Subrahmanyan) Hara et Chihara, Raphidophyceae, are fish-killing flagellates causing serious damage to fish farms in Japanese coastal waters, especially to yellowtail culture in the Inland Sea during summer. These organisms have an overwintering benthic stage in their life cycle [1], and cysts have recently been found from surface sediments in Suo-Nada, western Seto Inland Sea [2]. Discrimination of the cysts between *C. antiqua* and *C. marina* is impossible based on the morphological features (Imai and Itoh, in preparation).

The cysts of *Chattonella* spp. indicate clear seasonality in germination ability in sediments of Suo-Nada [3]. They acquire germinability during the winter season with a low temperature of around 10°C. From April to mid July, the number of germinable cysts maintained a high level, and it rapidly decreased in August accompanied by the rise of bottom water temperature. Dormancy and maturation are thus affected by temperature regime in the Inland Sea. In the present paper, we investigate the effects of different temperatures on dormancy and maturation in the cysts of *Chattonella* spp. in sediments. Secondary dormancy of the cysts is also examined, and its implications are discussed in relation to carrying over of the cysts and initiation of the next summer red tides.

RED TIDES: BIOLOGY, ENVIRONMENTAL SCIENCE, AND TOXICOLOGY
Okaichi, Anderson, and Nemoto, Editors

MATERIALS AND METHODS

For the maturation experiment, sediment samples were collected with a gravity core-sampler at St.H-12 located in southern Harima-Nada, in October 1985. This station is situated within the dense seed bed [1,4]. In autumn, most of the cysts of Chattonella spp. are in a state of spontaneous dormancy [3]. Collected sediments (top 1-cm depth) were mixed thoroughly, and divided for storage at seven different temperatures (5°, 11°, 15°, 18°, 20°, 22°, and 25°C). Sequential change of germinability of cysts in sediment stored at each temperature in darkness was determined by the extinction dilution method (MPN method) [5] during 7 month period. Incubation for enumeration of germinable cysts in each sediment sample was made for 8-12 days at 22°C (optimal temperature for germination) [1] and 3,500 lx with 14hL:10hD photo-cycle.

For the dormancy experiment, sediment samples (top 1-cm depth) were collected from St.S-19 located in the central part of Suo-Nada, in November 1985. Sediments with cysts were stored at 11°C for 5 months in darkness before the experiment to obtain the germinability. After that period, sediments were placed under seven different temperatures in darkness mentioned above. The change in the number of germinable cysts at each storage temperature was measured by the extinction dilution method.

For the secondary dormancy experiment, sediment samples (top 1-cm depth) were collected from St.H-16 in Harima-Nada [3] and St.S-17 in Suo-Nada [3] in April 1985. These samples contain many cysts with germination capacity. The sediment samples were initially stored at 22°C in darkness for 2 months to induce dormancy and then transferred to 11°C and kept for 5 months to recover the germinability. The same treatment was repeated once again. The numbers of germinable cysts in sediments were monitored sequentially by the extinction dilution method.

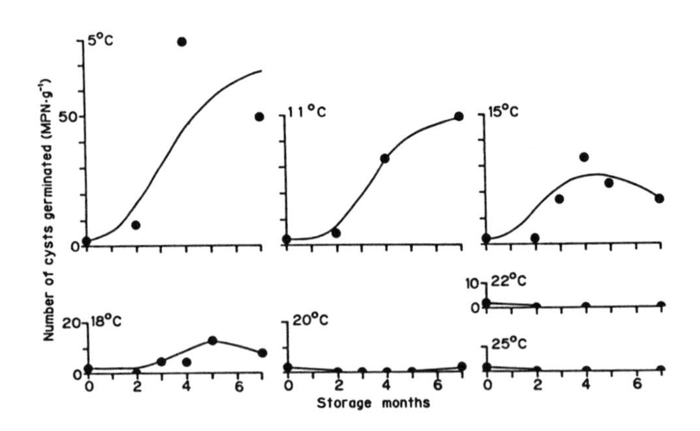

FIG. 1. Effects of different storage temperatures on maturation of the cysts of Chattonella spp. in sediments.

RESULTS AND DISCUSSION

Figure 1 shows the effects of storage temperatures on maturation of the cysts of Chattonella spp. in sediments. No maturation was observed for at least 7 months under the storage temperatures of 20°C or higher. In contrast, many cysts obtained germinability within 7 months when sediments were kept at 11°C or lower. Temperatures of 15° and 18°C seemed to be critical for maturation. These results indicate the necessity of low temperatures for breaking dormancy in the cysts of Chattonella spp., as was reported in the cysts of a dinoflagellate Peridinium cunningtonii [6]. During winter, temperatures of 11°C or lower are commonly observed in the Inland Sea.

The germinable cysts in sediment maintained this ability at storage temperatures of 11°C or below (Fig. 2). When sediments are stored at 5°C, the cysts of Chattonella spp. maintain germinability for more than 5 years [7]. The cysts lost germinability gradually at 15° and 18°C, and rapidly at 20°C or higher (Fig. 2). From summer to fall, bottom water temperatures of 20°C or higher are common in the Inland Sea.

Figure 3 represents the effects of shifts of storage temperature on dormancy and maturation of the cysts in sediments. The cysts with germination capacity lost it rapidly within 2 months when stored at 22°C. Total number of the living cysts of Chattonella spp. in a sediment sample, collected from Suo-Nada in March 1987, dose not change significantly at 22°C for at least 75 days (Imai and Itoh, unpublished data). After the lowering of the storage temperature to 11°C, these cysts gradually recovered germinability. Some of the cysts in sediment from St.H-16 could again recover the germinability after twice losing of the ability. These results imply that the cysts buried underneath the sediment surface cannot germinate even at optimal temperature for germination and they are induced into secondary dormancy to lose germinability. In copepods, it is known that

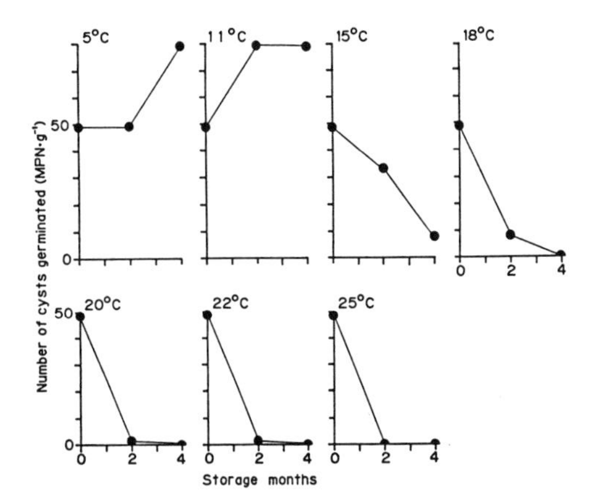

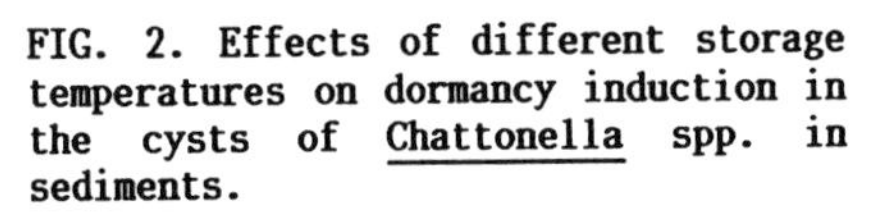
FIG. 2. Effects of different storage temperatures on dormancy induction in the cysts of Chattonella spp. in sediments.

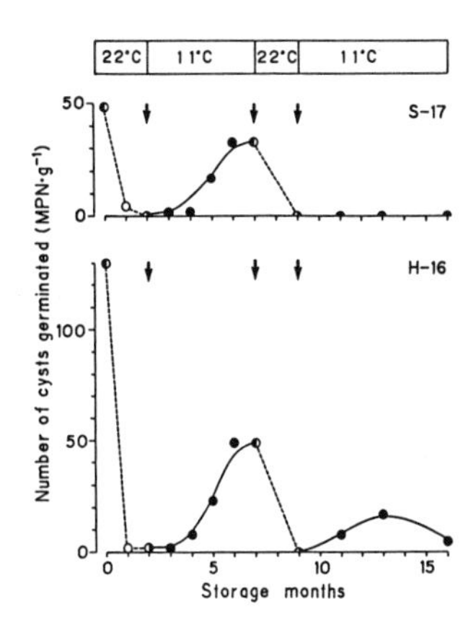

FIG. 3. Effects of shifts of storage temperature on dormancy and maturation in the cysts of Chattonella spp. in sediments. Arrows indicate the point of time of temperature shifts.

eggs do not hatch when buried underneath sediment [8,9]. Also in a toxic dinoflagellate *Gonyaulax tamarensis*, large populations of benthic cysts remain in sediments through blooming season, and they are carried over year-to-year without germination [10].

In bottom sediments of the Inland Sea, following behavior can be inferred about the cysts of *Chattonella* spp. during summer. Many cysts in sediment have germination capacity early in the summer. The cysts on the sediment surface germinate when bottom water temperature rises to adequate level for germination. The cysts underneath sediment surface, on the other hand, cannot germinate even at the optimal temperature. They are induced into secondary dormancy during summer, and carried over to the next year. Secondary dormancy is well known in higher plant seeds such as common ragweed [11]. In late summer, newly formed cysts sink to bottom. Overwintering cyst populations, both new and old (carried over) groups, acquire the germinability during winter, and will play a key role in initiating the next summer red tides in the Inland Sea of Japan.

ACKNOWLEDGEMENTS

We are grateful to the officers and crew of R/V "Shirafuji-Maru", Nansei Regional Fisheries Research Laboratory, for their cooperation of the sampling at sea. This study was supported partly by a grant from the Environment Agency of Japan.

REFERENCES

1. I.Imai, K.Itoh and M.Anraku, Bull. Plankton Soc. Japan, 31:35-42 (1984)
2. I.Imai and K.Itoh, Bull. Plankton Soc. Japan, 33:61-63 (1986)
3. I.Imai and K.Itoh, Mar. Biol., 94:287-292 (1987)
4. I.Imai and K.Itoh, Bull. Nansei Reg. Fish. Res. Lab., No.19:43-52 (1985)
5. I.Imai, K.Itoh and M.Anraku, Bull. Plankton Soc. Japan, 31:123-124 (1984)
6. Y.Sako, Y.Ishida, H.Kadota and Y.Hata, Bull. Japan. Soc. Sci. Fish., 51:267-272 (1985)
7. M.Yoshida, I.Kitakado, K.Saiura and T.Akizuki, Personal communication.
8. S.Uye, S.Kasahara and T.Onbe, Mar. Biol., 51:151-156 (1979)
9. N.H.Marcus and J.Schmidt-Gengenbach, Limnol. Oceanogr., 31:206-210 (1986)
10. D.M.Anderson, S.W.Chisholm and C.J.Watras, Mar. Biol., 76:179-189 (1983)
11. J.M.Baskin and C.C.Baskin, Ecology, 61:475-480 (1980)

NUTRIENT AND TEMPERATURE EFFECTS ON GROWTH AND SEXUAL PHASES OF FOUR MARINE DINOFLAGELLATES

Carmelo R. Tomas,* Marina Montresor,** and Elisabetta Tosti**
*Bureau of Marine Research, Florida Department of Natural Resources, St. Petersburg, FL 33701 USA; **Stazione Zoologica 'A. Dohrn', Villa Comunale, 80121 Naples, Italy.

ABSTRACT

Growth of the marine dinoflagellates Amphidinium carterae, Coolia monotis, Ceratium schrankii and Ceratium tripos, isolated from the Gulf of Naples, was observed in clonal cultures at 15, 20 and 25 °C. Nutrient additions as nitrate, ammonium, urea, orthophosphate and glycerophosphate gave differing growth rates and final yields. A. carterae and Coolia monotis grew particularly well with urea, ammonium and nitrate while final yields varied directly with temperature and nitrogen source. Both Ceratium species showed ammonium toxicity at levels above 16 uM, inability to utilize glycerophosphate for growth and slightly different temperature optima. C. tripos showed no temperature preference with rates between 0.25 to 0.7 div. d^{-1} while the larger C. schrankii had best growth at 20-25 °C with rates $\geq$0.25 div. d^{-1}. Clonal cultures of C. tripos and C. schrankii produced gamete like swarmers at all temperature and most nutrient additions.

CONDENSED REPORT

Growth of the epibenthic dinoflagellate species Amphidinium carterae and Coolia monotis at temperatures from 15 to 25 °C were compared to that of the two pelagic species Ceratium tripos and Ceratium schrankii. Nitrogen supplied either as nitrate, ammonium or urea and phosphorus as ortho or glycerophosphate gave differing growth patterns. A. carterae had growth rates which varied between 0.32 to 0.71 div. d^{-1} which increased with temperature for all nutrients except nitrate. Urea and glycerophosphate gave best growth with values of 0.42 and 0.71 div. d^{-1} respectively. Growth at 20 °C with nitrate was significantly different ($P<0.01\%$) than that at 15 and 25 °C while growth on ammonium showed no difference at each temperature. C. monotis also had mean growth rates which increased with temperature varying from 0.05 to 0.82 div. d^{-1}. All nutrients gave growth which increased directly with temperature while growth on ammonium yielded lower rates than all the other treatments.

Ceratium tripos showed no temperature preference for growth from 15 to 25 °C with values from 0.27 to 0.70 div. d^{-1} for the nutrients tested. Ammonium levels above 16 uM were toxic and glycerophosphate failed to support growth. Urea and nitrate gave the best growth results respectively. Ceratium schrankii had significantly greater growth at 20 and 25 °C with means of 0.27 and 0.25 div. d^{-1}. The highest growth rates recorded exceeded 0.30 div. d^{-1}.

Both species of Ceratium were observed to form smaller, lightly pigmented cells which survived for short periods of time. These cells were not considered aberrant forms but definite "swarmers" resembling Ceratium lineatum similar to the gametes described by von Stosch [1]. Both C.

RED TIDES: BIOLOGY, ENVIRONMENTAL SCIENCE, AND TOXICOLOGY
Okaichi, Anderson, and Nemoto, Editors

tripos and *C. schrankii* had these gametelike swarmers at all temperatures and nutrients although nitrate appeared to give the greatest production of swarmers than the other sources. These swarmers were formed from a rapid series of divisions in which vegetative cells divided to give "pregametes" which within 24 h divided again to form the gametelike swarmers. Although numerous observations were made, no fusion of these swarmers with heterothallic counterparts were observed. Since a limited number of clones of *Ceratium* were available, further mating experiments were not possible.

Urea and nitrate gave best results relative to growth while glycerophosphate could sustain growth only in the benthic or epibenthic species. Ammonium at various levels was toxic or gave inferior growth to other nitrogen sources except for *A. carterae*. The combination of nutrient and temperature regime did effect *life cycle* and the formation of gametelike swarmers.

ACKNOWLEDGMENTS

This work was supported by funds from Stazione Zoologica 'A. Dohrn'. Technical assistance was given by Mr. G. Forlani and G. Gariulo.

1. von Stosch, H.A., (1964) *Helgo. wiss. Meeresunters.*, 10, 140.

NOTE

A full account of this work is in press in Journal of Phycology.

DINOFLAGELLATE CYST MORPHOLOGY: AN ANALYSIS BASED ON LABORATORY OBSERVATIONS OF ENCYSTMENT

F.J.R. TAYLOR,* and G. GAINES**
*Departments of Oceanography and Botany, University of British Columbia, Vancouver, B.C. V6T 1W5, Canada; **2001 N. Adams Street, #903, Arlington, Virginia 22201, U.S.A.

ABSTRACT

Resting cyst formation has been studied in cultures of Gonyaulax spinifera and Polykrikos kofoidii. The sequence of events and variation in the final form of the cyst reveal serious taxonomic consequences in the case of G. spinifera and provide a basis for interpreting features found in other dinoflagellate cysts.

INTRODUCTION

Classification requires that judgements be made about which morphological characters are taxonomically the most meaningful. Among dinoflagellates, this is a special problem because the motile cells have been classified independently of the cysts. As a result, nineteen cyst species in six genera have been attributed to Gonyaulax spinifera (Claparede & Lachmann) Diesing or "the spinifera group." Are these cysts overclassified or is the motile stage underclassified? This is equivalent to asking: what characters are environmentally modified? Answers to this question can help coordinate the classification of cysts and motile stages, and can also help paleontologists infer paleoecological conditions from cyst morphologies.

Although encystment in culture has been observed before, it has only involved species that possess morphologically simple cysts. We have obtained encystment in two species, not closely related, that produce fossilizable cysts with complex wall morphologies.

EXPERIMENTAL

Batch cultures of Gonyaulax spinifera and Polykrikos kofoidii Chatton encysted spontaneously. Cysts were examined with light and scanning electron microscopy. Details of the procedures are given in Taylor and Gaines in press) and Gaines and Taylor (in prep).

RESULTS

In Gonyaulax spinifera, encystment begins with a planozygote casting off its flagella and plates, becoming spherical and exuding mucus. The outer membrane then lifts off the surface of the cell, apparently by the secretion of another kind of mucus underneath it. Spines form in this transparent mucoid layer under the outer membrane by a sinking down of the outer membrane at the centers of the plates. At the plate junctions, spines are formed apparently on invisible supports within the mucoid layer, like tentpoles supporting a tent. The central capsule is thus the size of the motile cell, and the fossiblizable cyst wall consists of two layers. More variation is found in the cysts than in the motile stages.

RED TIDES: BIOLOGY, ENVIRONMENTAL SCIENCE, AND TOXICOLOGY
Okaichi, Anderson, and Nemoto, Editors

Polykrikos kofoidii encystment also begins with the casting off of flagella, rounding of the cell and exudation of mucus. Again, the outer membrane lifts off, but instead of then sinking back down except where spines formed, the outer membrane is lost. Before that, however, spines form within the transparent mucoid layer. The spines are transparent at first, but later take on a brown colour. They form all at once; they do not grow inward or outward. Spines are made of fibrils, and constitute the outer layer of the cyst wall. There is, in addition, an inner, smooth layer. Again, the cyst variability exceeds that of the motile cell.

DISCUSSION

The transparent mucoid layer is the most important component of the encystment process. Complex wall structures form in this layer. Although visible structures do not form in it until late, there must be invisible structures present early on.

Cyst morphology varies more than the morphology of the motile cell. The length of spines is determined by the thickness of the mucoid layer, which must depend upon the availability of the precursor material. The completeness of the outer layer also depends on the manufacture of hardening material. Thus, the presence and number of perforations (claustra) in the outer layer is also variable. Therefore, both of these structures are related to the health and nutritional state of the cell. In short, these dinoflagellate cysts, and others in which ornamentation is a major taxonomic criterion, appear to be overclassified, relative to the motile forms. Ornamentation is less reliable for taxonomic purposes than is tabulation, which reflects the basic cell organization regardless of the environment.

More questions are raised by these findings than are answered. For example, are the spines of *G. spinifera* cysts homologous with the spines of *P. kofoidii* cysts, or are they the products of two different structural systems? The outer layer of *G. spinifera* cysts differs in several respects from that of *P. kofoidii*: it is smooth, thin and apparently sinks down everywhere except where the spines form, while that of *P. kofoidii* is fibrillar, thick and apparently forms entirely in place. In other cysts, processes that originate from the centers of plates appear to be fibrillar, while those on the margins are smooth. A few forms (e.g., *Callaiosphaeridium*) have both types of processes, and other forms (e.g., *Kisselovia*) have three wall layers, all of which suggests two separate systems for the development of processes.

What is the relationship between cultured cysts and field-produced cysts? Is the variation in cultures greater? Our cells and cysts are smaller than those normally found in the field, but this is a common phenomenon. Changes in nutrient conditions are generally thought to be more rapid in batch cultures than in nature. These changes produce greater differences in the microenvironments of different cells in a culture. If so, one can view the cultures as demonstrating the broad range of possible morphotypes, and any single field sample will reflect but a narrow band of that range.

REFERENCES

1. G. Gaines and F. J. R. Taylor, (in prep)
2. F. J. R. Taylor and G. Gaines, *J. Phycol.*, (in press)

MESODINIUM RUBRUM (LOHMANN) HAMBURGER & BUDDENBROCK - NOT ONLY A TAXONOMIC PROBLEM?

Lindholm, T.

Department of Biology, Åbo Akademi, SF-20500 Åbo, Finland

CLASSIFICATION OF *MESODINIUM RUBRUM*

The photosynthetic ciliate *Mesodinium rubrum* is a cosmopolitan marine plankton organism with a reduced cryptophyte as an endosymbiont (3, 8, 10). It is a very fast-swimming and fragile plankter and red tide organism. Among the ciliates it is classified with the Kinetophragminophora (order Haptorida, family Didiniidae) (1). It moves backwards with respect to the oral end, with jumping movements typical of the genus *Mesodinium*. Bifurcated oral tentacles seem to occur in most (all?) populations but not in all individuals. Forms lacking oral tentacles have been called *Myrionecta rubra* (4, 7) but the description of the genus *Myrionecta* is vague. Earlier taxonomical discussions are reviewed in several papers (2, 3, 8, 10). Studies of living material and the ultrastructure of many more populations seem to be needed before the species or species complex has been properly classified.

A SPECIES COMPLEX?

Mesodinium rubrum is usually pear-shaped and fairly rounded. During blooms short and irregular cells frequently occur (8). The size varies greatly (length 15-70 um, diameter 10-60 um) as does the number of the chloroplasts (6-20 in small forms, 20-100 in large forms). Small forms often have 36 cirri, large forms have more than 50 cirri. The ciliary belt is very dense. Bifurcated oral tentacles occur in fresh samples but not in all cells. The tentacles are usually lacking in preserved material. They have a unique ultrastructure with cylinders of 14 microtubules (identifiable as rings of 14 microtubules in cross sections). The tentacles may contain one or several extrusomes (toxicysts?) and the furca may consist of two extrusomes (9). Most fixatives destroy the cilia and the cirri and break the body of *M. rubrum* into two parts. Often fixation (or the escape reaction) turns all cirri over the oral end. The chloroplasts are rich in starch and they stain black with iodine. The colour of living specimens is mostly brick red or brown. Dead cells turn green if the cells are broken and the water soluble pigment phycoerythrin is lost. During bloom conditions the water may look red, wine-red or maroon and the chlorophyll a values are high (8, 10).

DISTRIBUTION AND BLOOMS

Mesodinium rubrum is a common cause of red water (distribution maps in 8 and 10). Blooms are especially frequent in the coastal waters off the west coast of South America. During the 1980s *M. rubrum* has been reported from many "new" areas, e.g. Kamtchatka, China, The Black Sea, and Brasil. There are still (1987) very few records from the Mediterranean Sea, the Indian Ocean and Arctic and Antarctic waters. Blooms have occurred in many fjords, estuaries and coastal lagoons in Europe and North America. *M. rubrum* occurs in salinities down to 3-5 permille.

RED TIDES: BIOLOGY, ENVIRONMENTAL SCIENCE, AND TOXICOLOGY
Okaichi, Anderson, and Nemoto, Editors

NONTOXIC?

Toxin production has not been reported for *Mesodinium rubrum*. Its physiology is, however, still poorly known and e.g. the function of the tentacles and their extrusomes (toxicysts?) is unknown (9). Indirectly, blooms may be noxious by deteriorating the water quality and dense blooms may be detrimental to fish. *M. rubrum* may also cause shellfish discolouration (6). The increased number of dense *M. rubrum* bloom reports from most parts of the world may reflect a general eutrophication of the coastal waters.

ACKNOWLEDGEMENTS

The author has received unpublished reports on *Mesodinium* blooms, as well as pictures of blooms, from a great number of planktologists all over the world. This help and enthusiasm is gratefully appreciated and acknowledged.

REFERENCES

1. Corliss, J. O. 1979. "The ciliated Protozoa". Pergamon, Oxford.
2. Grain, J., Puytorac, P. de, and Groliere, C-A. 1982. Protistologica 18, 7-21.
3. Hibberd, D. J. 1977. J. Mar. Biol. Ass. U. K. 57, 45-61.
4. Jankowski, A. W. 1976. Pages 167-168. In "Materials of the II All-Union Conference of Protozoologists, Part I, General Protozoology". Ed. A. P. Markevich et al., Naukova Dumka, Kiew.
5. Jacques, G. and Sournia, A. 1978. Vie Milieu, Ser. AB, 28-29, 175-187.
6. Kat, M. 1984. Aquaculture 38, 375-377.
7. Lee, J. J., Hutner, S. H. and Boree, E. C. (Eds). 1985. "An Illustrated Guide to the Protozoa". Allen Press, Lawrence, Kansas.
8. Lindholm, T. 1985. Adv. Aquatic Microbiol. 3, 1-48.
9. Lindholm, T., Lindroos, P. and Mörk, A-C. 1988. BioSystems 21, 141-149.
10. Taylor, F. J. R., Blackbourn, D. J. and Blackbourn,J. 1971. J. Fish. Res. Bd. Canada 28, 391-407.

A NEW METHOD FOR INTER- AND INTRA-SPECIES IDENTIFICATION OF RED TIDE ALGAE CHATTONELLA ANTIQUA AND CHATTONELLA MARINA BY MEANS OF MONOCLONAL ANTIBODIES

SHINGO HIROISHI,* ARITSUNE UCHIDA,** KEIZO NAGASAKI,**
AND YUZABURO ISHIDA**
*Division of Research and Development, Shiraimatsu Shinyaku Co., Ltd., Shiga 528, Japan;**Laboratory of Microbiology, Department of Fisheries, Faculty of Agriculture, Kyoto University, Kyoto 606, Japan

ABSTRACT

Four monoclonal antibodies were obtained from the culture supernatant of hybridomas which had been established by cell fusion between myeloma and spleen cells of mice immunized with *Chattonella antiqua* and *C. marina*. Two monoclonal antibodies were reactive species-specifically with *C. antiqua* and *C. marina*, respectively. The third antibody was reactive with *C. antiqua* and a strain of *C. marina*. The fourth antibody was reactive with *C. marina* and a strain of *C. antiqua*. These results show the possibility of inter- and intra-species identification of *C. antiqua* and *C. marina* using the monoclonal antibodies.

INTRODUCTION

Chattonella antiqua and *C. marina* are the major organisms causing red tids in the coastal sea of Japan. There have been several morphological studies on identification of these species (1-10). At present, criteria showing the differences between them were their size, shape and inner structure by using photomicroscope and transmission electronmicroscope (10). However, the cell size in a population is variable and their shapes change easily according to changes of environmental conditions because they lack rigid cell walls over the cell membranes.

Therefore, we have approached the problem in an immunological way, by preparing specific monoclonal antibodies to the organisms.

EXPERIMENTAL

The antibodies were obtained by immunizing BALB/c mice with whole cells of each strain of *Chattonella* according to the method described by Köhler and Milstein (11) and by Hiroishi et al. (12).

Reactions of these antibodies to some strains of *C. antiqua* and *C. marina* were determined by the following method: a hundred microliters of the culture (around 2 x 10^3 cells/ml) were pipetted into wells of 96 well plate and to them were added equal volume of 0.5 M manitol in phosphate buffered saline (Buffer 1). After shaking the plate, it was left for 3 min to precipitate the organisms and then 150 µl of a monoclonal antibody was added and incubated at 20°C for 30 min. After the incubation of the cells with 50 µl of fluorescein isothiocyanate conjugated goat anti-mouse IgG antibody, they were washed three times with Buffer 1. The cell suspension was pipetted out onto a glass slide and observed with compound light microscope or fluorescence microscope. In the case of the fluorescence microscope, B-exciting light and a -550IF objective filter were used.

RED TIDES: BIOLOGY, ENVIRONMENTAL SCIENCE, AND TOXICOLOGY
Okaichi, Anderson, and Nemoto, Editors

RESULTS AND DISCUSSION

Using the cell fusion procedure, four hybridoma cell lines producing monoclonal antibodies (AT-83, AT-86, MR-18 and MR-21) were obtained. Yellowish fluorescein corona around the cells of *Chattonella* were observed when the antibodies reacted with the cells. These results suggest that the antigens recognized by these antibodies are distributed on the surface region of the cells. The corona became clearer when using a -550IF objective filter because red fluorescence emitted from chlorophyll in the cells was eliminated by the filter.

Reactivity patterns of the obtained monoclonal antibodies were different from each other. The results are summerized in Table 1.

TABLE I. Reactivities of four monoclonal antibodies against various strains of *C. antiqua* and *C. marina* by means of indirect fluorescent assay.

Strain	Sampling place and year	Reactivity AT-83	AT-86	MR-18	MR-21
C. antiqua					
NIES-85	Harima-Nada 1978	+++	+++	-	++
NIES-86	Uranouchi Bay 1980	+++	+++	-	-
C. marina					
NIES-118	Harima-Nada 1983	-	-	+++	+++
NIES-121	Kagoshima Bay 1982	+++	-	+++	+++

a)+++, positive cell ratio 81-100%; ++, 51-80%; +, 11-50%; -, 0%

AT-86 was reactive specifically with two strains of *C. antiqua* and MR-18 was reactive specifically with two strains of *C. marina*. AT-83 was reactive with two strains of *C. antiqua* and NIES-121 strain of *C. marina* but not with NIES-118 strain. MR-21 was reactive with two strains of *C. marina* and NIES-85 strain of *C. antiqua* but not with NIES-86 strain. These results show the possibility of inter- and intra-species of identification *C. antiqua* and *C. marina* by the monoclonal antibodies.

REFERENCES

1. B. Biecheler, Arch. Zool. Exp. Gen. 78, 79-83 (1936).
2. R. Subrahamnyan, Indian J. Fish 1, 182-203 (1954).
3. A. Hollande and M. Enjumet, Bull. Trav. Publ. Stn. Aquicult. et Peche Castiglione, N.S. 8, 273-280 (1957).
4. B. Fott, Algenkunde. Gustv Fisher, Jena 439-443 (1971).
5. Y. Hada, Rep. Hiroshima Shoka-Daigaku 12, 27-57 (1972).
6. Y. Hada, Bull. Plank. Soc. Japan 20, 112-125 (1974).
7. H. Takano, In Kankyou to Seibutsu-Shihyo 2, Suikai-Hen, Kyouritsu Shuppan Co., Tokyo p234-242 (1975).
8. J. P. Mignot, Protistologica 12, 279-293 (1976).
9. C. Ono and H. Takano, Bull. Tokai Reg. Fish. Res. Lab. 102, 93-99 (1980)
10. Y. Hara and M. Chihara, Japan J. Phycol. 30, 47-56 (1982).
11. G. Köhler and C. Milstein, Nature 256, 495-497 (1975).
12. S. Hiroishi, S. Matsuyama, T. Kaneko, Y. Nishimura and J. Arita, Tissue Antigens 24, 307-312 (1984).

A NEW DISCOVERY OF CYSTS OF PYRODINIUM BAHAMENSE VAR. COMPRESSUM FROM THE SAMAR SEA, PHILIPPINES

Matsuoka, K.[1], Y. Fukuyo[2] and C. L. Gonzales[3]
1: Department of Geology, Faculty of Liberal Arts, Nagasaki University, Nagasaki, 852, Japan; 2: Department of Fisheries, Faculty of Agriculture, University of Tokyo, Tokyo, 113, Japan; 3: Fisheries Research Division, Bureau of Fisheries and Aquatic Resources, Quezon City, Philippines

ABSTRACT

Surface sediments collected from Maqueda Bay in the Samar Sea, Philippines contained many cysts which are identical to Pyrodinium bahamense var. compressum based on the characteristic epicystal archeopyle, and intratabular, hollow and slender processes. Many thecate cells of this species also occurred in the same sample. This is the first reliable record of the cysts of P. bahamense var. compressum in Southeast Asia.

An orthographic change of the biological name is made: from var. compressa to var. compressum.

INTRODUCTION

Pyrodinium bahamense was originally described from Waterloo Lake in the Bahamas by Plate[1]. Later Böhm[2] recognized a form of this species based on specimens obtained from the Persian Gulf. Recently Steidinger et al.[3] carefully observed thecate cells collected from the coasts of Brunei and Papua New Guinea in comparison with the specimens of Tampa Bay in the U.S. and of Oyster Bay in Jamaica. They recognized several morphological differences among the specimens and revised the taxonomy of P. bahamense, establishing two varieties, P. bahamense var. bahamense for the Atlantic specimens and P. bahamense var. compressum for the Indo-Pacific specimens. However, Balech[4] found no consistent differences for distinguishing the varieties.

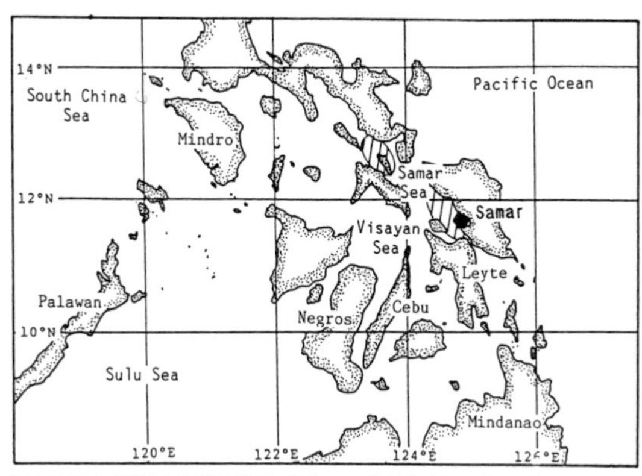

Fig. 1: Map showing the sampling location (●) and ares affected by the blooms of Pyrodinium bahamense var. compressum (◍) (after Estudillo and Gonzales[5]).

RED TIDES: BIOLOGY, ENVIRONMENTAL SCIENCE, AND TOXICOLOGY
Okaichi, Anderson, and Nemoto, Editors

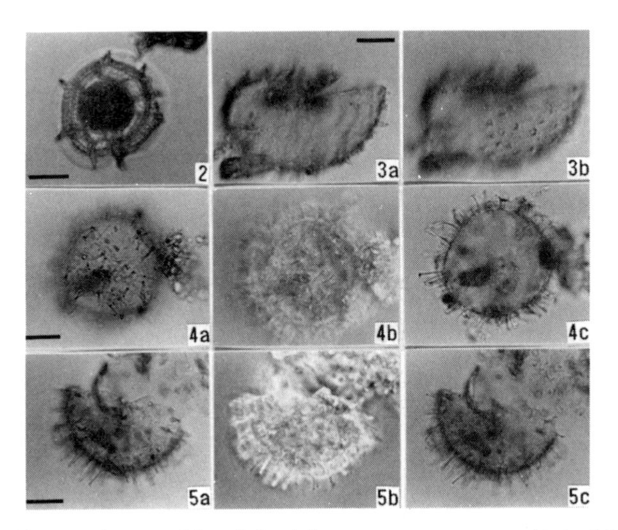

Figs. 2-5: Thecae and cysts of Pyrodinium bahamense var. compressum obtained from the surface sediments in the Samar Sea, Philippines. 2; Apical view of thecae, 3a, 3b: Hemispherical empty cyst showing the epicystal archeopyle and reduced processes. 4a, 4b, 4c; Spherical empty cyst showing the adnate archeopyle sutures and normal slender processes. 5a, 5b, 5c; Hemispherical empty cyst showing the epicystal archeopyle. Scale bar: 20μm.

From the view point of shellfish toxicity, Maclean[5] suggested that P. bahamense var. compressum is possibly responsible for PSP in Papua New Guinea and Harada et al.[6] confirmed the toxicity of this variety. Since then, toxic red tides caused by P. bahamense var. compressum have expanded into Borneo, Philippines, Palau and Fiji according to Maclean[7]. In Maqueda Bay of the Samar Sea, Philippines, Estudillo and Gonzales[8] reported that a major red tide outbreak accompanied by serious paralytic shellfish poisoning occurred in 1983. Gonzales et al.[9] recognized the PSP to be caused by P. bahamense var. compressum and expected the cysts of this species to be deposited in this region.

Wall and Dale[10] carried out cyst incubations of P. bahamense var. bahamense based on the material of Smiths Sound in Bermuda, Bahia Fosforescente and Bahia de Jobos in Puerto Rico and Isla de Vieques in the tropical Atlantic. In contrast, there is no record of a P. bahamense var. compressum cyst in the above-mentioned areas of the West Pacific except for a brief note given by Steidinger et al.[3]

This is the first report with description and illustrations of the modern cyst of P. bahamense var. compressum from Southeast Asia.

MATERIAL AND METHODS

The filmy surface layer of the sea floor was scooped with a 200 mℓ, wide-mouthed glass bottle while skin diving at Maqueda Bay in the Samar Sea. At the beginning of the study, raw samples including many cysts were inspected without any processing. Later, in order to observe more detailed morphology of the processes, standard palynological processing was employed.

RESULTS AND DISCUSSION

In the raw material, there were many empty cysts of P. bahamense var. compressum with a few other protoperidiniacean cysts, diatoms and other palynomorphs such as fern spores and pollen grains. The cysts of P. bahamense var. compressum are spherical to subspherical (62~72µm in diameter) and comprise a bi-layered colorless wall covered with many short, hollow, intratabular and slender cylindrical processes with capitate, or rarely bifurcate distal extremities (6~16µm in length). The processes, however, are sometimes much reduced to small nodes or low cones. (Fig. 3). Most specimens are spherical and do not show a distinct epicystal archeopyle (Fig. 4), but with careful observation of the surface, some parasutural slits are recognized. Some specimens have a typical epicystal archeopyle with several zigzag principal archeopyle sutures (Figs. 3, 5). Based on the literature, the morphology of the cysts of P. bahamense var. compressum is basically similar to those of P. bahamense var. bahamense described by Wall and Dale[10] excluding the cyst diameter and length of process. The cysts of P. bahamense var. compressum have larger body and relatively shorter processes.

The same material also contained some thecate cells with contracted protoplasm identical to P. bahamense var. compressum (Fig. 2). Although observations have not yet been done in detail, the thecate cells seem to be characterized by the absence of a conspicuous apical horn and anterio-posterior compression.

In the sample studied, the cyst assemblage contained almost exclsively the cysts of P. bahamense var. compressum except for very rare occurrences

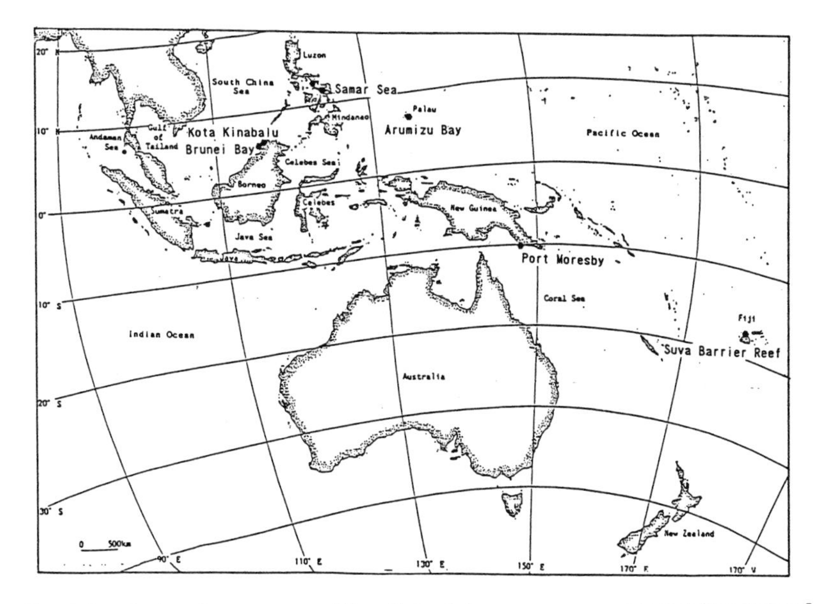

Fig. 6 Map showing the occurrence of Pyrodinium bahamense var. compressum (after MacLean[7] and Steidinger et al.[3])

of protoperidiniacean cysts. Most cysts were completely empty and associated with thecate cells. In comparison with the modern cyst assemblage around Japan reported by Matsuoka[11], this extrodinary cyst assemblage seems to be the aftermath of blooms of P. bahamense var. compressum.

For taxonomical discussion of P. bahamense, we need further detailed and statistical observation on the morphology of both thecate cells and cysts of P. bahamense var. bahamense and P. bahamense var. compressum. However, from the view-point of the PSP problem, the discovery of the cyst is very helpful for predicting future red tides of P. bahamense var. compressum.

Appendix:

The gender of the intraspecific epithet should agree gramatically with the generic one (I.C.B.N. 24.2) and the names published with an incorrect Latin termination are to be changed to accord with Art 24 of I.C.B.N., and without change of the author's name (I.C.B.N. 32.5). Thus as the gender of Pyrodinium is neuter, the name of variety should be spelled as compressum, and the formal name is as follows; Pyrodinium bahamense Plate var. compressum (Böhm) Steidinger, Tester and Taylor, 1980.

REFERENCES

1. Plate, L., Arch. Protistenk. 7, 411-428, (1906).
2. Böhm, A., Bernice P. Bishop Mus., Bull. 87, 1-46, (1931).
3. Steiginger, K.A., L.S. Tester and F.J.R. Tayler, Phycologia, 19, 329-337, (1980).
4. Balech, E. Rev. Palaeobotan. Palynol. 45, 17-34 (1985).
5. Maclean, J.L., Papua New Guinea Agric. J., 24, 131-138, (1973, distributed in 1975).
6. Harada, T., Y. Oshima, H. Kamiya and T. Yasumoto, Bull. Japan. Soc. Sci. Fish., 48, 821-825 (1982)
7. Maclean, J.L., In White, A.W. et al. (ed.) Toxic Red Tide and Shellfish Toxicity in Southeast Asia, 92-102, (1984).
8. Estudillo, R.A. and C.L. Gonzales, In White, A.W. et al. (ed.) Toxic Red Tide and Shellfish Toxicity in Southeast Asia, 52-79, (1984).
9. Gonzales, C.L., J.A. Ordoñez and A.M. Maala, Abstract of International Symposium on Red Tides, Takamatsu, 14, (1987).
10. Wall, D. and B. Dale, J. Phycol., 5, 140-149, (1969).
11. Matsuoka, K., Bull. Fac. Liberal Arts, Nagasaki Univ., 25(2), 21-115, (1985).

POSSIBLE SITE FOR THE CYST OF PERIDINIUM PENARDII (LEMM.) LEMM. TO EXCYST IN A RESERVOIR

ZEN'ICHIRO KAWABATA AND MASAYUKI OHTA
Department of Environmental Conservation, Ehime University, Tarumi 3-5-7, Matsuyama 790, Japan

ABSTRACT

To find a possible site for the cysts of *Peridinium penardii* (Lemm.) Lemm. to excyst in a reservoir, culture bottles containing the lake bottom mud which contained the cysts were suspended at several depths at four stations in the reservoir. After incubation for 7 days, vegetative cells were observed in all bottles suspended 0.5 m deep. A rapid decrease in the number of the cells was observed at 5.0 m and 10.0 m deep and almost none below 20.0 m. These results were repeatedly obtained during a three-month period in winter. The place where *P. penardii* first appears annually was constantly predicted to be at the head of the reservoir, using the data of the distribution of the cysts and the morphology of the lake basin.

INTRODUCTION

Water blooms, so-called freshwater red tides, have been recently observed especially in reservoirs from western Japan [1]. In most cases the freshwater "red tide" is caused by a dinoflagellate *Peridinium* species. Large populations of the organisms degrade water quality creating a nuisance in the filtration processes in making drinking water, and sometimes cause serious odor problems. Therefore, it is of practical importance in terms of water quality management to elucidate the mechanisms of occurrence of this freshwater "red tide" *Peridinium* species and to develop methods for suppression of the blooms. The life cycle of *Peridinium* species contains a cyst stage [2, 3]. The excystment of the cyst precedes an increase in numbers of vegetative cells. Therefore, a prevention of excystment is important to effectively suppress the occurrence of freshwater "red tide".

The purpose of this study was to find the site for the excystment of *Peridinium penardii* in the reservoir, and to predict the place where *P. penardii* first appears annually using cyst distribution data and the morphology of the lake basin.

THE STUDY AREA

The Ishitegawa Reservoir (N33° 53', E 132° 50') was constructed in Matsuyama City, Ehime Prefecture, Japan, in 1973 by damming up the Ishite River for the supply of drinking and irrigation water and for flood control. The reservoir's area, volume, mean depth and length are 0.50 km^2, 1.28 x 10^7 m^3, 25.6 m and 2.3 km, respectively, when full (Fig. 1-A). Freshwater "red tides" of *Peridinium* species has been observed at the head of the reservoir every winter since 1975 [1, 4].

MATERIALS AND METHODS

Four experimental stations (A-D) were placed in the reservoir (Fig. 1-A). Cultures were carried out at the depths of 0.5 m at station A, 0.5 m,

RED TIDES: BIOLOGY, ENVIRONMENTAL SCIENCE, AND TOXICOLOGY
Okaichi, Anderson, and Nemoto, Editors

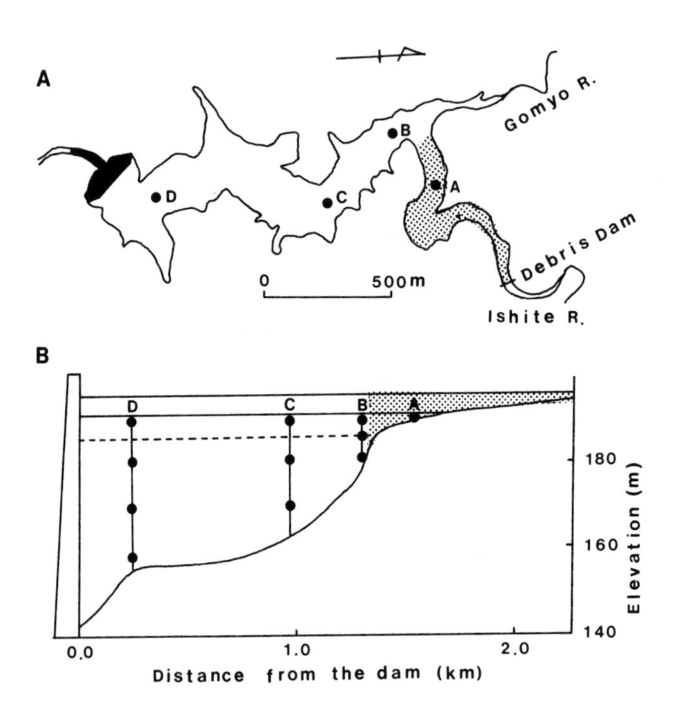

Fig. 1. An overview and cross section of the Ishitegawa Reservoir. The shaded area indicates the hypothesized area of excystment when the water level is high. A: an overview with the experimental stations (A-D). B: a cross section with culture depths (●) at each station. Solid and dotted lines indicate the water surfaces when the water levels are high and low, respectively.

5.0 m and 10.0 m at station B, 0.5 m, 10.0 m and 20.0 m at station C and 0.5 m, 10.0 m, 20.0 m and 33.6 m at station D (Fig. 1-B).

Surface lake bottom mud, about 2 cm deep, was collected a few days before incubation, from several places in the area between the debris dam and station A which was exposed because of low water level. After complete mixing, 25 cm^3 of the mud mixture was transferred to glass bottles with a diameter of 4.0 cm and a height of 12.0 cm. Lake water was collected at each depth at each station and filtered by a nylon plankton net with a pore size of 10 μm in order to remove the vegetative cells of *Peridinium penardii*. The culture bottles containing the mud were gently filled with the filtered lake water. After covering the mouths of the bottles with a nylon net with a pore size of 10 μm, sets of three bottles were each put in a large-mesh nylon net bags which were tied to ropes. The ropes were fixed to anchors on the lake bottom and suspended by buoys. These bottles were incubated at each depth at each station. In order to ascertain the existence of cysts of P. penardii which could excyst in the lake mud used for the in situ culture, six bottles containing the lake bottom mud were filled with distilled water. Three were placed in an incubator at 10 °C under the illumination of 3000 lux by fluorescent lamp on a 12 hr. LD cycle and three were placed in the dark. All cultures both in situ and in vitro were carried out for 7 days and repeated 12 times from November 25, 1985 to February 24, 1986.

Field measurements of water temperature and underwater irradiance as illuminant flux density (lux) were measured at each station using a thermistor (YSI Model 33 S-C-T Meter) and an irradiance meter (WL-2, Murayama Denki, Ltd.), respectively.

All the water in the bottles was collected and centrifuged at 3000 rpm. The sediment was transferred to plankton counting chamber and all the vegetative cells of P. penardii were counted microscopically.

RESULTS

Vegetative cells were almost always found in all bottles which were cultured in vitro with few vegetative cells found in dark bottles. An

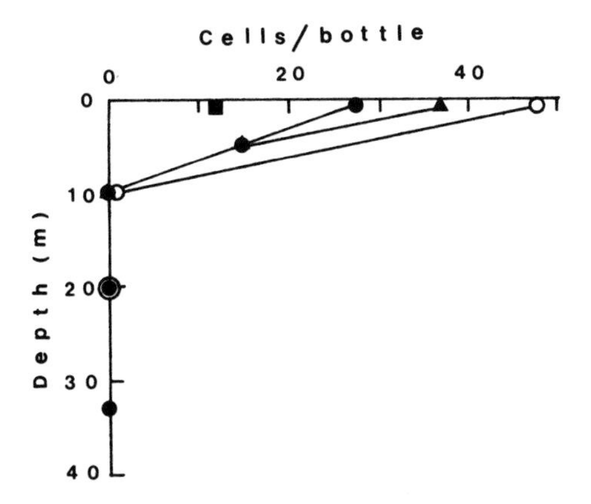

Fig. 2. The number of vegetative cells (mean value) per bottle incubated in situ from Feb. 3 to 10, 1986, at each depth at each station (■ , stn. A; ▲, stn. B; ○ , stn. C; ●, stn.D).

example of the number of vegetative cells found in the bottle incubated in situ at each depth at each station is shown in Fig. 2. The vegetative cells were observed in all bottles suspended 0.5 m deep. A rapid decrease in the number of the vegetative cells was observed in the bottles at 0.5 m and 10.0 m deep and almost none at deeper than 20.0 m. These tendencies were almost always observed regardless of position within the reservoir during the period of experiment.

No noticeable change in water temperature was found vertically. The water temperature at station A was usually 1-2 °C lower than at the other stations. Water temperature decreased from 14.5 °C to 5.0 °C through the period of the experiment. One per cent of relative irradiance usually reached the depth of between 10.0 m and 20.0 m.

DISCUSSION

From the result that the vegetative cells of *P. penardii* were observed in all bottles cultured in an incubator at 10 °C, it is evident that the lake bottom mud used for the culture *in situ* usually contained cysts of *P. penardii*. Generally speaking, water temperature is the primary environmental factor controlling excystment in dinoflagellate resting cysts [3, 5]. The cyst of *P. penardii* excysted at the water temperatures recorded during the period of this experiment. Therefore, according to the range of our experiment, the water temperature had little influence on excystment.

Light conditions have generally been found to exert little effect on excystment of dinoflagellates [3, 6, 7]. In contrast to these findings, it was reported [8] that light was required for excystment of a marine dinoflagellate. It was observed in our present culture *in vitro* that the excystment of *P. penardii* was not prevented, but significantly reduced, in the absence of light.

No appreciable effect of the nutrient condition on the excystment of dinoflagellate cysts [3, 7] was observed. No noticeable difference was observed in water quality among each depth during the period of this experiment (unpubl. data). Therefore, the fact that no excystment was observed in the lake water deeper than 20.0 m indicated that water depth was an important environmental factor in preventing excystment.

Cysts able to excyst under light are ubiquitous on the lake bottom of the reservoir (unpubl. data). There is a shallow area at the head of the reservoir which changes depending on the water level. Therefore, it was suggested that the area where the cyst of *P. penardii* can excyst is restricted to the head of the reservoir when the water level is high (Fig. 1-A, B). It was also predicted that no excystment would be found even at the head of the reservoir when the water level is low (Fig. 1-B), because the shallow area dries out and only deep water remains.

ACKNOWLEDGMENTS

The authors appreciate the permission granted by the Ishitegawa Reservoir Office to conduct our survey in the reservoir. We are also grateful to Miss K. Hino and Mr. Otsuka of our laboratory who shared in the labor of sampling and helped with enumeration.

REFERENCES

1. T. Ito, Bull. Plankton Soc. Japan 26, 113-116 (1979) (In Japanese).
2. L. A. Pfiester and J. J. Skvarla, Phycologia 18, 13-18 (1979).
3. Y. Sako, Y. Ishida, H. Kadota and Y. Hata, Jap. Soc. Sci. Fish. 51(2), 267-272 (1985).
4. H. Kagawa, Y. Iseri and T. Ito, Jap. J. Wat. Pollut. Res. 7(6), 375-383 (1984) (In Japanese with English summary).
5. L. A. Pfiester and D. M. Anderson in: The Biology of Dinoflagellates F. J. R. Taylor, ed. (Blackwell Scientific, Oxford 1987) pp. 611-648.
6. T. Endo and H. Nagata, Bull. Plankton Soc. Japan 31(1), 23-33 (1984) (In Japanese).
7. D. M. Anderson and D. J. Wall, J. Phycol. 14(2), 224-234 (1978).
8. B. J. Binder and D. M. Anderson, Nature 322(14), 659-661(1986).

DISTRIBUTION OF ALEXANDRIUM EXCAVATUM RESTING CYSTS IN A PATAGONIC SHELF AREA (ARGENTINA)

OROZCO, F.E. AND J.I. CARRETO
Instituto Nacional de Investigación y Desarrollo Pesquero,
P.O. Box 175, 7600 - Mar del Plata, Argentina

ABSTRACT

During winter 1985-1986, sediment samples were collected in a patagonic shelf area to determine the abundance and distribution of *Alexandrium excavatum* resting cysts. A defined distributional pattern was not observed, though resting cysts were distributed along the area studied (42-47°S). Cyst concentration varied between 0-9,000 cysts cm^{-3} in top sediment. Vertical distribution in some samples indicated the occurrence of maximun values near the 2-3 cm depth. These results show the great toxic potential of the study area.

INTRODUCTION

During 1980, *A. excavatum* blooms in Argentine shelf waters were restricted to the area off Península Valdés (1). From this initial outbreak, the toxic area expanded through successive stages to cover nearly all the Argentine Sea (2). Although *A. excavatum* resting cysts were previously documented (1,3), nothing was known about their distribution. The importance of resting cysts in the initiation of *A. excavatum* and other dinoflagellate blooms has been extensively documented (4,5,6). Anderson (7), has recently discussed their role as a dispersion agent, as a source of toxicity and as means for genetic recombination through sexuality. Generally, distribution of cysts is not homogeneous. Transport and sedimentary processes, result in dilution and concentration zones ("seed-beds") (4).

We hope that knowledge of the distribution and abundance of resting cysts in the Argentine Sea can be employed in helping to decide the location of shellfish monitoring and aquaculture projects. Therefore, we report the first semi-quantitative analysis of *A. excavatum* resting cysts in sediments of a patagonic shelf area.

EXPERIMENTAL

Fifty nine samples were obtained during three oceanographic cruises (July, October 1985 and January 1986). A "Picard" type dredge and a gravity corer (1.8 cm I.D.) were used. After ninety minutes of sedimentation, samples were stored at -30°C. In the case of the dredge samples, a 2.5 cm^3 wet sediment subsample was used. Samples taken with corer were transversally severed taking the first sediment centimeter and the water of the floculent layer. In some cases, the PVC tube containing the subsample was severed in different 1 cm length pieces. The subsample was filled to 20 ml volume with filtered sea water, and later sonicated (8). Particles under 25 µm were discarded by a mesh sieve. The remaining sediment was quantitatively transferred to a glass vessel making successive washes with filtered sea water. Three aliquots were taken from each subsample with an automatic pipette, while the suspension was shaken to assure its homogeneity. The volume of the aliquot ranged between 0.2-0.4 ml, depending on the granularity of the sediment. Cysts were counted using an inverted microscope. In all cases, the total area of the optical field was examined. The results presented correspond to the average obtained

RED TIDES: BIOLOGY, ENVIRONMENTAL SCIENCE, AND TOXICOLOGY
Okaichi, Anderson, and Nemoto, Editors

from three aliquots.

RESULTS

As can be seen in Figs. 1 and 2, the area of cyst distribution is as wide as the area sampled, and extends from Pla. Valdes(42 S) to the southern extreme of San Jorge Gulf (47 S). The highest concentrations occur in the first few centimeter, and reach values of 9,000 cysts cm^{-3} wet sediment. These values are similar to those found by White and Lewis (8) in the sediments of Fundy Bay. However, the spatial distribution is highly heterogeneous, and concentration values found were very different between near stations. This spatial heterogeneity seems to be a characteristic common for

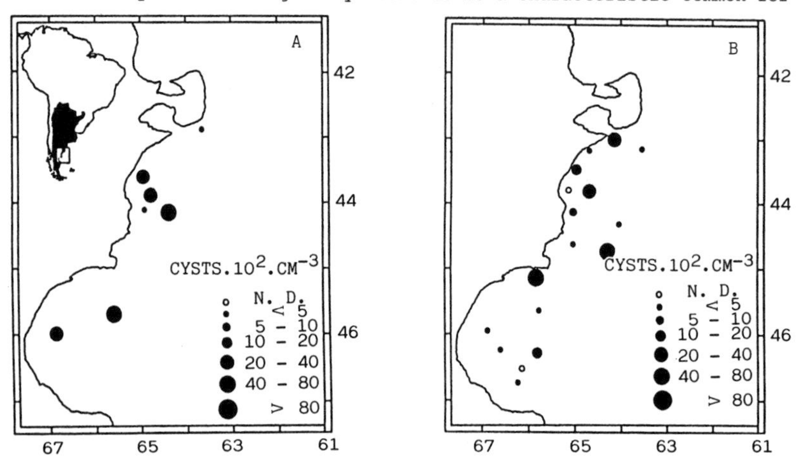

FIG. 1. Distribution of A. excavatum resting cysts (dredge samples): (A) July 1985; (B) October 1985.

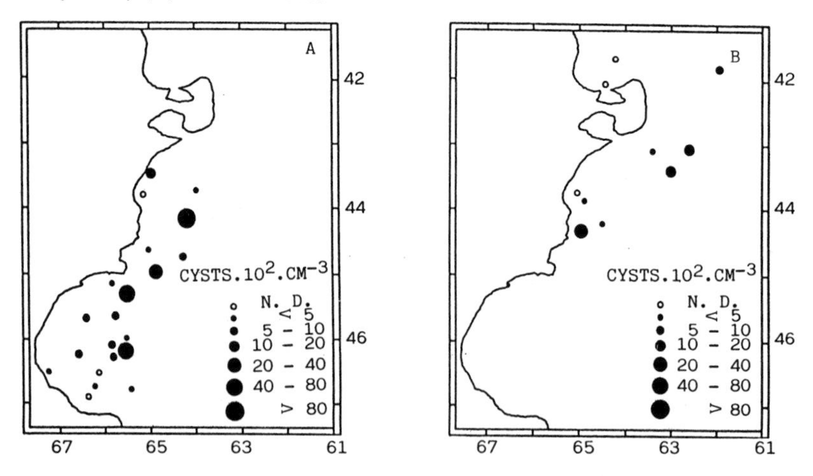

FIG. 2. Distribution of A. excavatum resting cysts (corer samples): (A) October 1985; (B) January 1986.

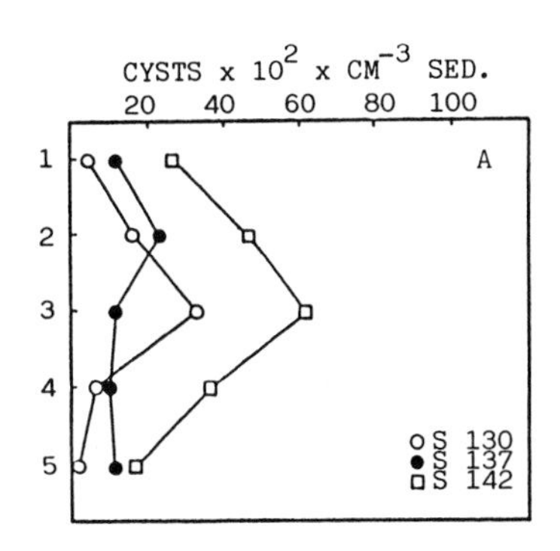

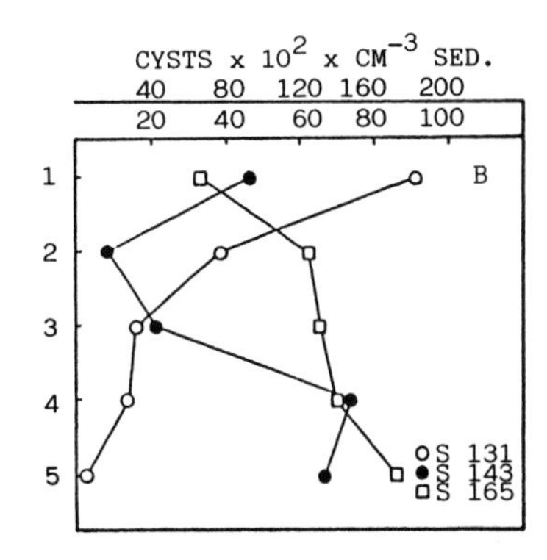

FIG. 3. (A) and (B) Vertical distribution of A. excavatum cysts in some stations.

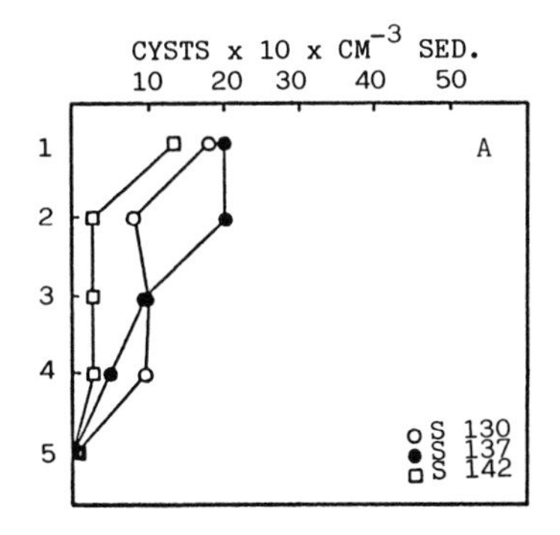

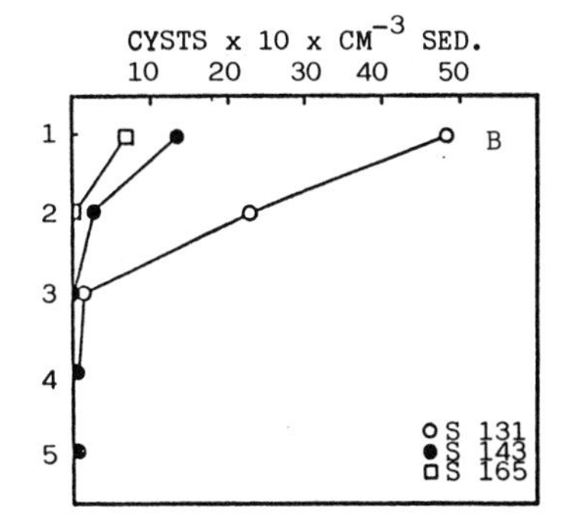

FIG. 4. (A) and (B) Vertical distribution of P. schwartzii cysts in some stations.

both coastal (about 40 m) and deep (about 100 m) samplings. The data obtained are insufficient for estimating temporal variability of cyst concentrations. Although high values were observed during spring, summer and winter, the highest ones correspond to spring. The vertical distribution studied during spring allowed us to separate the stations into two groups. The first one (Fig. 3 A) includes those that present a common distribution pattern, in which a concentration maximum in the second or third centimeter is evident. The second group (Fig. 3 B) significantly differs from the first one. The distribution observed in these stations do not present common features. In station 131, the maximum occurs at the surface layer, with 9,000 cysts cm^{-3} and values decreasing to the detection limit in the fifth centimeter. In station 165, the maximum occurs in the fifth centimeter, with values as high as 17,000 cysts cm^{-3}. Finally, in station 143, two maxima are observed. It is interesting to compare the vertical distribution of A. excavatum cysts, with that of the phagotrophic dinoflagellate Polykrikos schwartzii (Fig. 4). For the latter, the distribution pattern is the same in all stations, with a maximum observed in the surface layer. On the other hand, the observed concentrations are lower than those of A. excavatum, and their proportions range between 1% and 30%.

DISCUSSION

The first PSP outbreak recorded in the Argentine Sea was caused by A. excavatum and was restricted to the frontal system of Pla. Valdés (1). From this initial outbreak, the toxic area expanded through successive stages to cover nearly all the Argentine Sea (2). Our results indicate the presence of A. excavatum cysts in the whole area studied. Furthermore, cysts were recently observed in front of Mar del Plata coast (39 S) (De Marco, pers. comm.) and in the San Julián Bay (49 S) (Benavides, pers. comm.). This indicates that the spread of the toxic area, whether through the transport of vegetative cells or cysts, finally resulted in colonization by this species. The high concentration values found (more than 9,000 cysts cm^{-3}) are similar to those reported by White and Lewis (8) in Fundy Bay, where the phenomenon is endemic (9). Thus, a similar condition could be expected on the Argentine shelf. The highest concentrations are generally found in offshore stations. However, the observed distribution is irregular, probably in patches. The biological and physical mechanisms underlying surface and subsurface cyst concentration are not known, and seem to be dependent on several factors (6) complicating the interpretation of the results. The heterogeneous distribution of A. excavatum cysts is quite different from that observed for P. schwartzii cysts. Therefore, a specific behaviour probably related to planktonic stages of both species may be suggested, and this could be highly interesting, since a well known predator-prey relationship exists (10).

REFERENCES

1. J.I. Carreto, M.L. Lasta, R.M. Negri and H.R. Benavides, Contrib. (INIDEP, Mar del Plata) 399 (1981).
2. J.I. Carreto, R.M. Negri, H.R. Benavides and R. Akselman in: Toxic Dinoflagellates, D.M. Anderson, A.W. White and D.G. Baden, eds. (Elsevier 1985) pp. 147-152.
3. H.R. Benavides, R.M. Negri and J.I. Carreto, Physis (Buenos Aires) Secc. A 41, 135-142 (1983).
4. K.A. Steidinger in: Proc. 1st Internat. Conf. on Toxic Dinoflagellate Blooms, V.R. LoCicero, ed. (Mass. Sci. Tech. Found., Wakefield, Mass. 1975) pp. 153-162.
5. C.M. Lewis, C.M. Yentsch and B. Dale in: Toxic Dinoflagellate Blooms, F. J.R. Taylor and H.H. Seliger, eds. (Elsevier North Holland 1979) pp. 235-238.
6. D.M. Anderson, D.G. Aubrey, M.A. Tyler and D.W. Coats, Limnol. Oceanogr. 27 (4), 757-765 (1982).
7. D.M. Anderson in: Seafood Toxins, E.P. Ragelis, ed. (Amer. Chem. Soc. Symposium Series 262, Wash. D.C. 1984) pp. 126-138.
8. A.W. White and C.M. Lewis, Can. J. Fish. Aquat. Sci. 39 (8), 1185-1194 (1982).
9. A. Prakash, J.C. Medcof and A.D. Tennant, Bull. Fish. Res. Bd. Can. 168, 68 pp. (1971).
10. J.I. Carreto, H.R. Benavides, R.M. Negri and P.D. Glorioso, J. Plankton Res. 8, 15-28 (1986).

CELL CYCLE AND GROWTH RATE OF *CHATTONELLA ANTIQUA* (HADA) ONO DURING RED TIDES

CHITARI ONO
Akashiwo Research Institute of Kagawa Prefecture, Yashima Higashimachi, Takamatsu 761-01, Japan

ABSTRACT

The cell cycle of *C. antiqua* was examined in culture and in nature. In all cases, the cell division occurred at night or during dark periods but did not occur in daytime or during light periods. Since the cell division required about 9 hours, it seems that the maximum growth rate of *C. antiqua* in nature did not exceed 1 div./day.

INTRODUCTION

The growth rates of red tide phytoplankton have been estimated based on cell density variations in natural water. However, it is very difficult to measure cell density reliably for a fixed population in nature, and grazing losses must be considered as well with such methods. In the present study, the cell cycle of *C. antiqua* was observed during 24-hr continuous-sampling of red tides during the summers of 1986 and 1987. The population growth rate of *C. antiqua* in nature was estimated based on these data.

MATERIALS AND METHODS

Natural Specimens: *C. antiqua* was collected by 24 hour continuous sampling at a station in the yellowtail culture ground off Hiketa in the south coastal area of Harima Nada during the summer, 1986 and at a station in Yashima Bay during the summer, 1987. Two and four liter surface water samples were filtered with 20 μm mesh nylon screening, and suspended matter was collected and fixed with Carnoy's solution. Fixed samples were washed with distilled water and stored in 70% ethanol.

Cultured Specimens: Colonies of *C. antiqua* (strain 53 Harima, isolated from the red tide which occurred in the summer of 1978) were cultured under a 14/10 L/D cylce with 12 Klux light intensity at 23°C. Samples were taken from the colonies every hour and fixed with Carnoy's solution. The fixed samples were stored in 70% ethanol.

Observation of Cell Cycle: Samples stored in 70% ethanol were stained with an aceto-iron-heamatoxylin chloral hydrate mixture [1]. Slides were prepared using one drop of sample and one drop of stain. This solution was heated on the slide and observed with a photomicroscope.

RESULTS

Cell Cycle of Cultured Specimens: Cells with prophase nuclei and chromatids could be observed about an hour after the beginning of the dark period. Double-structured chromosomes were visible about an hour later. Metaphase nuclei were observed midway through the dark period in most cells, and at that time, chromosomes gathered on the equatorial plate of the cells and divided into two groups. Each of the two groups of chromosomes was pulled to a pole, but the cell plate did not form. The equatorial plate formed in the longitudinal plane of the cells. The invagination of the cell wall, which is equivalent in formation to the cell

Published 1989 by Elsevier Science Publishing Co., Inc.
RED TIDES: BIOLOGY, ENVIRONMENTAL SCIENCE, AND TOXICOLOGY
Okaichi, Anderson, and Nemoto, Editors

plate in higher plants, occurred on the same plane where the equatorial plate formed. At the same time that invagination of the cell wall occurred, cytoplasmic division started. The cytoplasmasic division began at the flagellar pores in the anterior part of the cell and finished at the posterior end. Cell division was complete by the end of the dark period and was observed in the light. The cell division rate reached about 70% under 12 Klux light intensity.

Natural Specimens: The prophase nuclei appeared at 18:00, about half an hour before sunset and both nuclear and cytoplasmic division was complete by 04:00, about one and a half hours before sunrise during the red tide in the summer, 1986. Cell division was not observed at all in the daytime, as in the culture experiment, and the cell division rate reached about 30%. In contrast, the cell division of C. antiqua in the red tide which occurred during 1987 started at 03:00 and finished at 10:00, peaking at 06:00. Moreover, the cell division rate peaked at 15%.

DISCUSSION

The mean generation times of C. antiqua in nature were reported by Yoshimatsu [2] to be 25-27 hrs. These estimates were based on cell density variations during the red tide, which occurred during the summer of 1978. The growth rate of this flagellate has typically been estimated in terms of either mean generation time or the number of cell divisions per day. These estimates have been primarily based on cell density variations observed at one or a few fixed stations in the areas concerned. There are obvious concerns as to whether this method provides valid estimates of the growth of a natural population, especially one subject to grazing pressure. In this study, cell cycles of laboratory cultured C. antiqua, which is the major red tide species in Harima Nada, were examined as were naturally occurring cells of C. antiqua collected from the red tides which occurred off Hiketa and in Yashima Bay. The nuclear division of the cultured specimens began about one hour after dark and finished before light. The peak occurrence of nuclear division appeared 4-6 hrs. after dark. During interphase, C. antiqua chromomers were distributed homogeneously and 2-3 nucleoli were seen in the nucleus. About an hour after dark, chromatids became visible and prophase started. As prophase progressed, the duplicated, long thread-like chromatids became visible. In the metaphase, chromatids grew shorter and thicker and each chromatid was clearly discernable. The chromatids were separated into two groups and each moved toward a pole. Simultaneously with the chromatid separation, the new cell wall formed and the division of cytoplasm started. On the other hand, nuclear division of cells in natural samples were observed only at night. One exception was the C. antiqua red tide that occurred in 1987. In this red tide, nuclear division took place during the 03:00-10:00 time period, although the duration was nearly the same as the usual nighttime division. In several cases, large short term population density variations took place without any accompanying nuclear division.

REFERENCES

1. W. Wittmann, Stain Tech. 40, 161-164 (1965).
2. S. Yoshimatu, Scientific reports on Hornelian red tide occurred in 1978, Okaichi eds. (Kagawa Prefecture, Takamatsu 1979) pp. 34-37.

IV BIOCHEMISTRY

ROLE OF ALKALINE PHOSPHATASE ACTIVITY IN THE GROWTH OF RED TIDE ORGANISMS

KAZUTSUGU HIRAYAMA, TAKASHI DOMA, NOBUAKI HAMAMURA AND TSUYOSHI MURAMATSU
Faculty of Fisheries, Nagasaki University, Nagasaki, 852, Japan

ABSTRACT

A field survey, conducted in Nagasaki Bay in summer 1982 showed chlorophyll *a* levels to be positively correlated with alkaline phosphatase (AP) activity, but negatively correlated with orthophosphate concentration. The highest activity was observed in a bloom of *Prorocentrum* sp. We cultured *Gymnodinium nagasakiense* axenically in 200 ml flasks to examine the role of alkaline phosphatase on P uptake. When only inorganic phosphate was used as a P source, cells growing exponentially did not exhibit AP activity. After exhaustion of P from the medium, however, extremely high AP activity was observed. When glycerophosphate was used as the only P source, AP activity appeared from the first day of culture and gradually increased with cell growth, while AP activity per cell decreased gradually with growth and the P content fluctuated in the range (400-600) x 10^{-15} g at. per cell during the exponential growth stage. When complete consumption of P restrained growth, internal P content was maintained at (350-380) x 10^{-15} g at. per cell.

INTRODUCTION

In eutrophic water, even after inorganic phosphate is exhausted by phytoplankton uptake, phytoplankton can still multiply to a bloom. This is considered to be due to the ability of unicellular algae to store phosphorus as internal polyphosphate in the cells. In addition, the role of alkaline phosphatase activity which is produced by microorganisms cannot be ignored [1, 2].

In order to investigate the role of alkaline phosphatase activity in red tide organisms on the outbreak of red tides, we conducted a field survey in Nagasaki Bay, and also cultured *Gymnodinium nagasakiense* in the laboratory.

MATERIALS AND METHODS

Field survey

In Nagasaki Bay, surrounded by Nagasaki city center, discoloration due to blooms of diatoms or dinoflagellates frequently occurs every summer. We collected seawater samples from depths of 0, 1, 2, 3, and 4 m at a sampling station which is located in the center of Nagasaki Bay (FIG. 1), once a week at about noon, from June to September, 1982. We then mixed equal volumes of all water samples from every depth as sample water for analysis. Immediately after mixing, the sample was divided into three parts, two of which were filtrates made by filtering through Milipore filters of 0.45 µm and 5 µm pore size and the last part was left unfiltered. Analysis of alkaline phosphatase (AP) activity was conducted on all three parts by a fluorometrical method according to Perry [3]. We also analyzed chlorophyll *a* in unfiltered samples and obtained the concentrations of the various

RED TIDES: BIOLOGY, ENVIRONMENTAL SCIENCE, AND TOXICOLOGY
Okaichi, Anderson, and Nemoto, Editors

forms of dissolved phosphorus in the filtrate through the 0.45 μm pore filter.

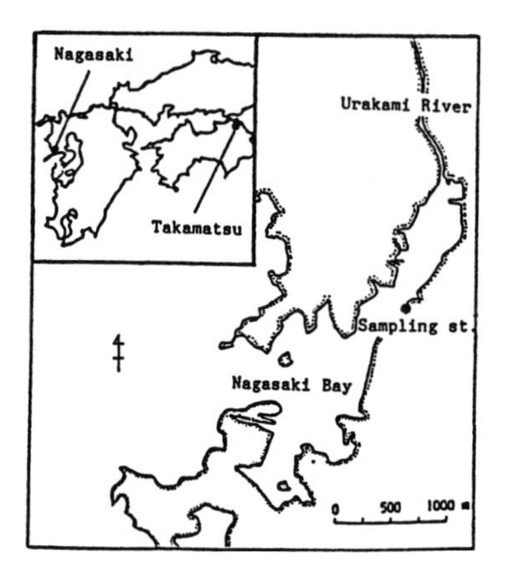

FIG. 1. Map of sampling station.

Laboratory culture

We prepared 200 ml flasks each containing 100 ml artificial culture medium, NH-15 plus soil extract. P source was added axenically into each flask after autoclaving. *G. nagasakiense* was inoculated into each flask at 120 cells/ml and then axenically cultured at 23° C, 6000 lux, 14L:10D. Then, one flask was used every day for analyzing cell density, AP activity and various forms of phosphorus. The AP activities were revised into the values corrected according to the contents of inorganic phosphate and glycerophosphate present in the medium, because the activity was underestimated by their existence.

RESULTS

Field survey

The changes in chlorophyll *a*, levels of various forms of dissolved phosphorus and AP activities in the three parts are shown in FIG. 2. The larger particles such as phytoplankton accounted for the high proportion of the AP activity in Nagasaki Bay, compared with the AP level of the parts associated with particles less than 5 μm in diameter. On 12 August, we observed a dense red tide of *Prorocentrum* sp. which shows as a peak of chlorophyll *a* in FIG. 1. At that time, total AP in the unfiltered sample was also recorded at its highest value, 121 nM/h, while inorganic orthophosphate could not be detected, and the activities of two filtrates through 0.45 and 5 μm pore size filters were at almost equal level. There was no detectable activity associated with small particles such as bacteria which passed through the 5 μm pore size filter. During the survey period except for 15 July, which was just after heavy rain, the relationship between total AP activity and inorganic orthophosphate content in the samples was found to be significantly correlated, with a linear inverse relationship.

Laboratory culture

In the case where inorganic phosphate had been supplied at 3 μM as the only P source, the activity per cell started increasing rapidly after the 9th day of culture, when inorganic phosphate was consumed down to 0.1 μM (FIG. 3A). Internal phosphorus in the cell remained at very high level from the start of culture, while *G. nagasakiense* was exponentially growing and inorganic phosphate remained in the culture medium. By the 9th day of culture, growth stopped and internal phosphorus in the cells had declined to (350-380) x 10^{-15} g at./cell.

The culture experiments were conducted with the media into which organic phosphate monoester, sodium b glycerophosphate had been supplied at 10 μM as the only P source (FIG. 2B). The AP activity/l could be detected

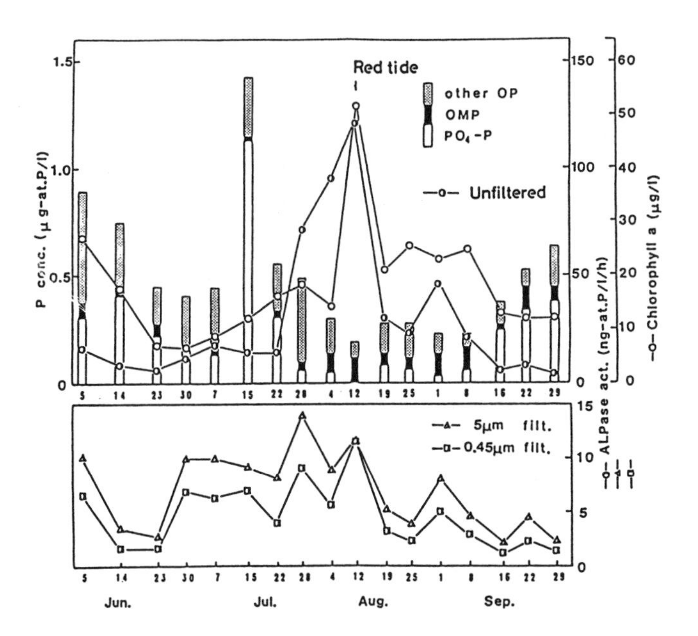

FIG. 2. Changes in alkaline phosphatase activity associated with three parts, chlorophyll a and various forms of phosphorus in Nagasaki Bay, 1982. OMP, organic phosphate monoester.

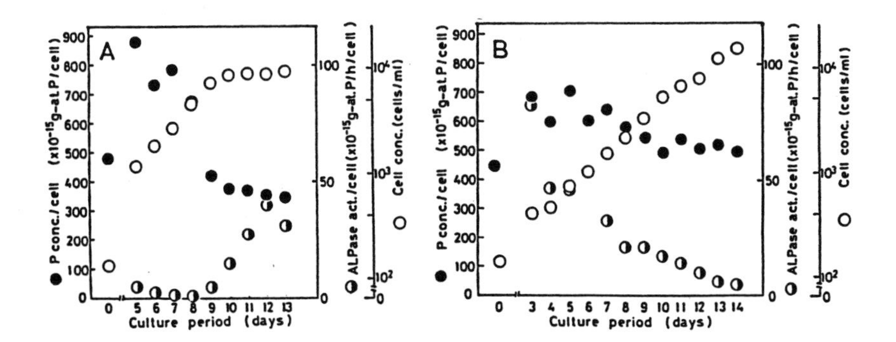

FIG. 3. Change in Gymnodinium cell density, alkaline phosphatase activity/cell and internal phosphorus in a cell of G. nagasakiense in axenic culture. A, 3 μM orthophosphate; B, 10 μM sodium b glycerophosphate as the only P source.

just after the start of culture, and gradually increased with exponential growth. However, the activity per Gymnodinium cell gradually decreased with the growth. Internal phosphorus content in a cell was maintained almost constant at (400-600) x 10^{-15} g at./cell during exponential growth stage. It appears that during exponential growth, G. nagasakiense utilized phosphomonoesters as a P source by means of the Alkaline phosphatase it produced itself. The daily change in AP activity per cell suggested that Gymnodinium density and/or organic phosphorus content present in medium had some correlation with the activity per cell. We tested these relationships separately by means of experimental culture work. We found that the higher the cell density is, the lower the AP activity per cell is, while the activity has no relationship with the glycerophosphate content.

DISCUSSION AND CONCLUSION

The field survey on Nagasaki Bay revealed that the alkaline phosphatase activity associated with some red tide organisms is related to outbreak of red tides by allowing further multiplication after exhaustion of the inorganic phosphate in environmental seawater.

In axenically cultures, G. nagasakiense consumes and stores excess phosphate in the cells as luxury uptake. If the inorganic phosphate is exhausted from the medium, stored P in the cells is consumed for further multiplication, and then AP activity becomes detectable. If the medium contains phosphate monoester as the P source, G. nagasakiense can further grow exponentially by uptake of phosphorus from the monoester by means of alkaline phosphatase now produced by the Gymnodinium cells. The activity per cell decreases with increase in cell density, while the internal phosphorus content per cell remains almost constant at (400-600) x 10^{-15} g at./cell. This represents a minimum limit for its normal division rate during the exponential growth phase. When the growth stops from exhaustion of both inorganic and organic P sources in the medium, the alkaline phosphatase activity per cell rapidly increases, while internal P content per cell remains at only (350-380) x 10^{-15} g at./cell. This P level is considered to be the minimum limit for G. nagasakiense to keep the cell alive.

REFERENCES

1. E. J. Kuenzler, J. Phycol. 1, 156-164 (1965).
2. N. Taga and H. Kobori, Mar. Biol. 49, 223-229 (1978).
3. M. J. Perry, Mar. Biol. 15, 113-119 (1972).

^{31}P-NMR STUDY OF POLYPHOSPHATE METABOLISM ASSOCIATED WITH DIEL VERTICAL MIGRATION BY *HETEROSIGMA AKASHIWO* UNDER SALINITY AND PHOSPHATE STRATIFICATIONS

MASATAKA WATANABE,* KUNIO KOHATA,* AND M.KUNUGI
*Laboratory of Marine Environment,
The National Institute for Environmental Studies,
Onogawa 16-2, Tsukuba, Ibaraki 305, Japan

ABSTRACT

A large scale axenic culture tank (2 m high by 1 m diameter) was used to observe diel vertical migration (DVM) of *H. akashiwo*. Vertical stratification, with low salinity and low orthophosphate (P_i) concentration in the surface waters and high salinity and high P_i concentration in the bottom waters were simulated in the tank. During night this species moved to the bottom waters where P_i was rich and took up P_i rapidly which will be accumulated as polyphosphate(PP_i). During daytime this species moved to the phosphate-depleted surface waters and utilized the accumulated PP_i for photophosphorylation. The combination of DVM and PP_i accumulation gives this species a strong ecological advantage over coastal diatoms which have no migratory ability.

INTRODUCTION

Common conditions during summer in the Seto Inland Sea (Avg. depth: 30m) include heavy rainfall followed by long periods of sunny weather. Consequently, stable salinity and temperature stratifications are observed at depths of 5-10 m. Inorganic phosphate concentrations are low in the surface waters and high in the bottom waters. *Heterosigma akashiwo* exhibits clear diel vertical migration behavior, swimming approximately $1.0 m \cdot h^{-1}$ [8]. It seems likely that *H. akashiwo* migrates a depth range of the order of 10-15m during a 24 hr period and thus that it can move between the stratified layers. Although the ecological importance of diel vertical migration on the nutrition and accumulation of flagellates has been discussed many times [1-7], the detailed analysis of the phosphate metabolism associated with such migration has not been performed. In this paper, a large axenic culture tank [8] was used to simulate diel vertical migration of *H. akashiwo* under vertically stratified conditions of salinity, temperature and orthophosphate. These are analogous to conditions observed in the Seto Inland Sea in certain summer periods when red tides occur. The phosphate metabolism of *H. akashiwo* during this vertical migration was studied by ^{31}P-NMR spectroscopy of perchloric acid extracts of this species. The ecological advantages of diel vertical migration and phosphate accumulation in the phosphate-rich bottom waters are demonstrated.

EXPERIMENTS

A large culture tank (2 m high by 1 m int. diam., working volume of about 1 m^3 and an airspace of about 0.4 m^3) was used to grow *H. akashiwo* under axenic conditions [8]. The clonal axenic culture strain of *H. akashiwo*(Hada) Hada (OHE-1, isolated from Tanigawa fishing port, Osaka Bay, Japan and maintained in the Microbial Culture Collection of the NIES) was

RED TIDES: BIOLOGY, ENVIRONMENTAL SCIENCE, AND TOXICOLOGY
Okaichi, Anderson, and Nemoto, Editors

inoculated into the tank to give an initial cell concentration of 173 cells·mL^{-1}. Initial concentration of orthophosphate (P_i) was 1.0μ M and salinity was 30.0 ‰. An irradiance level was 530μ E·m^{-2}·s^{-1} on a 12:12 h LD regime and temperature was kept as 20.5°C. The 7th day after the inoculation, the cell concentration of H. akashiwo in the tank (a fully mixed condition) reached 1.4 x 10^4 cells·mL^{-1} (measured at 0900 h) in the exponential growth phase and P_i in the medium was completely depleted. The specific growth rate was μ =0.68 (day^{-1}). At 1100 h of the 8th day, H. akashiwo accumulated on the surface due to active vertical migration when aeration was stopped. Then, 100 L of the bottom water was discharged. Immediately, 100 L of the new medium, which was enriched with P_i (19μ M) in higher salinity (33.4‰), was introduced to the bottom of the tank. Salinities were 30 ‰ in the upper layer (st. 1, 2, 3, 4, in which st. 1 located at the surface) and 33.4 ‰ in the lower layer (st. 5, which located at the bottom). Initial P_i concentrations were 0 μM in the upper layer and 19μ M in the lower layer. Large salinity difference of Δs=3.4‰ prevented vertical mixing and a stable stratification of P_i was maintained throughout the experiment. Samplings were repeated alternately at 1300 h (in the surface) and at 2300 h (in the bottom) for ^{31}P-NMR and phosphate measurements [9]. For each ^{31}P-NMR measurement, 3 L of the sampled water were immediately centrifuged at 3600g for 15 min at 5 °C. The extracts were prepared by rapid mixing the cell pellet with 18 mL of ice-cold perchloric acid and by ultrasonication during PCA addition [10]. The ^{31}P-NMR spectra were recorded at 161.8 MHz observing frequency on a JEOL GX-400 NMR spectrometer operating in the Fourier Transform mode. Typically 9000 scans were accumulated for each spectrum with a pulse repetition time of 1.2 s and 45 to 60 flip angle with 16 kHz spectral width. ^{31}P-NMR spectra were recorded using 0.1 M triphenylphosphate as the external reference.

RESULTS

^{31}P-NMR spectra of metabolites associated with vertical migration and phosphate uptake at the bottom by H. akashiwo cells are summarized in Table 1. In the P starved cells in the surface waters (1300 h, the 1st day), dominant peaks are from sugar P, P_i and phosphodiester (Table 1,(I)). When the cells migrated to the phosphate-rich bottom waters (2300 h, the 1st day), the increase in the P cell quota from 53 to 95 fmol·cell^{-1} was attributed to the increase in P_i and no other intermidiates were synthesized (Table 1,(II)). Since the cells were in the P starved condition, probably no extra energy was stored to promote synthesis of the phosphate intermediates, even with orthophosphate uptake. When the cells migrated to the light-sufficient surface waters(1300 h, the 2nd day), a dramatic increase in polyphosphates (PP_i) was the most pronounced change (Table 1,(III)). Since the orthophosphate concentration in the surface waters was zero, it was suggested that accumulated P_i in the cells was utilized for photophosphorylation to generate ATP, and then PP_i were synthesized at the expense of ATP. When the cells again migrated to the phosphate-rich bottom waters, orthophosphate was taken up, which was evident from the increase in the P cell quota. However, cell quota of P_i did not increase but rather cell quota of PP_i increased from 39.7 to 52.0 fmol·cell^{-1}. Also, the average chain lengths of PP_j increased from ca. 12 to 20 phosphate residues. These facts indicate that the cells synthesized PP_i as soon as P_i was taken up at the bottom (2300 h, the 2nd day)(Table 1,(IV)).

When the cells migrated again to the upper layer (1300 h, the 3rd day)(Table 1(V)), cell quota of PP_i decreased from 52 to 40.1 fmol·cell^{-1} and the average chain lengths of PP_i decreased from ca. 20 to 14 phosphate residues. Since the orthophosphate concentration in the surface waters was zero, it was suggested that the accumulated PP_i were degradated to regulate the level of P_i, which was used for photophosphorylation.

Table 1 Variation of cell quota of P intermediates in *H. akashiwo* cells [9]

	sugar P	Pi	Phospho diester	PP	Lpp	P cell quota
(I) 13:00 1st day surface	11.6 (36.4%)	6.1 (19.2%)	4.0 (12.6%)	—	—	53
(II) 23:00 1st day bottom	16.2 (28.5%)	22.8 (40.0%)	5.0 (8.8%)	—	—	95
(III) 13:00 2nd day surface	14.4 (15.7%)	12.4 (13.5%)	6.4 (7.0%)	39.7 (43.3)	12	153
(IV) 23:00 2nd day bottom	13.6 (13.4%)	13.3 (13.1%)	6.6 (6.5%)	52.0 (51.3%)	20	169
(V) 13:00 3rd day surface	13.5 (14.9%)	14.4 (15.9%)	5.3 (5.8%)	40.1 (44.3%)	14	151

1)unit: fmol•cell^{-1} 2)Lpp : average chain length

DISCUSSION

Now several important points can be summarized; 1) that *Heterosigma akashiwo* is capable of migrating through a very sharp salinity gradient, as well as temperature gradient [8]; 2) following phosphorus starvation, *H. akashiwo* can uptake phosphate in the bottom waters at night, predominantly into the inorganic phosphate pool in the cell; 3) inorganic phosphate is converted to polyphosphates (PP_i) by increasing the chain length of PP_i; and 4) there are subsequent migrations to nutrient depleted surface waters, where internal polyphosphate pools are used for photophosphorylation.

Excessive amounts of polyphosphate were accumulated in the bottom waters at night, and therefore their formation is promoted by increased ATP synthesis by photosynthetic phosphorylation, or by the accumulation during photosynthesis of large amounts of sugars and other easily oxidisable substrates.

This experiment shows clearly that the polyphosphate fractions maintain a constant, low level of orthophosphate in the cell(Table 1). The accumulation of significant amounts of orthophosphate in the cell would result in a considerable change in its osmotic pressure and pH. The accumulation of polyphosphates appears to be an efficient means of detoxifying free orthophosphate [11].

ACKNOWLEDGEMENT

The authors wish to thank Mr. M. Yoshida of Ibaraki Environmental Technical Center for skilled technical assistance.

REFERENCES

1. R.W. Eppley, O. Holm-Hansen, and J.D.H. Strickland, J. Phycol. 4,330-340 (1968).
2. R.W. Eppley, and W.G. Harrison, in: Toxic Dinoflagellate Blooms, V.R. LoCicero, ed. (Mass. Sci. Tech. Found. Wakefield 1975) pp. 11-22.

3. D. Kamykowski, and S.J. Zentara, Limnol. Oceanogr. 22, 148-151 (1977).
4. S.I. Heaney, and R.W. Eppley, J. Plankton Res. 3, 331-344 (1981).
5. D. Kamykowski, Mar. Biol. (Berl.) 62, 57-64 (1981).
6. J.J. Cullen, and S.G. Horrigan, Mar. Biol. (Berl.) 62, 81-89 (1981).
7. J.J. Cullen, in: Migration: Mechanisms and Adaptive Significance, M.A. Rankin, ed. (Supplement to contributions in Mar. Sci. 1985) 27, 135-152.
8. K. Kohata, and M. Watanabe, J. Exp. Mar. Biol. Ecol. 100, 209-224 (1986).
9. M. Watanabe, K. Kohata, and M. Kunugi, J. Phycol. 24, 22-28 (1988).
10. M. Watanabe, K. Kohata, and M. Kunugi, J. Phycol. 23, 54-62 (1987).
11. I.S. Kulaev, in: The Biochemistry of Inorganic Polyphosphates, (John Wiley & Sons., Chichester 1979) pp. 255.

ELECTROPHORETIC ANALYSIS OF ISOZYMES IN RED TIDE DINOFLAGELLATES (*GYMNODINIUM NAGASAKIENSE, PROTOGONYAULAX CATENELLA, & PERIDINIUM BIPES*)

YOSHIHIKO SAKO, ARITSUNE UCHIDA, AND YUZABURO ISHIDA
Faculty of Agriculture, Kyoto University, Kyoto 606, Japan

ABSTRACT

Isozyme patterns of red tide dinoflagellates were examined to distinguish these organisms at the species level or lower. Isolates of *P. bipes*, *G. nagasakiense*, and *P. catenella* from many different regions were subjected to polyacrylamide gel electrophoresis (PAGE) and banding patterns were examined for MDH, LDH, EST, ME, TO, ACP and NADP(R). As a result there were obvious differences in isozyme patterns under various culture conditions in *P. bipes* and especially complicated differences between isolates of *P. catenella*. On the contrary, there were few differences in all strains and culture conditions in *G. nagasakiense*. Isozyme analyses of MDH and EST of *P. bipes* and *G. nagasakiense* were very useful for determining the differences of locality and population. On the other hand, ME and TO showed a high similarity value under various conditions in all strains. This suggests that *P. bipes* has a higher level of genetic variation than *G. nagasakiense* and that *P. catenella* reveals a quite higher degree of genetic polymorphism than other dinoflagellates.

INTRODUCTION

The freshwater dinoflagellates belonging to the genus *Peridinium* bloom regularly in many reservoirs in western Japan, sometimes with tremendous cell densities in chocolate color waters [1]. In coastal regions, the marine dinoflagellate *Gymnodinium nagasakiense* causes large scale red tides every summer and *Protogonyaulax catenella* is the primary cause of paralytic shellfish poisoning.

In our recent work with clonal cultures in synthetic media, we made clear the life cycle and environmental conditions which induce encystment and excystment in several species of genus *Peridinium* [1-5] and *Protogonyaulax catenella* [1]. However, taxonomy based on the morphology of dinoflagellates remains controversial and confusing because some morphological features happen to vary, perhaps with environmental conditions. In order to distinguish these dinoflagellates at the species level, or lower, and examine their genetic variability, we have used electrophoretic analysis of isozymes.

EXPERIMENTAL

Organisms

Field samples of *Peridinium* were collected by plankton net from the blooms in four different reservoirs (Kamiji, Kuroda, Iwaya, Takihata), because such blooms were typically unialgal (99 % <). *P. bipes* was isolated from Ananai Reservoir in Kochi Prefecture and Kamiji Reservoir in Mie Prefecture. *G. nagasakiense* was isolated from Harima-Nada, Omura Bay, Suo-Nada and Gokasho Bay. Wild mating types (+,-) of *P. catenella* were the kind gift of Mr. S. Yoshimatsu of the Akashiwo Research Institute, Kagawa Prefecture. Using cross experiments, F1 strain mating types were isolated from

RED TIDES: BIOLOGY, ENVIRONMENTAL SCIENCE, AND TOXICOLOGY
Okaichi, Anderson, and Nemoto, Editors

germinated cells. All cultures used in the experiments were clonal and *P. bipes* and *G.nagasakiense* were bacteria free.

Culture and harvest

P. bipes was grown in MW1 medium [2]. *G. nagasakiense* was grown in MES, f/2, NH15 and ASP-7 media. *P. catenella* was grown in ESJ (ES + Jamarin S). Standard cultures of all species were maintained at 20℃ on 14 : 10 h light-dark cycle at 5,000lux. Culture conditions, for example temperature, light and nutrient, were modified by design. Cells were harvested at late exponential, stationary phase, and as cyst (only *P.bipes*) by centrifugation at 4000xg for 5min at 4℃,then washed with buffer A (20mM Tris-HCl,1mM EDTA, 5mM β ME, PH7.5) and stored at -90℃.

Enzyme extraction and electrophoresis

Cells were suspended in cold buffer A and disrupted by sonication at 0-4℃. The sonicates were centrifuged at 15,000xg for 30min. The supernatant was applied to the gels. Electrophoresis was carried out with 7.5% poly-acrylamide gels as described by Gabriel [6]. Gels were stained for the following enzymes : NAD-dependent malate dehydrogenase (MDH), lactate dehydrogenase (LDH), esterase (EST), NADP-dependent malic enzyme (ME), tetrazolium oxidase (TO), acid phosphatase (ACP), NAD(P) reductase (NAD(P)R). The migration distance for each band was converted into a relative Rf value by reference to a standard strain.

RESULTS AND DISCUSSION

Each isozyme band was scored as a phenotype and zymogram band patterns were compared pairwise by use of the similarity coefficient of Jaccard [7]. We summarized linkage dendrograms using unweighted pair-group arithmetic average. Based on the coincidence of isozyme patterns of all species, multiple molecular forms of all enzymes were identified.

In *P. bipes*, there were many common bands, especially in the case of TO and NAD(P)R. However, isozyme patterns of MDH and EST were different depending on population and culture conditions. The dendrogram showed that similarity values of isozymes from each strain formed clusters regardless of culture conditions, although similarity values were not so high. All field samples formed a cluster different from that of the cultured samples, regardless of different source locations. The average similarity values of

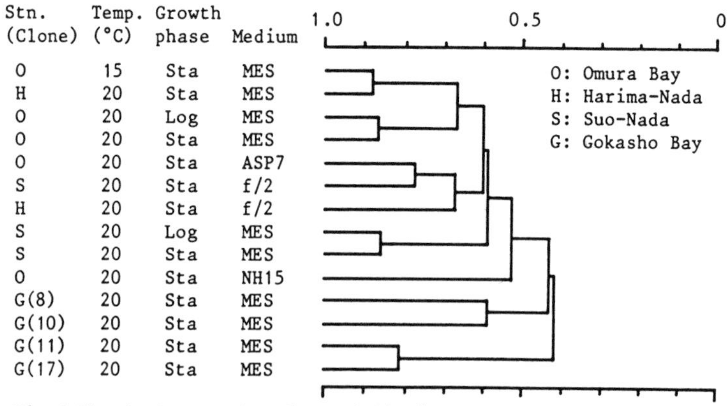

Stn. (Clone)	Temp. (°C)	Growth phase	Medium
O	15	Sta	MES
H	20	Sta	MES
O	20	Log	MES
O	20	Sta	MES
O	20	Sta	ASP7
S	20	Sta	f/2
H	20	Sta	f/2
S	20	Log	MES
S	20	Sta	MES
O	20	Sta	NH15
G(8)	20	Sta	MES
G(10)	20	Sta	MES
G(11)	20	Sta	MES
G(17)	20	Sta	MES

1.0 0.5 0

O: Omura Bay
H: Harima-Nada
S: Suo-Nada
G: Gokasho Bay

Fig. 1 The dendrogram based on similarity values of Malate Dehydrogenase of *Gymnodinium nagasakiense*.

each strain in *P. bipes* were higher than those of *P. willei* [8].

In *G. nagasakiense*, the isozyme patterns of MDH and EST showed some differences depending on locality and culture conditions. The similarity values of MDH showed clearly the difference of locality since strains from the same region were clustered together (Fig.1). On the other hand, isozyme patterns of ME and TO were nearly the same in all strains regardless of culture conditions. As a result, similarity values of TO (Fig.2) and ME in all strains were very high regardless of culture conditions. The similarity values of all strains of *G. nagasakiense* were higher than those of *P. bipes*.

However, similarity values between isolates of *P. catenella* were lower than those of other two species studied. Both parent strains and their F1 strains which were mating type + or - showed very different isozyme patterns.

Stn. (Clone)	Temp. (°c)	Growth phase	Medium
O	15	Sta	MES
O	20	Sta	MES
O	20	Sta	MES
O	20	Sta	NH15
O	20	Sta	ASP7
S	20	Log	MES
S	20	Sta	MES
H	20	Sta	f/2
H	20	Sta	MES
G(8)	20	Sta	MES
G(10)	20	Sta	MES
G(11)	20	Sta	MES
S	20	Sta	f/2
G(17)	20	Sta	MES

1.0 0.5 0

O: Omura Bay
H: Harima-Nada
S: Suo-Nada
G: Gokasho Bay

Fig. 2 The dendrogram based on similarity values of Tetrazolium Oxidase of *Gymnodinium nagasakiense*.

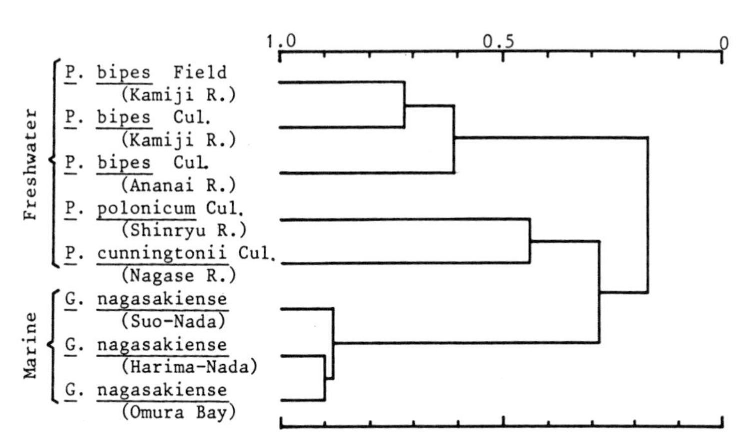

Fig. 3 Summarized dendrogram based on similarity values in the typical electrophoretic isozyme patterns (MDH, TO, NAD(P)R) of *Peridinium* spp. and *G. nagasakiense*.

This low degree of similarity among *P. catenella* was similar to the observation of the *P. tamarensis/catenella* species complex from Northwest Canada and the United States [9].

Figure 3 gives a summarized dendrogram based on 3 enzymes of several species of *Peridinium* and *G. nagasakiense*. The strains of *P. bipes* and *G. nagasakiense* formed clusters respectively. Different species, which were clearly distinguished by morphological observation show very low similarity values. This study suggested that *P.bipes* had a higher level of genetic variation than *G. nagasakiense*. P. catenella reveals the highest level of genetic variation.

REFERENCES

1. Y. Sako and Y. Ishida, Microbial Ecology 14. Gakkai Schuppan Center, p. 99-113, (1986).
2. Y. Sako, Y. Ishida, H. Kadota and Y. Hata, Bull. Japan. Soc. Sci. Fish., 50, 743, (1984).
3. Y. Sako, Y. Ishida, H. Kadota and Y. Hata, Bull. Japan. Soc. Sci. Fish., 51, 267, (1985).
4. Y. Sako, M. Nakanishi, T. Konda, Y. Ishida, H. Kadota, K. Shrestha, H. R. Bhandary and R. L. Shrestha, Bull. Jpn. Soc. Microb. Ecol., 1, 19, (1986).
5. Y. Sako, Y. Ishida, T. Nishijima and Y. Hata, Nippon Suisan Gakkaishi, 53, 473, (1987).
6. O. Gabriel, Methods Enzymol., Vol.22, p.565, (1971).
7. P.H.H. Sneath and R.R. Sokal, Numerical taxonomy p.1, Freeman (1973).
8. B.A. Hayhome and L.A. Pfiester, Amer. J. Bot., 70, 1165, (1983).
9. A.D. Cembella and F.J.R. Taylor, Biochem. System Ecol., 14, 311, (1986).

DIEL CHANGES IN THE COMPOSITION OF PHOTOSYNTHETIC PIGMENTS IN CHATTONELLA ANTIQUA AND HETEROSIGMA AKASHIWO (RAPHIDOPHYCEAE)

KUNIO KOHATA,* AND MASATAKA WATANABE*
*Laboratory of Marine Environment,
National Institute for Environmental Studies,
Onogawa 16-2, Tsukuba, Ibaraki 305, Japan

ABSTRACT

Pigment analyses by HPLC show clear diel periodicity of cellular Chl *a* and *c*, and carotenoid composition in *C. antiqua* and *H. akashiwo* grown on a 12:12 LD cycle. Cellular pigment contents increased generally during the light period and decreased during the dark period with increase of cell concentration by phased cell division. Fucoxanthin, violaxanthin, β-carotene were observed as the dominant carotenoids of both the species. Cellular carbon and nitrogen showed similar diel periodicity to pigments. The cellular contents of Chl *a*, fucoxanthin and carbon increased in a parallel manner during the light period. However, the increase of violaxanthin was restricted to only a few hours at the beginning of the light period both in *C. antiqua* and *H. akashiwo*. The ratio of violaxanthin to Chl *a* showed diel periodicity in the contrary phase to that of fucoxanthin. This suggested a different function of violaxanthin from that of fucoxanthin which was considered as a light-harvesting pigment.

INTRODUCTION

Many unicellular algae undergo phased cell division under synchronized conditions [1]. During a cell division cycle, it has been reported that cellular contents such as C, N and photosynthetic pigments showed clear diel changes in *Heterosigma akashiwo* [2] and in *Chattonella antiqua* [3] grown on a 12:12 h light:dark cycle. Cellular carbon and pigment contents increased during the light period and decreased during the dark period with an increase of cell concentration by phased cell division. In the present study, detailed analyses were performed regarding the diel change in carotenoid composition of both species.

EXPERIMENTAL

Clonal axenic cultured strains of *Chattonella antiqua* (Hada) Ono and *Heterosigma akashiwo* (Hada) Hada (NIES-1 and OHE-1, respectively, the Microbial Culture Collection of NIES) were grown in the NIES tank [4] filled with 1 m^3 of f/2 medium [5]. Algal cultures used to inoculate the tank were prepared in the same way as those previously reported [4]. Before the experiments, the tank and filters for air and water were sterilized with steam. Illumination was provided by a 5 kW xenon lamp to give an average quantum flux density of 565 $\mu E \cdot m^{-2} \cdot s^{-1}$ over the water surface in the tank. The light period was controlled by an electric timer providing a 12:12 h LD cycle; lights were turned on at 6:00 and off at 18:00. Mixing conditions were maintained by introducing sterilized air into the bottom of the tank throughout the experiments. Water temperature in the tank was kept constant at 20 ± 1 °C.

When the cell concentration reached ca. 300 $cells \cdot mL^{-1}$ for *C. antiqua*

RED TIDES: BIOLOGY, ENVIRONMENTAL SCIENCE, AND TOXICOLOGY
Okaichi, Anderson, and Nemoto, Editors

and ca. 3100 cells·mL^{-1} for H. akashiwo in the exponential growth phase, intensive measurements were started and continued for 72 h at 3 h intervals (the intensive measurement period = the IM period). Cell concentrations and mean cell volumes were measured using a cell counting system which included a Coulter Counter model TA-II. Particulate carbon and nitrogen on the Whatman GF/C 47 mm glass fiber filters were measured by elemental analyzer (MT-3, Yanaco, Japan). For reversed-phase liquid chromatographic analysis of chlorophyll(Chl) and carotenoids, 200-500 mL of sea water was filtered through a 47 mm Whatman GF/C with 2mL of $MgCO_3$ solution. Analytical conditions for HPLC are described in Fig. 1 legend. Visible spectroscopy was also applied to pigments scraped off from reversed-phase TLC plates (Whatman, C-8) developed with aqueous methanol to confirm pigment identification [3].

RESULTS AND DISCUSSION

Figure 1 shows a typical HPLC chromatogram of each C. antiqua and H. akashiwo. Dominant carotenoids for the two species were found to be fucoxanthin, violaxanthin, and β-carotene. Chlorophylls a and c were also found in the two species as well as trace amounts of chlorophyllide a and isomers of Chl a. Small peaks Nos. 4 and 5 of H. akashiwo are probably antheraxanthin and zeaxanthin, respectively. These two xanthophylls were not observed in C. antiqua under this experimental condition. The specific growth rate μ(ln unit of increase·day^{-1}) of C. antiqua was 0.67 day^{-1} at the beginning of the exponential growth phase and 0.45 day^{-1} during the IM period. The rate of H. akashiwo was observed 0.80 day^{-1} and 0.62 day^{-1}, at the beginning and during the IM period, respectively. Irradiance level in the tank was reduced by self-shading of algal cells at the end of culture experiments. This reduced light condition affected the growth rate of C. antiqua. The reversible epoxization of zeaxanthin via antheraxanthin to violaxanthin has been investigated with regard to light-induced conversions of the xanthophylls. [7,8] The growth under light-limited conditions may account for the absence of antheraxanthin and zeaxanthin in C. antiqua.

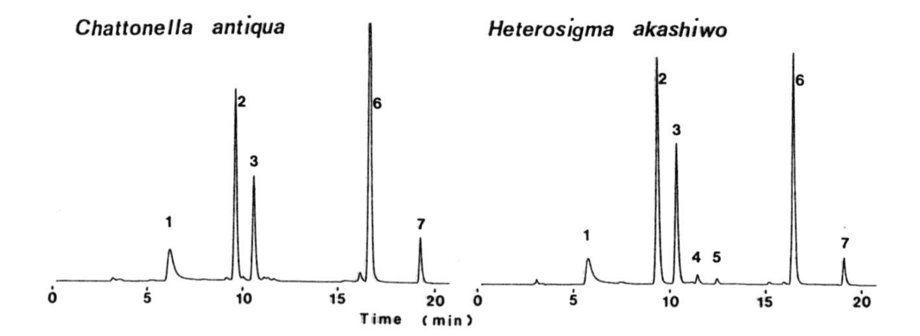

FIG. 1. Typical HPLC chromatogram from absorbance at 440 nm. An ODS column (5 μm) ODS-3, Whatman) was used with linear gradient elution (Shimadzu, LC-6A) of solvent A (ion-pair solution [6]: water: methanol, 5: 5:90) to solvent B (ethyl acetate), over a 20 min period. Flow rate was held at 1.0 mL·min^{-1}. Peak identities are (1) Chl c, (2) fucoxanthin, (3) violaxanthin, (6) Chl a, (7) β-carotene.

Cellular C, N and pigments increased during the light.

These cellular contents were observed to be minimum at beginning of the light period and maximum at start of the dark for C. antiqua (FIG. 2). Almost the same diel patterns as those for C. antiqua were observed in the cellular contents of H. akashiwo. Ratios between maximum and minimum levels of carbon and Chl a, which were obtained consecutively for three division cycles, behave in the same manner as those for cell concentrations. Hence, specific growth rate in the field should be estimated directly from diel changes of the ratios. The mean cell volume of H. akashiwo oscillated according to growth rates. However, the ratio for mean cell volume in C. antiqua was higher than that expected from daily growth rate. This indicated that mean cell volume in C. antiqua reflected not only the maturation of carbon assimilation of cells but also other factors, such as regulation of cell shape and density by control of osmotic pressure.

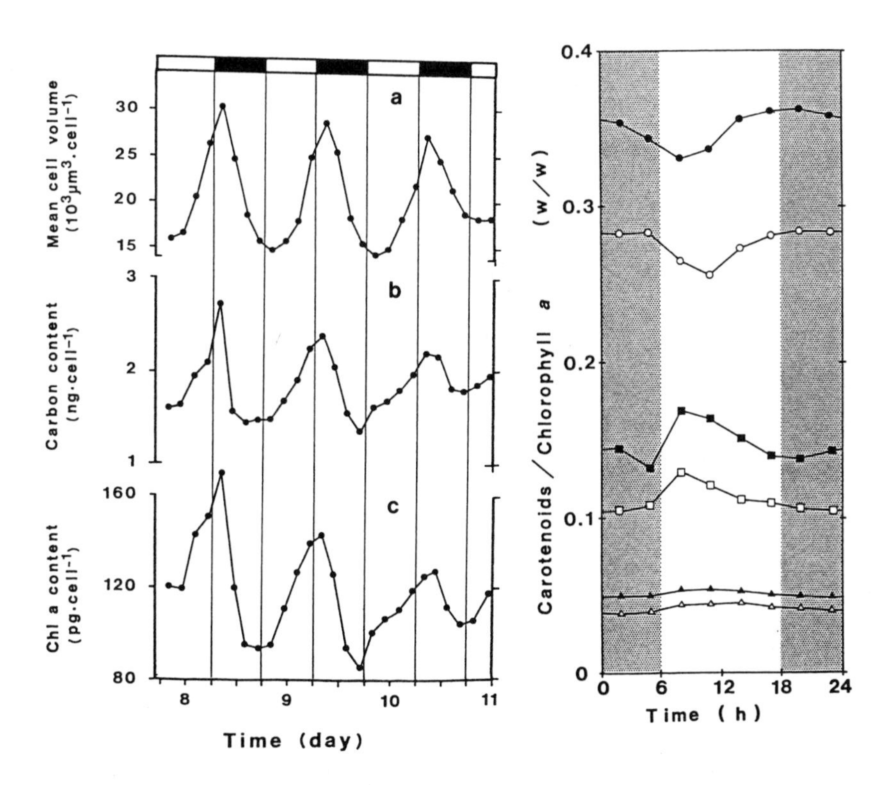

FIG. 2(a)-(c). (Left) Diel changes in cellular components of C. antiqua grown on a 12:12 h LD cycle [Ref. 3]. Black bars at the top of the figure denote the dark period. Data obtained at the three levels in the NIES tank during the IM period (see Text) have been averaged to give one trace.

FIG. 3. (Right) Diel changes in the carotenoids/Chl a (w/w) ratio of C. antiqua (Open marks) and of H. akashiwo (Solid marks). Carotenoids are (●, ○) fucoxanthin; (■,□) violaxanthin; (▲,△) β-carotene.

Excellent correlation was observed between cellular carbon and Chl a.

Quantitative relationship between cellular carbon and Chl a content was observed to be r^2=0.988 (n=81) in C. antiqua [3] and r^2=0.995 (n=75) in H. akashiwo. The carbon/Chl a ratio (w/w) was 18.4 and 19.1 in C. antiqua and H. akashiwo, respectively. The excellent correlation indicated that the ratio of carbon/Chl a did not change significantly during cell division cycles. It appears that Chl a concentration can be converted to biomass and also to cell concentration. This conversion should have ecological applications , but nutrients and growth condition must be considered.

Violaxanthin was produced for only a few hours.

The synthesis rates of pigments were calculated by differentiation as function of time (1/C dC/dt). During the light period, all pigments increased, but clearly in different ways. Violaxanthin was produced for only a few hours, starting at the beginning of the light period, and remained constant in amount thereafter. Carotenoids are elements of the carotenoid cycle in oxygen evolution [8] and are light-harvesting components of chloroplasts. They are important to photosynthetic electron transport and the protection of photosynthetic tissue against photosensitized oxidation. Each carotenoid component in C. antiqua and H. akashiwo was synthesized at different stages of the cell division cycle. This suggests that the components do not have simultaneous multiple functions but rather that each has at least one specific function. The light-harvesting function of fucoxanthin has been confirmed by direct energy transport measured by fluorescence excitation and emission spectra in isolated pigment-protein complexes [9]. Pattern similarity between fucoxanthin and Chl c indicates fucoxanthin may exist as the fucoxanthin-Chl a and c complex, which is generally considered to be a light-harvesting complex.

The pigment-ratio violaxanthin/Chl a was in a reversed phase to that of fucoxanthin/Chl a (FIG. 3). Thus, these two xanthophylls may be synthesized by different biosynthetic processes and be included in separate pigment-protein complexes. Violaxanthin, a member of the carotenoid cycle mentioned above, may act as a photoprotective rather than a light-harvesting pigment. A sufficient amount of light-protectant appears to be synthesized at the beginning of the light period when cell division is complete.

ACKNOWLEDGEMENTS

The authors wish to thank Mr. M. Yoshida, Ibaraki Environmental Technical Center, and Ms. M. Kato for their technical assistance.

REFERENCES

1. eg., S.W.Chisholm in: Physiological Bases of Phytoplankton Ecology, T. Platt, ed. (Can. Bull. Fish. Aquat. Sci 1981) 210, pp. 150-81.
2. K. Kohata and M. Watanabe, Res. Rep. Natl. Inst. Environ. Stud. 80, 23-32 (1985) (in Japanese).
3. K. Kohata and M. Watanabe, J. Phycol. 24, 58-66 (1988).
4. K. Kohata and M. Watanabe, J. Exp. Mar. Biol. Ecol. 100, 209-224 (1986).
5. R.R.L. Guillard and J.H. Ryther, Can. J. Microbiol. 8, 229-239 (1962).
6. R.F.C. Mantoura and C.A. Llewellyn, Anal. Chim. Acta 151, 297-314 (1983).
7. A. Hager, Ber. Deutsh. Bot. Ges. 88, 27-44 (1975).
8. S. Liaaen-Jensen in: Marine Natural Products, P.J.Scheuer ed. 2, 1-73 (1978).
9. eg., T.G. Owens and E.R. Wold, Plant Physiol. 80, 732-738 (1986).

UV-ABSORBING PIGMENTS IN THE DINOFLAGELLATES ALEXANDRIUM EXCAVATUM AND PROROCENTRUM MICANS. EFFECTS OF LIGHT INTENSITY

CARRETO, J.I., S.G. DE MARCO AND V.A. LUTZ
Instituto Nacional de Investigación y Desarrollo Pesquero,
P.O. Box 175, 7600 - Mar del Plata, Argentina

ABSTRACT

The concentration of UV-absorbing pigments in *Alexandrium excavatum* and *Prorocentrum micans* cultures depends on the light intensity of growth (PAR). The greatest concentrations are observed at high light intensities. Changes in irradiance from low (20 μEin m^{-2} s^{-1}) to high (250 μEin m^{-2} s^{-1}) light intensity and viceversa induce a rapid and reversible response (hours for completion). Synthesis regulation on these UV photoprotector substances is interpreted as a genetic adaptation that confers these organisms a competitive advantage at high light intensity and short wavelengths. These conditions seem to prevail during the initiation and early development of near surface blooms.

INTRODUCTION

The migration patterns of various dinoflagellates species have shown that in abundant nutrient conditions, cells accumulate at relatively high irradiance levels (1,2). In these conditions, they are also exposed to the deleterious effect of UV radiation (3,4,5). Some red tide dinoflagellates show "in vivo" absorption spectra characterized by the rapid absorption increase in the near UV region (6,7,8). Although the nature of the substances that cause this absorption is unknown, they are related to those observed by Shibata (9) in cyanobacteria and symbiotic dinoflagellates. Their photoprotector filtering effect has been noted for some tropical phytoplankton communities (5), symbiotic dinoflagellates (9,10) and some algal communities associated to coral reefs (11). Therefore, it can be suggested that their presence in dinoflagellates, as well as in other near surface blooms species, is an adaptive mechanism to high light intensities and short wavelengths.

EXPERIMENTAL

Cultures of *P. micans* and *A. excavatum* were isolated in the frontal area off Pla. Valdés (12). Cultures were grown in F/2 medium without silicon addition, alternating 12:12 light/dark cycle at 18°C, and maintained in sufficient nutrient conditions at rather constant cell density (5 x 10^3 cell/ml) by periodic addition of fresh medium. Cells grown at 250 μEin $m^{-2}s^{-1}$ (high) and 20 μEin $m^{-2}s^{-1}$ (low) were used. In the first experiment, cultures were transferred from low to high light intensity and viceversa. In another experiment, aliquots of the low light culture were transferred to various higher light intensities, using a suitable incubator. Light intensities were measured with a LI-185 Lambda-Licor. Cells were counted with a Coulter Counter TA II. "In vivo" absorption spectra were made in a manner similar to that described by Yentsch (13), using GF/C glass fiber filters. The absorption spectra of 90% acetone extracts were recorded, and chlorophyll concentrations were calculated (14). Carotenoids percent composition was determined after TLC on cellulose. In all cases a Shimadzu 210 A dual beam spectrophotometer was used.

RED TIDES: BIOLOGY, ENVIRONMENTAL SCIENCE, AND TOXICOLOGY
Okaichi, Anderson, and Nemoto, Editors

RESULTS

"In vivo" absorption spectra of A. excavatum and P. micans (Fig 1) show that low light adapted cells present a characteristic spectrum of peridinin-containing photosynthetic dinoflagellates (8). In addition, high light adapted cells present a sharp absorption increase in the near UV region. Furthermore, in the acetonic extracts, an absorption maximum between 365-342 nm (Fig 2) shows that high light cells have high concentrations of additional pigments which strongly absorb in the UV region. Other minor differences in the A 435/A 672, A 490/A 672 and A 535/A 672 absorption ratios are also evident (Fig 1). Chlorophyll cell contents are higher in low light cells. However, at high light intensity total carotenoids/chlorophyll a ratio is higher. These differences are accompanied by important variations in the carotenoid composition. The highest peridinin percentages are observed at low light intensities, diadinoxanthin percentage being higher at high light intensities (Table 1). This result agrees with the above mentioned relative increase of photoprotector carotenoids in high light conditions (15,16). Changes from low to high light intensity (Fig 3) result in a rapid increase of the A 490/A 672 "in vivo" absorption ratio, and in the content of UV absorbing pigments (A 365/A 672). The time needed to reach the steady state of the process is below the cell division time. The response is partially reversible, the process being induced by the high-to-low light transfer (Fig 3). The magnitude of the observed change at constant time (6 h) depends on the light intensity to which the low light cells are transferred. The observed response (Fig 4) shows that after an initial range without evident changes, the A 365/A 663 and A 476/A 663 ratios linearly increase with the light intensity, to reach a plateau near the 700 µEin $m^{-2}s^{-1}$. The increase

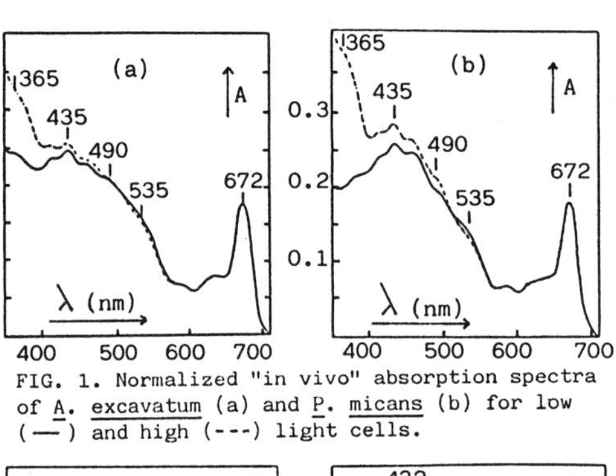

FIG. 1. Normalized "in vivo" absorption spectra of A. excavatum (a) and P. micans (b) for low (—) and high (---) light cells.

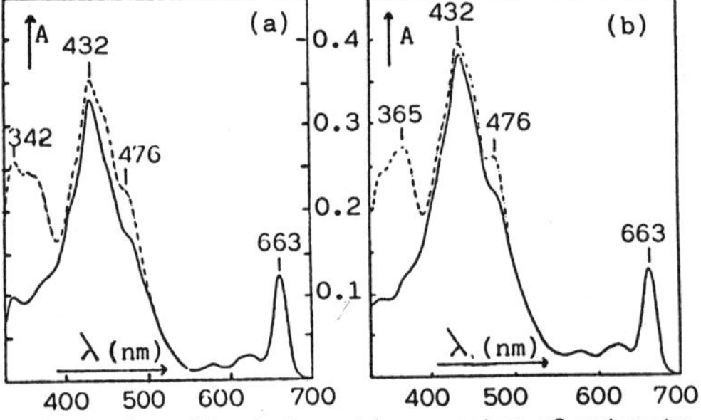

FIG. 2. Normalized absorption spectra of extracts (90% acetone) of A. excavatum (a) and P. micans (b) for low (—) and high (---) light cells.

TABLE I. Pigment composition and absorbance ratios of A. excavatum and P. micans for low (L.L.) and high (H.L.) light cells. a) Diadino + dinoxanthin.

		Chl. a pg $cell^{-1}$	A 476 / A 663	Peridinin (%)	Diadinoxanthin (%)a)	β-carotene (%)
A. excavatum	L.L.	64.1	1.43	83.0	13.0	4.0
	H.L.	51.1	1.85	69.0	27.2	3.8
P. micans	L.L.	42.7	1.71	82.0	15.3	2.8
	H.L.	36.0	2.01	69.0	25.1	5.9

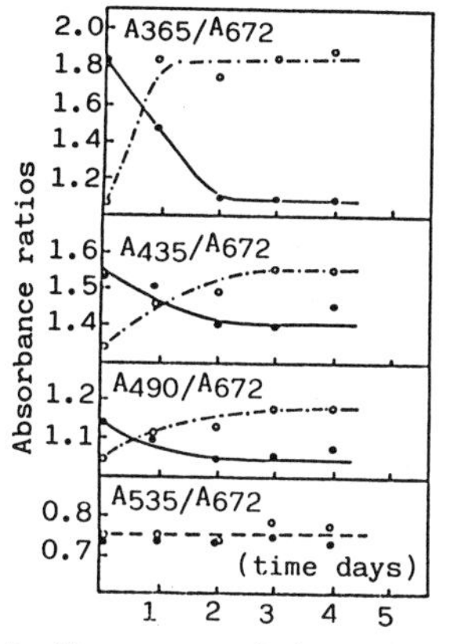

FIG. 3. Time course of the transfer of P. micans from low to high light intensities (O) and viceversa (●).

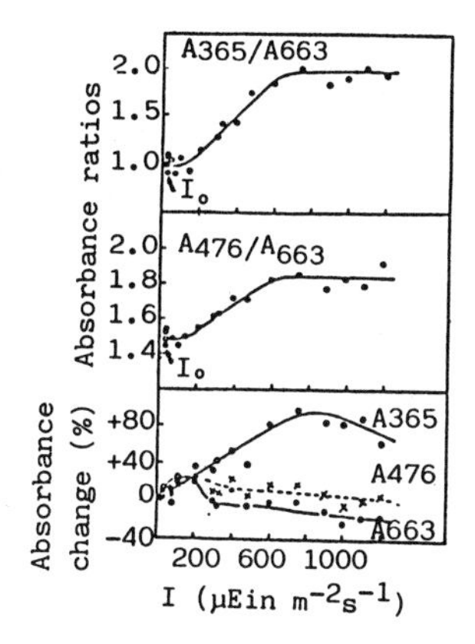

FIG. 4. Response of A. excavatum low cells (I_0= 20 μEin $m^{-2}s^{-1}$) to different light levels.

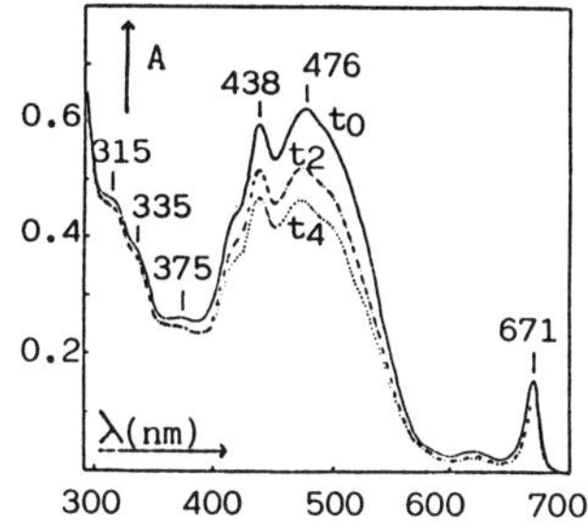

FIG. 5. Photooxidation of PCP of P. micans at 1200 μEin $m^{-2}s^{-1}$ (O): 476nm; (●): 438 nm.

in the A 365/A 663 ratio is a result of the increment in the synthesis of the UV absorbing pigments, rather than of the decrease of the chlorophyll a cell contents (Fig 4). Finally, "in vitro" photooxidation of the unpurified peridinin-chlorophyll a-protein complex (PCP) isolated from P. micans, is not accompanied by formation of UV absorbing pigments (Fig 5). This result indicates that the synthesis of these pigments is the metabolic product of a well regulated enzymatic system that includes a photochemical step, rather than a consequence of a non-specific photooxidation.

DISCUSSION

Dinoflagellates are considered among the most primitive eukaryotes (17) upon which the ozone layer might have not acted as an effective filter to the high incident UV radiation. Their motility allowed them to select the depth of the water column to filter most of the lethal UV radiation. This strategy significantly reduces PAR availability and requires a very efficient energy antennae system (18). This seems to be a characteristic strategy of dinoflagellates, which present relatively high photosynthesis and growth rates at low light intensities (18). However, some of these species, e.g. Amphidinium carterae, present photoinhibition at very low light intensities (18) and lack near UV-absorbing pigments (Haxo,F.T.,pers. comm.). An alternative strategy could have been the synthesis of selective filters to UV radiation, like those present in other primitive algal groups (9,19). Some red tide dinoflagellates present similar substances (6,7,8), and our results indicate their functional significance. In A. excavatum and P. micans, the abundance of these pigments is regulated by the irradiance levels (PAR) and accompanied by a parallel response of the photoprotector/photocollector carotenoids ratio (15,16). The response time is also consistent with the rapid changes of intensity and spectral composition of light during their vertical migration. The migration patterns of various red tide dinoflagellate species have shown that in abundant nutrient conditions, cells accumulate at relatively high light intensities (1, 2,20). In short term experiments, at high natural radiation levels, about 50% of the observed photoinhibition is caused by UV radiation (3). However, the growth photoinhibition seems to be almost entirely produced by UV radiation (5). The wide range of adaptive responses to UV radiation observed in microalgae seems to be directly related to the normal environmental conditions in which the species grows (5,10). The coastal water species of high or medium latitudes, show little or no resistence to an increased dose of UV radiation (4,21). In those environments, the incident near surface radiation and the penetration of UV radiation is weak. One of the most conspicuous characteristics that accompanies red tide development is high insolation and calm periods (20,22). Although there are no available data, it is possible that with those meteorological conditions, the increase of UV radiation, in combination with other factors (nutrients and stability), may alter the specific composition of the phytoplanktonic community (4,5), thus favouring the development of the best adapted species. This hypothesis -although speculative- deserves to be studied, considering the increment of these red tide phenomena (12) and their possible relation to climatic changes observed in the planet (23).

REFERENCES

1. J.J. Cullen and S.G. Horrigan, Mar. Biol. 62, 81-89 (1981).
2. D.M. Anderson and K.D. Stolzenbach, Mar. Ecol. Prog. Ser. 25, 39-50 (1985)
3. R.C. Smith, O. Holm-Hansen and R. Olson, Photochem. Photobiol. 31, 585-592 (1980).
4. R.C. Worrest, Physiol. Plant. 58, 428-434 (1983).
5. P.L. Jokiel and R.H. York Jr., Limnol. Oceanogr. 29, 192-199 (1984).
6. F.T. Haxo in: Comparative Biochemistry of Photoreactive Systems, M.B. Allen, ed. (Academic Press 1960) pp. 339-360.
7. C.S. Yentsch and C.M. Yentsch, J. Mar. Res. 37, 471-483 (1979).
8. W.M. Balch and F.T. Haxo, J. Plankton Res. 6, 515-525 (1984).
9. K. Shibata, Plant and Cell Physiol. 10, 325-335 (1969).
10. P.L. Jokiel and R.H. York Jr., Bull. Mar. Sci. 32, 301-315 (1982).
11. R.C. Carpenter, Limnol. Oceanogr. 30, 784-793 (1985).
12. J.I. Carreto, H.R. Benavides, R.M. Negri and P.D. Glorioso, J. Plankton Res. 8, 15-28 (1986).
13. C.S. Yentsch, Limnol. Oceanogr. 7, 207-217 (1962).
14. S.W. Jeffrey and G.F. Humphrey, Biochem. Physiol. Pflanzen 167, 191-194 (1975).
15. S. Shimura and Y. Fujita, Mar. Biol. 33, 185-194 (1975).
16. J.B. Soo Hoo, D.A. Kiefer, D.J. Collins and I.S. McDermid, J. Plankton Res. 8, 197-214 (1986).
17. F.J.R. Taylor in: Toxic Dinoflagellates, D.M. Anderson, A.W. White and D.G. Baden, eds. (Elsevier 1985) pp. 11-26.
18. K. Richardson, J. Beardall and J.A. Raven, New Phytol. 93, 157-191 (1983).
19. P.M. Sivalingam, T. Ikawa, Y. Yokohama and K. Nisizawa, Bot. Mar. 17, 23-29 (1974).
20. B. Kimor, A.G. Moigis, V. Dohms and C. Stienen, Mar. Ecol. Prog. Ser. 27, 209-215 (1985).
21. J. Calkins and T. Thordadottir, Nature 283, 563-566 (1980).
22. J.I. Beaulieu and J. Menard, ibid 17, pp. 445-450 (1985).
23. J.C. Farman, B.G. Gardiner and J.D. Shanklin, Nature 315, 207-210 (1985).

ACKNOWLEDGEMENTS

Special thanks to Dr. Clarice M. Yentsch for reviewing and commenting on this work.

PURIFICATION AND CHARACTERIZATION OF CALMODULIN (Ca^{2+} BINDING PROTEIN) FROM *CRYPTHECODINIUM COHNII* AND *PERIDINIUM BIPES*

YOSHIHIKO SAKO,[1] SACHIKO MACHIDA,[1] HIROKO TODA,[2] AND YUZABURO ISHIDA[1]
[1] Faculty of Agriculture, Kyoto University, Kyoto 606, Japan
[2] Institute for Protein Research, Osaka University, Suita, Osaka 565, Japan

ABSTRACT

Calmodulin (Ca^{2+} binding protein) which has very conservative amino acid sequence and activates cAMP phosphodiesterase by binding with Ca^{2+} was purified from dinoflagellates *Crypthecodinium cohnii* and *Peridinium bipes*. Purified calmodulin of these dinoflagellates could activate cAMP phosphodiesterase in the presence of Ca^{2+} and was inhibited by trifluoperazine and EGTA. Both calmodulins had a molecular weight of about 16,000 and showed a different migration pattern under presence of Ca^{2+} or EGTA in polyacrylamide gel electrophoresis. These characters were very similar to those of eucaryote calmodulin. But calmodulin from *C. cohnii* had trimethyllysine and unknown amino acids; the evolutionary status of this dinoflagellate is interesting.

INTRODUCTION

Calmodulin is a small, acidic, heat-stable, Ca^{2+}-binding protein, which modulates a large number of Ca^{2+}-dependent events in eukaryotes [1]. Calmodulin contains four homologous structural domains and is an activator of several enzymes, including Ca^{2+}-dependent cyclic nucleotide phosphodiesterase [2], Ca^{2+}-stimulated ATPase [3], NAD-kinase [4] and myosin light-chain kinase [5]. Calmodulin is widely distributed in eukaryotes, being isolated from animal, plant, green alga, protozoa, and slim mold. These studies demonstrate that calmodulin is a highly conserved protein in terms of both structure and function. Calmodulin is one of the proteins evolving with slowest speed in molecular evolution.

The dinoflagellates possess numerous unusual and primitive characteristics such as few histone, lack of nucleosomes, condensed chromosome, and unusual mitosis [6]. Accordingly, the evolutionary position of dinoflagellates has been very interesting. In this study, we have purified and characterized calmodulin for the first time from marine dinoflagellate *C.cohnii* and freshwater dinoflagellate *P. bipes* causing freshwater red tide in Japanese reservoirs.

EXPERIMENTAL

Organisms

C. cohnii was provided by Nihon Yushi company. The cells of *P. bipes* were collected by plankton net from the bloom in Iwaya reservoir in Gifu prefecture. The bloom was unialgal (99% *P. bipes*).

Culture and harvest

C. cohnii was grown in MLH medium [7] at 30℃ in dark and cells were harvested at the late-exponential growth phase. Both algae were washed with buffer A (20mM Tris-HCl, 5mM EDTA, pH7.5) and stored at -90℃.

Analytical procedure

Calmodulin was determined according to its ability to activate the cal-

RED TIDES: BIOLOGY, ENVIRONMENTAL SCIENCE, AND TOXICOLOGY
Okaichi, Anderson, and Nemoto, Editors

modulin-deficient brain phosphodiesterase [5]. Protein was measured by the Lowry method [8]. Electrophoresis of calmodulin was done in 15% polyacrylamide gels in the buffer system of Davis [9] containing either $CaCl_2$ or EGTA at a final concentration of 0.2mM and in 0.1% SDS in a Laemmli buffer system [10].

RESULTS AND DISCUSSION

Purification of calmodulin

Purification schemes of calmodulin from both species were basically the same. All operations were carried out at 0-4℃. Cells of *C. cohnii* or *P. bipes* were suspended in 2vol of buffer A and disrupted by ultrasonication. The suspension was centrifuged at 16000xg for 45min and the supernatant collected. The supernatant was heated at 100℃ for 5min and the denatured proteins were removed by centrifugation. The supernatant was brought to 80% saturation of ammonium sulfate. After centrifugation, the pellet was dissolved in buffer A and dialysed against the same buffer (Preparation I). Preparation I was applied to a column of DEAE-cellulose equilibrated with buffer A. After the sample application, the column was washed by buffer containing 0.2M NaCl. Many proteins were eluted by this wash. Calmodulin was eluted by raising the concentration of NaCl to 0.3M in *C. cohnii* or 0.28M in *P. bipes*. Active fractions were pooled and dialysed against buffer B (50mM Tris-HCl, 0.3M NaCl, 0.1mM $CaCl_2$, 1mM β ME, pH7.5) (Preparation II). Preparation II was applied to Affi-gel phenothiazine (Bio-Rad) equilibrated with buffer B. After most proteins were eluted with continued washing of buffer B, calmodulin was eluted with buffer C (50mM Tris-HCl, 5mM EGTA, pH7.5). Active fractions were pooled and dialysed against buffer A (Preparation III). The purity of calmodulins from both species were determined by PAGE. Electrophoresis revealed a single protein band which corresponded to cAMP phosphodiesterase-activating activity in both species.

Characterization of calmodulin

Calmodulins obtained from *C. cohnii* and *P. bipes* were homogenous using PAGE with addition of either Ca^{2+} or EGTA and their mobilities decreased in the presence of Ca^{2+}.

Each of these purified samples also showed a single band on SDS-PAGE. Molecular weights were 16,200 and 15,000 for *C. cohnii* and *P. bipes* proteins, respectively. Molecular weight of *C. cohnii* calmodulin was standard size, but that of *P. bipes* was smaller than those of other species and similar in size to *Tetrahymena* calmodulin [11].

Purified calmodulins of *C. cohnii* and *P. bipes* were capable of activating the activity of calmodulin-deficient brain phosphodiesterase in the presence of Ca^{2+}. The amount of calmodulin required for half-maximal activation of enzyme were different from each other. The ability of *C. cohnii* and *P. bipes* calmodulins to activate phosphodiesterase were examined as a function of free Ca^{2+} concentration by use of a 10-fold excess of calmodulin. Consequently *C. cohnii* and bovine heart calmodulins were active at low ratios of $[Ca^{2+}]/[EGTA]$ exhibiting half-maximal activity at ratios of 0.71 and 0.75, respectively. On the other hand, ratios of *P. bipes* calmodulin required for half-maximal activity was 0.93. This ratio was similar to those of trypanosome and *Tetrahymena* calmodulin [12].

The activation of phosphodiesterase activity by each calmodulin was blocked by trifluoperazine and EGTA.

The amino acid composition of *C. cohnii* calmodulin was similar to these determined previously for animal and plant calmodulins, for example an absence of Trp and Cys, and presence of ε-N-trimethyllysine and relatively large amounts of Glu and Asp. However,it is very interesting that *C.cohnii* calmodulin contained unknown amino acids.

REFERENCES

1. W.Y. Cheung, Science, 207, 19, (1980).
2. W.Y. Cheung, J. Biol. Chem., 246, 2859, (1971).
3. R.M. Gopinath and F.F. Uincenzi, Biochem. Biophys. Res. Commun., 77, 1203, (1977).
4. J.M. Anderson and M.J. Cormier, Biochem. Biophys. Res. Commun., 84, 595, (1978).
5. K. Yagi, M. Yazawa, S. Kakiuchi, M. Ohshima and K. Uenishi, J. Biol. Chem., 253, 1338, (1978).
6. D.L. Spectcor, Dinoflagellates, Academic Press, p.1, (1984).
7. R.C. Tuttle and A.R. Laeblich, Phycologia, 14, 1, (1975).
8. O.H. Lowry, N.J.Posebrough, A.L. Farr and R.J. Randoll, J. Biol. Chem., 193, 265, (1951).
9. B.J. Davis, Ann. NY Acad. Sci., 121, 404, (1964).
10. U.K. Laemmli, Nature, 227, 680, (1980).
11. G.A. Jamieson,Jr., T.C. Vahaman and J.J. Blum, Proc. Natl. Acad. Sci. USA., 76, 6471, (1979).
12. L. Ruben, C. Egwuagu and C.L. Patton, Biochem. Biophys. Acta., 758, 104, (1983).

COMPLETE AMINO ACID SEQUENCE OF FERREDOXIN FROM THE FRESH-WATER DINOFLAGELLATE, *Peridinium bipes*

ARITSUNE UCHIDA*, SHINYA EBATA*, KEISHIROU WADA**, HIROSHI MATSUBARA**, and YUZABURO ISHIDA*
*Laboratory of Microbiology, Department of Fisheries, Faculty of Agriculture, Kyoto University, Kyoto 606; ** Department of Biology, Faculty of Science, Osaka University,Toyonaka, Osaka 560, Japan

ABSTRACT

The amino acid sequence of the major ferredoxin component isolated from a dinoflagellate, *Peridinium bipes*, was completely determined. The sequence was as follow: Phe-Lys-Val-Thr-Leu-Asp-Thr-Pro-Asp-Gly-Lys-Lys-Ser-Phe-Glu-Cys-Pro-Gly-Asp-Ser-Tyr-Ile-Leu-Asp-Lys-Ala-Glu-Glu-Glu-Gly-Leu-Glu-Leu-Pro-Tyr-Ser-Cys-Arg-Ala-Gly-Ser-Cys-Ser-Ser-Cys-Ala-Gly-Lys-Val-Leu-Thr-Gly-Ser-Ile-Asp-Gln-Ser-Asp-Gln-Ala-Phe-Leu-Asp-Asp-Asp-Gln-Gly-Gly-Asp-Gly-Tyr-Cys-Leu-Thr-Cys-Val-Thr-Tyr-Pro-Thr-Ser-Asp-Val-Thr-Ile-Lys-Thr-His-Cys-Glu-Ser-Glu-Leu. It was composed of 93 amino acid residues with seven cysteine residues. The cysteine residue at the 18th position usually found in other chloroplast-type ferredoxins was found at the 16th position in *Peridinium* ferredoxin. Calculation of the numbers of amino acid differences among chloroplast-type ferredoxins indicates that the *Peridinium* ferredoxin is far divergent not only from higher plant ferrodoxins but also from blue-green algal ferredoxins.

INTRODUCTION

The dinoflagellates constitute a Class of unicellular algae that includes both photoautotrophic and heterotrophic species and represents an important group of marine and freshwater plankton. They are well known as the causative plankton of red-tides. The evolutionary position and systematics of the dinoflagellates have been widely investigated. They were shown to be a distinctive group of protistan organisms, having a mixture of plant and animal characeristics, with no close relatives(1). Since they have features of both procaryotes and eucaryotes they were at one time classified as "mesocaryotes"(2). However, the cellular organization of the dinoflagellates is typically eucaryotic, showing membrane bound organelles, therefore, it was suggested that they are not an intermediate group, but rather a group of eucaryotes which retained some primitive features(3). In the attempts to formulate classification schemes of the dinoflagellates and to assess their evolutionary position, emphasis has been placed almost solely upon morphological characteristics(4), nuclear organization(5) and the development of the plates that form the cellulotic theca(6), while the use of biochemical characters has been very limited(7-9). Recent molecular data on the structure and organization of the DNA(10,11), RNA(12,13) and protein(14) have provoked renewed interest in dinoflagellate evolution.

RED TIDES: BIOLOGY, ENVIRONMENTAL SCIENCE, AND TOXICOLOGY
Okaichi, Anderson, and Nemoto, Editors

Ferredoxins are iron-sulphur electron transfer proteins ubiquitous in bacteria, algae, higher plants and animals. Chloroplast-type ferredoxins, each containing two atoms of iron and sulphur, are found in all oxygenic photosynthetic organisms from blue-green algae to higher plants. Their functional, structural and evolutionary characteristics are widely investigated(15,16).

In view of the widespread use of ferredoxin in evaluating phylogeny, it was of interest to characterize these proteins in a species of Peridinium, one of the most prevalent dinoflagellate groups. Peridinium bipes used in this experiment is freshwater photosynthetic dinoflagellate. Large-scale freshwater blooms by this organism have often occurred in reservoirs and rarely in natural lakes in Japan during April through early June.

In this article we report the isolation and complete amino acid sequence of ferredoxin from P. bipes and compare it with sequences of other chloroplast-type ferredoxins.

EXPERIMENTAL

Cells of P. bipes were collected by net from water blooms in Takihata Reservoir, Osaka Prefecture. The cells were kept at -90 C until use.

P. bipes ferredoxin was prepared basically as described by Hase et al.(17). After precipitation of ferredoxin with trichloroacetic acid the apoprotein was reduced with 2-mercaptoethanol and carboxymethylated with iodoacetic acid. S-carboxymethyl(Cm)-ferredoxin was digested with Staphylococcal V8 protease or with Trypsin.

Purification of Peptides was accomplished by staphylococcal protease digestion or tryptic digestion, separately fractionated by reverse C18 column using high-performance liquid chromatography (HPLC) system.

Amino acid composition and sequence determination of Cm-ferredoxin and purified peptides were determined with an amino acid analyzer. The amino (N)-terminal sequences of purified peptides were determined by the manual Edman degradation procedure. Phenylthiohydantoin derivatives(PTH) were identified by HPLC and TLC.

RESULTS and DISCUSSION

The total number of residues was 93, giving a molecular weight of 9,959, excluding iron and sulfur. These values are in fairly good agreement with those obtained by direct analysis. The complete amino acid sequence is shown in Fig. 1. In particular, although the resemblances among other chloroplast-type ferredoxins are significant, there are deletions of small peptides in N- and C-terminals. The 93 amino acid residues was the smallest number in the known chloroplast-type ferredoxins while the content of seven cysteine residues was the largest. Dinoflagellate algae are assumed to be a distinctive group of protistan organisms. These organinsms occupy an important position in the study of the evolutionary processes of the plant kingdom. An amino acid sequence homology method(18) was used to construct a phylogenetic tree of 48 chloroplast-type ferredoxins determined. The calculations were based on the homology distance of amino acid sequence at corresponding loci in the various ferredoxins, and a deletion or insertion was counted based on

Dayhoff's mutation data(19) by means of a computational method. One of the possible phylogenetic treesis present in Fig. 2. The similarity of higher plant ferredoxins is obvious and they form a cluster with a marked exception of genus Equisetum. The green algal ferredoxin is in a position relatively near higher plant ferredoxins. The blue-green algal ferredoxins are even more remote from other higher plant ferredoxins. It is interesting to note that P. bipes ferredoxin is remote from blue-green algal ferredoxins and earliest branching of the higher plant lineage leads to dinoflagellates. It is relatively closer to Marchantia polymorpha or Gleichenia japonica ferredoxins.

```
Peridinium bipes
        10        20        30        40        50        60        70        80        90
 FKVTLDTPDGKKSFECPGDSYILDKAEEEGLELPYSCRAGSCSSCAGKVLTGSIDQSDQAFLDDDQGGDGYCLTCVTYPTSDVTIKTHCESEL
 ::::: :: :          : :::: ::: :: :::::::: ::::::::  :  ::::  ::::::     : :::  ::::: :: :: :  :
TFKVTLNTPTGQSVIDVEDDEYILDAAEEAGLSLPYSCRAGACSSCAGKVTAGEVDQSDESFLDDDQMDEHYVLTCIAYPTSDLTIDTHQEEALI
        10        20        30        40        50        60        70        80        90
Marchantia polymorpha

    DISTANCE = -333
```

Fig. 1. Parallel between the alignments of ferredoxin of P. bipes and Marchantia polymorpha

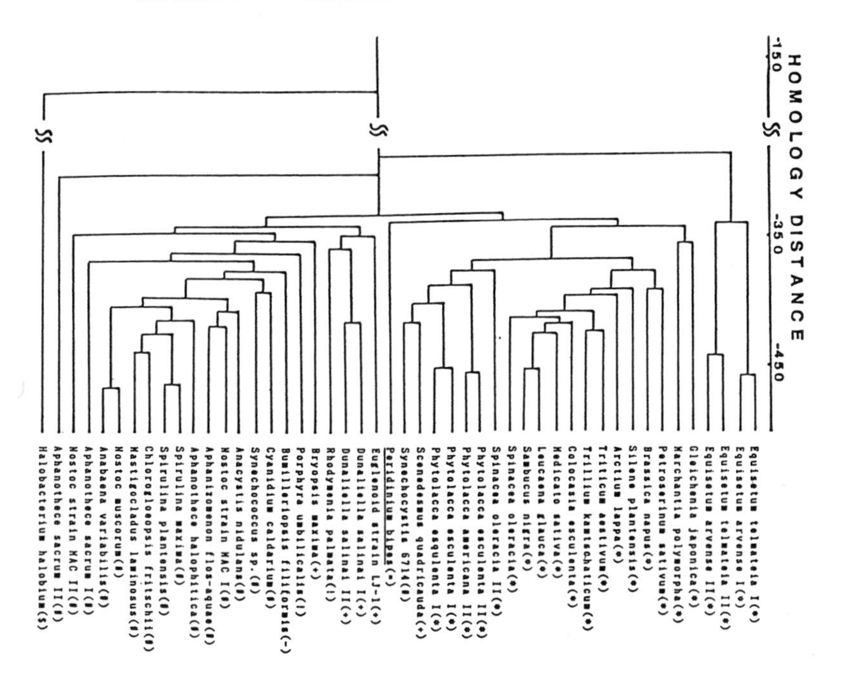

Fig. 2. A phylogenetic tree of chloroplast-type ferredoxins *, higher plants; +, green algae; =, dinoflagellate; -, brown alga; !, red algae; #, blue-green algae; $, bacterium

Maroteux et al.(11) investigated the secondary structures of 5.8 S rRNA of dinoflagellates and described that an early divergence of the dinoflagellate lineage occurred from the typical eucaryotes. Based on the peculiar characteristics of dinoflagellates such as presence of hydroxymethyluracil in the DNA, and existence of the specific pigment peridinin it has recently been suggested that this group is monophyletic and has evolved independently of the other non-procaryotic organisms. As Herzog et al.(5) suggested, dinoflagellates may be considered as their sister group. This assumption seems to be supported by our present amino acid sequence study of ferredoxin. If indeed the dinoflagellate lineage has diverged early from the eucaryotic lineage as initially proposed by Loeblich(7), this implies that the split occurred before the eucaryotic chromatin was organized in nucleosomal structures, but after (i) the appearance of repeated DNA sequences(13),(ii) the splitting of the polymerases which have been characterized in C. cohnii(20), and (iv) the appearance of six capped small nuclear RNAs described in C. cohnii which are similar in several respects to the U1 - U6 snRNAs of eucaryotes(13).

REFERENCES

1. Dodge, J.D.(1983) Brit. Phycol. J. 18, 335-356.
2. Dodge, J.D.(1965) Second International Conference on Protozool. Exc. Med. Inter. Congr. Ser. No. 91, 339
3. Kubai, D.F., and Ris, H.(1969) J. Cell Biol.40, 508-528.
4. Taylor, F.J.R.(1985) In Toxic Dinoflagellates:Proceedings of the Third International Conference on Toxic Dinoflagellates. pp11-26.
5. Herzog, M., Von Boltzky, S., and Soyer, M.-O.(1984) Origin of Life 13, 205-215.
6. Taylor, F.J.R.(1979) In Toxic Dinoflagellate Blooms. pp47-56.
7. Loeblich, A.R.III.(1976) J. Protozool. 23, 13-28.
8. Hayhome, B.A., and Pfiester, L.A.(1983) Amer. J. Bot. 70, 1165-1172.
9. Hochman, A.(1982) Arch. Microbiol., 133, 62-65.
10. Hinnebusch, A.G., Klotz, L.C., Immergut, E., and Loeblich, A.R.III.(1980) Biochemistry, 19, 1744-1755.
11. Maroteaux, L., Herzog, M., and Soyer-Gobillard, J.-O.(1985) BioSystems, 18 307-319.
12. Hinnebusch, A.G., Klotz, L.C., Blanken, R.L., and Loeblich, A.R.III.(1981) J. Mol. Evol., 17, 334-347.
13. Reddy, R., Spector, D., Henning, D., Liu, M.H., and Busch, H. (1983) J. Biol. Chem., 258, 13965-13969.
14. Hochman, A., Berman, T., Plotkin, B., and Schejter, A.(1985) Arch. Biochem. Biophys., 243, 161-167.
15. Rao, K.K., and Cammack, R.(1981) In Thirty-second Symposium of the Society for General Microbiology.pp175-213.
16. Matsubara, H., Hase, T., Wakabayashi, S., and Wada, K.(1980) In Evolution of Protein Structure and Function. pp.245-266.
17. Hase, T., Wada, K., and Matsubara, H.(1976) J. Biochem. 79, 329-343.
18. Goad, W.B., and Kanehisa, M.(1982) Nucleic Acids Res. 10, 247-263.
19. Dayhoff, M.O., Scjwartz, R.M., and Orcutt, B.C.(1978) In Atlas of Protein Sequence and Structure. pp345-352
20. Rizzo, P.J.(1979) J. Protozool. 26, 290-294.

A METALLOTHIONEIN-LIKE PROTEIN INDUCED IN A DIATOM, PHAEODACTYLUM TRICORNUTUM

Maita,Y., S.Kawaguchi and K.Tada

Faculty of Fisheries, Hokkaido University, Minatocho 3-1-1, Hakodate 041, Japan

ABSTRACT

The inducing ability and chemical nature of metallothionein like protein were investigated using the protoplasm of a marine diatom, Phaeodactylum tricornutum, under the presence of trace amount of heavy metals. The metallothionein-like protein was prepared from cells exposed to Cd using gel filtration (Sephadex G-75). Cadmium content of a peak corresponding to metallothionein-like protein accounted for 61% of the Cd incorporated into cultured cells. Estimated molecular weight of this component was ca. 12,000 dalton. The amino acid residues showed the typical composition of metallothionein found in several terrestrial phyla, i.e. high abundance of half cystine and the trace amounts of aromatic amino acids. However, a difference between the metallothionein-like protein from diatom species and from mammals was found, i.e. there was more glutamic acid and less serine in diatom species. Results of time series experiments suggest - that this substance plays an important role in the detoxification and transport of heavy metal incorporated into phytoplankton cells.

INTRODUCTION

It is well known that organisms living in aquatic environments polluted by heavy metals have tolerance against them (Foster,1977; Butler et al.,1980). Recent studies have shown that metal-binding proteins called metallothionein, in which cystein is relatively abundant, play an important role in detoxification and transport of heavy metals in living cells. While the proteins are found in phyla from bacteria to mammals, their presence in marine phytoplankton is yet to be studied (e.g. Cloutier-Mantha and Brown,1980).

This paper reports on the induction of a metallothionein-like protein by a species of marine phytoplankton, Phaeodactylum tricornutum.

MATERIALS AND METHODS

A diatom, P. tricornutum, used in this study was cultured in 50 ml of Mutsu medium (modified HESNS medium) at 20°C under the illumination of 3,000 lux (Light:Dark=16:8). Prior to the experiment, 1 ml of pure cultured P. tricornutum was inoculated into 50 ml of the medium and cultured for 4 days until they reached exponential growth phase. Aliquots of pre-cultured P. tricornutum (ca. 10^6 cells) were inoculated into 100 ml of experimental medium containing metal concentrations adjusted at 0-10 ppm using reagent grade metal chloride and cultured further for 10-14 days under the same light and temperature conditions. At logarithmic growth phase, the living cells were harvested by centrifugation at 1,000 x g for 10 min. Pellets were rinced

RED TIDES: BIOLOGY, ENVIRONMENTAL SCIENCE, AND TOXICOLOGY
Okaichi, Anderson, and Nemoto, Editors

twice with metal free filtered sea water in order to remove any metal on the cell surface, and then resuspended in a small quantity of 10 mM Tris buffer, adjusted to pH 8.6.

GEL FILTRATION

The cell walls of phytoplankton were destroyed by three successive two minute exposures to a homogenizer (Ultra Turrax Homogenizer). The homogenates were centrifuged at 1,200 x g for 10 min., then 12,000 x g for 30 min. and 105,000 x g for 60 min. to remove the precipitates. The supernatants were further incubated for 2 min. at 70°C and then centrifuged at 36,000 x g to give a clear solution. Finally, the supernatants were submitted to gel filtration by Sephadex G-75 using a column of 2.6 x 60.9 cm in size and eluted with 10 mM Tris buffer adjusted to pH 8.6. Fractions were taken as 5ml aliquots.

ANALYSIS OF AMINO ACIDS

UV absorption at 260 and 280 nm for proteinous component and atomic absorption for cadmium content in the fractionated samples were examined. Subsequently, the fractions corresponding to the protein peaks were combined, concentrated and hydrolyzed under vacuum less than 10^{-2} torr at 105°C for 22 hrs. Hydrolyzed samples were evaporated and diluted with citrate buffer adjusted to pH 2.2, and then injected into HPLC for amino acid analysis (Hitachi Seisakusho Co., High Performance Liquid Chromatograph, Model 835).

RESULTS

GROWTH CURVE

Fig.1 shows the growth curve of P. tricornutum cultured with various concentration of metals and without metal. The growth was not affected by the addition of Cd in the any concentration level under the given conditions, but the growth was inhibited by Zn and Cu at concentrations of 10 ppm.

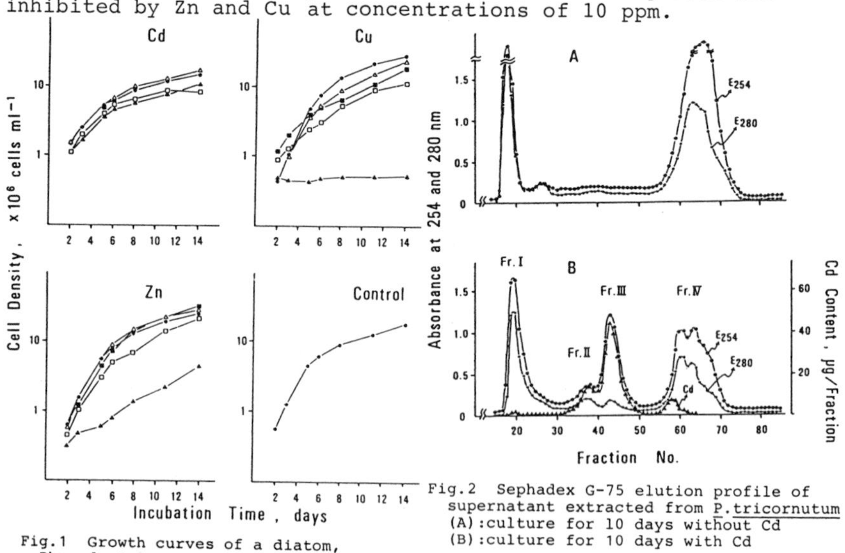

Fig.1 Growth curves of a diatom, Phaeodactylum tricornutum, cultured in metal containing medium, ●:0.05 ppm ○: 1 ppm ■: 2 ppm □: 5 ppm △: 10 ppm

Fig.2 Sephadex G-75 elution profile of supernatant extracted from P.tricornutum (A):culture for 10 days without Cd (B):culture for 10 days with Cd

GEL FILTRATION BY SEPHADEX G-75

Fig.2 shows the elution diagram of the samples cultured with (loaded sample) and without Cd (Control). In the case of the control, U.V. absorption peaks appeared at void volume near 100 ml eluate (Fraction I) and at 350 ml eluate (Fraction IV). On the other hand, one peak having small shoulder was detected at 200 ml eluate (Fraction II and III) in a loaded sample. These U.V. peaks completely coincided with those of Cd content.

Using standard proteins (i.e. bovine albumin; MW=66,000, α-chymotrypsinogen; MW=25,700, cytochrome-c; MW=12,400) as indicators, the molecular weights of Fr. II and Fr. III were estimated to be about 14,000 and 12,000 dalton, respectively.

AMINO ACID COMPOSITION OF FRACTION II AND III

Amino acid compositions of these fractions are shown in Table 1. Characteristic amino acid composition was found in Cd rich fractions: cysteine (about 30%) was markedly increased in Fr. III. It is considered that the enrichment of sulfur containing amino acid bears a marked resemblance to the metallothionein found in higher plants and animals.

TIME COURSE EXPERIMENT

The results of a time course experiment showed that a Cd containing peak appeared near the void volume (corresponding to the high molecular weight fraction) just after Cd inoculation at day 1 and then this fraction did not increase further. However, at days 3 and 5, the Cd content of both the medium molecular weight fraction and low molecular weight fraction increased gradually (Fig.3). Both fractions were estimated to contain about 61.7% and 29.8% of the total Cd in the supernatant.

Table 1 Amino acid composition of each fractions separated by gel filtration (expressed as g/100g protein)

	Fr.I 70,000	Fr.II 14,000	Fr.III 12,000	Fr.IV 3000
Asp	12.6	10.0	4.2	12.1
Thr	6.7	4.3	2.3	1.7
Ser	7.0	5.5	2.7	1.4
Glu	9.8	22.1	33.5	34.3
Gly	9.3	11.0	10.3	2.2
Ala	11.7	7.6	2.9	10.5
Cys/2	1.7	14.1	30.8	0.5
Val	8.1	4.2	2.0	2.0
Met	0.8	-	1.0	0.2
Ile	4.7	2.7	1.3	1.0
Leu	7.9	3.9	2.2	1.2
Tyr	0.7	-	0.7	0.2
Phe	3.9	1.9	0.7	0.3
Lys	5.6	5.3	2.6	6.2
His	1.3	0.8	0.3	0.2
Arg	3.6	3.1	1.2	7.6
Pro	4.6	3.4	1.6	18.5

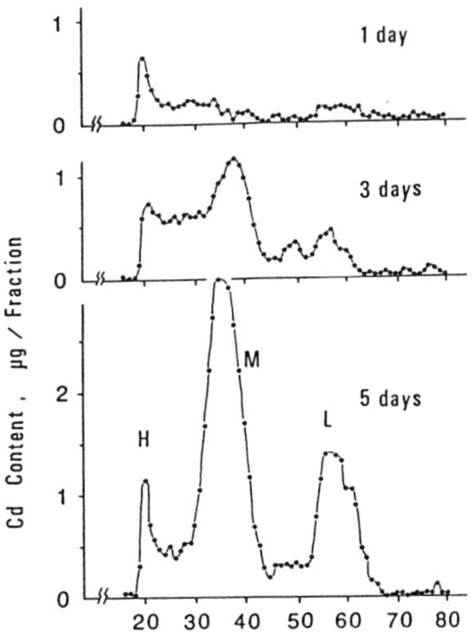

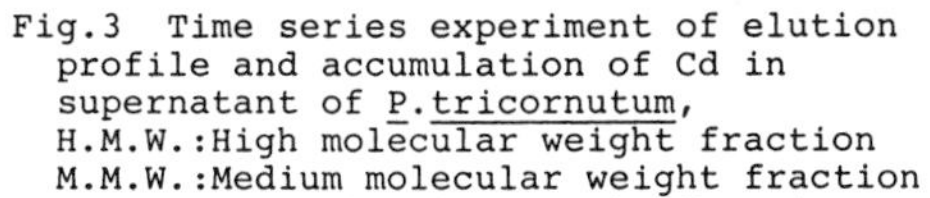
Fig.3 Time series experiment of elution profile and accumulation of Cd in supernatant of P.tricornutum,
H.M.W.:High molecular weight fraction
M.M.W.:Medium molecular weight fraction
L.M.W.:Low molecular weight fraction

DISCUSSION

In this study, proteins coexisting with Cd were first separated from a diatom species, P. tricornutum cultured in Cd containing media. Gel filtration showed that the molecular weight compositions were about 14,000 and 12,000 daltons for Fr. II and III, respectively. These values are somewhat higher than those of metallothionein in mammals. However, the high cystein content (30.8%) and the lack of aromatic amino acids are characteristics of metallothionein. The difference between the Cd-containing fraction of our diatom species and the metallothionein of mammals are in the amino acid composition (i.e. high glutamic acid and low serine). This tendency is not only found in P. tricornutum but also in higher plants such as cabbage (Wagner, 1984), agrostis (Rauser et al., 1984) and Rauvolfia serpentina (Grill et al., 1985). The name phytochelatin is prososed for these types of natural products in higher plants.

The results of our our time course experiment showed that Cd accumulated predominantly into the medium and low molecular weight fractions, not in the high molecular weight compounds. Generally, it is known that metals located at similar or adjacent periodic table positions have similar characteristics as ligands. For example, if a metalloenzyme (generally MW>70,000) contains zinc, the affinity of mercury or cadmium to that enzyme will be stronger than that of zinc. Therefore, the zinc in the enzyme will be replaced by mercury or cadmium. Due to the presence of mercury or cadmium, the enzyme activity will be lost (Wada, 1875). We suggest that a precurser to the metallothionein-like protein is first produced by P. tricornutum in the presence of Cd. The precursor acts as an acceptor in order to isolate the toxic metal from high molecular weight compounds with important biological functions, such as metalloenzymes.

In conclusion, a metallothionein-like protein is induced in P. tricornutum in the presence of Cd. The protein probably plays an important role in the detoxification and transport of heavy metals incorporated into phytoplankton cells. Further experiments are needed to clarify the molecular weight of the Cd binding protein, its amino acid sequence, and the ecological significance of such a biochemical system in the natural environment.

REFERENCES

1. M. Butler, A. E. J. Haskew and M. N. Young, Plant, Cell and Environment, 3, 119, (1980)
2. L. Cloutier-Mantha and D. A. Brown, Botanica Marina 23, 53, (1980)
3. P. L. Foster, Nature, 269, 322 (1977)
4. E. Grill, Ernst-L. Winnacker and M. H. Zenk, Science, 230, 674, (1985)
5. W. E. Rauser, J. Plant Physiol., 115, 143, (1984)
6. O. Wada, Taisha (Metabolism), 12,219, (1975)
7. G. J. Wagner, Plant Physiol., 76, 797, (1984)

PHOTOREDUCTION OF FE(III)-EDTA COMPLEX AND ITS AVAILABILITY TO THE COASTAL DIATOM *THALASSIOSIRA WEISSFLOGII*

Takeda,S.* and A.Kamatani**
*Faculty of Agriculture, University of Tokyo, Tokyo 113; **Tokyo University of Fisheries, Tokyo 108, Japan

ABSTRACT

The iron availability to *Thalassiosira weissflogii* was studied with media containing Fe(III)-EDTA complex at different concentrations in polycarbonate and glass flasks. The threshold of iron concentration in media maintaining the maximum growth was around 10^{-7} M, corresponding to $\sim 10^{-16}$mol Fe•cell^{-1}. In the case of lower iron concentrations ($10^{-8.5}$-$10^{-7.9}$ M), the growth rates in glass flasks were 1.5-2.0 times higher than those in polycarbonate ones. This was ascribed mostly to photoreduction of Fe(III)-EDTA to Fe(II) which was available to phytoplankton, because effective UV light passed in abundance through the walls of glass flasks compared with that of the polycarbonate ones.

INTRODUCTION

There is no evidence whether natural levels of iron act as a limiting factor for the growth of phytoplankton[1,2]. This uncertainty stems from questions regarding metal contamination and/or available form of iron in seawater[3,4,5,6]. In addition, photochemical reactions of metal-organic compounds complicate the speciation of iron in seawater[7,8,9]. To minimize the contamination by metals in culture medium, polycarbonate equipment is better than glass[3,10], while in the sense of light transparency the latter should be better than the former.

The purpose of this work was to study the iron availability to phytoplankton and to assess the importance of photoreduction of Fe(III)-EDTA to Fe(II) using polycarbonate and glass flasks.

EXPERIMENTAL

Stock cultures of *Thalassiosira weissflogii* were grown using polycarbonate flasks. All experiments were performed under sterile conditions, but it was observed microscopically that the cultures were contaminated to some degree.

All culture experiments were conducted in a cabinet at 20±1°C under the 14:10 h light:dark cycle at a light intensity of 5000 lux supplied by cool white fluorescent lamps.

Seawater from Sagami Bay was used for preparing culture medium. After being passed through 0.8 µm membrane filters and UV irradiation in order to decompose organic matter, the seawater was enriched with nutrients: 1.0×10^{-4} M NH_4NO_3, 1.0×10^{-5} M KH_2PO_4, 4.0×10^{-5} M Na_2SiO_3. Trace metals and vitamins were added as follow: 1.0×10^{-7}M $MnCl_2$, 1.0×10^{-7} M $ZnSO_4$ and 0.5 µg vitamin B_{12}•liter^{-1}. This enriched seawater was used as basal medium (pH = 8.1), in which iron was not added but was present at 1.8×10^{-9} M as background. This basal medium was autoclaved at 121°C for 20 minutes in a polycarbonate bottle.

The glass and polycarbonate ware used in the experiments were cleaned with detergent, soaked in 2N HCl for several days, and finally rinsed several times with Milli-Q water.

RED TIDES: BIOLOGY, ENVIRONMENTAL SCIENCE, AND TOXICOLOGY
Okaichi, Anderson, and Nemoto, Editors

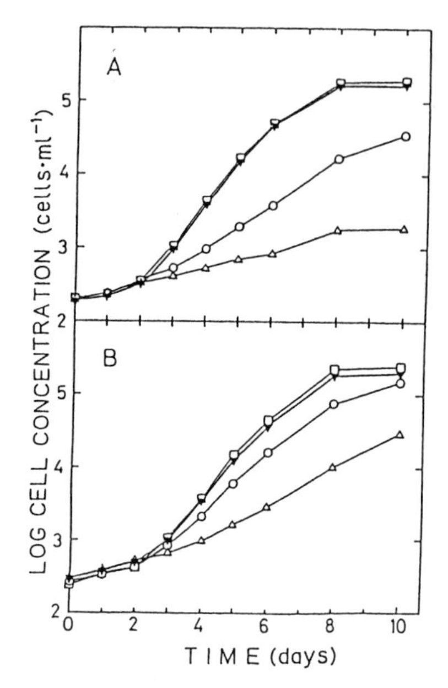

FIG. 1. Time course of cell growth in polycarbonate(A) and glass(B) vessels. EDTA=10^{-5} M. Total iron concentrations: △, 2.8×10^{-9}M; ○, 1.2×10^{-8}M; □, 1.0×10^{-7}M; ▼, 1.0×10^{-6}M.

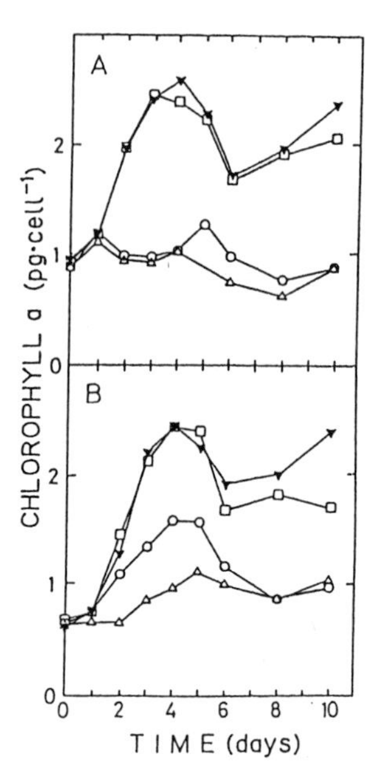

FIG. 2. Time course of chlorophyll a content in a cell in polycarbonate (A) and glass(B) vessels. Symbols as Fig. 1.

In order to study the iron effect on the growth of phytoplankton, iron-starved cells subcultured in the basal medium, were inoculated into 8 liters of basal medium at concentrations of 100-200 cells • ml^{-1}. About 800-ml portions of this inoculum culture was then poured into 1-liter polycarbonate(Nalgene) and glass(Pyrex) flasks, and various levels of Fe(III)-EDTA were added(1.0×10^{-9} M, 1.0×10^{-8} M, 1.0×10^{-7} M, and 1.0×10^{-6} M). The final concentration of EDTA was adjusted to 1.0×10^{-5} M in all experiments.

Cell concentrations were measured with a Coulter Counter model Z_B. Growth rates were calculated from linear regressions of the natural logarithm of cell number against time during the exponential phase of growth. Chlorophyll a concentration was determined by the fluorometric *in vitro* method using an Hitachi 650-10 fluorometer.

The per cent transmittance of light (200-800 nm) through the wall of polycarbonate and glass flasks was measured with an Hitachi 220A spectrophotometer.

Photoreduction of Fe(III)-EDTA to Fe(II) was determined using bathophenanthroline disulfonate(BPDS). Seawater containing 1.0×10^{-4} M EDTA, 1.0×10^{-5} M $FeCl_3$, and 4.0×10^{-5} M BPDS, was divided into polycarbonate and glass flasks. After standing at 20°C for 12 hours under light(5000 lux) or dark conditions, the $Fe(II)\text{-}(BPDS)_3^{2-}$ formed in each flask was extracted with isoamyl alcohol and measured spectrophotometrically.

RESULTS

The growth of *T. weissflogii* strongly depends on the iron concentration in the culture medium(Fig. 1). Iron levels greater than $\sim10^{-7}$ M allow

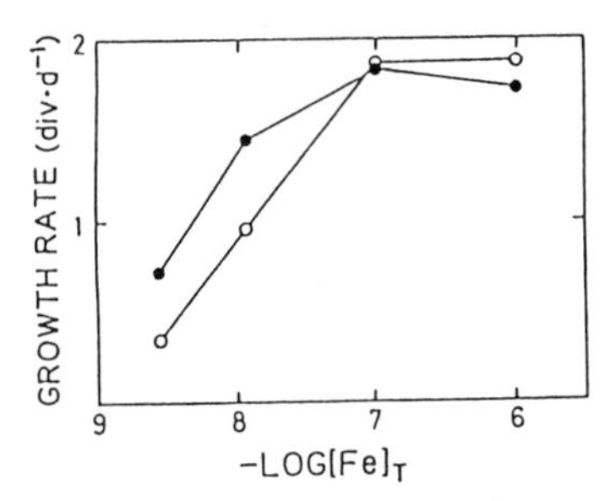

FIG. 3. Growth rate plotted against the negative log of the total iron concentration in medium. o, polycarbonate; •, glass.

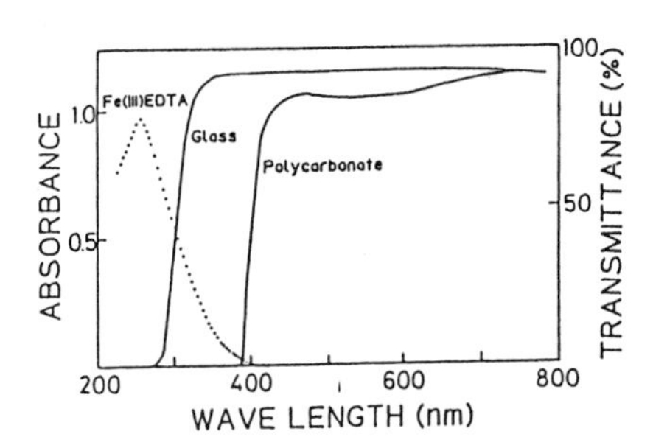

FIG. 4. Transmittance(%) of the Pyrex glass and polycarbonate(solid line), and absorption spectrum of Fe(III)-EDTA complex solution(dotted line).

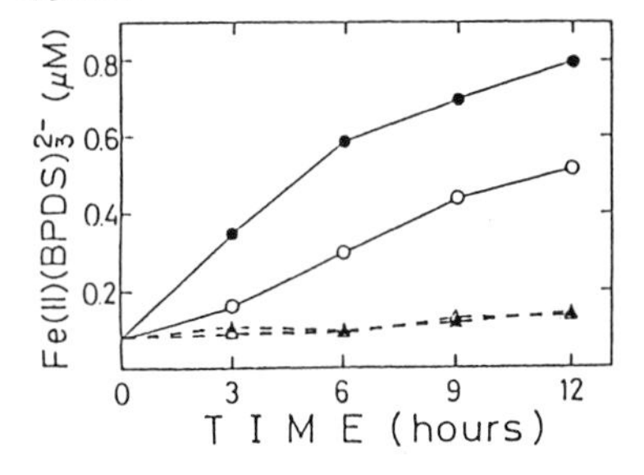

FIG. 5. Time course of $Fe(II)-(BPDS)_3^{2-}$ formation in seawater solutions of Fe(III)-EDTA. $Fe(III)=10^{-5}$ M, EDTA = 10^{-4} M, BPDS=$4x10^{-5}$ M. o, polycarbonate in light; Δ, polycarbonate in dark; •, glass in light; ▲, glass in dark.

maximum growth rate. If iron is less than $\sim 10^{-7}$ M the growth rate is depressed. Therefore, the threshold of iron concentration maintaining the maximum growth rate is around 10^{-7} M, corresponding to $\sim 10^{-16}$ mol Fe•cell^{-1}. This is supported by Figure 3 which shows the relationship between the growth rate and the negative log of the total iron concentration. In the early exponential phase of growth under optimum conditions for iron, the chl-a content was estimated at 2.4 pg•cell^{-1}(Fig. 2).

In the experiments in which the medium contained low levels of iron (2.8 - $12x10^{-9}$ M), the growth rates and chl-a contents were greater in glass flasks than in polycarbonate ones(Figs. 1&2). This result suggests that when the total amount of iron is restricted, glass flasks have some mechanism of accelerating the supply of iron available to phytoplankton relative to polycarbonate flasks.

Next, the light transparency of the walls of flasks used in the culture experiments was studied in the wavelength region of 200 to 800 nm. As can be seen in Fig. 4, the polycarbonate completely absorbed light shorter than 385 nm, while the transparency of glass extends to 280 nm. Absorption spectrum of Fe(III)-EDTA in the medium is characterized by having a maximum at 260 nm (Fig. 4). The overlapping range between the UV region passing through the glass and the absorption band of Fe(III)-EDTA may have some significance for supplying available iron to phytoplankton. To clarify this point, photoreduction of Fe(III)-EDTA to Fe(II) was studied and the results are shown in Fig. 5. The glass flask had a greater photoreduction capability than the polycarbonate one. The photoreduced iron after 12 hours' illumination was, on the basis of iron concentration added in the medium, estimated at 3.8 and 6.5 % for polycarbonate and glass flasks, respectively. The photoreduction rates for the light conditions used here were calculated to be $3.7x10^{-8}$ and $8.5x10^{-8}$ M•h^{-1} for polycarbonate and glass vessels in the first 6 hours' experimental period.

DISCUSSION

The availability of iron to phytoplankton is dependent upon its chemical speciation in aquatic systems; ferric and ferrous ion activity controls biological availability of iron[5,11].

Using medium-restricted iron concentrations($<10^{-8}$ M), kept at the same level as the iron ion activity, the growth rates and chl-a contents were significantly enhanced in glass flasks compared to polycarbonate ones (Figs. 1&2). It can determined from Figure 3 that the growth rates at total iron concentrations of 2.8×10^{-9} and 1.2×10^{-8} M in glass flasks correspond to those at 6.7×10^{-9} and 3.8×10^{-8} M in polycarbonate flasks, respectively. These results suggest that the total amount of the available form of iron in glass flasks should be supplied to phytoplankton at a rate of 2.4-3.2 times higher than in polycarbonate ones. This discrepancy may be caused by the desorption/adsorption of iron on the walls of flasks used in the experiments, because glass vessels have been shown to contain relatively high impurity levels of trace metals[12] and to adsorb larger quantities of iron from aqueous solution than polycarbonate does[10]. Therefore, we studied further the wall characters of the polycarbonate and glass flasks used here, in the presence and absence of EDTA. No significant desorption/adsorption of iron was observed for the both flasks.

The apparent photoreduction rate($\sim10^{-8}$ $M\cdot h^{-1}$) in glass flasks was observed to be about 2.3 times faster than in polycarbonate. Furthermore, the free ferric ion activity calculated from thermodynamic equilibrium was only about 10^{-19} M when the ferric ion coexisted with plenty of EDTA. Accordingly, these results strongly suggest that photoreduction of Fe(III)-EDTA to Fe(II) must be one of the important processes supplying available iron to phytoplankton.

As a result, it should be emphasized that sunlight-induced photoreduction of iron complexes with organic ligands may enhance iron availability to phytoplankton in surface waters, particularly where the iron concentration is very low and acts as a limiting factor for phytoplankton growth.

REFERENCES

1. F.M.M.Morel and N.M.L.Morel-Laurens, in: Trace Metals in Seawater C.S.Wong, E.Boyle, K.W.Bruland, J.D.Burton and E.D.Goldberg, eds. (Plenum, New York 1983) pp. 841-869.
2. L.E.Brand, W.G.Sunda and R.R.L.Guillard, Limnol. Oceanogr. 28, 1182-1198 (1983).
3. F.M.M.Morel, J.G.Rueter, D.M.Anderson and R.R.L.Guillard, J. Phycol. 15, 135-141 (1979).
4. M.L.Wells, N.G.Zorkin and A.G.Lewis, J. Mar. Res. 41, 731-746 (1983).
5. M.A.Anderson and F.M.M.Morel, Limnol. Oceanogr. 27, 789-813 (1982).
6. C.G.Trick, R.J.Andersen, N.M.Price, A.Gillam and P.J.Harrison, Mar. Biol. 75, 9-17 (1983).
7. D.G.Hill-Cottingham, Nature 175, 347-348 (1955).
8. C.J.Miles and P.L.Brezonik, Environ. Sci. Technol. 15, 1089-1095 (1981).
9. R.G.Zika and W.J.Cooper, Photochemistry of Environmental Aquatic Systems (American Chemical Society, Washington 1987).
10. S.E.Fitzwater, G.A.Knauer and J.H.Martin, Limnol. Oceanogr. 27, 544-551 (1982).
11. M.A.Anderson and F.M.Morel, Mar. Biol. Letters 1, 263-268 (1980).
12. D.E.Robertson, Anal. Chem. 40, 1067 (1968).

Present address: Abiko Research Laboratory, Central Research Institute of Electric Power Industry, 1646 Abiko, Abiko City Chiba, 27o-11, Japan (Shigenobu Takeda).

THE ROLE OF IRON IN THE OUTBREAKS OF *CHATTONELLA* RED TIDE

T. OKAICHI, S. MONTANI, J. HIRAGI AND A. HASUI
Faculty of Agriculture, Kagawa University, Kagawa
761-06 Japan

ABSTRACT

Iron is known as one of the micro-nutrients which enhances the growth of *Chattonella* spp., Raphydophyceae. In sea water of Harima Nada in the eastern part of the Seto Inland Sea, where *Chattonella* red tides occur and cause fishkills, soluble iron is usually found in the range from 0.04 to 0.1µM. The concentration of total iron ranges from 0.7 to 1.4µM, which is comparable with Ks(0.5µM) for the uptake of EDTA-Fe. The cells of *Chattonella* spp. are covered with a glycocalyx layer composed of polysaccharides. Colloidal and particulate iron seem to be absorbed by the glycocalyx layer and transfered by siderophores inside the cells.

INTRODUCTION

Iron is known as one of the micro-nutrients enhancing the growth of marine phytoplankton. Chelated iron is also required for the culture of *Chattonella* spp., Raphydophyceae. The requirement for maximum growth in batch cultures is about 1µM of EDTA-Fe.

Chattonella red tides appear frequently in Harima Nada in the eastern part of the Seto Inland Sea, Japan. In the sea water of Harima Nada soluble iron was usually found in the range from 0,04 to 0.1 µM. Total iron ranges from 0.7 to 1.4 µM. On the occasion of the red tide in July, 1983, 24-32% of the total iron was found in *Chattonella marina*. The high utilization of iron by *C. marina* caused us to speculate that there might be a peculiar mechanism for absorbing iron in *Chattonella* spp. To ascertain whether this mechanism does indeed exist, the presence of siderophores was checked, and the role of the glycocalyx on the wall surface of *Chattonella* spp. was examined.

EXPERIMENTAL

Materials

During the red tide, sea water with *Chattonella* spp. was continuously centrifuged. Mass culture was also carried out with 100L tanks and 10L glass vessels.

Extraction of siderophore

The scheme of extraction is shown in Fig.1. After final treatment with Sephadex G-25, thinlayer chromatography (using Kiesen Gel with BuOH:AcOH:H_2O(60:15:25) mixtures as solvents)was conducted.

RED TIDES: BIOLOGY, ENVIRONMENTAL SCIENCE, AND TOXICOLOGY
Okaichi, Anderson, and Nemoto, Editors

Extraction and fractionation of polysaccharide

The extraction of polysaccharide was carried out by the method shown in Fig.2. Alkali soluble polysaccharides were fractionated with DEAE cellulose(Cl^-).

Staining of glycocalyx layer

After fixation of Chattonella spp., iron cacodylate was added and Chattonella spp. were stained with ferrocyanate. The composition of the fixative reported by Yokote and Honjo [1] was as follows; conc. Formalin 100ml, EtOH 20ml, $Ficol_{400}$ 5g, $CaCl_2$ 4g. One ml of cold fixative was added to 1-5ml of medium.

RESULTS AND DISCUSSION

Chattonella marina obtained from the red tide which occurred in 1983 was used as material for the isolation of siderophores. Two fractions with iron chelating activity were isolated after final chromatography with Sephadex G 25 (Fig. 1).

The active fraction was purified with thinlayer chromatography. One of the components was determined to be deferrioxamine B by NMR. The presence of siderophores in Chattonella antiqua was also confirmed. Further chemical studies are still required, however.

The presence of a glycocalyx layer surrounding Chattonella spp. was first reported by Yokote and Honjo[1]. They estimated histochemically that mucopolysaccharide was the main component of the glycocalyx. But its physiological function was not clear.

With the method shown in Fig.2. the alkali soluble polysaccharide was separated into 4 fractions. The glucose and iron content in each fraction was analyzed. The results are shown in Table 1. In fraction IV the iron concentration per polysaccharide was the highest among the fractions. It is interesting that about 90% of the iron was rather easily washed out from the polysaccharide in fraction IV with EDTA(refer to alkali soluble fr.IV and IV' in Table 1). It seems to be reasonable to say that colloidal and particulate iron is absorbed on the glycocalyx layer and then transferred to the iron chelater to be conveyed inside the cell.

Ks uptake values for the absorption of iron by three red tide species was estimated by the present authors (Table 2). There were no significant differences among Ks values in these species, but the time to attain the maximum uptake rate was shortest in Chattonella antiqua. This result is reasonable if one assumes that iron is absorbed by the glycocalyx of C. antiqua. Glycocalyx layers on the other two species were tiny compared with that of C. antiqua.

The route of iron absorption in Chattonella spp. seems to be similar to the iron taxi route due to Exochelin and Mycobaction shown by K.N. Raymond and C.J. Carrano[2].

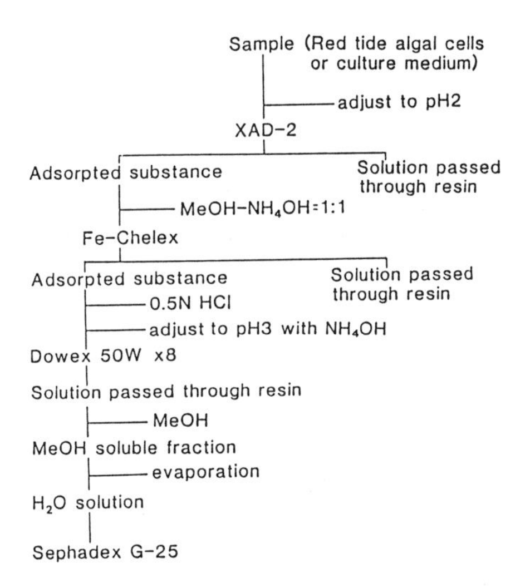

Fig. 1 Extraction method of iron chelating substance from *Chattonella* spp.

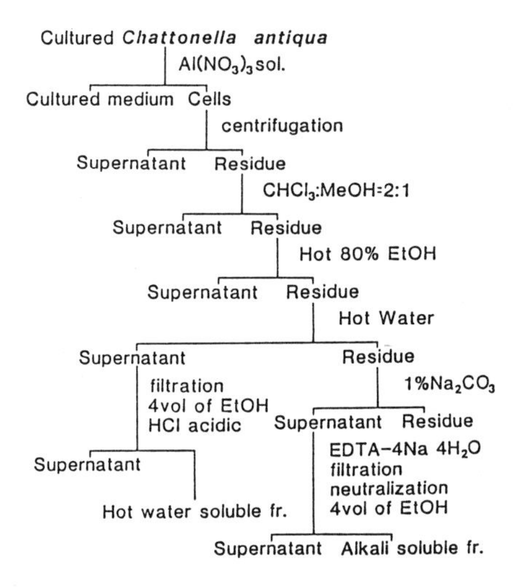

Fig. 2 Flow diagram of extraction and fractionation procedure of polysaccharides

Table 1 Crude polysaccharides from *Chattonella antiqua*[1]

Fraction		Hot water soluble fr.	Alkali soluble fr.				
			I	II	III	IV	IV'[2]
Yield	(mg)	15	5	6	14	46	5
Saccharide (as glucose)	(mg)	0.18	1.3	1.9	0.52	3.4	0.50
Iron	(μg) (%)	2.0 (2.5)	1.3 (1.7)	7.2 (8.5)	4.8 (4.8)	64 (82.5)	1.0
Saccharide-C/Fe (atomic ratio)		170	1900	500	200	99	890

(1) 3.7×10^7 cells from 14L cultured medium.

(2) Alkali soluble fr. IV (10mg) washed by EDTA.

Table 2 ^{59}Fe(as EDTA-Fe) uptake of red tide organisms

	Chattonella antiqua	*Prorocentrum minimum*	*Heterosigma akashiwo*
K_s (μg/l)	27.8	27.0	23.5
Maximum uptake rate ($\times 10^{-9}$μg/cell·hr.)	87.7	8.26	1.40
Uptake quantity ($\times 10^{-8}$μg/cell) 24hrs.	61.0	12.0	1.3
Time attained to maximum uptake rate (hr.)	2	12	8

REFERENCES

1. M. Yokote and T. Honjo, Experementia 41, 1143-1145 (1985).
2. K.N. Raymond and C.J. Carrano, Accounts Chem. Res., 12, 183-190 (1979).

THE SELENIUM REQUIREMENT OF GYMNODINIUM NAGASAKIENSE

ISHIMARU, T.*, T. TAKEUCHI**, Y. FUKUYO+ AND M. KODAMA++
*Ocean Res. Institute, Univ. Tokyo, Nakano, Tokyo, 164 Japan
**Wakayama Pref. Fisheries Exp. Sta., Kushimoto, Wakayama, 649-35 Japan
+Faculty of Agriculture, Univ. Tokyo, 113 Japan
++School of Fisheries Sci., Kitasato Univ., Sanriku, Iwate, 022-01 Japan

ABSTRACT

Gymnodinium nagasakiense reaches concentrations of more than 100,000 cells/ml during red tides. However, it did not exceed 20,000 cells/ml when cultured in medium based on offshore seawater containing nitrate, phosphate, vitamins (B_{12}, biotin, thiamin) and metals (Fe,Mo,Co,Cu). Maximum cell concentration increased when selenium (H_2SeO_3) was added. Growth rates of *G. nagasakiense* at various concentrations of selenium were measured using an artificial seawater medium treated with a chelating resin, Chelex 100. A maximum growth rate of 0.79 div./day and a half-saturation constant of 0.075 nM were obtained.

INTRODUCTION

A red tide of *Gymnodinium nagasakiense* Takayama & Adachi [1] (Gymnodinium sp. type-'65) was first reported in 1965 at Omura Bay [2] and has been observed in various regions of Japan thereafter. In 1984, red tide covered about 150 km of shoreline on the east coast of Kii Peninsula and caused serious damage to aquaculture. Culture studies on *G. nagasakiense* have been conducted on the effects of pH and salinity [3], bottom mud extract [4] and dissolved organic matter near fish farms [5], but the physiological characteristics of *G. nagasakiense* have not been detailed. A strain of *G. nagasakiense* isolated from 1984 red tide water grew well in enriched seawater medium T1 [6], a modification of f/2 [7]. However, during our study of its growth under various conditions of temperature, salinity and irradiance (results to be published elsewhere), the maximum cell yield was ca. 10,000 cells/ml, which is one order of magnitude less than that in the red tide water. Large amounts of nitrate and phosphate remain in the medium when the culture is in stationary phase. This suggests that some other nutrient not contained in the T1 medium is limiting the growth of *G. nagasakiense.*

In this study, as a result of an extensive survey for the limiting nutrient by additions of trace metals, significant stimulation of the growth of *G. nagasakiense* was observed following the addition of selenium.

MATERIALS AND METHODS

A unialgal strain of *G. nagasakiense* isolated from Temma Bay in Wakayama Prefecture was used in this study. Cultures were grown in 25 mm diameter flat bottom test tubes containing 25 ml of medium at about 100 $\mu Em^{-2}s^{-1}$ (measured by a quantum sclar irradiance meter; Biospherical Instruments QSL-100) under a 12:12 hour LD cycle at 20° C. An enriched seawater medium T1 [6] was used for the spike experiments to determine the limiting element. Various trace elements (10 nM) were added to each culture after it reached the stationary growth stage. An artificial seawater medium shown in Table I was used with varying concentrations of Se (H_2SeO_3) for further experiments. Seawater salts (identical to ASP7; [8]) and major nutrients were treated with a chelating resin, Chelex 100 (Bio-Rad) to

RED TIDES: BIOLOGY, ENVIRONMENTAL SCIENCE, AND TOXICOLOGY
Okaichi, Anderson, and Nemoto, Editors

reduce the trace metal contaminants [9]. Growth was monitored by the *in vivo* fluorescence method [10]. Growth rate (div./day) during the exponential growth phase was calculated by the following formula:

$$\mu_2 = \ln F_1 - \ln F_0 / (T_1 - T_0)\ln 2$$

where F_0 and F_1 indicate the relative fluorescence intensity at times T_0 and T_1 respectively.

Population density at stationary phase was determined using a microscrope and a Sedgwick-Rafter chamber. Each experiment to determine the growth rate and maximum cell concentration was carried out in triplicate.

Table I. The artificial seawater medium

Compound	Concentration (M)
NaCl	3.42×10^{-1}
$MgSO_4$	2.92×10^{-2}
$CaCl_2$	6.0×10^{-3}
KCl	7.5×10^{-3}
$NaNO_3$	5.0×10^{-4}
NaH_2PO_4	5.0×10^{-5}
Fe-EDTA	1.0×10^{-8}
$ZnSO_4$, $MnCl_2$, $NaMoO_4$, $CoCl_2$, $CuSO_4$, KBr, LiCl, $SrCl_2$, RbCl, KI	1.0×10^{-8}
Na_2-EDTA	1.0×10^{-8}
Thiamine HCl	5.93×10^{-7}
Biotin	4.1×10^{-9}
Cyanocobalamin	7.38×10^{-10}
Tris (pH 8.0)	5.0×10^{-3}

RESULTS AND DISCUSSION

In the spike experiments, among various trace elements (Fe, Mn, Co, Zn, Cu, Mo, I, Br, Cr, B, Li, Sr, Rb, F, Zr, Ba, V, W and Se) added at the stationary phase, only Se in the form of H_2SeO_3 significantly enhanced the growth of *G. nagasakiense* (Fig. 1). Cell concentrations in Se spiked cultures on day 12 and 25 were about 10,000 cells/ml and 88,000 cells/ml, respectively. In the spike experiments with Co (Fig. 1) and some other elements (eg. Cu; data not shown), slight increases in population density were observed after day 20. Lindstrom [11] reported that Se [VI] caused a long lag period (15-20 days) for *Peridinium cinctum.* The increase in population density after 1 week of Co addition might have resulted from selenate in the seawater medium.

The relationship between Se concentrations and the maximum cell density is shown in Fig. 2. The maximum cell concentration was limited by Se at concentrations <2 nM, and was limited by nitrate at Se >2 nM. From the initial slope of Fig. 2 (Se <2 nM), the cellular requirement of Se is calculated to be 4.4×10^{-17} moles/cell. Se limits the maximum cell yield when the ratio of Se to N is less than 1:28,000.

The exponential growth rate is shown to be a function of Se concentration in a manner similar to a Michaelis-Menten type hyperbola (Fig. 3). However, slow growth of *G nagasakiense* was observed in the medium without addition of Se, after three succesive transfers to the artificial seawater medium without Se addition. Hence, a Lineweaver-Burk plot for Se concentrations (assuming no Se contamination) and growth rate did not yield a linear relationship (Fig. 4; Δ). The concentration of contaminating Se was estimated (using a computer simulation with varying Se concentrations) as the concentration which gives the minimum value of the sum of squares of the errors about the regression line between the reciprocals of Se concentration and growth rate. The Se contamination was estimated to be 0.012 nM, and a maximum growth rate of 0.79 div./day and a K_s value of 0.075 nM were obtained.

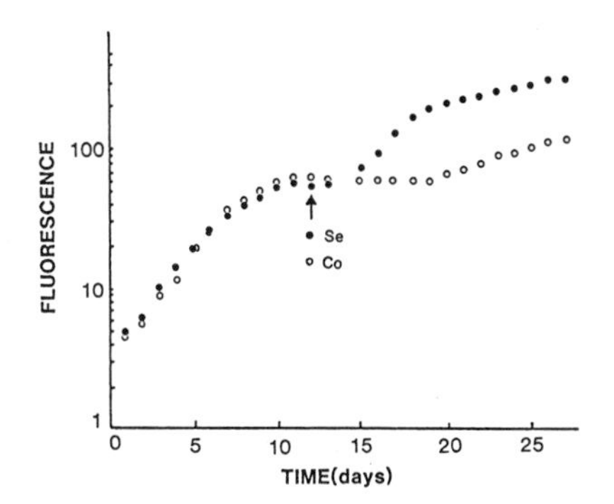

FIG. 1. Growth enhancement of G. ngasakiense by Se and Co addition at the stationary growth phase.

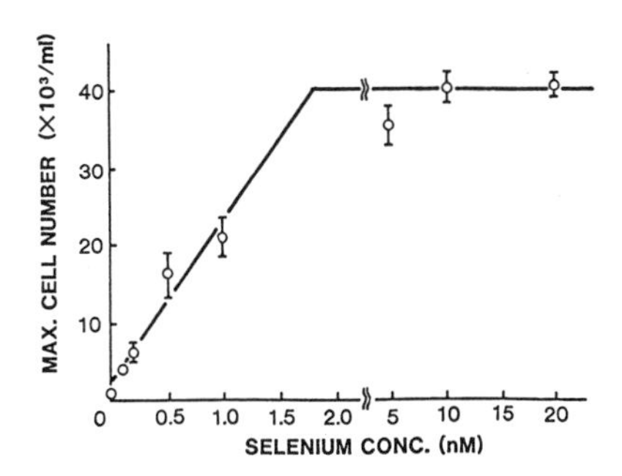

FIG. 2. Relationship between maximum cell concentration of G.nagasakiense and Se concentration added to the medium.

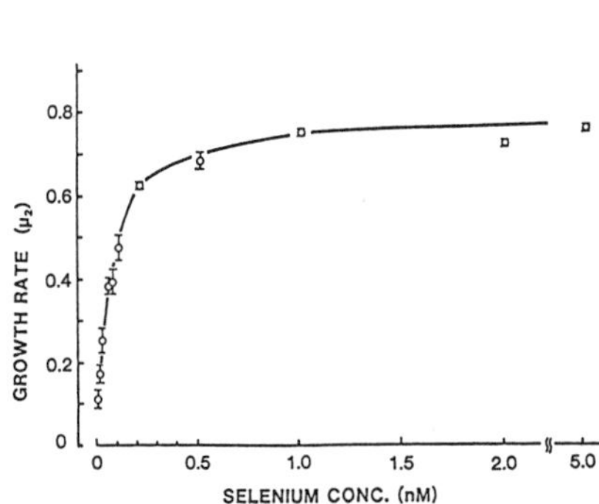

FIG. 3. Growth rate of G. nagasakiense as a function of Se concentration added to the medium.

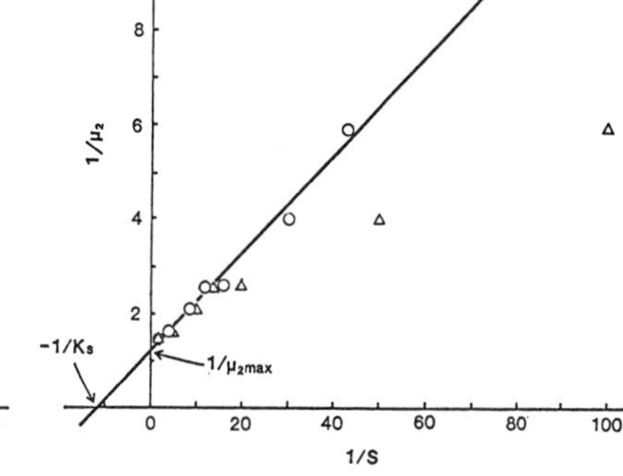

FIG. 4. Linweaver-Burk plot for growth rates and Se concentration. Se contamination of 0 (Δ) and 0.012 nM (o) were assumed.

Selenium was recently reported to be an essential element for phytoplankton such as the dinoflagellate Peridinium cinctum [12], the fresh water Chrysomonad, Chrysochromulina breviturrita [13], and a marine diatom, Thalassiosira pseudonana [14]. Although algal species in various taxa require Se as an essential element, the specific effect of selenium differs depending on the species of algae [15]. The effect of selenium on growth kenitics was only reported for P. cinctum, in which a K_s value of 0.095 nM was obtained. That is slightly higher than the value for G. nagasakiense that we determined in this study.

The concentration of total Se (selenite, selenate and organic form) in natural water is generally >0.5 nM and averages approximately 1 nM [16,17]. It is also reported that Se concentration is high in the coastal region, and low in river waters and in the open ocean. Our K_s for Se (0.075 nM) for the growth of G. nagasakiense is low compared to the average ambient Se concentration. It has therefore been suggested that G. nagasakiense is not limited by Se in natural waters.

Wrench and Measures [18] observed a preferential utilization of selenite by natural phytoplankton assemblages, suggesting that the level of selenium can vary during phytoplankton blooms. In coastal areas where phytoplankton populations are high and environmental conditions are variable, Se concentration would thus be expected to fluctuate. Although little is known of the oxidation state of selenium in natural waters, it seems to us that growth of G. nagasakiense is at times limited by Se in coastal seas, despite its low K_s value. Further information on Se in natural waters and on the response of G. nagasakiense to Se in various oxidation states is needed in our studies of the mechanisms of red tide formation.

REFERENCES

1. H. Takayama and R. Adachi, Bull. Plankt. Soc. Japan 31, 7-14 (1984).
2. S. Iizuka in : Toxic Dinoflagellate Blooms, D.L. Taylor and H.H. Seliger, eds. (Elsevier, North-Holland, Amsterdam 1979) pp. 111-114.
3. K. Numaguchi and K. Hirayama, Bull. Fac. Fisheries, Nagasaki Univ. 33, 7-10 (1972).
4. K. Hirayama and K. Numaguchi, Bull. Plankt. Soc. Japan 30, 139-146 (1972).
5. A. Nishimura, Bull. Plankt. Soc. Japan 29, 1-7 (1982).
6. T. Ogata, T. Ishimaru and M. Kodama, Mar. Biol. 95, 217-220 (1980).
7. R.R.L. Guillard in: Culture of Marine Invertabrate Animals, W.L.Smith and M.H. Chanley, eds. (Plenum, N.Y., 1975) pp. 29-60.
8. L. Provasoli in: Proc. 4th Int Seaweed. Symp., D. DeVirville and J. Feldmann, eds. (Pergamon Press, Oxford, 1963) pp. 9-17.
9. F.M.M. Morel, J.G. Reuter, D.M. Anderson and R.R.L. Guillard, J. Phycol. 15, 135-141 (1979).
10. L.E. Brand, R.R.L. Guillard and L.S. Murphy, J. Plankt. Res. 3, 193-201 (1981).
11. K. Lindstrom, Hydrobiologia 101, 35-48 (1985).
12. K. Lindstrom and W. Rhodhe, Mitt. Internat. Verein. Limnol. 21, 168-173 (1978).
13. J.D. Wehr and L.M. Brown, Can. J. Aquat. Sci. 70, 77-85 (1985).
14. N.M. Price, P.A. Thompson and P.J. Harisson, J. Phycol. 23, 1-9 (1987).
15. A.E. Wheeler, R.A. Zingaro, K. Irgolic and N.R. Bottino, J. Exp. Mar. Biol. Ecol. 57, 181-194 (1982).
16. K. Hiraki and Y. Nakaguchi, Kaiyou Kagaku (Marine Sciences) 17, 540-547 (1985):in Japanese.
17. Y. Sugimura, Y.Suzuki and Y. Miyake, J. Oceanogr. Soc. Japan 32, 235-241, (1976).
18. J.J. Wrench and C.I. Measures, Nature 299, 431-433 (1982).

V TOXICOLOGY

SAXITOXIN-PRODUCING BACTERIUM ISOLATED FROM *PROTOGONYAULAX TAMARENSIS*

MASAAKI KODAMA, TAKEHIKO OGATA AND SHIGERU SATO
Laboratory of Marine Biological Chemistry, School of Fisheries Sciences, Kitasato University, Iwate 022-01, Japan

ABSTRACT

Saxitoxin was identified in the culture media of a bacterium isolated from *Protogonyaulax tamarensis* grown in the medium containing antibiotics. This bacterium was not detected in the medium of *P. tamarensis*, suggesting that it originated from inside cells of *P. tamarensis*. A positive relationship between toxicity of *P. tamarensis* and number of bacterial colonies derived from *P. tamarensis* was observed. All the toxic strains of *P. tamarensis* gave bacterial colonies whereas nontoxic strains did not. These facts suggest the possible association of the bacterium with toxin production of *P. tamarensis*.

INTRODUCTION

Paralytic shellfish toxins (PSP toxins) are produced by several species of dinoflagellates [1] and a blue green alga [2]. There have been many reports on the toxin itself and biology of the toxin producing organisms. However, little is known about the physiological roles or biosynthetic mechanism of the toxin in these organisms. In a previous paper [3], we showed the difference of toxicity among various clones of *Protogonyaulax tamarensis* which were grown under the same conditions. The presence of various clones with different characteristics could be explained by the heterothalism of this species [4]. However, subclones derived from a single clone also showed the different toxicities when they were grown under the same conditions [5], suggesting that toxin production is not a hereditary characteristic of *P. tamarensis*.

We report here the saxitoxin-producing bacterium which was isolated from *P. tamarensis* and discuss the association of this bacterium with toxin production of *P. tamarensis*.

MATERIALS AND METHODS

Isolation and culture of the bacterium from *P. tamarensis*

A toxic clone of *P. tamarensis* (OF84432D-3)[3] was cultured in 1 L of T1 medium [3] containing a mixture of 10mg of gentamicin sulfate, 150mg of sodium ampicillin and 150mg of sodium cloxacillin. The mixture of antibiotics was repeatedly added 3 times to the culture during the exponential phase every 2 days. The cells were harvested by centrifugation (1 500 x g, 10 min) 3 days after the last addition of antibiotic mixture. The harvested cells were washed several times with sterilized seawater, lightly homogenized, inoculated onto agar plates of nutrient medium (Eiken Chemical Co., Ltd.), and then incubated for 4 days at 25°C. Bacterial isolates were cultured in a liquid nutrient medium [10g meat extract, 10 g peptone (Mikuni Kagaku Co. Ltd.), 2g NaCl and 2g agar in 1000 ml of distilled water, pH 5.5]. The culture was shaken for 4 days at 25 °C and then left for 4 days at 4 °C. The cultured media were centrifuged at 3000 x g for 20 min and the supernatant solutions were kept frozen until use.

RED TIDES: BIOLOGY, ENVIRONMENTAL SCIENCE, AND TOXICOLOGY
Okaichi, Anderson, and Nemoto, Editors

Purification and Identification of the toxin

About 70 L of the supernatant was dried in vacuo. The residue was extracted twice with 2 L of methanol containing 1% acetic acid. The extracts were applied on a column of Amberlite CG-50 (Na^+ form, 5 x 50 cm). The column was washed with 2 L of 1M acetate buffer (pH 4) and 2 L of water. The toxin adsorbed was eluted with 3 L of 10% acetic acid. The crude toxin was treated with activated charcoal [6], and then passed through a TSK G-3000S column (1.5 x 2 cm). The filtrate was further fractionated by rechromatography on Amberlite CG-50. The toxin was eluted with 0.5 M acetic acid. The toxic fractions were combined, evaporated to dryness, dissolved in a small amount of 0.03M acetic acid and then filtered through a Sepak C-18 cartridge. The filtrate was chromatographed on a Develosil ODS-5 column (8 x 250 mm) using an aqueous solution containing 0.03M ammonium acetate (pH 5.0), 0.005M heptafluorobutyric acid and 5% acetonitrile. The eluate was monitored by the absorbance at 206 nm. Purified toxin was analyzed by TLC [7], cellulose acetate membrane electrophoresis [8] and HPLC-fluorometric analysis [9] with the authentic specimen of saxitoxin (Calbiochem-Behring Corp., La Jolla, CA). Throughout the procedures, toxin was monitored by mouse bioassay and HPLC-fluorometric analysis [9]. The toxicity was expressed in mouse units (MU) using a dose-death time table of saxitoxin [1] where one MU represents a dose to kill a 20-g male mouse in 15 min.

Toxicity of P. tamarensis and number of colonies derived from P. tamarensis cultured in media containing antibiotic mixtures

P. tamarensis clone was cultured in the medium containing different concentration of antibiotic mixture. The cells were harvested at the end of exponential phase. The harvested cells were counted under a microscope. The homogenates of known numbers of the cells were inoculated to the agar plates and incubated for 4 days. Bacterial colonies appeared were counted. At the same time, the toxicity of P. tamarensis cells was measured.

Isolation of bacteria from toxic species of dinoflagellates

Other strains of P. tamarensis than OF8443D-3 collected from the same area and nontoxic strains of P. tamarensis [13] were cultured in the medium containing antibiotics as described above. The homogenates of these strains were inoculated to agar plates and incubated for 4 days at 25°C. As controls, newly found toxic species such as P. cohorticula [11] and Gymnodinium catenatum [12], and a nontoxic strain of P. catenella were treated in the same manner.

RESULTS

Isolation of the bacterium from P. tamarensis

After the cell harvest, cultured medium of P. tamarensis was mixed with a nutrient medium for bacteria, and incubated at 25°C for a week. No bacteria were grown in the medium, showing that the culture of P. tamarensis was bacteria-free. On the other hand, nutrient agar plates inoculated with the homogenate of P. tamarensis cells which were washed thoroughly with sterilized seawater showed colonies of bacteria after 4 days incubation. All the colonies were of the same shape and color (yellowish). The isolates were Gram-negative rod-shaped bacteria which were motile with flagella.

Toxin in the cultured media

About 100 MU of toxin as saxitoxin was recovered after activated charcoal treatment. The effects on mice were indistinguishable from those of tetrodotoxin and PSP toxins. Upon cellulose acetate membrane electrophoresis, purified toxin showed a blue fluorescent spot with identical mobility to saxitoxin by spraying with 1% H_2O_2 and heating for 5 min. TLC developed with solvent systems of pyridine- ethyl acetate- acetic acid- water (75:35:15:30) and t-butanol- acetic acid- water (2:1:1) showed blue fluorescent spots with identical Rf values to those of saxitoxin (0.53 in the former system and 0.31 in the latter). Analysis by HPLC-fluorometry [9] gave a peak with an identical retention time as saxitoxin (Fig. 1). The ratio of peak height to toxicity coincided with that of standard saxitoxin.

Relationship between toxicity of P. tamarensis and number of colonies derived from P. tamarensis.

In Table I is shown the relationship between toxicity of P. tamarensis and number of colonies derived from P. tamarensis. As the concentration of antibiotic mixture increased, the number of colonies decreased. In contrast, the toxicity of P. tamarensis decreased with increase of antibiotic mixture concentration. These results indicate the positive relationship between toxicity of P. tamarensis and number of colonies derived from P. tamarensis.

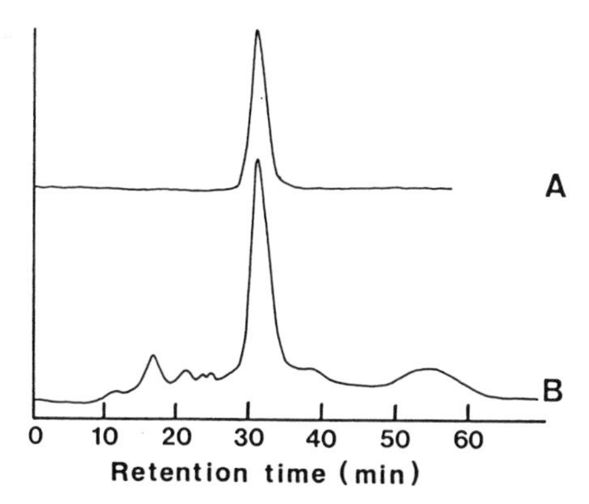

FIG. 1. Identification of bacterial toxin as saxitoxin by HPLC-fluorometric analysis. A: saxitoxin standard (0.2 MU), B: bactrial toxin (0.3 MU).

TABLE I. Relationship between amounts of bacteria derived from P. tamarensis and toxicity of P. tamarensis.

Culture*	Number of colonies from 10^4 cells	Toxicity of the cell MU/10^4 cells
A	15,900	0.578
B	1,500	0.097
C	330	0.057

* Culture was grown in the medium without antibiotics (A), containing gentamicin sulfate, sodium ampicillin and sodium cloxacillin in concentrations of 5, 50, 50 mg/L, respectively (B) and 15, 150, 150 mg/L, respectively (C).

Bacteria derived from other strains of P. tamarensis and other species of toxic dinoflagellates

Bacteria were obtained from the cultures of only toxic strains such as five strains of P. tamarensis, two strains of P. cohorticula, and Gymnodinium catenatum. No bacteria were obtained from the cultures of nontoxic strains of P. tamarensis and P. catenella.

DISCUSSION

Saxitoxin was detected in the cultured media of the bacterium isolated from the homogenate of *P. tamarensis* which was cultured in the medium containing antibiotics mixture. This bacterium, the identification of which is under investigation, seemed to originate from the inside of *P. tamarensis* cells, because it was not detected in the culture medium of *P. tamarensis*. When the antibiotic concentration in the medium of *P. tamarensis* increased, number of bacterial colonies derived from *P. tamarensis* decreased. At the same time, the toxicity of *P. tamarensis* also decreased, indicating the positive relationship between amount of bacteria and toxicity of *P. tamarensis*. In addition to these, bacteria were obtained only from toxic strains. They were not obtained from nontoxic strains of toxic dinoflagellates.

The primary producers of PSP toxins have been attributed to several species of dinoflagellates [1] and a blue-green alga [2]. On the other hand, Silva [14] found the presence of intracellular bacteria in *P. tamarensis* and presented a hypothesis that the toxin is produced not by dinoflagellates but by the intracellular bacteria. We also suggested that toxin production of *P. tamarensis* is not a hereditary character [5]. The results of the present study support both Silva's hypothesis and our previous results.

The main toxin components of *P. tamarensis* from which our bacterium was isolated, are sulfated derivatives of saxitoxin (gonyautoxins)[3]. The procedures which we applied to the isolation of bacterial toxin were not suitable for the isolation of gonyautoxins. Thus gonyautoxins could not be identified in the present study, if any. However, the presence of saxitoxin suggests the occurrence of gonyautoxins, because bioconversion between these toxins is reported [15]. The low toxin productivity of our bacterium seems to depend on the culture condition of the bacterium, or symbiosis may enhance the toxin production of this bacterium.

REFERENCES

1. E.J. Schantz, ANN. N.Y. Acad. Sci. 479, 15-23 (1986).
2. M. Alam, M. Ikawa, J.J. Sasner Jr. and P.J. Sawyer, Toxicon 11, 65-72 (1973).
3. T. Ogata, T. Ishimaru and M. Kodama, Marine Biology 95, 217-220 (1987).
4. D.H. Turpin, P.E.R. Dobell and F.J.R. Taylor, J. Phycol. 14, 235-238 (1978).
5. T. Ogata, M. Kodama and T. Ishimaru, Toxicon 25, 923-928 (1987).
6. Y. Kotaki, Y. Oshima and T. Yasumoto, Nippon Suisan Gakkaishi 47, 943-946 (1981).
7. T. Yasumoto, Y. Oshima, M. Hosaka and S. Miyakoshi, Nippon Suisan Gakkaishi 47, 929-934 (1981).
8. W.E. Fallon and Y. Shimizu, J. Environ. Sci. Health A12, 455-464 (1977).
9. Y. Oshima, M. Machida, K. Sasaki, Y. Tamaoki and T. Yasumoto, Agric. Biol. Chem. 48, 1707-1711 (1984).
10. H. Sommer and K.F. Meyer, Arch. Pathol. 24, 560-598 (1937).
11. Y. Fukuyo, K. Yoshida, T. Ogata, T. Ishimaru, M. Kodama, P. Pholpunthin, S. Wissesang and T. Piyakarnchana, This volume.
12. T. Ikeda, S. Matsuno, S. Sato, T. Ogata, M. Kodama, Y. Fukuyo and H. Takayama, This volume.
13. M. Kodama, T. Ogata, Y. Fukuyo, T. Ishimaru, P. Pholpunthin, S. Wisessang, K. Saitanu, V. Panikiyakarn, T. Piyakarnchana, Nippon Suisan Gakkaishi 53, 1491 (1987).
14. E.S. Silva, International IUPAC Symposium on Mycotoxins and Phycotoxins, Lausane (Pahotox Publication)(1979).
15. Y. Kotaki, Y. Oshima and T. Yasumoto, in: Toxic Dinoflagellates, D.M. Anderson, A.W. White and D.G. Baden eds. (Elsevier, New York, Amsterdam, Oxford 1985) pp. 287-292.

HEMOLYTIC TOXINS IN MARINE CYANOBACTERIUM SYNECHOCOCCUS SP.

Mitsui, A.*[1], D. Rosner[1], A. Goodman[1], G. Reyes-Vasquez[1],
T. Kusumi[1 2], T. Kodama[2], and K. Nomoto[2]
1) School of Marine and Atmospheric Science, University of Miami, Miami, Florida, USA
2) Suntory Fundamental and Bioorganic Research Institutes, Osaka, Japan

ABSTRACT

In tropical and subtropical marine environments, not only dinoflagellates, but also cyanobacteria (blue-green algae) should be considered as the major toxic photosynthetic microorganisms which implicate marine life, food poisoning public health. Many marine cyanobacterial strains from Atlantic sub-tropical marine environments which produced biologically active substances have been isolated in our laboratory at the University of Miami. Among these strains Synechococcus sp. Miami BGII 6S, which was isolated from a mass fish mortality site at a bay in South Florida, produced hemolysins. The toxin production depended on environmental factors. Maximum production was found in an early stationary growth phase. Chemical structure of one of the hemolysin of this strain was determined with cooperative study of above institutes as Galactopyranosyl lipid.

INTRODUCTION

Comparatively, less information is available on marine toxic cyanobacteria than fresh water toxic cyanobacteria. It is known in marine cyanobacteria that Gomphosphaeria aponina produces compounds active against the toxigenic dinoflagellate Ptychodiscus brevis[1], Oscillatoria erythraea has been linked to fish mass mortalities in the southern Caribbean[2], Phormidium corallyticum cause the black band disease of Atlantic reef coral[3], and shrimp mass mortalities have been attributed to toxic cyanobacteria[4].

Toxic activity either against mice or red blood cells or both have been demonstrated in aqueous and lipophilic extracts from marine cyanobacteria belonging to the Oscillatoriaceae.[5, 6] Some of the isolated toxins exhibit strong activity against P-388 lymphocytic mouse leukemia[7] and tumor promoters.[8, 9] It is also reported that marine cyanobacteria, Lyngbya gracilis, Oscillatoria nigroviridis and Schizothrix calcicala produced Debromoaplysiatoxine group.[10]

For several years our laboratory at the University of Miami Marine School has been surveying the tropical and subtropical marine environment for bioactive and toxic cyanobacteria.[11]

During 1972 and 1973, fish mortalities occurred in Biscayne Bay, Florida, but the responsible agent was not clearly identified. A sampling program was established immediately after every fish kill was reported. Water samples from surface, midwater and bottom water were collected from 12 stations located in affected and nearby unaffected areas. From them many strains of cyanobacteria were isolated by Mitsui and Rosner. Those isolates were screened for toxicity using the hemolysis test. Activity against human red blood cells was found in several of the isolates from affected areas, while those from unaffected ones showed no activity (Goodman & Mitsui).

One strain, designated as Synechococcus sp. strain Miami BGII 6S, was found to exhibit strong hemolytic activity. Preliminary investigations to establish the nature of its toxins have been carried out by Goodman and Mitsui. Partially purified an ether soluble toxic compound which appears to be lipoide and heat stable in nature, but the chemical structure was not elucidated. Studies carried out for the following years showed that the specific toxic activity varied significantly from one batch culture to another. In some cases high activity was present, while other cultures exhibited no activity.

---

*Person to whom correspondence should be addressed

RED TIDES: BIOLOGY, ENVIRONMENTAL SCIENCE, AND TOXICOLOGY
Okaichi, Anderson, and Nemoto, Editors

However, it was then found that hemolysin production occurred at a particular culture age and the particular culture conditions (Reyes-Vasquez and Mitsui). Finally, one of the hemolysin was isolated and its chemical structure was determined (Kusumi, Kodama, Nomoto, Reyes-Vasquez and Mitsui). This paper briefly reports such results. Detailed reports will be published elsewhere.

EXPERIMENTAL

Organism

The strain Miami BG II 6S occurs as paired spherical (1.5 to 2.0 µm in diameter) to ovoidal cells (long axis seldom exceed 2.3 µm), without sheath. They are pale blue-green under phase optics. Cells divide by binary fission in one plane only. In old cultures or when subjected to non-optimal conditions, short chains (3 to 8 cells) are formed. Such morphological characteristics are in agreement with those assigned to *Synechococcus naegeli*.[12]

Culture condition

An artificial medium (Medium A)[13] was used as a basal medium to culture the strain in 3.5L water-jacketed glass columns, or 2L capacity aspirator bottles, under controlled environmental conditions. When large amounts of biomass were required, 20L plastic carboys were used. Illumination was provided continuously, from bottom or sides, by cool white 40W fluorescent lamps. The light intensities employed were indicated in the results section.

Hemolysin extraction

Cells were harvested either by centrifugation or by filtration, and extracted with n-buthanol under constant stirring overnight, then allowed to separate into two phases, and the buthanol fraction was collected and evaporated to dryness under vacuum. The dry residue was used to prepare the water soluble and the ethanol soluble fractions for hemolysis tests.

Hemolysin assay

A red blood cell working suspension (RBCS), equivalent to 5 x 10^6 cells/ml, was prepared. For each assay 4.5 ml RBCS was added to 0.5 ml of sample. The mixture was incubated for 1 hr. at 37°C and then centrifuged at 510 x g for 5 min. Absorbance of the supernatant at 410 nm was then measured.

RESULTS

Growth conditions for hemolysis production

Figure 1 shows the relationship between the age of the culture and hemolysin production. Hemolysin was not found in cells during the exponential growth phase and suddenly appeared during the stationary phase and maximized at the latter stationary phase.

Various light intensities, temperatures and salinities on the growth were tested for hemolysin production. Optimum conditions for hemolysin production were found at 12 to 16 ‰ salinity, 25°C temperature and 50 µE/m^2/sec. irradiation (data not shown).

Isolation and chemical structure of hemolysin

For isolation of hemolysin, *Synechococcus* sp. Miami BG II 6S was cultured in the optimum conditions mentioned above and cells were harvested at late stationary phase (9 day culture age). Dried butanol extract (500 mg) of the cells was extracted again with $CHCl_3$-$MeOH$-H_2O (7: 13: 8). Then by the guide of hemolytic activity, the hemolysin was isolated as outlined in the flow sheet. The lower layer of the $CHCl_3$-$MeOH$-H_2O extract showed a strong hemolytic activity. Evaporation of the lower layer gave a residue. Chromatography of the residue over silica-gel, elution with $CHCl_3$-MeOH (10:1), and evaporation afforded compound I as colorless solids (38 mg).

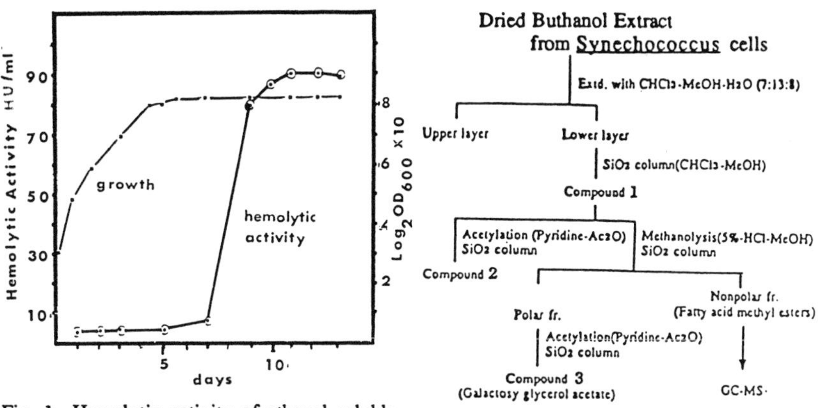

Fig. 1 Hemolytic activity of ethanol soluble fraction from BuOH extract of the cells during batch culture growth of Synechococcus sp. Miami BG II 6S.

Flow sheet of hemolysin isolation procedures.

Compound I, $[\alpha]_D$ +16.7° (in benzene), showed IR absorption due to hydroxyls (3425, 1180, 1070 cm-1), ester (1735, 1240 cm-1), and double bonds (1650 cm-1). The ^{1}H-NMR spectrum of I exhibits a 6-proton broad triplet (0.97 ppm, J=6), the signals equivalent to 27 methylene protons (between 1.10 and 2.95 ppm), the signals for 12 protons on carbons carrying the oxygen atoms (between 3.70 and 4.85 ppm), and the signals for 6 protons due to the double bonds (ca. 5.55 ppm). Acetylation of I (15 mg) with acetic anhydride in dry pyridine gave tetraacetate 2 (4.2 mg) as colorless powder. Compound 2 has no absorption due to hydroxyl group in IR spectrum. In the ^{1}H-NMR spectrum of five signals for five hydrogens (protons) on carbons carrying the acetoxyl groups are visible at 3.92, ca. 4.18, ca 4.25, 4.97 and 5.67 ppm, the presence of one primary OH group and three secondary OH groups I being proved.

Methanolysis of I (9 mg) with 5% HCl-MeOH afforded two products. The polor fraction (6 mg) of the methanolysis products was acetylated with acetic anhydride in dry pyridine to give hexaacetate 3 (4.1 mg) as colorless solids. The structure of 3 was confirmed by the ^{1}H-NMR and mass spectra, as shown in Fig. 2.

The nonpolor fraction was analyzed by GC-Mass spectrum. The result indicates that the nonpolar fraction is a mixture composed of four fatty acid methyl esters as shown in Fig. 3. However, C was a minor component and only a trace amount was detected.

Therefore as shown in Fig. 4 compound I with hemolytic activity have been deduced to be a mixture that C-2' and C-3' positions of glycerol moiety in I are substituted for the fatty acids as shown in Fig. 4.

Fig. 2

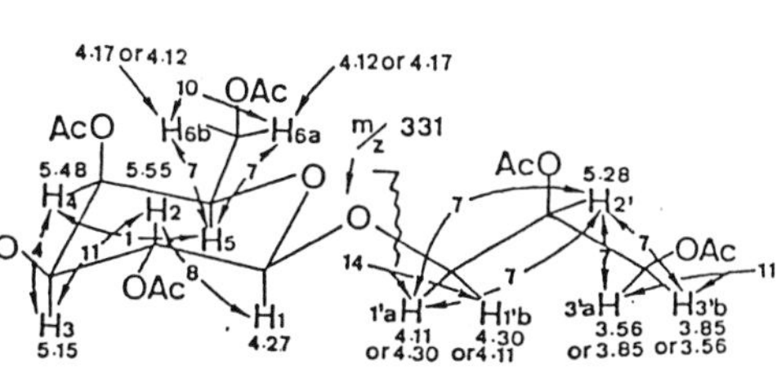

Fig.3

A—— $CH_3\text{-}(CH_2)_5\text{-}CH{=}CH\text{-}(CH_2)_7\text{-}COOCH_3$

B---- $CH_3\text{-}(CH_2)_{14}\text{-}COOCH_3$

C—— $CH_3\text{-}(CH_2)_m\text{-}CH{=}CH\text{-}(CH_2)_n\text{-}COOCH_3$ (m+n=13)

D---- $CH_3\text{-}(CH_2)_4\text{-}CH{=}CH\text{-}CH_2\text{-}CH{=}CH\text{-}(CH_2)_7\text{-}COOCH_3$

Fig. 4

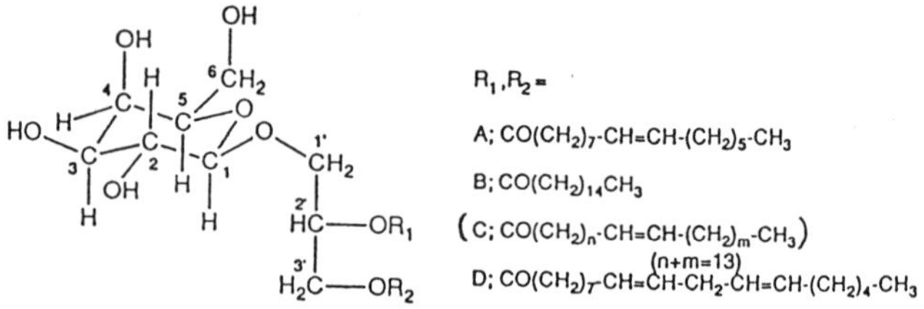

DISCUSSION

As the result of chemical analysis, it elucidated that our isolated compounds include four galactosyl diacyl glycerol compounds. These compounds change the moiety of fatty acids, cis-9-hexadecenic, hexadecanic, cis-cis-9-12-octadienic and uncertain one. The structurally related compounds were found as mono- and digalactosyldiglycerides in higher plant, mono-diglycerides acylated with highly unsaturated fatty acids in a dinoflagellate Glenodinium sp., digalactosyl monoglyceride with palmitic acid in a green alga Ulva pertusa[14], and β-galactosidic digalactosyl monoglycerifes with two unsaturated fatty acids in phytoflagellate Prymnesium parvum.

Phytoflagellate Prymnesium parvum is well known to produce toxins with ichthyotoxic and hemolytic activities and thus cause mass mortality of fish in blackish water culture ponds. The chemical properties of the major toxins of this and dinoflagellates were clarified[15,16],1'-O-octadecatetraenoyl-3'-O-ß-D-galactopyranosyl)-glycerols and 1'-O-octadecapentaenoyl-3'-O-(6-O-ß-D-galactopyranosyl)-glycerols.

Our cyanobacterial hemolysins are closely related in higher plants. But from fatty acids moiety some of hemolysin are different in higher plants. Our cyanobacterial hemolysin appeared only at the late stationary growth phase (Fig. 1). This suggests that hemolysin might originate from a cellular component. However, as a similar hemolytic compounds were isolated from sponge[17], the further experiments are needed for this conclusion.

REFERENCES

1. D.L. Eng-Wilmont, et al. Toxic Dinoflagellate blooms. Elsevier, New York, 355-360. (1979).
2. C.van Buurt. Gulf and Carib. Fish. Inst. Ad-Hoc. Symp., Puerto Rico, 9-13. (1981)
3. K. Rutzler and D.L. Santavy. P.S.Z.N.I., Mar. Ecol. 4, 301-319. (1983).
4. D.W. Lightner. Invert. Pathol. 32, 139-150. (1978).
5. Y.Hashimoto, et al. Animal, Plant and Microbial Toxins. Vol I, Plenum Press, New York, 33-338. (1976).
6. Y.Hashimoto, et al. Proc. 7th Int. Seaweed Symp., Sapporo, 1971, 569-572. (1972)
7. J.S. Mynderse, et al. Science 196, 538-540. (1977).
8. R.E. Moore. Pure and Appl. Chem., 54: 1919-1934. (1982).
9. H. Fujiki, et al. Jpn. J. Cancer Res. (Gann), 76, 257-259. (1985).
10. R.E. Moore. The Water Environment. Algal Toxins and Public Health, Plenum Press, 15-23. (1981).
11. A. Mitsui, et al. Biotechnology and Bioprocess Engineering, United India Press, 119-155. (1985).
12. R.J. Ripka, et al. J. Gen. Microbiol., 111: 1-61. (1969).
13. S. Kumazawa and A. Mitsui. Int. J. Hydrogen Energy, 6, 339-348. (1981).
14. N. Fusetani and Y. Hashimoto. Agr. Biol. Chem., 39, 2021-2025. (1975).
15. H. Kozaki, et al. Agr. Biol. Chem. 46, 233-236. (1982).
16. T. Yasumoto, et al. Biol. Bull. 172: 128-131. (1987).
17. H. Kikuchi, et al. Chem. Pharm. Bull. 30, 3544-3547. (1982).

SEPARATION OF TOXINS FROM HARMFUL RED TIDES OCCURRING ALONG THE COAST OF KAGOSHIMA PREFECTURE

ONOUE, Y. AND K. NOZAWA
Faculty of Fisheries, Kagoshima University, Kagoshima 890, Japan

ABSTRACT

Three toxic fractions, neurotoxic, hemolytic and hemagglutinative, were separated from red tides of *Chattonella marina, Cochlodinium* type'78 Yatsushiro and *Gymnodinium* type '84 K occurring along the coast of Kagoshima Prefecture. A variety of organic solvents were employed for fractionation of neurotoxin, whereas aqueous phenol was used for extraction of hemolysins and hemagglutinins. Ichthyotoxicity of these three toxins to juvenile red seabream was examined. The test fish died within 4-10 minutes after exposures to the neurotoxin fractions (0.01-0.02%) from the three types of red tides, causing conspicuous edema on their secondary lamellae. On exposures to the hemolysin and hemagglutinin fractions (0.02%) fish also died with a marked mucous release on their gill filaments. The survival time of the fish fell between 20-50 minutes. The hemolytic activity on sheep blood cells was estimated to be 0.18 h.u./mg for *Chattonella* and *Gymnodinium* and 0.50 h.u./mg for *Cochlodinium*. The hemagglutinating titres on the same blood cells were 12,500 for *Chattonella* and *Gymnodinium,* and 6,200 for *Cochlodinium*. The hemagglutinating fraction induced violent convulsions followed by respiratory paralysis on intravenous injection into mice.

INTRODUCTION

In Kagoshima region three species of phytoflagellates have been incriminated in massive fish kills. *Chattonella marina,* which is most commonly found in coastal waters of Kagoshima Prefecture becomes dangerous on propagation in young yellowtail farming areas [1]. *Cochlodinium* type'78 Yatsushiro occurs in Yatsushiro Sea, southwest of Kyushu Island, posing a threat to cultured red seabream and young yellowtail [2]. The first occurrence of *Gymnodinium* type'84 K was observed in Kagoshima Bay in 1984 [3]. This dinoflagellate has been found highly toxic to cultured fish. In the present work three types of toxins were separated from each of the above three red tide organisms, and were examined on their potency and effects on fish.

EXPERIMENTAL

Red tide organisms

The algal cells of *Chattonella marina* were collected by continuous centrifugation (800-1,000 rpm, 500 ml/min) from 300 l of the 1985 red tide (80,000-100,000 cells/ml) in Kagoshima Bay, and then freeze-dried. The other two plankton cells were collected similarly from 1,000 l of the 1984 red tide (1,000-1,500 cells/ml) of *Cochlodinium* type'78 Yatsushiro in Yatsushiro Sea and 1,000 l of the 1985 red tide (15,000-20,000 cells/ml) of *Gymnodinium* type'84 K in Kagoshima Bay. The number of algal cells per g of dry weight was 1×10^9 in *Chattonella,* 3×10^8 in *Cochlodinium* and 2×10^8 in *Gymnodinium*.

Separation of toxic fractions

Three toxic fractions were separated from plankton cells by a combination of the method of Baden and Mende [4] with that of Bass *et al.* [5].

RED TIDES: BIOLOGY, ENVIRONMENTAL SCIENCE, AND TOXICOLOGY
Okaichi, Anderson, and Nemoto, Editors

Algal cells were shaken with three volumes of chloroform and ethanol (2:1), and centrifuged at 3,500 rpm for 15 minutes. The chloroform layer was evaporated, dissolved in 90% methanol, and extracted with petroleum ether. The methanol layer was evaporated and reextracted with a small amount of acetone to obtain neurotoxin fraction. The yield of neurotoxin fraction was 340 mg (1.20%) in *Chattonella*, 103 mg (3.03%) in *Cochlodinium* and 128 mg (1.03%) in *Gymnodinium*. The defatted algal cells were homogenized, extracted with 2 volumes of 40% phenol, and centrifuged. The phenol layer was dialyzed against water for 3 days, centrifuged, and lyophilized to recover hemolysin fraction. The recovery of hemolysin fraction was 55.4 mg (0.20%) in *Chattonella*, 18.6 mg (0.55%) in *Cochlodinium* and 40.9 mg (0.33%) in *Gymnodinium*. The aqueous layer was defatted with chloroform, dialyzed, and lyophilized to produce hemagglutinin fraction. The yield of hemagglutinin fraction was 1.50 g (5.30 %) in *Chattonella*, 180 mg (5.29%) in *Cochlodinium* and 120 mg (0.97%) in *Gymnodinium*.

Ichthyotoxicity

Toxic fractions (0.01-0.03%) were added to 200-ml beakers with seawater (temp. 26-27°C, DO 8-9 ppm) and a couple of juvenile red seabream (body length 30-40 mm, body weight 500-600 mg), and were tested.

Histological observation

The gills of juvenile red seabream exposed to toxic fractions were removed, fixed in a Bouin's fluid, treated with 2% trichloroacetic acid, and embedded in paraffin. Sections of 5-10 µm were cut crosswise and horizontally and stained with periodic acid-Schiff (PAS).

Hemolytic activity

Hemolytic index. Hemolysin fraction (5 mg) was dissolved in 1 ml of isotonic phosphate buffer and a series of dilution (0.001-0.25%) prepared. An equal volume of 2% sheep blood cell suspension was added to each dilution and left at 15-16°C for 10 hours. The hemolytic index (H.I.) or dilution factor was calculated from the concentration at which a complete hemolysis was attained.

50% hemolytic dose (HD_{50}). Hemolysin fraction (2 mg) was dissolved in 1.4 ml of 0.2 M borate buffer containing 0.1 ml of 1 M $CaCl_2$ and added with an equal volume of 1% sheep blood cell suspension. After being kept in a 37°C water bath for 30 minutes, the suspension was centrifuged. Absorbance of the supernatant at 550 nm was measured by a Hitachi 101 spectrophotometer.

Hemagglutinating activity

Hemagglutinative fraction (5 mg) was dissolved in 1 ml of isotonic phosphate buffer. Activity of the test solution was measured similarly as described in the hemolytic activity test.

Mouse toxicity

Intraperitoneal injection. Hemagglutinative fraction (5 mg) was dissolved in 5 ml of physiological saline. A 1-ml portion of the test solution was injected intraperitoneally into male mice (ddY) weighing approximately 20 g.

Intravenous injection. Hemagglutinative fraction (2-20 µg) in 0.2 ml of physiological saline was injected into the tail vein of a 20-g mouse.

Chemical composition

The following methods were used for analysis of hemagglutinative fraction. The ascorbic acid method of Chen *et al.* [6] for phosphorus, the phenol-sulfate method of Dubois *et al.* [7] for hexose, a Wako reagent kit with a color development of GPO-p-chlorphenol for triglyceride, and the Cu-Folin method of Lowry *et al.* [8] for protein.

RESULTS

Three lethal factors, neurotoxin, hemolysin and hemagglutinin, were fractionated from red tides of *Chattonella marina*, *Cochlodinium* type'78 Yatsushiro and *Gymnodinium* type'84 K.

Neurotoxin fraction

When exposed to 0.02% neurotoxin fraction, a test fish subsided, as if it were anesthetized, while color change (whitening) occurred in the body. This was followed by labored breathing and respiratory arrest. The survival time was shortened in *Gymnodinium* (less than 4 min), but prolonged in *Chattonella* (7-9 min) and *Cochlodinium* (8-10 min). The branchial edema produced in the secondary lamella of exposed fish accounted for more than 40% in *Chattonella* and *Cochlodinium*.

Hemolysin fraction

On exposure to hemolysin fraction (0.02%), fish evoked vomiting and violent convulsions with occasional jumping, which was followed by loss of balance, labored breathing and respiratory arrest. *Cochlodinium* allowed a slightly shorter survival time (20-30 min), as compared with *Chattonella* (25-37 min) and *Gymnodinium* (35-49 min). The edema in exposed fish fell between 15-30% in these three phytoflagellates. In this case, hemorrhage and mucous release also seemed to occur in the secondary lamella. Hemolytic units (HD_{50}) per mg of fraction were estimated to be 0.18 in *Chattonella* and *Gymnodinium*, and 0.50 in *Cochlodinium*.

Hemagglutinin fraction

At 0.02% of this fraction, fish provoked similar signs as seen in hemolysin fraction. Death ensued within 20-50 minutes in three species of red tide organisms. Besides edema, deformations of the secondary lamella became evident in *Chattonella* and *Gymnodinium*. The hemagglutinating titres on sheep blood cells were 12,500 for *Chattonella* and *Gymnodinium* and 6,200 for *Cochlodinium*.

Mice did not exhibit any abnormal signs on intraperitoneal injection (2.5 mg), but they died with respiratory paralysis on intravenous injection (20 μg). The lethal dose (LD_{50}) in mice was found to be 2-4 mg/kg on the fractions from these three species of plankton.

The chemical analysis of hemagglutinin fraction resulted in: (as μg per mg) phosphorus 102-234; sugar (as hexose) 408-575; triglyceride 30-127; protein 11-50. A fairly large difference in composition was noted between *Gymnodinium* and the other two species. The phosphorus content was about three times higher in *Gymnodinium* than in the other two, but *vice versa* the triglyceride content.

DISCUSSION

Fish gills have large surface areas and a high permeability to various substances. They therefore are important sites of entry for toxins. Opercular lesions might occur in contact with toxins on the gill surfaces. Toxins entering the blood stream through the gill membranes could affect many parts of the fish. According to Ungell *et al.* [9], stress stimuli would exert very severe effects on gills and other target organs of fish. Primary effects are believed to involve the branchial vascular resistance induced by hypoxia and the hypothalamic control via the autonomic nervous system and the corticosteroid and catecholamine (chromaffin) secreting tissues. This may secondarily affect ion transfer and adrenergic nerve activity (branchial and systemic vasculatures).

Extensive branchial edema occurred in fish gills which were exposed to red tide toxins, suggesting that such toxins might pose an intense stress on fish with a severe impairment in the transfer of oxygen and other compounds across the gills.

Pärt *et al.* [10] have found in their experiments of gill transfer of rainbow trout that the structural deterioration of the gills could be prevented by adding adrenaline in the perfusion fluid.

All these findings will give an insight into the mechanism of massive fish kills.

ACKNOWLEDGEMENTS

We express our gratitude to Mr. K. Kumanda, the Director of Kagoshima Prefectural Fisheries Experimental Station for his assistance in collecting red tide organisms. This study was supported in part by grant-in-aid for scientific research and for publication of scientific research result from the Ministry of Education, Science and Culture, and by a grant from the Ministry of Agriculture, Forestry and Fisheries.

REFERENCES

1. T. Noro and K. Nozawa, Jap. J. Phycol. 29, 73 (1980).
2. Y. Onoue, K. Nozawa, K. Kumanda, K. Takeda and T. Aramaki, Bull. Japan. Soc. Sci. Fish. 51, 147 (1985).
3. Y. Onoue, K. Nozawa, K. Kumanda, K. Takeda and T. Aramaki, Bull. Japan. Soc. Sci. Fish. 51, 1567 (1985).
4. D.G. Baden and T.J. Mende, Toxicon 20, 457 (1982).
5. E.L. Bass, J.P. Pinion and M.E. Sharif, Aquatic Toxicology 3, 15 (1983).
6. P.S. Chen, Jr., T.Y. Toribara and H. Warner, Anal. Chem. 28, 1756 (1956).
7. M. Dubois, K.A. Gilles, J.K. Hamilton, P.A. Rebers and F. Smith, Anal. Chem. 28, 350 (1956).
8. O.H. Lowry, N.J. Rosebrough, A.L. Farr and R. J. Randall, J. Biol. Chem. 193, 265 (1951).
9. A.L. Ungell, A. Kiessling and S. Nilsson in: Toxins, Drugs and Pollutants in Marine Animals, L. Bolis, J. Zadunaisky and R. Gilles, eds. (Springer-Verlag, Berlin 1984) p. 114.
10. P. Pärt, H. Tuurala and A. Soivio, Comp. Biochem. Physiol. 71C, 7 (1982).

ICHTHYOTOXINS IN A FRESHWATER DINOFLAGELLATE *PERIDINIUM POLONICUM*

YASUKATSU OSHIMA, HARUTAKA MINAMI, YUTAKA TAKANO, AND TAKESHI YASUMOTO
Department of Food Chemistry, Faculty of Agriculture, Tohoku University
Tsutsumi-dori, Sendai, Miyagi 980, Japan

ABSTRACT

High ichthyotoxicity was detected in a freshwater dinoflagellate *Peridinium polonicum*, but not in *P. penardii* and *P. bipes*. Three toxins, polonicumtoxins A, B and C (PT-A, B and C), were isolated from cultured cells and natural bloom of *P. polonicum*. PT-A, B and C killed killifish *Oryzias latipes* within 40 min at concentrations of 13, 33 and 300 ppb, respectively. PT-A and C also had mouse lethalities of 1.5 and 2.0 mg/kg by ip injection. Neither purified toxins nor crude methanol extracts of the dinoflagellates had mutagenicity in Ames and umu tests. Novel structures of the toxins with unique tetrahydropyridine moiety were elucidated by spectroscopic analyses.

INTRODUCTION

In recent years, blooms of freshwater dinoflagellates *Peridinium* spp. have been reported at many man-made lakes in Japan. The blooms cause serious environmental problems, since most of the affected lakes are used as reservoirs for water supply [1]. Occasional fish kills accompanying the bloom events made the problem worse. Among several species reported in Japan, *P. polonicum* is the only species known to produce ichthyotoxin(s). Hashimoto *et al.*[2] isolated a toxin called glenodinine from the dinoflagellate bloom occurred at Lake Sagami, Kanagawa in 1962 and reported its biological and chemical properties. However the structure of glenodinine remains unknown.

The present study deals with the isolation of three closely related ichthyotoxins from *P. polonicum* and elucidation of their chemical structures. Toxicity to mice and mutagenicity of the toxins were also investigated to evaluate their health risks.

EXPERIMENTAL

Specimens

Natural bloom of *P. polonicum* cells were concentrated with a plankton-net (20 um pore size) at Lake Shinryu, Hiroshima Prefecture in November, 1985 and transported to the laboratory in a plastic bottle. Unialgal culture of *P. polonicum* isolated at the lake was grown in No. 10 medium of Chu [3] supplemented with thiamine (100 ug/L), vitamin B_{12}(2 ug/L), biotin (1 ug/L), streptomycin (2 mg/L), dihydrostreptomycin (50 mg/L), neomycin (2 mg/L) and P-II metal mixture [4](5 mL/L), at 17±2°C under illumination of white fluorescent lights at 3000 lux of 16-8 hrs light-dark cycles. The cultured cells were harvested by continuous flow centrifugation at 7500 rpm and 100 mL/min. *P. polonicum* and *P. bipes* from Lake Okutama, Tokyo, were also cultured under the same conditions. Bloom of *P. penardii* was collected from Lake Kanna, Gunma Prefecture, in November, 1984.

Isolation of toxins

The toxins in cultured cells were extracted with MeOH, whereas the extraction was done with $CHCl_3$ for the lake water containing organisms. The

RED TIDES: BIOLOGY, ENVIRONMENTAL SCIENCE, AND TOXICOLOGY
Okaichi, Anderson, and Nemoto, Editors

extracts were purified by liquid-liquid partition between $CHCl_3$ and water followed by column chromatography on Develosil ODS (15-30 mesh, Nomura Chemical) with the following two solvents: 0.05 N AcOH-MeOH (4:6) and (8:2). Final purification was achieved on a silicagel 60 column in $CHCl_3$-MeOH (9:1) or (7:3). Purity of the toxins was tested by TLC using silicagel 60 plates with a solvent of $CHCl_3$-MeOH-H_2O (65:25:4).

Spectroscopy

PMR and CMR spectra were taken with JEOL FX-100 and Varian XL-200 spectrometers, IR spectra with JASCO IR-1 spectrometer, and mass spectra with Hitachi M-52 and JEOL JMS-01SG-2 spectrometers.

Bioassays

Three killifish *Oryzias latipes* were placed in 50 ml of test solution prepared by dissolving the sample in 0.04 M Tris-HCl buffer (pH 8.8). Toxins suspended in 1% Tween 60 were injected intraperitoneally into male mice of ddY strain weighing 16-18 g.

Mutagenicity tests

Pure toxins and methanol extracts of cultured cells were dissolved in DMSO and applied for Ames test using *Salmonella typhimurium* strains TA98 and TA100 with or without S9 mix [5]. PT-A, B and C were also tested for mutagenicity using a kit (UMU-RAC kit, Otsuka Assay Laboratory) for *umu* reaction [6].

RESULTS

Ichthyotoxicity of *Peridinium* spp.

Toxicities of culture and natural bloom of *Peridinium* spp. are summarized in TABLE I. All *P. polonicum* specimens showed high lethalities to the fish although toxicity levels differed between the two strains. No toxicity was detected either in *P. bipes* or *P. penardii*.

Isolation and structures of toxins

Bioassay-directed fractionation of the extract from 700 L culture of *P. polonicum* Hiroshima strain led to the isolation of one major toxin named polonicumtoxin A (PT-A, 14 mg) and one minor toxin, polonicumtoxin B (PT-B, 1 mg). Along with 36 mg of PT-A, another active component, polonicumtoxin C (PT-C, 43 mg) was obtained from 18 L of concentrated bloom water. No PT-B was detected in the natural organisms. On TLC, they were detected as characteristic grayish-green spots with ninhydrin and as orange spots with Dragendorff reagent at Rf values of 0.74, 0.69 and 0.30, for PT-A, B and C,

TABLE I. Ichthyotoxicity of freshwater *Peridinium* spp.

Specimens	Locality	Toxicity*
P. polonicum (bloom)	Lake Shinryu, Hiroshima	24,000
P. polonicum (culture)	Lake Shinryu, Hiroshima	17,000
P. polonicum (culture)	Lake Okutama, Tokyo	2,400
P. penardii (bloom)	Lake Kanna, Gunma	>5,000,000
P. bipes (culture)	Lake Okutama, Tokyo	>1,000,000

*Expressed in the minimum cell number required to kill a killifish.

respectively. Production of PT-A and B by the Tokyo strain was also confirmed by TLC.

Molecular formula of PT-A, $C_{15}H_{23}NO_2$, was deduced from elemental analysis and mass spectrum (m/z 249.1447,M). Infrared spectrum indicated ester (1735, 1165 cm^{-1}) and olefin (1660 cm^{-1}) and no OH or NH in the molecule. CMR and PMR spectra implied a vinylacetic acid ester, a trisubstituted olefin and a methyl on olefin in the PT-A molecule. Medium intensity IR absorption at 1640 cm^{-1} and CMR signal at 178.4 ppm indicated that the nitrogen atom was present as a part of imine and the structure 1 possessing unique tetrahydropyridine moiety was deduced from decoupling experiments. Imine nature of the nitrogen was also evidenced by the observation of rapid exchange of four methylene protons (*) with solvent deuterium (CD_3OD)in PMR, presumably through imine-enamine equilibration. The PMR spectrum of PT-B was very much similar to that of PT-A except for presence of an acetate signal at 2.1 ppm and absence of vinylacetate signals, indicating the structure 2. IR (3300 cm^{-1}), mass spectrum (m/z 181, M^+), CMR and PMR spectra indicated PT-C to be an alcohol 3. Actually PT-C was obtained by alkaline hydrolysis of PT-A and PT-B. Further confirmation of the structures was achieved by the following reactions. Reduction of PT-C with $NaBH_4$ yielded a 2-substituted piperidine 4 and acetylation of PT-C with acetic anhydride in pyridine gave two products, an enamine 5 and ketone 6.

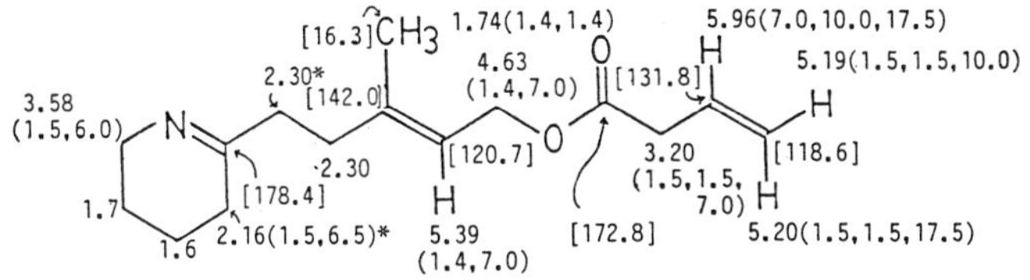

1

PMR Chemical shift in ppm with coupling constant ( ) in $CDCl_3$.
CMR Chemical shift in ppm [ ]in CD_3CN.

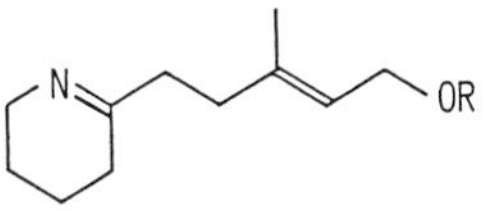

2 R = Ac
3 R = H

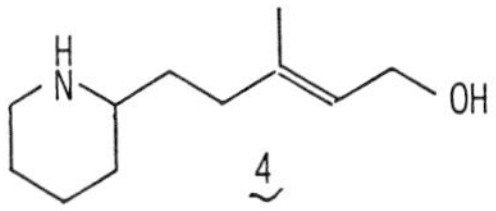

4

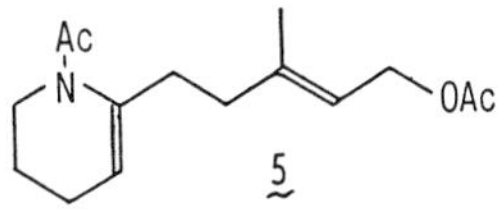

5

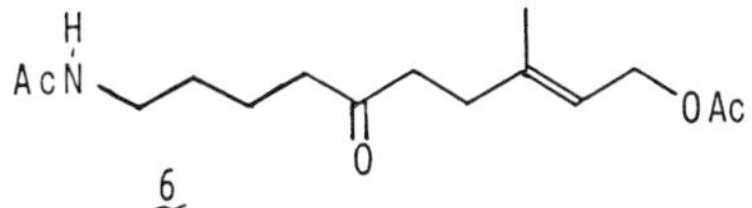

6

Biological activity of toxins

Ichthyotoxicities and mouse lethalities of purified toxins are shown in TABLE II. Although the potencies of toxins varied widely, the test fish exposed to PT-A, B and C showed the same symptoms; a slight excitement followed by stiffness of the pectoral fins, decrease of response to stimuli, loss of balance and occasional abrupt jumping before death. The fish died in 30-40 min at the minimum lethal concentration. Test solution pH affected the toxixity greatly, as reported for glenodinine[3]. LC_{99} of PT-A went up 4 times by decreasing pH from 8.8 to 6.5. The subjected mice died quickly, within 5 min, after convulsive jumps and spasmodic breath. In contrast to ichthyotoxicity, PT-A and PT-B had comparable levels of toxicity to mice. Ames and umu tests of the purified toxins were all negative with or without S9 mix. Also no mutagenicity was detected in the crude methanol extracts of cultured P. polonicum and P. bipes by Ames test.

TABLE II. Biological activities of polonicumtoxins A, B and C.

Activity			A	B	C
Ichthyotoxicity to O. latipes		at pH 8.8	13	33	300
(LC_{99} in ppb)		at pH 6.5	52	-	-
Lethality to mice (LD_{99} in mg/kg)			1.5	-	2.0
Mutagenicity	by Ames test	with S9 mix	ND	-	ND
		without S9 mix	ND	-	ND
	by umu test	with S9 mix	ND	ND	ND
		without S9 mix	ND	ND	ND

(ND) : not detected. (-) : not tested.

DISCUSSION

This will be the first to clarify the toxin structures from a freshwater dinoflagellate. In comparison with glenodinine, the first toxin reported for P. polonicum by Hashimoto *et al.*[2], the present structures differed from their descriptions of sulfur-containing alkaloid, although the symptoms of test animals and solubility of the toxins were very much similar It is difficult to conclude the identity of the toxins without direct comparison.

Ichthyotoxicity of the major toxin PT-A was very potent and comparable to those of brevetoxins isolated from marine dinoflagellate *Gymnodinium breve* [7]. However mouse lethal potency was very low in comparison with other dinoflagellate toxins causing health hazard to human, such as paralytic and diarrhetic shellfish toxins. Although no mutagenicity was detected in this study, studies on chronic toxicity and pathology may be required to evaluate the potential health risks of the toxins. Systematic studies to assess the safety of other bloom-forming species of *Peridinium* is strongly advisable since they related to water supply.

ACKNOWLEDGEMENTS

The authors wish to express sincere thanks to Dr. Yasuwo FUKUYO, Tokyo University; Mr. Kentaro IMAMURA, Mihara Public Health Center, Hiroshima Prefecture; Mr. Mitsutsugu HOSAKA, the water-supply department of Tokyo for their collaboration on collecting samples and identification of the organisms.

REFERENCES

1. T. Itoh, Bull. Plankton Soc. Japan 26, 113 (1979).
2. Y. Hashimoto, T. Okaichi, L.D. Dang, and T. Noguchi, Bull. Jpn. Soc. Soc. Fish. 34, 528 (1968).
3. S.P. Chu, J. Ecol. 30, 284 (1942).
4. L. Provasoli in: Proceedings US-Japan Cont. Hakone, Japanese Society Plant Physiology, 63 (1968).
5. B.N. Ames, T. Yahagi, Y. Seino, T. Sugimura, and N. Ito, Mutation Res. 42, 335 (1975).
6. Y. Oda, S. Nakamura, I. Oki, T. Kato, and H. Shinagawa, Mutation Res. 147, 219 (1985).
7. H. Chou and Y. Shimizu, Tetrahedron Lett. 23, 5521 (1982).

PHARMACOLOGICAL ACTIONS OF BREVETOXIN FROM PTYCHODISCUS BREVIS ON NERVE MEMBRANES

JAMES M. C. HUANG AND CHAU H. WU
Department of Pharmacology, Northwestern University Medical School, 303 E. Chicago Avenue, Chicago, Illinois 60611 U.S.A.

ABSTRACT

Pharmacological actions of brevetoxins (PbTXs) isolated from the Florida red-tide dinoflagellate, Ptychodiscus brevis, were studied using invertebrate axonal preparations. Application of PbTX-3 to the crayfish and squid giant axons caused a dose-dependent depolarization of the axonal membranes and eventual loss of membrane excitability. The extent of the maximum depolarization was 30 mV and the ED_{50} was 1.7 nM. Voltage-clamp experiments provided direct evidence that the primary action of the toxin is to induce an opening of the sodium channel at the resting membrane potential. The induced opening of the channels can be antagonized by tetrodotoxin and procaine. Thus, the electrophysiological evidence shows that brevetoxin induces the sodium channel to open, resulting in the depolarization of the axon membrane. Depolarization of the nerve terminal and the resulting enhanced release of transmitters can account for a number of pharmacological actions of brevetoxin on various organ systems as well as clinical symptoms of intoxication.

INTRODUCTION

A catastrophic episode of red tide in the Gulf of Mexico in 1946-1947 affected a wide area along the coast of Florida. Tons of dead fish littered the beaches with disastrous consequences to the economy and public health of the region. More than 80 episodes of red tide have since been recorded there. The responsible organism was subsequently identified as a new species of unarmored dinoflagellate and named Gymnodinium breve (1). This organism has been reclassified as Ptychodiscus brevis (2).

A total of eight toxins has been isolated from the dinoflagellate and purified to homogeneity. These toxins are named brevetoxins, and a notation system, PbTX-1 through PbTX-8, based on the numbering system of Shimizu (3), has been proposed to designate the various brevetoxins isolated by several laboratories (4). The brevetoxins can be divided into two subgroups according to their chemical structures; PbTX-2, -3, -5, -6 and -8 belong to one group, and PbTX-1 and PbTX-7 to the other. We have examined the pharmacological action of PbTX-3 (formerly designated T17) on the membrane excitability of single giant axons of the crayfish and squid. The results indicate that PbTX-3 acts on the sodium channel and induces the channel to open at the resting membrane potential and thus depolarizes the nerve.

MATERIALS AND METHODS

Giant axons in the circumesophageal connectives of the crayfish, Procambarus clarkii, were used for experiments on membrane excitability. The axon was impaled with a glass microelectrode filled with 3 M KCl solution. The electrical resistance of the microelectrodes was 5-10 megohm. The axon was continuously perfused with van Harreveld's solution which

RED TIDES: BIOLOGY, ENVIRONMENTAL SCIENCE, AND TOXICOLOGY
Okaichi, Anderson, and Nemoto, Editors

contained: 205 mM NaCl, 5.4 mM KCl, 2.6 mM $MgCl_2$, 10 mM $CaCl_2$, and 3 mM HEPES buffer, pH 7.55. The temperature was at 20°C.

Giant axons of the squid, Loligo pealei, were used for voltage clamp experiments. The basic arrangement of the axial-wire voltage clamp was essentially the same as that described by Wu and Narahashi (5). To record only sodium currents, potassium currents were eliminated by perfusing the axon with potassium-free solutions as follows: (a) External solution: 450 mM NaCl, 50 mM $CaCl_2$, and 10 mM HEPES buffer, pH 8.00; (b) Internal solution: 250 mM Cs glutamate, 20 mM NaF, 30 mM Na phosphate buffer, and 400 mM sucrose, pH 7.30. The temperature of the preparation was maintained at 10°C.

Purified samples of PbTX-3 were generously supplied by Dr. Daniel G. Baden of the University of Miami. Tetrodotoxin (TTX) was obtained from Sankyo Company (Tokyo, Japan) and procaine HCl from Sigma Chemical Company (St. Louis, MO).

RESULTS AND DISCUSSION

Depolarization of Axonal Membranes

Brevetoxin depolarizes nerve membranes in a dose-dependent manner. External application of PbTX-3 to a crayfish axon at a concentration of 1.1 μM induced a depolarization which attained a maximum of 30 mV within 30 sec, and the depolarization continued for 30 min. Application of the toxin to the inside of the axon gave similar results. Data from 22 crayfish axons were fitted to a dose-response curve which gave an ED_{50} of 1.7 nM. This value is in reasonable agreement with the K_D of about 3 nM obtained in the binding study of rat brain synaptosomes (4). The maximum depolarization obtained with the highest doses tested was 30 mV. The depolarization was prevented or reversed when the axon was treated with 300 nM TTX. The depolarization was highly dependent on the external sodium concentration; decreasing the sodium concentration from 205 mM to 1 mM completely reversed the depolarization. Taken together, these results indicate that PbTX-3 causes depolarization by inducing the sodium channel to open. We therefore conducted the following voltage clamp experiments on squid axons to observe this effect directly.

Sodium Currents Induced by PbTX-3

Isolated squid giant axons were voltage-clamped by the axial wire technique. The axon was perfused with potassium-free physiological solutions to eliminate the potassium currents. Figure 1 shows a typical record of the effect of PbTX-3 on the holding current of an axon. Prior to toxin application, the axon was voltage-clamped at -130 mV, and no test pulses were applied. At this potential the holding current (labeled I_H) was negligible, indicating that few channels were open. At the first arrow PbTX-3 (1 μg/ml) was applied internally, and an increase in the holding current in the inward direction was recorded. Since the potassium currents had been eliminated, the increase in the holding current must have come from sodium currents induced by the toxin. This is better demonstrated when TTX (300 nM) was applied (second arrow). The holding current was completely blocked by TTX. Since TTX is a specific blocker of the sodium channel, the results indicate that the channels induced to open are indeed those of sodium channels. Analysis of the current-voltage relationship indicates that the brevetoxin-modified channel can be activated in the potential range from -160 to -80 mV where normally sodium channels do not respond to open.

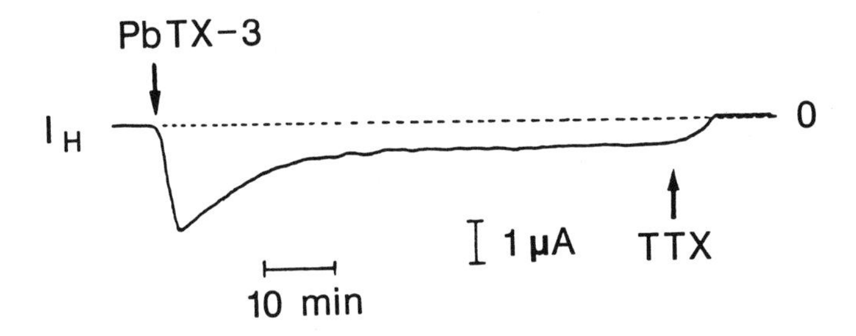

FIG. 1. Induction of sodium current by PbTX-3 in a squid axon. At the first arrow, 1 µg/ml of PbTX-3 was applied, and the holding current (I_H) increased in amplitude. At the second arrow, 300 nM of TTX was applied which completely blocked the current. Dashed line indicates the zero baseline.

Antagonism by Procaine

Previous experiments by Huang et al. (6) suggest that local anesthetics can antagonize the action of brevetoxin. We therefore carried out the following experiment to assess the antagonism. The axon was voltage-clamped at -120 mV and perfused with potassium-free physiological solutions (Fig. 2). Application of PbTX-3 (1 µg/ml at the first arrow) immediately increased the holding current by 3.3 µA, indicating simultaneous opening of a large number of sodium channels induced by the toxin. This increase in the holding current was fairly well maintained during the 30-min period of toxin treatment. Following this, procaine (30 mM) was applied to both inside and outside of the axon (second arrow), causing a time-dependent suppression of the holding current. It is noteworthy that the suppression of the induced current by procaine, even at a relatively high dose of 30 mM, was incomplete as compared with the action of TTX, which abolished the induced current.

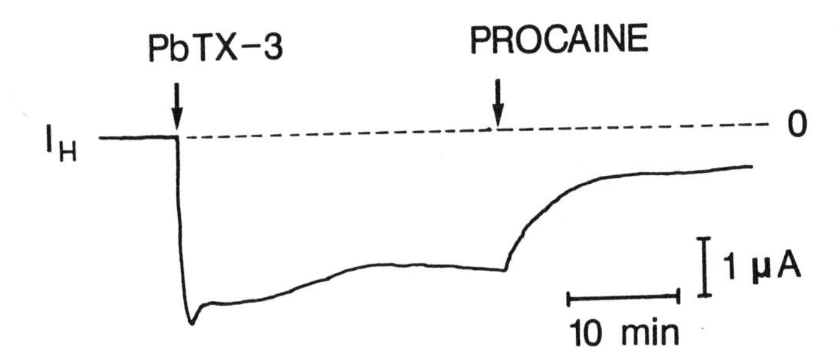

FIG. 2. Antagonism of the brevetoxin-induced increase in the sodium current by procaine. PbTX-3 (1 µg/ml) was applied to a squid axon (first arrow) for 30 min, resulting in a prolonged increase in the holding current (I_H). Procaine (30 mM) was then administered both externally and internally (second arrow), causing a partial block of the induced current. Dashed line indicates the zero baseline.

This finding is consistent with the previous result with the microelectrode recording technique showing that, if the axon was first treated with brevetoxin, the effectiveness of the local anesthetic in reversing the toxin-induced depolarization was reduced. These results suggest that the antagonism between brevetoxin and procaine may arise from a competitive interaction between the two agents acting on the same binding site or an allosteric interaction between separate binding sites bound with brevetoxin and procaine.

SUMMARY

We have demonstrated that brevetoxin induces sodium channels to open at the resting membrane potential, resulting in depolarization of the nerve membrane. At the nerve terminal, depolarization of the membrane would lead to an enhanced release of transmitters. This basic action can account for a number of pharmacological actions of brevetoxin on various organ systems as well as clinical symptoms of intoxication. The induced opening of sodium channels may be blocked effectively by TTX or partially by procaine.

ACKNOWLEDGEMENTS

The authors wish to thank Dr. Daniel G. Baden of the University of Miami for the gift of crystalline PbTX-3. C.H.W. was a recipient of the Research Career Development Award (K04-AM00928) from the National Institute of Health.

REFERENCES

1. C.C. Davis, Bot. Gaz. 109, 358-360 (1948).
2. K.A. Steidinger, In: Toxic Dinoflagellate Blooms, D.L. Taylor and H.H. Seliger, eds. (Elsevier/North Holland, New York 1979) pp. 435-442.
3. Y. Shimizu, Pure Appl. Chem. 54, 1973-1980 (1982).
4. M.A. Poli, T.J. Mende, and D.G. Baden, Mol. Pharmacol. 30, 129-135 (1986).
5. C.H. Wu and T. Narahashi, J. Pharmacol. Exp. Ther. 184, 155-162 (1973).
6. J.M.C. Huang, C.H. Wu, and D.G. Baden, J. Pharmacol. Exp. Ther. 229, 615-621 (1984).

TISSUE CULTURE ASSAY METHOD FOR PSP AND RELATED TOXINS

KAZUHIRO KOGURE,* MARK L. TAMPLIN,** USIO SIMIDU,* AND RITA R. COLWELL***
*Ocean Research Institute, University of Tokyo, Minamidai, Nakano, Tokyo 164, Japan; **Center of Marine Biotechnology, University of Maryland, 600 East Lombard Street, Baltimore, MD 21202, USA; ***Department of Microbiology, University of Maryland, College Park, MD 20742, USA

ABSTRACT

A simple and sensitive assay method for saxitoxin (STX), tetrodotoxin (TTX) and related toxins, was developed using tissue culture cells. In the presence of ouabain (1mM), veratridine (0.05mM) causes influx of sodium ion into the mouse neuroblastoma cell line, Neuro-2A, resulting in the cellular swelling and subsequent death. STX, or other sodium channel blockers counteracts this effect and enables cells to grow. Cells thus treated in a 96-well microtiter plate were observed using an inverted microscope. The minimum detectable limit was c.a. 200pg or 10^{-3} mouse unit. With less than c.a. 10ng of STX, the toxin concentration and percent of living cells showed a good correlation, which was used for the quantitative measurement of the toxin.

INTRODUCTION

Paralytic shellfish poisons (PSP) are a group of potent toxins which originate in unicellular algae and are accumulated into various marine shellfish through the food chain [1]. Since some shellfish are of commercial importance, a continuous and close watching of toxin levels are necessary. The same is also true for tetrodotoxin, which is known as "puffer fish toxin". The monitoring of these toxins are also indispensable for the basic research in laboratory. The standard mouse bioassay [2] and chemical approach using HPLC [3, 4] are well known and commonly used, but these are not always suitable, mainly because of sensitivity and/or simplicity. The immunoassay [5] looks promising, but the antibody is not easily produced in animals due to the small molecular size of these toxins. Therefore, the development of a superior substitute has been urgently required.

Catterall and Nirenberg [6] observed the effect of several compounds on the sodium channel of neuroblastoma cell lines. They showed that in the presence of ouabain, a specific inhibitor of Na^+-K^+ATPase, the alkaloid veratridine and some other chemicals stimulated sodium influx and depolarized the action potential of excitable membranes. Further, this function was counteracted by TTX or STX [7]. We found that these processes were recognizable directly under the microscope as a morphological change in the cells. This was used as a basis of a new assay method for STX, TTX and other sodium channel blocking agents. The objective of the present investigation was to

RED TIDES: BIOLOGY, ENVIRONMENTAL SCIENCE, AND TOXICOLOGY
Okaichi, Anderson, and Nemoto, Editors

develop a suitable assay protocol which permits quantitative measurements of these toxins.

MATERIALS AND METHODS

Mouse neuroblastoma cell line, Neuro-2A was obtained from ATCC (No. CCL 131)(Maryland, USA) and Dainippon Seiyaku Co., Tokyo, Japan. Cells were cultured in RPMI 1640 medium (Gibco, New York, USA) supplemented with 10% fetal bovine serum (Armour Pharmaceutical Co., Tokyo, Japan) at 37°C. The experimental conditions have been described elsewhere [8]. In brief, for the assay, 200 µl of medium containing 1.0 - 1.5 x 10^5 cells/ml were transfered to each well on a 96-well microtiter plate. Prior to the assay, the plate was further incubated overnight or at most for two days. For quantitative work, triplicate wells were prepared for each sample. At least 200 cells were randomly chosen and the viability was judged under the inverted microscope. The sodium influx induced by veratridine (0.05mM) and ouabain (1mM) causes swelling of cells, weakening the intracellular organelles structure, and loss of typical neuronal shape. Those which lost rigid membrane structures were judged to be dead. A slight change of focus helped in seeing thickness or rigidity of membranes.

Ouabain, veratridine and TTX were purchased from Sigma (Missouri,USA). STX was a generous gift from Dr. M. Kodama, Kitasato University, Iwate, Japan.

RESULTS AND DISCUSSION

Veratridine (up to 0.1mM) or ouabain (up to 2mM) alone did not cause any morphological change of Neuro-2A cell. The concomitant addition of both chemicals was necessary to elicit substantial sodium influx [6], and subsequent morphological change. At least 0.02mM of veratridine and 0.5mM ouabain were necessary to clearly demonstrate the effects of STX or related sodium channel blockers. Although higher concentrations elicited faster change, more than 0.075mM veratridine or 2mM ouabain had a harmful effect on the cells and caused significant mortality which then interfered with the assay. After checking, the combination of 0.05mM veratridine and 1mM ouabain is recommended as the standard protocol.

With 0.05mM veratridine and 1mM ouabain, the swelling and morphological change of the Neuro-2A cells were recognized within one or two hours. After about three hours, the difference with and without STX became apparent. With the toxin, cells maintained the neuronal shape, or the extent of the swelling was much less. Without the toxin, the number of dead cells increased rapidly with time. After 6 hours, normally 80-90% of the cells died. They were swollen and looked faint, lacking intracellular structure. Longer incubation time up to one day did not always enhance the difference with and without the toxin under this condition [8].

With less than c.a. 10ng/200µl of STX or TTX, the percentages of living cells increased in parallel with the toxin concentration. Fig. 1 shows the case for TTX assay. STX showed a quite similar pattern. The minimum detectable amounts were about 200pg/200µl for both toxins. A preliminary test for gonyautoxin also gave similar results. It should be noted that it is quite possible to detect smaller amounts of toxins simply by reducing the quantity of medium in a microtiter plate well. These results are in good agreement with the direct observation of $^{22}Na^{+}$ influx by Catterall and Nirenberg [6]. Using the same cell line, they found that the veratridine-dependent $^{22}Na^{+}$ influx was regulated by TTX within a range from c.a. 1 to 100nM.

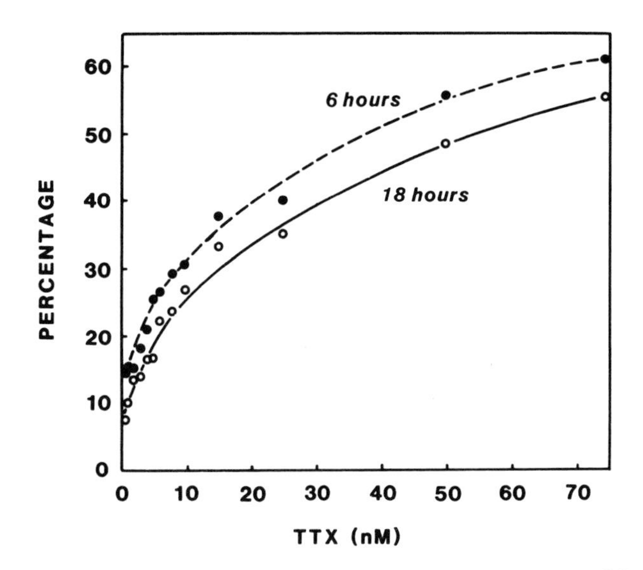

Fig. 1. Percentage of living neuroblastoma cells in the presence of tetrodotoxin.

From the results shown above, it is clear that this system can be a novel quantitative assay method for STX, TTX and related sodium channel blockers. The sensitivity obtained is quite high compared with HPLC [3, 4] or immunoassay methods [5]. Although the assay needs a one half-day incubation, it is easy to handle many samples at one time. On the other hand, it seems possible that there are other better chemicals and/or cell lines for the present purpose. So far, however, we found that aconitine instead of veratridine, and human neuroblastoma cell line, SK-N-MC (ATCC, HTB-10) did not work well.

To utilize this assay, some equipment and techniques for tissue culture are required, including access to the tissue culture laboratory facilities. This should not be very difficult at least in many universities or institutions. The technique can be acquired from a few-days course or by personal direction. In the beginning, some practice will be necessary, especially for judging the viability of cells. The judging is not completely objective, and there may be some differences among workers. However, this did not cause a fundamental problem as long as each worker has his own standard. Other problems inherent to this method are, first, this cannot differentiate any toxin from others. Depending on the question, it is recommended that other chemical or immunological methods be performed also. Second, in some crude samples, there may be an impurity which has a harmful effect on Neuro-2A cells, resulting in an underestimation of toxins. Although this is not always the case, we usually use the smallest possible amount of sample for the assay.

In conclusion, the morphological change of mouse neuroblastoma cell line, Neuro-2A, after treatment with ouabain and veratridine provides a new assay method for STX, TTX and related sodium channel blockers. The method is simple, inexpensive, and sensitive compared with other conventional methods. This method can be a powerful tool for basic research on these toxins in laboratories, and also for routine monitoring of their levels in natural foodstuff.

ACKNOWLEDGEMENT

We thank Dr. M. Kodama for generous gift of STX. This work was supported by Center of Marine Biotechnology, University of Maryland, Naval Res. Grant No. N00014-86-K-0669, and the Nippon Life Insurance Foundation, C-86110-173.

REFERENCES

1. D.M. Anderson, D.M. Kulis, J.A. Orphanos, and A.R. Ceurvels, Estuar. Coast. Shelf Sci. 14, 447-458 (1982).
2. T. Kawabata, in: The Manual for the Methods of Food Sanitation Tests (Japan Food Hygieneic Association, Tokyo 1978) pp. 232-244.
3. J. Sullivan, and M. Iwaoka, J. Assoc. Off. Anal. Chem. 66, 297-304 (1983).
4. T. Yasumoto, and T. Michishita, Agric. Biol. Chem. 49, 3077-3080 (1985).
5. S.R. Davio, J.F. Hewetson, and J.E. Beheler, in: Toxic Dinoflagellates, D.M. Anderson et al., eds. (Elsevier Publish. 1985) pp. 343-348.
6. W.A. Catterall, and M. Nirenberg, Proc. Nat. Acad. Sci. 70, 3759-3763 (1973).
7. M.M. Tamkun, J.A. Talvenheimo, and W.A. Catterall, J. Biol. Chem. 259, 1676-1688 (1984).
8. K. Kogure, M.L.Tamplin, U. Simidu, and R.R. Colwell, Toxicon 26, 191-197 (1988).

DETECTION OF PARALYTIC SHELLFISH POISON BY A LOBSTER NERVE-MUSCLE PREPARATION

K. DAIGO,* T. NOGUCHI,* N. KAWAI,** A. MIWA,** AND K. HASHIMOTO*
*Faculty of Agriculture, University of Tokyo, Bunkyo, Tokyo 113, Japan;
**Department of Neurobiology, Tokyo Metropolitan Institute for Neurosciences, 2-6 Musashidai, Fuchu, Tokyo 183, Japan

ABSTRACT

A highly sensitive detection method for paralytic shellfish poison (PSP) has been developed which is based on its inhibitory action on the excitatory postsynaptic potential (EPSP) of a lobster nerve-muscle preparation.

Amplitude of EPSP was decreased in a dose-dependent manner by PSPs. The decrease of EPSP was parallel with the reduction in the amplitude of the presynaptic spike, indicating that PSPs affected the Na channels in the presynaptic axon.

Saxitoxins reduced the amplitude of EPSP by 40% at a dose of 10^{-14}mol, whereas gonyautoxins completely reduced the amplitude at 10^{-16}mol. The method developed here could be useful to screen many samples for PSP.

INTRODUCTION

Paralytic shellfish poison (PSP) is assayed by the official method [1] using ddY strain mice in Japan. This method is convenient but of low sensitivity [limit of detection; 2 MU, equivalent to 0.4 μg saxitoxin (STX)]. In addition, it cannot give any information on toxin composition. Recently, several types of PSP analyzer [2-4] using a high performance liquid chromatograph (HPLC) [limit of detection [4]; 0.01-0.5 MU for gonyautoxin 1-4 (GTX_{1-4}), STX and neoSTX] have been developed. However, a simple, sensitive method is also expected to be developed which can screen many samples for PSP.

PSP, as well as tetrodotoxin, is a useful tool in neurobiology, because of the specific and strong blocking action on the sodium channel. In this connection, the nerve-muscle preparation [5] from the walking leg of crustacea is well studied, and axons can be isolated with comparative ease. In such a situation, attempts were made to develop a highly sensitive detection method for PSP which is based on its inhibitory action on the excitatory postsynaptic potential (EPSP) in a lobster nerve-muscle preparation.

MATERIALS AND METHODS

PSP samples

STX-rich sample: A STX-rich sample (7,500 MU, 4 μmol) was prepared from Philippine specimens of a toxic crab *Zosimus aeneus* by the method reported previously [6]. Most toxicity of the sample was accounted for by STX, as demonstrated by cellulose acetate membrane electrophoresis [6].

GTX-rich sample: A GTX-rich sample (5,000 MU, 6 μmol) was prepared by the previously reported method [6] from the toxic digestive glands of scallop *Patinopecten yessoensis* specimens which were collected from Ofunato Bay, Iwate Prefecture. This sample was composed mainly of roughly equal

RED TIDES: BIOLOGY, ENVIRONMENTAL SCIENCE, AND TOXICOLOGY
Okaichi, Anderson, and Nemoto, Editors

amounts of GTX_1 and GTX_4, as analyzed by HPLC [6].

Protogonyautoxin-rich sample: A Protogonyautoxin(PX)-rich sample (35 MU, 2 μmol) was prepared from digestive glands of toxic oyster *Crassostrea gigas* specimens from Senzaki Bay, Yamaguchi Prefecture, by the method reported previously [7]. This sample consisted of PX_1 (GTX_8 epimer) and PX_2 (GTX_8), since its acid hydrolyzate gave rise to GTX_2 and GTX_3, in HPLC analysis [8].

Oyster toxin: Toxic oyster digestive glands were extracted with three volumes of distilled water. Toxins were partially purified from the extract by the method reported previously and dissolved in distilled water [6]. To a portion of the toxin solution, was added an equal volume of 0.1 N HCl, and the mixture was heated at 100°C for 5 min. The acid-treated sample was submitted to HPLC analysis for PSP, resulting in detection of GTX_2, GTX_3, STX and neoSTX.

Gymnodinium toxin: *Gymnodinium catenatum* which was a responsible plankton for toxic oysters in Senzaki Bay in Dec. 1986, was cultured under the conditions reported before [9], and the cells harvested by centrifugation. After addition of distilled water, the cells were ultrasonicated and ultrafiltered through a Diaflo YM-2 membrane (Amicon). A portion of *Gymnodinium* toxin was treated with 0.1 N HCl in the same manner as was the oyster toxin. When analyzed by HPLC, the non-treated toxin showed PX_1 and PX_2, but both PXs were changed into GTX_2 and GTX_3 on acid hydrolysis.

Measurement of inhibitory action of PSP on EPSP of a lobster nerve-muscle preparation

Excitatory axons of lobster *Panulirus japonicus* were isolated from the meropodite of a walking leg according to Miwa and Kawai [5], and stimulated by suction electrodes. For recording, the dorsal surface of the stretcher muscle was exposed by cutting a round hole in the exoskeleton and overlying connective tissue so as to make a pool. Microelectrodes filled with 3 M KCl were used for intracellular recording, and also for current passage. In some experiments, extracellular recording was made by using 2 M NaCl electrode positioned at the synapse.

Normal saline was composed of 468 mM NaCl, 10 mM KCl, 20 mM $CaCl_2$, 8 mM $MgCl_2$, and 2 mM Tris-HCl (pH 7.4). The preparation was placed in a chamber with a groove, made of silicon.

A toxin solution was applied on the surface of the stretcher muscle, and the changes in EPSP were recorded 5 min later. The preparation was thoroughly washed with the normal saline until the EPSP was completely recovered, and used for the next experiment and so on. Thus, the effect of each toxin at 5-10 concentrations on EPSP was examined on one and the same preparation.

RESULTS AND DISCUSSION

The toxins tested showed a blocking action on EPSP of the lobster nerve-muscle preparation at an extremely low dose. In this case, the decline in amplitude of EPSP was almost parallel with that of nerve terminal spike (figure not shown). Consequently, PSPs affected the fine axonal branches and reduced Na currents, resulting in a decrease of transmitter release. In addition, the effect of toxin was suggested to be presynaptic since resistance of the muscle was not affected by PSPs (figure not shown).

As TABLE I shows, EPSP was completely blocked at a dose of 10^{-12} mol of STX-rich sample, and blocked by 40% even at 10^{-14} mol. The preparation was more sensitive to GTX-rich sample: EPSP was completely blocked at a dose of

TABLE I. Blocking action of various toxins on EPSP of lobster nerve-muscle preparation.

Toxin \ Mol applied	Reduction of EPSP amplitude (%)					
	10^{-17}	10^{-16}	10^{-15}	10^{-14}	10^{-13}	10^{-12}
STX-rich sample				40		100
GTX-rich sample		100				
PX-rich sample			30	50		
Oyster toxin	70					
Gymnodinium toxin			45			
Gymnodinium toxin[a]				70		

[a]After heating in 0.05 N HCl for 5 min.

10^{-16} mol. A similar blocking was caused by PX-rich sample: A dose of 10^{-14} mol was enough to block EPSP completely.

On the other hand, oyster toxin strongly blocked EPSP of the nerve-muscle preparation at a dose of 10^{-17}mol (TABLE I). *Gymnodinium* toxin completely blocked EPSP at a dose of 10^{-15}mol. Hydrolysis did hardly affect the action of this toxin.

As described above, any of STX-, GTX- and PX-rich samples, as well as toxic oyster and planktonic extracts, completely or almost completely blocked EPSP of the nerve-muscle preparation at a minute dose from 10^{-14}~ 10^{-16}mol. These doses of PSP are roughly equivalent to 10^{-4}~10^{-8} MU. In this connection, it was noted that the method can detect PXs, low-toxic components, with a comparable sensitivity to highly toxic PSP members.

This method, however, cannot distinguish PSP from tetrodotoxin and related substances (Noguchi *et al.*, unpublished). Therefore, it is desirable to run simultaneously other assay methods to confirm the toxin(s) as PSP.

REFERENCES

1. T. Kawabata, Shokuhin Eisei Kensa Shishin [Manual of Methods for Food Sanitation Testing], Vol. 2 (Japan Food Hygienic Association, Tokyo 1978) pp. 240.
2. J. J. Sullivan and W. I. Iwaoka, J. Assoc. Off. Anal. Chem. 66, 297-303 (1983).
3. Y. Oshima, M. Machida, K. Sasaki, Y. Tamaoki, and T. Yasumoto, Agric. Biol. Chem. 48, 1707-1711 (1984).
4. Y. Nagashima, J. Maruyama, T. Noguchi, and K. Hashimoto, Nippon Suisan Gakkaishi 53, 819-823 (1987).
5. A. Niwa and N. Kawai, J. Neurophysiol. 47, 353-361 (1982).
6. T. Noguchi, O. Arakawa, K. Daigo, and K. Hashimoto, Toxicon 24, 705-711 (1986).
7. Y. Onoue, T. Noguchi, J. Maruyama, K. Hashimoto, and H. Seto, J. Agric. Food Chem. 31, 420-423 (1983).
8. Y. Nagashima, Y. Sato, T. Noguchi, Y. Fuchi, K. Hayashi, and K. Hashimoto, Marine Biology (in press).
9. Y. Onoue, T. Noguchi, J. Maruyama, Y. Ueda, K. Hashimoto, and T. Ikeda, Nippon Suisan Gakkaishi 47, 1347-1350 (1981).

DETOXIFICATION MECHANISMS OF FLORIDA'S RED TIDE DINOFLAGELLATE PTYCHODISCUS BREVIS

DANIEL G. BADEN,* THOMAS J. MENDE**, and LAURIE E. ROSZELL*
University of Miami *Rosenstiel School of Marine and Atmospheric Science Division of Biology and Living Resources, 4600 Rickenbacker Causeway, Miami FL. 33149 U.S.A. and **School of Medicine, Department of Biochemistry and Molecular Biology, P.O. Box 016129, Miami FL. 33101 U.S.A.

ABSTRACT

Florida's red tide dinoflagellate Ptychodiscus brevis produces at least eight different polyether toxins based on two different structural backbones. Cultures yield different toxin profiles depending upon growth phase of the cultures. Total mass amount of toxin does not vary, regardless of growth phase. In logarithmic growth phase, the two aldehyde toxins, PbTx-1 and PbTx-2 predominate. When cultures reach stationary phase, the profiles are composed of six toxins including: the C-27,28 epoxide, the C-37 O-acetate, and the C-42 primary alcohol forms of PbTx-2; the C-43 primary alcohol of PbTx-1; and the two predominant toxins PbTx-1 and PbTx-2. Detoxification pathways account for the multiplicity of toxins, and for the reduced potency of the derivatives.

INTRODUCTION

Toxic marine dinoflagellates produce a multiplicity of potent materials, often based on a single structural backbone. The classic illustration is the toxin suite isolated from Protogonyaulax, where all potent materials are derivatives of one of the more potent toxins in the suite, Saxitoxin [1]. A similar multiplicity of toxins exists in Florida's red tide dinoflagellate Ptychodiscus brevis, which are based on two similar polyether lipid-soluble structures [2] (Fig 1.).

Using the diploid strain of Ptychodiscus brevis isolated by W.B. Wilson in 1953 [3], we isolated and quantified toxins from multiple 10 liter cultures and attempted a correlation of toxin profile with culture density and growth phase. The potency of each individual toxin isolated was also determined by Gambusia affinis (mosquito fish) bioassay [4].

EXPERIMENTAL

Culture Growth and Extraction of Toxins

Cultures of P. brevis were grown in 10 L quantities under continuous illumination (Gro-Lux WS, Sylvania Corp.) at 4000 lux and 24°C. Cells were sampled throughout culture growth

RED TIDES: BIOLOGY, ENVIRONMENTAL SCIENCE, AND TOXICOLOGY
Okaichi, Anderson, and Nemoto, Editors

and counts were performed on aliquots suspended in Lugoll's solution. Cultures were extracted at mid-logarithmic (4 x 10 L, n=50) or in stationary (4 x 10 L, n=25) phase using techniques we have described previously [5].

Quantification of Toxins

Toxins were separated using C-18 reverse phase HPLC (isocratic 85% aqueous methanol, uv detection at 215 nm), and were quantified by comparison with brevetoxin standards. The identity of each toxin was ascertained using Fourier transform infrared spectrometry.Toxin content on a per cell basis was calculated based on cell counts and on toxin mass.

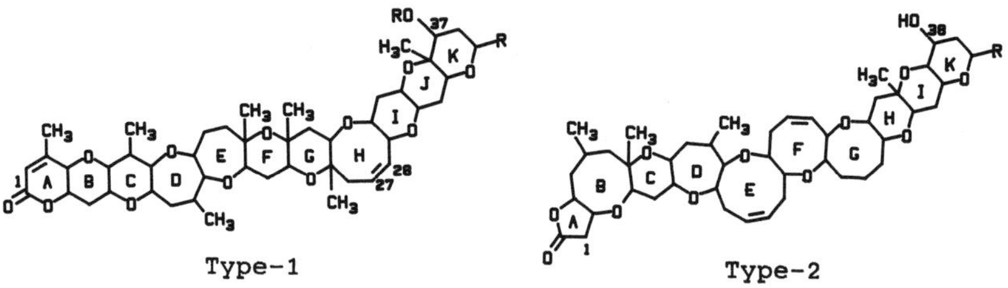

Type-1 Type-2

	R_1	R_2		R
PbTx-2	H	CH2C(=CH2)CHO	PbTx-1	CHO
PbTx-3	H	CH2C(=CH2)CH2OH	PbTx-7	CH2OH
PbTx-5	Ac	CH2C(=CH2)CHO		
PbTx-6	H	CH2C(=CH2)CHO (27,28 epoxide)		
PbTx-8	H	CH2COCH2Cl		

Figure 1. Structures of the brevetoxins.

RESULTS

Cultures

At the time of dilution, cell concentrations of P. brevis ranged from 0.2- 5 x 10^6 per L. At mid-logarithmic phase, culture concentrations reached 2-3 x 10^7 per L; stationary phase yielded cell counts ranging from 4-4.8 x 10^7 per L. Both PbTx-1 and PbTx-2, the alpha beta unsaturated toxins of each structural class, represented the major toxins isolated in either logarithmic or stationary cultures. The amounts of PbTx-2 always exceeded the amounts of PbTx-1 obtained from cultures, and during logarithmic growth phase well over 90% of the toxin was associated with P. brevis cells, as determined by radioimmunoassay of Percoll density gradient fractions of cultures. About 30% of the total toxin isolated was extracellular in stationary phase cultures. Determination of the profile of intracellular versus extracellular toxins is in progress.

Toxin Profiles

The relative proportions of the individual toxins show a

marked difference between logarithmic and stationary phases, however. The proportion of toxin present in the alpha beta unsaturated forms of each structural backbone progressively decreases from 95 and 98 % for types-1 and -2 respectively in logarithmic phase, to about 20 and 50 % respectively in stationary phase. The principal conversion of the alpha beta unsaturated toxins appears in each case to be a reduction of the aldehyde functionality to the less potent primary alcohol toxins PbTx-3 and PbTx-7, respectively (Table I). A typical tracing of the relative proportions of Type 1 toxin aldehyde versus alcohol toxins throughout culture growth is shown in Figure 2.

TABLE I. Toxin Yields from Cultures: Type 1 Versus Type 2 Toxins.

Notation	Type 1 Yield (pg/cell) L	S	Notation	Type 2 Yield (pg/cell) L	S	Note (1)
PbTx-2	8.7	7.50	PbTx-1	1.7	0.25	(2)
PbTx-3	0.42	0.90	PbTx-7	0.02	0.04	(3)
PbTx-5	0.06	----	PbTx-9*	0.01	----	(4)
PbTx-6	0.04	0.12	PbTx-10*	0.01	0.02	(5)

*Toxins underlined have not been demonstrated in cultures. By analogy with Type 1 toxins, they are proposed to exist, and the yields expected are in proportion to analogous Type 1 structures. Notes: (1) L=logarithmic, S=stationary; (2) alpha beta unsaturated aldehyde; (3) alpha beta unsaturated primary alcohol; (4) O-acetate; (5) epoxide; (6) methylene-reduced primary alcohol.

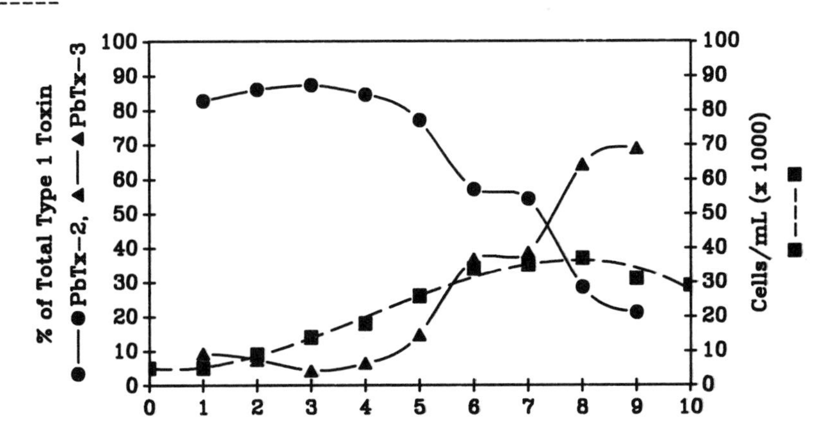

Figure 2. Relative Proportions of PbTx-2 and PbTx-3 with Respect to Culture Growth Phase. Ten liter cultures were extracted at weekly intervals following initial inoculation at time 0, and were assayed for both PbTx-2 (circles) and PbTx-3 (triangles) for 9 weeks. Lag phase weeks 0-1; Log phase weeks 1-4; Stationary phase weeks 4-7; culture demise weeks 7-10.

DISCUSSION

Shimizu et al. [2] and others have shown that the two predominant toxins in P. brevis cultures are PbTx-1 and PbTx-2. In our logarithmic phase cultures, we also find the predominant toxins to be these two materials. Through examination of over 3000 liters of culture, we have discovered that our cultures vary in toxin profile throughout the growth phase from lag to log to stationary growth. The predominant profile change is in the relative proportion of the alpha beta unsaturated toxin PbTx-2 in favor of the primary alcohol toxin PbTx-3, but we also detected increases in relative proportions of epoxide and acetylated derivatives of PbTx-2, and also of the primary alcohol derivative of PbTx-1. Changes in toxin profile can be explained on the basis of detoxification reactions, primarily the NADH/NADPH-dependent alcohol dehydrogenase system, phase I reactions, and O-acetyltransferase enzymology acting on PbTx-1 and PbTx-2. Preliminary experiments utilizing purified enzymes and co-factors, and pure brevetoxins illustrate these proposals to be plausible, and that the toxins are interconvertible using classical in vitro enzymology.

It is important to point out the correlation in the two structural backbones and their respective metabolism, both to illustrate the identical detoxification schemes and to call attention to the increased water solubility of each derivative (and hence excretability), and over 10-fold decreased potency upon derivatization (data not shown and [4]). We believe the brevetoxins are toxic to P. brevis itself, and that the organism is continually battling the accumulation of PbTx-1 and PbTx-2. Its principal weapon is logarithmic growth, but under conditions of decreased growth, classical detoxification reactions are also invoked to minimize the effects of these potent materials on the dinoflagellate itself. The implications of such an hypothesis include a toxigenic entity within the dinoflagellate.

ACKNOWLEDGEMENTS

This work was supported in part by the US Army Medical Research and Development Command, Contract Numbers DAMD17-85-C-5171 and DAMD17-87-C-7001.

REFERENCES

1. S. Hall and P.B. Reichardt in: Seafood Toxins, E.P. Ragelis, ed.(ACS Symp. Ser. 262, Washington D.C., 1984) pp. 113-124.
2. Y. Shimizu, H.N. Chou, H. Bando, G. VanDuyne and J. Clardy, J. Am. Chem. Soc. 108, 514 (1986).
3. W.B. Wilson, Fla. Bd. Conserv. Prof. Pap. Ser. 7.(1966).
4. D.G. Baden, T.J. Mende, A.M. Szmant, V.L. Trainer, R.A. Edwards and L.E. Roszell, Toxicon 26, 97.
5. D.G. Baden, T.J. Mende, M.A. Poli and R.E. Block in: Seafood Toxins, E.P. Ragelis, ed. (ACS Symp. Ser. 262, Washington D.C. 1984) pp. 359-368.
6. M.A. Poli, T.J. Mende and D.G. Baden, Molec. Pharmacol. 30, 129, (1986).

MORTALITY OF FISH LARVAE FROM EATING TOXIC DINOFLAGELLATES OR ZOOPLANKTON CONTAINING DINOFLAGELLATE TOXINS

A.W. WHITE, O. FUKUHARA* and M. ANRAKU* **
Sea Grant Program, Woods Hole Oceanographic Institution, Woods Hole, Massachusetts, U.S.A. 02543; *Nansei Regional Fisheries Research Laboratory, Japanese Fisheries Agency, Ohno, Hiroshima 739-04, Japan

ABSTRACT

First-feeding red sea bream (Pagrus major) and Japanese anchovy (Engraulis japonica) larvae were fed the toxic dinoflagellate Gonyaulax excavata. Older larvae were fed zooplankton (mostly copepods) that had eaten G. excavata. Despite low toxin content of the dinoflagellates relative to field conditions, effects of the toxins were apparent. The mortality rate of first-feeding red sea bream larvae feeding on Gonyaulax was about three times that of starved controls. First-feeding Japanese anchovy larvae fed poorly on Gonyaulax, and no difference in mortality between treatments and controls was observed. Older larvae of both species showed symptoms typical of "paralytic shellfish poisoning" within a few hours after eating zooplankton that contained Gonyaulax toxins; 20 to 30% of the larvae died. Results indicate that fish larvae, like adult fish, are sensitive to paralytic shellfish toxins and suggest that blooms and red tides of G. excavata and its toxic relatives cause kills of larval, as well as adult, fish.

INTRODUCTION

Blooms and red tides of the toxic dinoflagellate Gonyaulax excavata (tamarensis) have caused mass kills of adult fish in nature [1,2], resulting in losses to traditional fisheries and mariculture. Kills of Atlantic herring, sand lance, menhaden, and cage-cultured Atlantic salmon and rainbow trout have been reported [3,4,5]. Laboratory experiments have shown that Gonyaulax toxins are lethal to various marine fishes, including Atlantic herring, Atlantic salmon, winter flounder, American pollock, and cod and that the sensitivity of these fishes to the toxins is similar to that of warm-blooded animals [6]. Fish acquire the toxins either by direct ingestion of Gonyaulax or, as has been well documented, by ingestion of zooplankton that have fed on Gonyaulax and accumulated the toxins [4,6,7,8]. It is not known whether other routes of toxin acquisition are important for fish, such as ingestion of contaminated shellfish or direct uptake of toxins from solution.

Of even greater concern in terms of fish populations is whether larval fish are killed by Gonyaulax toxins. Year-class strength is determined in large part by the success of early life stages of fish. Fish larvae in the midst of a toxic bloom have little choice but to eat either the toxic dinoflagellates or toxin-containing zooplankton. The one previous study on this topic showed increased mortality of winter flounder larvae upon exposure to G. excavata [9].

We investigated the effects of G. excavata on first-feeding larvae of two commercially important fishes in Japan, red sea bream (Pagrus major) and Japanese anchovy (Engraulis japonica), and the effects of zooplankton containing G. excavata toxins on older larvae.

**Present address: Overseas Fishery Cooperation Foundation, Akasaka Twin Tower, 17-22, Akasaka 2, Minato-Ku, Tokyo, Japan

RED TIDES: BIOLOGY, ENVIRONMENTAL SCIENCE, AND TOXICOLOGY
Okaichi, Anderson, and Nemoto, Editors

MATERIALS AND METHODS

Gonyaulax excavata (clone 7 from eastern Canada) was grown bacteria-free in 1.5 L of enriched seawater medium "f/2" in 2.8-L Fernbach flasks at 18°C and 4,000 lux illumination on a 16:8-hr LD cycle. Seawater (32-33 ppt salinity) was obtained from Hiroshima Bay and glass-fiber filtered before use. Cells in mid-log phase growth were used for feeding fish larvae and zooplankton. Single cells were 25-40 μm in length, duplets were twice as long. Cells looked normal and maintained normal swimming behavior during the experiments; in fact, they doubled in number after several days in the tanks.

Wild zooplankton were collected from Hiroshima Bay, using either 300 μm net tows or continuous plankton traps. In the laboratory the material was screened through 600 μm netting to remove large animals. Screened material was split between two, 10-L plastic buckets, gently aerated, and fed either G. excavata or the non-toxic diatom Phaeodactylum tricornutum. Zooplankton composition varied between experiments, but generally the dominant organisms were the copepods Acartia clausii and Labidocera japonica. Fortunately, the fish larvae showed a clear feeding preference for copepods. Zooplankton were allowed to feed for 2-3 hours on G. excavata (when their intestines became full) and were then gently collected on a piece of netting and added to the experimental tanks containing fish larvae.

Toxin content of G. excavata and zooplankton which ate G. excavata was measured by modifications of the mouse bioassay described before [7].

Experiments were done in cylindrical, 9-L glass tanks (15 cm high, 28 cm diam.) covered externally, except on top, with black plastic to provide a good background for larval feeding. Tanks were filled with 7 L of filtered seawater (31.5 ppt salinity) and kept in a walk-in incubator at 18°C (for red sea bream) or 20°C. Tanks received light from cool-white fluorescent lamps at 3,000 lux at the surface for 16 hr/day and at 25 lux from an incandescent bulb at night. Water was gently aerated with glass pipettes.

In experiments with first-feeding larvae, 5 treatments were made in duplicate: starved controls; G. excavata; G. excavata in a cage made of 10-μm netting; rotifers (Brachionus plicatilis); and rotifers plus G. excavata. Gonyaulax was supplied at an initial concentration of 300 cells/mL. Rotifers (fed Chlorella and screened through 100-μm netting) were supplied at 5/mL.

The experiment with first-feeding red sea bream larvae was started by adding 100, 2-day-old larvae in yolk-sac stage to each tank after Gonyaulax and rotifers had been already been added. Since Japanese anchovy larvae were so delicate, eggs (100) were added to each tank and 3 days later (2 days after hatching), when larvae were ready to begin feeding, Gonyaulax and/or rotifers were added.

Twice daily, dead larvae were removed and counted and observations on larval behavior were made. Temperature, pH, and oxygen in the tanks were checked periodically; temperature remained constant, pH remained at 7.8-7.9 and oxygen remained at near saturation levels in all tanks throughout the tests.

RESULTS AND DISCUSSION

First-feeding red sea bream larvae began feeding behavior (S-posturing and striking) one day after the experiment was started (3-day-old larvae). They fed well on rotifers and moderately well on Gonyaulax. After 4 days about 95% of the larvae exposed directly to Gonyaulax alone had died, whereas only about 20-40% of the larvae in the other treatments, including starved controls, had died (Fig. 1). Larvae which died in the Gonyaulax treatments usually had Gonyaulax cells observable in their stomachs; most contained only several cells, but some contained a mass of yellow-brown material representing the remnants of about 12 or more cells. These larvae lay

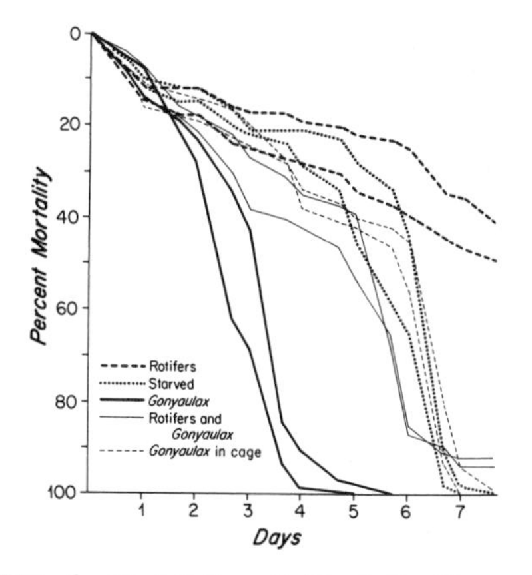

FIG. 1. Mortality of first-feeding red sea bream larvae

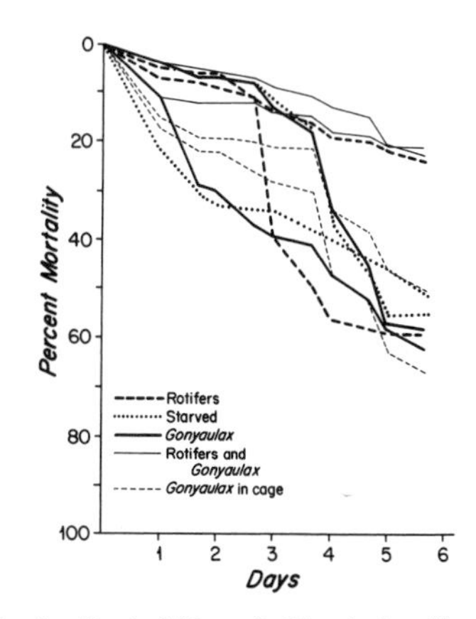

FIG. 2. Mortality of first-feeding Japanese anchovy larvae

paralyzed on the tank bottom, unable to move when prodded and with pumping of the heart being the only sign of life, until death ensued. Paralysis was not observed in larvae in the other treatments. There was a tendency for the larvae which were alone with *Gonyaulax* to have a more sluggish avoidance response (when prodded with a pipette) than larvae in the other treatments. Paralysis and reduced avoidance was also observed in winter flounder larvae fed *G. excavata* [9].

Survival was best in the rotifer treatments, although there was gradual mortality even in these treatments (Fig. 1). Effects of starvation on mortality were not clearly apparent until the fifth day when many larvae died. The pattern of mortality in the treatments with *Gonyaulax* separated from larvae by plankton netting was not different from that in the starved controls, indicating that excretion products from *Gonyaulax* were not responsible for the increased mortality in the treatments in which larvae were fed *Gonyaulax*. The data suggest that *Gonyaulax* caused a slight increase in mortality over starved controls even in the presence of rotifers. Indeed, some dead larvae in the tanks containing rotifers and *Gonyaulax* had *Gonyaulax* cells in their guts.

First-feeding Japanese anchovy larvae fed poorly on *G. excavata*. The rate of mortality of larvae exposed to *Gonyaulax*, either directly or indirectly, was similar to that for starved larvae (Fig. 2). Characteristic feeding behavior (searching, S-posturing, and striking) developed only in tanks containing rotifers, by the end of the first day. Intestines of larvae soon became packed with rotifers. After 5 days, at the termination of the experiment, 50-65% of the larvae in the *Gonyaulax* or starvation treatments had died, whereas only 20-25% of the larvae in the rotifer treatments (including rotifers plus *Gonyaulax*) had died, with the exception of one rotifer tank in which a sudden bacterial contamination caused rapid mortality (see Fig. 2).

On occasion in the *Gonyaulax* treatments one or two *Gonyaulax* cells were seen in the guts of dead anchovy larvae. An interesting exception was that one larva contained about 8 *Gonyaulax* cells, lined up in its intestine like beads on a string, and it was paralyzed except for the beating of its heart. Aside from this, however, there was no clear difference in avoidance response behavior among treatments.

Regarding older larvae (4-6 weeks), in several experiments in which either red sea bream or Japanese anchovy larvae were exposed to zooplankton which had fed on *Gonyaulax*, effects of the toxins were clearly apparent.

Within 1-2 hr after feeding on Gonyaulax-fed zooplankton had begun, some larvae of both species (22-38% of the fish tested) lost their equilibrium and swam on their sides or upside down or in circles, as was previously observed in adult fish of other species [6,8]. Symptoms progressed to include paralysis on the tank bottom, with little motion of the opercula. A few larvae recovered from this condition and resumed normal swimming, but most died within a few hours. In total, in the experiments in which pronounced symptoms of poisoning developed, 20% (11 of 55) red sea bream larvae and 33% (5 of 15) Japanese anchovy larvae died. Larvae of both species which ate Phaeodactylum-fed zooplankton showed no abnormal behavior and no deaths occurred.

Effects of G. excavata on first-feeding larvae and of zooplankton containing G. excavata on older larvae were observed despite low toxin content of the dinoflagellates and zooplankton relative to field conditions. In these experiments toxin content of G. excavata ranged from 1 to 2 x 10^{-5} μg STX equivalent/cell, and toxin content of zooplankton averaged 3.5 μg STX equivalent/g wet weight. Toxin levels 10 times greater than this often occur in G. excavata and zooplankton in nature [3,10].

At 2 x 10^{-5}μg STX equivalent/cell, and assuming the oral LD_{50} to be similar to that for adult fish [6], calculations suggest that a first-feeding fish larva would need to eat only 6-11 G. excavata cells to acquire a lethal dose.

This study shows that some fish larvae can be killed by eating G. excavata and zooplankton containing G. excavata and suggests that blooms of toxic dinoflagellates have impact on fish larvae, as well as adults, in nature.

ACKNOWLEDGMENTS

AWW gratefully acknowledges support from the Japanese Fisheries Agency and the Canadian Department of Fisheries and Oceans to conduct this work at the Nansei Regional Fisheries Research Laboratory in 1982. Support to prepare this paper and present it at the International Symposium on Red Tides was provided in part by NOAA National Sea Grant College Program Office, Department of Commerce, under Grant No. NA86-AA-D-SG090, WHOI Sea Grant Project Nos. M/O-2 and E/L-1. Woods Hole Oceanographic Institution Contribution No. 6647.

REFERENCES

1. A.W. White, In: Proc. Conf. on Impact of Toxic Algae Blooms on Mariculture, Aqua-Nor '87 Exhibition, Trondheim, Norway, (in press).
2. B. Dale, D.G. Baden, B.Mc. Bary, L. Edler, S. Fraga, I.R. Jenkinson, G.M. Hallegraeff, T. Okaichi, K. Tangen, F.J.R. Taylor, A.W. White, C.M. Yentsch and C.S. Yentsch, In: Problems of Toxic Dinoflagellate Blooms in Aquaculture, Sherkin Island Marine Station, Ireland, (1987).
3. A.W. White, In: Seafood Toxins, E.P. Ragelis, ed., Amer. Chem. Soc. Symposium Series 262, ACS, Washington, D.C., p. 171-180, (1984).
4. A.W. White, Can. J. Fish. Aquat. Sci., 37, 2262-2265, (1980).
5. A.M. Mortenson, In: Toxic Dinoflagellate Blooms, D.M. Anderson, A.W. White and D.G. Baden, eds., Elsevier, New York, p. 165-170, (1985).
6. A.W. White, Marine Biology, 65, 255-260, (1981).
7. A.W. White, Limnol. Oceanogr., 26, 103-109, (1981).
8. M.H. Yazdandoust, In: Toxic Dinoflagellate Blooms, D.M. Anderson, A.W. White and D.G. Baden, eds., Elsevier, New York, p. 419-424, (1985).
9. L.J. Mills and G. Klein-MacPhee, In: Toxic Dinoflagellate Blooms, D.L. Taylor and H.H. Seliger, eds., Elsevier, NY, p. 389-394, (1979).
10. A.W. White, Toxicon, 24, 605-610, (1986).

OCCURRENCE OF PROTOGONYAULAX TAMARENSIS AND SHELLFISH TOXICITY IN OFUNATO BAY FROM 1980-1986

KATSUSHI SEKIGUCHI,* NOBUYUKI INOGUCHI,* MICHIHIKO SHIMIZU,* SATORU SAITO,* SHOJI WATANABE,* TAKEHIKO OGATA,** MASAAKI KODAMA** AND YASUWO FUKUYO***
*Iwate Prefectural Fisheries Experimental Station, Iwate 026, Japan; **Laboratory of Marine Biological Chemistry, School of Fisheries Sciences, Kitasato University, Iwate 022-01, Japan; ***Faculty of Agriculture, Tokyo University, Tokyo 113, Japan

ABSTRACT

The abundance of *Protogonyaulax tamarensis* and the toxicity of scallops were monitored in Ofunato Bay from 1980-1986. *P. tamarensis* developed to high cell densities in 1980-1982, but was at lower densities from 1983-1986. The toxicity of scallops was well correlated with the abundance of *P. tamarensis*. When *P. tamarensis* was present in high quantities, environmental factors such as precipitation, irradiance, water temperature, salinity, concentrations of dissolved oxygen, chlorophyll *a*, PO_4-P and NO_3-N were within ranges of 0-190 mm/10 days, 40-90 hr/10 days, 5-10 °C, 33.0-33.7, 93-112%, 0.7-8.5 µg/L, 0.1-0.7 µg-at/L and 0.3-5.9 µg-at/L, respectively. These conditions continued for 60-90 days each year from 1980-1982, while they continued for less than 15 days each year from 1983-1986. These data suggest that the continuance of these environmental conditions is necessary for the growth of *P. tamarensis* in the field.

INTRODUCTION

Paralytic shellfish poisoning (PSP) is a major, global problem for the the fisheries industry and for public health. In a previous report [1], we indicated that *P. tamarensis*, which occurred from spring to early summer in Ofunato Bay, was responsible for the toxification of bivalves, whereas *P. catenella*, which occurred in the fall, was not. One fatal case of PSP occurred in 1961 [2]. Although intensive studies have been carried out on *P. tamarensis* as well as the toxin it produces [3], little is known of the environmental factors which affect the appearance and the growth of this species in natural waters.

In this report, we describe the seasonal variation of *P. tamarensis* abundance, as well as the environmental conditions and the toxicity of scallops in Ofunato Bay from 1980-1986. The environmental conditions best suited for the growth of *P. tamarensis* are also discussed.

MATERIALS AND METHODS

Monitoring Station

Ofunato Bay is a small bay with maximum width, length and depth of 1.5 km, 7 km and 40 m respectively, located at 39°N, 141°41'E. Two rivers enter the headwaters of the bay. The mouth of the bay is restricted by the breakwater against tsunami, with an opening of 200 m width and 16 m depth. Our sampling station was at Shizu (24 m depth) [1].

RED TIDES: BIOLOGY, ENVIRONMENTAL SCIENCE, AND TOXICOLOGY
Okaichi, Anderson, and Nemoto, Editors

Monitoring of P. tamarensis and environmental factors

Seawater samples were collected every 2nd meter through-out the 24 m deep water column 2 - 5 times each month. Enumeration of P. tamarensis cells was carried out microscopically as described previously [1]. Environmental factors such as water temperature, salinity, concentrations of dissolved oxygen, PO_4-P, NO_3-N and chlorophyll a were measured according to Strickland and Parsons [4]. The data from the ten day meteolorogical report of the Sendai district meteorological observatory were used for quantifying rainfall and irradiance.

Monitoring of scallop toxicity

Nontoxic scallops (Patinopecten yessoensis) were transplanted to the station at 10 m depth. Five scallop specimens were collected for a toxicity test whenever the seawater sampling was done. Digestive glands were excised, combined, and measured for PSP toxicity by the standard mouse bioassay method [5].

RESULTS AND DISCUSSION

Seasonal variation in P. tamarensis abundance and scallop toxicity

Fig. 1 shows the seasonal variation in the abundance of P. tamarensis and the toxicity of scallops in Ofunato Bay from 1980-1986. These data show that P. tamarensis appeared every year. In 1980, P. tamarensis first appeared in February or March, and disappeared in April. The species appeared again in May and increased to a high cell density. The bloom diminished by the end of June. A similar pattern of appearance was observed in 1982. However, the first peak was observed in March and was much larger than that of 1980. In 1981, P. tamarensis appeared in March and reached maximum density in April. After the peak, it disappeared by the end of May. No fluctuations were observed, showing that it bloomed only once that year. In contrast, P. tamarensis did not bloom in high densities from 1983-1986. Each year's maximum cell number during this period ranged from 160 to 5,820 cells, whereas it exceeded 70,000 cells each year from 1980-1982. These observations show that the pattern of appearance and the final abundance are quite variable from year to year.

Scallop toxicity increased in parallel to the occurrence of P. tamarensis (Fig. 1). Previously we reported that the ratio of toxicity to cell number for spring P. tamarensis was generally larger than that for the summer P. tamarensis in Ofunato Bay [1]. A similar pattern was observed every year of this study. Toxin production of P. tamarensis is reported to be highest at low temperatures [6]. Thus the higher ratio for the spring P. tamarensis is probably due to lower temperature during this period. The curious phenomenon which we pointed out before [1] was also observed every year, i.e., the maximum shellfish toxicity occurred 1-2 weeks after P. tamarensis reached maximum abundance. The toxicity of scallops was maximal when the population of P. tamarensis was decreasing. The time lag between both maximum peaks was about one week. This phenomenon is most easily seen from 1980-1982.

Optimum environmental conditions for the growth of P. tamarensis

P. tamarensis bloomed in high density from spring to early summer in 1980-1982, suggesting that the environmental conditions during these periods were favourable for growth. In order to isolate the optimum environmental conditions, values of environmental parameters observed during the period from the beginning of the increase to the maximum density were tabulated.

Table I shows the maximum and minimum values of each environmental factor. We tentatively defined this condition as an optimum condition for the growth of P. tamarensis. In Fig. 1, the periods during which the optimum condition continued are shown. In 1980-1982, when P. tamarensis bloomed in high density, the conditions were maintained for significantly long periods (60 days in 1980, 80 days in 1981, 80 and 50 days in 1982). On the other hand, they continued for less than 15 days in 1983-1986 when P. tamarensis did not bloom in high density. These facts suggest that the growth of P. tamarensis was controlled either by a single or a combination of environmental factors and that the persistence of those conditions for a relatively long period is essential for growth to high cell densities.

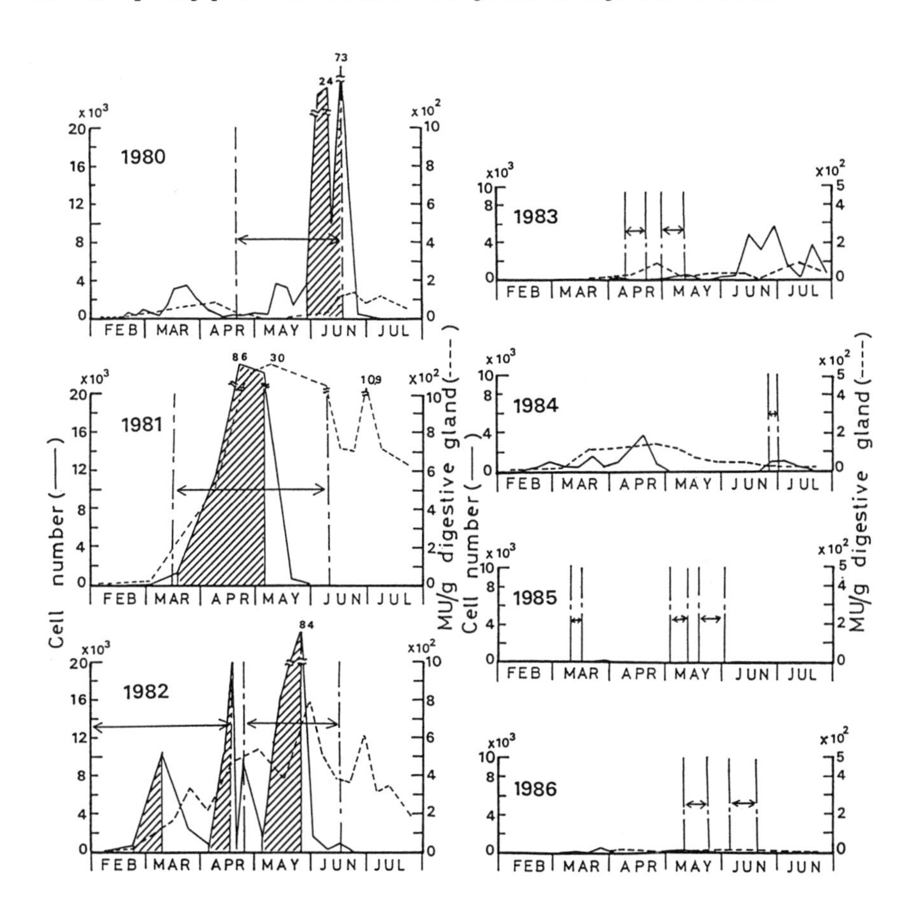

FIG. 1. Seasonal variation in abundance of P. tamarensis and scallop toxicity. Tentatively defined optimum conditions for the growth of P. tamarensis consist of ranges from minimum to maximum values of various environmental factors observed during the periods shwon by ////. This optimum condition is maintained during the periods shown by ⟷. The abundance of P. tamarensis is the sum of cell concentrations observed at discrete depths from surface to bottom.

TABLE I. Optimum environmental conditions for the growth of *P. tamarensis* in Ofunato Bay.

Environmental factors	Range	
	Minimum	Maximum
Precipitation (sum of 10 days)	0	190 mm
Irradiance (sum of 10 days)	40	90 hrs
Temperature*	5	10 °C
Salinity*	33.0	33.7
D.O.*	93	112 %
Chlorophyll *a**	0.7	8.5 μg/L
NO_3-N*	0.3	5.9 μg-at/L
PO_4-P*	0.1	0.7 μg-at/L

* Average value of 6 layers from 12 to 22 m.

ACKNOWLEDGEMENTS

This work was supported in part by a grant from Ministry of Agriculture, Forestry and Fishery in Japan. The toxicity of a part of the samples was analyzed by Iwate Prefectural Institute of public health.

REFERENCES

1. T. Ogata, M. Kodama, Y. Fukuyo, T. Inoue, H. Kamiya, F. Matsuura, K. Sekiguchi and S. Watanabe, Nippon Suisan Gakkaishi 48, 563-566 (1982).
2. T. Kawabata, T. Yoshida, and Y. Kubota, Nippon Suisan Gakkaishi 28, 344-351 (1962).
3. E.J. Schantz, ANN. N.Y. Acad. Sci. 479, 15-23 (1986).
4. J.D.H. Strickland and T.R. Parsons, A Practical Hand Book of Sea Water Analysis (Fisheries Research Board of Canada, Ottawa 1972) Bull. 167, 2nd edition.
5. S. Williams in : Official Methods of Analysis (Association of Official Analytical Chemists, Arligton, Virginia, 1984) pp.344-345.
6. T. Ogata, T. Ishimaru and M. Kodama, Marine Biology 95, 217-211 (1987).

SUSPECTED CAUSATIVE DINOFLAGELLATES OF PARALYTIC SHELLFISH POISONING IN THE GULF OF THAILAND

YASUWO FUKUYO,* KATSUMI YOSHIDA,** TAKEHIKO OGATA,*** TAKASHI ISHIMARU**** MASAAKI KODAMA,*** PORNSILOP PHOLPUNTHIN,***** SUCHANA WISESSANG,***** VIMOL PHANICHYAKARN,***** and TWESUKDI PIYAKARNCHANA*****
*Faculty of Agriculture, University of Tokyo, Tokyo 113, Japan; ** Aquatic Resources Laboratory, Tokyo 152, Japan; ***School of Fisheries Sciences, Kitasato University, Iwate 022-01, Japan; ****Ocean Research Institute, University of Tokyo, Tokyo 164, Japan; *****Faculty of Science, Chulalongkorn University, Bangkok 10500, Thailand

ABSTRACT

Four species of Protogonyaulax were found in the Gulf of Thailand. They were identified as P. cohorticula, P. fratercula, P. leei and P. tamarensis, respectively. Two strains of P. cohorticula and four of P. tamarensis were cultured and tested for toxicity. All the strains of P. cohorticula were toxic whereas those of P. tamarensis were non-toxic. These results show that P. cohorticula is at least one of the causative organisms of paralytic shellfish poisoning in Thailand.

INTRODUCTION

In 1983, paralytic shellfish poisoning (PSP) with 63 patients including one fatal case, occurred in Thailand [1]. In the phytoplankton samples collected one month after the incident, Protogonyaulax tamarensis cells were found, though the number was very small [1]. Toxin analysis of toxic mussels suggested the causative organism was Protogonyaulax [1]. However, there are no reports on PSP caused by Protogonyaulax in tropical waters. In order to identify the causative organisms, a monitoring survey for toxic dinoflagellates has been carried out in the Gulf of Thailand.

In this paper, we report the occurrence of four species of Protogonlyaulax in the Gulf of Thailand. Toxicities of two species were examined using cultures.

MATERIALS AND METHODS

Species Identification

Phytoplankton samples were collected from Pranburi and several areas in the innermost Gulf of Thailand from 1983 to 1986. Morphological characteristics of Protogonyaulax were observed under a light microscope (NIKON SX-NT21 with differential interference contrast) after dissecting thecal plates using 5% sodium hypochlorite solution and staining with a solution of chloral hydrate, potassium iodine and iodine [2]. Criteria used for the species differentiation are: shape of the apical pore complex (APC), position of a ventral pore (vp), position of anterior and posterior attachment pores (aap and pap), and shape of a sulcal anterior plate (sa) [3,4].

Toxin production

Cultures of Protogonyaulax were established and maintained in T1 medium [5] at 25°C under 4000 lux with 16:8 h L:D cycle using white fluorescent lamps. They were harvested at the end of the exponential growth stage by

RED TIDES: BIOLOGY, ENVIRONMENTAL SCIENCE, AND TOXICOLOGY
Okaichi, Anderson, and Nemoto, Editors

centrifugation. Harvested cells were extracted with an equal amount of water. The extracts were directly analyzed by an HPLC-fluorometric analyzer of PSP [6]. In order to determine the presence of acid-labile components such as gonyautoxin (GTX) VIII, a portion of the water extracts was treated with 0.1 N HCl to hydrolyze the carbamoyl-N-sulfate moiety and then analyzed as above. The toxin was partially purified as previously reported [1] to examine the toxic components.

RESULTS

Identification of Protogonyaulax spp.

From the plankton community study, 73 dinoflagellate species were recorded (Fukuyo, unpublished data). Four Protogonyaulax species, P. cohorticula, P. fratercula, P. leei and P. tamarensis, were collected as suspected toxic species among them. Pyrodinium bahamense var. compressa [7], a recognized toxic species in tropical waters, was not found.

Protogonyaulax cohorticula is round to subpentagonal in ventral view, and slightly wider than long. The sulcus is deeply impressed and fairly wide posteriorly. High wing-like lists develop at the right and left margins of the sulcus. Thecal plates are sparsely porulated. The APC is triangular, narrower ventrally, and sometimes truncated at the ventral end. An apical pore (ap) has a narrow drop-shape and is located in the left half of the APC of cells having the aap, or at center of the APC of cells without the aap. The aap is large, round to short ellipsoidal. The pap is round and located at center of a sulcal posterior plate (sp). The vp is clearly found near the right corner of apical 1' on the suture between apical 1' and 4'. A sulcal anterior plate (sa) has a triangular or rectangular anterior expansion which invades into the epitheca. Length of the cell is 30.3 to 50.0 µm, and width is 30.0 to 57.5 µm. Chains of more than eight cells are often found. In culture, chains of more than 64 cells are formed.

Protogonyaulax fratercula is round to subpentagonal in ventral view, and slightly longer than wide. The sulcus is deeply excavated and wide posteriorly. High wing-like lists develop at the right and left margins of the sulcus. Thecal plates are sparsely porulated. APC is fusiform and pointed both ventrally and dorsally. Ap is small, narrow and located in the left half of the APC. Aap is round and situated in the right half of the APC. Pap is round and located at center of the sp plate. Vp is clearly found in the middle of the suture between apical 1' and 4'. Length of cell is 27.5 to 42.5 µm, and width is 27.5 to 42.5 µm. Chains of more than eight cells are often found.

Protogonyaulax leei is round, slightly longer than wide. The epitheca is hemispherical. Shape of hypotheca is asymmetrical, with the height of the left half longer than that of the right. The sulcus is shallowly impressed, and curves in the middle toward the right. The cingulum is narrow in width, and excavated deeply. Thecal plates are porulated. The APC is narrow triangular to rectangular, prolonged dorso-ventrally and curves to the left in the middle. The large, drop-shaped ap occupies the center of the APC. Some pores are found in the marginal area of the APC. Two attachment pores, aap and pap, are rarely found. Aap cuts the APC in the middle of right margin. The pap is small and located in the right half of the sp plate. The vp is located inside apical 1' and does not contact the suture between apicals 1' and 4'. Length of the cell is 35.0 to 40.0 µm, and width is 35.0 to 42.5 µm. Chains of two cells are rarely found.

Protogonyaulax tamarensis is round and slightly longer than wide. The sulcus is weakly impressed, and wide posteriorly. The antapex is slightly concave and the cingulum deeply concave. Thecal plates are thin and sparsely porulated. The APC is triangular or rectangular, narrower ventrally, with large drop-shaped ap. The aap is round to short ellipsoidal. The pap is found in the right half of the sp plate near the margin between the 5th posterior plate. Both aap and pap are sometimes sealed. The vp is clearly found near the middle of the suture between apical plates 1' and 4'. Length of the cell is 25.0 to 32.5 μm, and width is 25.0 to 32.5 μm. Chains of two cells are rarely found.

Toxin production

Although some cells of *P. fratercula* and *P. leei* were obtained, they could not be cultured. Four strains of *P. tamarensis* and two of *P. cohorticula* were cultured under conditions described above. Mice injected with extracts of all *P. tamarensis* strains did not die nor show any signs of PSP, even when extracts equivalent to 10^7 cells were injected. In HPLC analyses, no toxin peak was observed. From these data, we conclude that these strains were non-toxic.

In contrast, extracts of both strains of *P. cohorticula* were toxic. The toxicity of the two strains was different from each other, but similar to that of strongly toxic strains of *P. tamarensis* in Japan [5]. Fig. 1 shows the results of HPLC analyses of the toxin of *P. cohorticula*.

The water extract shows peaks of GTX I+IV, II, III and saxitoxin (STX), with some unidentified peaks. When analyzed after HCl treatment, peak heights of GTX II, III and STX increased significantly, indicating the presence of carbamoyl-N-sulfate derivatives of these components. The peak height of GTX I+IV did not change after HCl treatment. Ecectrophoretic analysis of partially purified toxin showed a band of GTX IV, although it was very faint. No band of neoSTX was observed. These data indicate that about 80% of the toxin in *P. cohorticula* is GTX I.

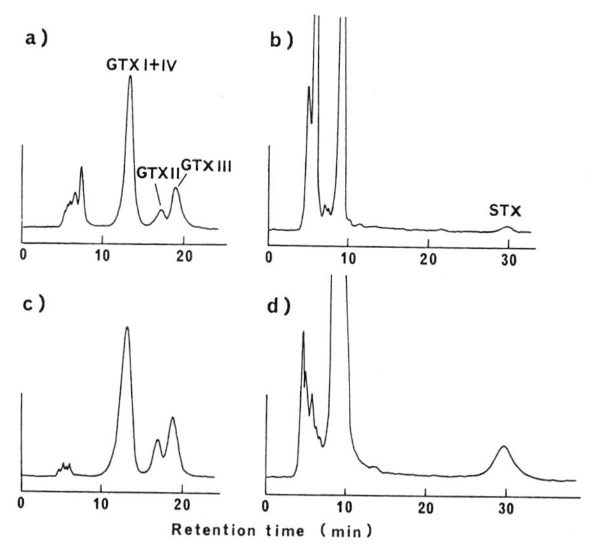

FIG. 1. Identification of *P. cohorticula* toxins by HPLC-fluorometric analyses according to Oshima *et al.* [6]. Mobile phase in a) and c): 0.15 M sodium citrate buffer pH 6.5, mobile phase in b) and d): 0.35 M sodium citrate buffer pH 5.5. a) and b): water extracts of *P. cohorticula* cells, c) and d): that after HCl treatment.

DISCUSSION

In our survey, Pyrodinium bahamense var. compressa [7], a known toxic species in tropical waters, was not found. Instead, four species of Protogonyaulax were identified as suspected causative species of PSP in the Gulf of Thailand. Among them, P. cohorticula was found to produce paralytic shellfish toxins. This fact shows that P. cohorticula is at least one of the causative organisms of PSP in the Gulf of Thailand. There are only two records on the occurrence of P. cohorticula. One is the original description by Balech [8]. He found this species for the first time in phytoplankton samples from the Gulf of Mexico. The second finding is Taylor's monograph on P. fratercula in the Andamann Sea [9], which was reclassified as P. cohorticula by Fukuyo et al. [4]. Our result which shows the occurrence of this species in the Gulf of Thailand, suggests a wide distribution in tropical waters. On the other hand, four strains of P. tamarensis isolated from the Gulf of Thailand were non-toxic. Although P. tamarensis is a well known toxic species, non-toxic strains are occasionaly reported [10, 11], Therefore, the possibility cannot be excluded that there are toxic and non-toxic strains of P. tamarensis in the Gulf of Thailand. Two other species could not be cultured and thus could not be examined for their toxin production.

Recently, PSP cases caused by Pyrodinium bahamense var. compressa have been reported from tropical areas [7,12]. There are no reports of PSP caused by Protogonyaulax spp. in tropical waters. However, PSP cases, the causative organisms of which were not elucidated, have been reported from tropical countries [13]. In such cases, species belonging to Protogonyaulax as well as Pyrodinium bahamense var. compressa, should be examined.

REFERENCES

1. S. Tamiyavanich, M. Kodama and Y. Fukuyo in: Toxic Dinoflagellates, D.M. Anderson, A.W. White and D.G. Baden eds. (Elsevier, N.Y. 1985) pp.521-524.
2. Y. Fukuyo and K. Imamura in: A Guide for Studies of Red Tide Organisms , Japan Fish. Resources Conserv. Ass. ed. (Syuwa, Tokyo 1987) pp. 54-73. (in Japanese).
3. Y. Fukuyo, K. Yoshida and H. Inoue in: Toxic Dinoflagellates, D.M. Anderson, A.W. White and D.G. Baden eds. (Elsevier, N.Y. 1985) pp. 27-32.
4. Y. Fukuyo, P. Pholpunthin and K. Yoshida, Bull. Plankton Soc. Japan, in press.
5. T. Ogata, T. Ishimaru and M. Kodama, Mar. Biol. 95, 217-220 (1987).
6. Y. Oshima, M. Machida, K. Sasaki, Y. Tamaoki and T. Yasumoto, Agric. Biol. Chem. 48, 1707-1711 (1984).
7. T. Harada, Y. Oshima, H. Kamiya and T. Yasumoto, Bull. Japan. Soc. Sci. Fish. 48, 821-825 (1982).
8. E. Balech, Rev. Mus. Argent. Cienc. Nat. Ber. Riv., Hidrobiol., 2, 77-129 (1967).
9. J.R. Taylor, Bibliotheca Bot., 132, 1-234 (1976).
10. R.J. Schmidt and A.R. Loeblich III, J. Mar. Biol. Ass. U.K. 59, 479-487 (1979).
11. C.M. Yentsch, B. Dale and J.W. Hurst, J. Phycol. 14, 330-332 (1978).
12. R.Q. Gacutan, M.Y. Tabbu, E.J. Aujero and F. Icatlo Jr. Mar. Biol. 87, 223-227 (1985).
13. A.W. White, M. Anraku and K. Hooi eds. Toxic red tides and shellfish toxicity in Southeast Asia. Proceedings of a consultative meeting held in Singapore, 11-14 September 1984 (The Southeast Asian Fisheries Development Center 1984).

PARALYTIC SHELLFISH POISONING BY THE MUSSEL MYTILUS EDULIS IN KOREA

O. ARAKAWA,* T. NOGUCHI,* D.-F. HWANG,* D.-S. CHANG,** J.-K. JEON,*** AND K. HASHIMOTO*
*Faculty of Agriculture, University of Tokyo, Bunkyo, Tokyo 113, Japan; **Department of Microbiology, National Fisheries University of Pusan, Pusan, Korea; ***Korea Advanced Institute of Science & Technology, Panwol Ind. 171-14, Korea

ABSTRACT

We have attempted to determine the toxin in the mussel Mytilus edulis, responsible for a food poisoning incident in Pusan, Korea, in April 1986. A few days after the outbreak, mussels were collected from the sea area where the victims obtained the causative shellfish. Mussel digestive glands showed a lethal potency of 490 MU/g as paralytic shellfish poison. Toxin was partially purified from the ethanolic extract by ultrafiltration, Bio-Gel P-2 and Bio-Rex 70 column chromatography. Electrophoresis, TLC and HPLC analyses demonstrated that the toxin consisted of protogonyautoxins (gonyautoxin 8 and its epimer), gonyautoxins and saxitoxins, with toxicity ratios of 2:84:14. From these results, it was concluded that the causative agent of the above incident was paralytic shellfish poison.

INTRODUCTION

Paralytic shellfish poison (PSP) is a potent neurotoxin consisting of more than ten components: protogonyautoxins (PXs), gonyautoxins (GTXs) and saxitoxins (STXs)[1]. PSP is originally produced by several planktonic organisms such as Protogonyaulax spp. [1,2]. Many food poisonings due to ingesting shellfish which have accumulated these toxins have been reported in North America, Japan and Europe [2]. Recently, the PSP appears to have spread to new areas, including some tropical regions [2].

In January 1986, purple clam Soletellina diphos caused food poisoning incidents in South Taiwan, and was found to contain PSP [3]. On April 1, 1986, a similar incident occurred in Pusan, Korea, in which 15 persons were poisoned due to ingesting the mussel Mytilus edulis. Two of them died. In this situation, we made some attempts to elucidate the responsible toxin.

EXPERIMENTAL

Materials

A few days after the outbreak of the poisoning incident, mussels were collected from the sea area near Pusan, where the victims obtained the causative shellfish. Digestive glands were excised from the mussels, immediately frozen, transported to the Laboratory of Marine Biochemistry, University of Tokyo, and kept below -20°C until use. In addition, several mussels were collected from the same area in early May 1986, and their digestive glands were kept frozen similarly.

Assay for Lethal Potency

Digestive glands (118 g) from the April specimens were extracted three

RED TIDES: BIOLOGY, ENVIRONMENTAL SCIENCE, AND TOXICOLOGY
Okaichi, Anderson, and Nemoto, Editors

times with 3 volumes of 80% ethanol adjusted to pH 2 with HCl. A portion of the combined extracts was assayed for lethal potency in mice by the official method for PSP [4], and the remainder was used to purify the toxin by the following method. Digestive glands from the May specimens were used exclusively for lethal potency assay.

Purification of Mussel Toxin

The extract of the digestive glands (58,000 MU) was defatted with dichloromethane, and ultrafiltered through a Diaflo YM-2 membrane. Half of the filtrate (27,000 MU) was chromatographed on a Bio-Gel P-2 column (2.0 x 95.5 cm) using 0.03 M AcOH as eluant. Toxic fractions (20,800 MU) were combined and freeze-dried. A portion of the toxin (7,000 MU) was dissolved in a small amount of water, and applied to a column (0.8 x 96 cm) of Bio-Rex 70 (H^+), followed by elution with water, 0.03 M AcOH and 0.01 N HCl. Toxic fractions obtained with each solvent [designated PX Fr. (130 MU), GTX Fr. (5,130 MU) and STX Fr. (830 MU), respectively] were freeze-dried and subjected to the following analyses. Authentic PXs (PX_2 and PX_1 corresponding to GTX_8 and its epimer, respectively), GTXs, and STXs were prepared by the methods reported previously [5-7], and used as reference standards.

Methods of Analysis

The three PSP fractions were analyzed for toxin composition by electrophoresis, TLC and HPLC, according to the methods previously reported [8].

Acid hydrolysis of PX Fr.

A portion (10 MU) of PX Fr. was dissolved in 3 ml of 0.1 N HCl and heated at 100°C for 5 min. After cooling, the solution was freeze-dried, dissolved in a small amount of water, and subjected to HPLC analysis.

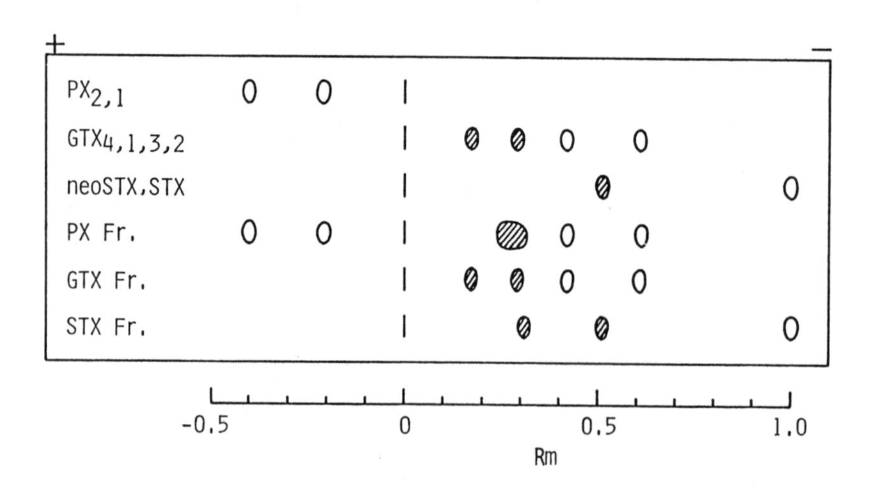

FIG. 1. Electrophoresis of PX, GTX and STX Frs., along with authentic PSPs.
O blue fluorescence; ⊘ greenish yellow fluorescence.

RESULTS

Lethal Potency of Mussel

The digestive glands from the April specimens showed a lethal potency of 490 MU/g, whereas those from the May specimens showed 19 MU/g, indicating a sharp decrease in lethal potency.

Purification of Mussel Toxin

After Bio-Gel P-2 column chromatography, 95 mg of a partially purified toxin was obtained, whose specific toxicity was 220 MU/mg. The toxin was separated into three fractions (PX Fr., GTX Fr. and STX Fr.) by Bio-Rex 70 column chromatography, with toxicity ratios of 2:84:14.

Identification of Mussel Toxin

In electrophoresis, PX Fr. gave five fluorescent spots, four of which corresponded to $PX_{1,2}$ and $GTX_{2,3}$ in both mobility and color (FIG. 1). GTX Fr. showed four spots which agreed well with those of GTX_{1-4}. STX Fr. exhibited three spots, two of which corresponded to neoSTX and STX. Essentially the same results were obtained in TLC (data not shown).

In HPLC analysis, PX Fr. gave rise to a fast sharp peak with a small preceding peak, followed by two small peaks whose retention times agreed well with GTX_2 and GTX_3 (FIG. 2). After acid hydrolysis, the two fast peaks disappeared, increasing the relative intensity of GTX_2 and GTX_3 peaks. In addition, two very small peaks appeared, whose retention times agreed well with those of GTX_1 and GTX_4. In this connection, PX Fr. showed a 17-fold enhanced lethal potency when hydrolyzed. GTX Fr. showed four peaks which corresponded to GTX_{1-4} (data not shown). STX Fr. showed a large peak whose retention time agreed with that of neoSTX, along with a trailing shoulder regarded as STX (data not shown).

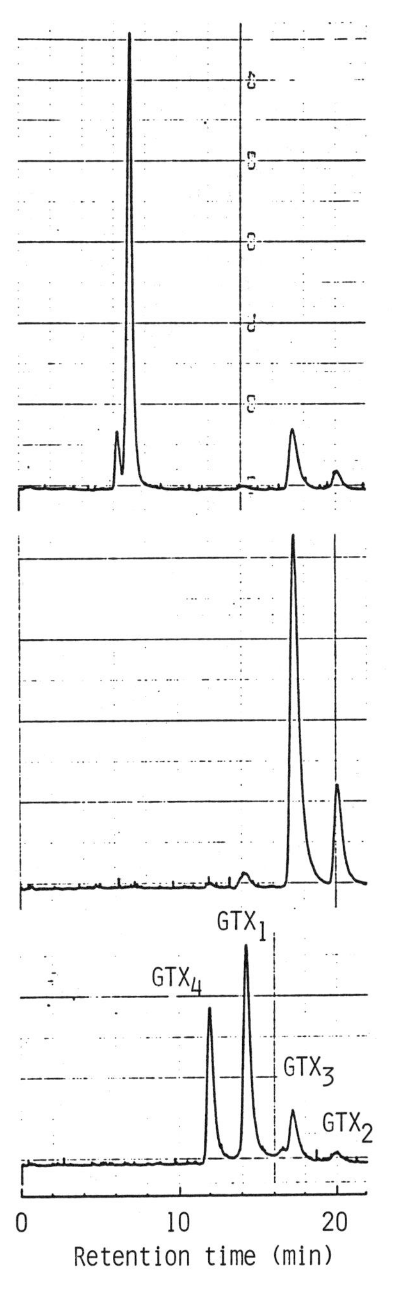

FIG. 2. HPLC of PX Fr. before (upper) and after (middle) acid hydrolysis, and of authentic GTXs (lower).

DISCUSSION

It was concluded from these results that the food poisoning incident which occurred in Pusan, in April 1986, was caused by PSP contained in mussels. The lethal potency of the mussels collected a few days after the outbreak was 490 MU/g digestive gland. Only six grams of such toxic digestive glands (or several specimens of mussel) could kill a man, since the MLD of this toxin is estimated to be 3,000 MU [2].

PSP of the mussel was composed mainly of GTX_{1-4}, along with $PX_{1,2}$ and small amounts of STXs. It has so far been reported [2] that most bivalves such as scallop, mussel and oyster, once infested, possess GTXs as the major toxins. This was the case with the Korean mussels.

In addition, the mussel specimens contained $PX_{1,2}$ at a significant level: both these components accounted for only 2% of the total potency. $PX_{1,2}$ or PXs are mildly toxic, but are converted into highly toxic analogues (GTX_{1-4}) when hydrolyzed [1,6], as demonstrated also by our experiments (FIG. 2). In our mussel specimens, $PX_{1,2}$ is calculated to account for more than 30% of the total toxicity after hydrolysis. However, no confirmative information is available on the *in vivo* behavior of those low-toxic PSP components in humans.

The causative organisms producing the PSP detected in the mussels have not yet been determined.

REFERENCES

1. T. Noguchi, in: Toxic Phytoplankton - Occurrence, Mode of Action, and Toxins, Suisan-Gaku Series, No. 42 (Nippon Suisan Gakkai, Tokyo 1982) pp. 88-101.
2. T. Noguchi and K. Hashimoto, Igaku No Ayumi 112, 861-870 (1980).
3. D.-F. Hwang, T. Noguchi, Y. Nagashima, I-C. Liao, and K. Hashimoto, Nippon Suisan Gakkaishi 53, 623-626 (1987).
4. T. Kawabata, Shokuhin Eisei Kensa Shishin [Manual of Methods for Food Sanitation Testing], Vol. 2 (Japan Food Hygienic Association, Tokyo 1978) p. 240.
5. T. Noguchi, Y. Ueda, K. Hashimoto, and H. Seto, Nippon Suisan Gakkaishi 47, 1227-1231 (1981).
6. Y. Onoue, T. Noguchi, J. Maruyama, K. Hashimoto, and H. Seto, J. Agric. Food Chem. 31, 420-423 (1983).
7. K. Daigo, A. Uzu, O. Arakawa, T. Noguchi, H. Seto, and K. Hashimoto, Nippon Suisan Gakkaishi 51, 309-313 (1985).
8. T. Noguchi, O. Arakawa, K. Daigo, and K. Hashimoto, Toxicon 24, 705-711 (1986).

FIRST REPORT ON PARALYTIC SHELLFISH POISONING CAUSED BY GYMNODINIUM CATENATUM GRAHAM (DINOPHYCEAE) IN JAPAN

TAKEHIKO IKEDA,* SUSUMU MATSUNO,* SHIGERU SATO,** TAKEHIKO OGATA,** MASAAKI KODAMA,** YASUWO FUKUYO*** AND HARUYOSHI TAKAYAMA****
*Yamaguchi Prefectural Inland Sea Fisheries Experimental Station, Yamaguchi 754; **School of Fisheries Sciences, Kitasato University, Iwate 022-01; ***Faculty of Agriculture, Tokyo University, Tokyo 113; ****Hiroshima Fisheries Experimental Station, Hiroshima 737-12, Japan

ABSTRACT

Gymnodinium catenatum Graham (Dinophyceae) was identified as a causative organism of paralytic shellfish poisoning (PSP) for the first time in Japan. The toxin composition of this species consisted mostly of N-sulfocarbamoyl derivatives of paralytic shellfish toxins. The toxin composition of oysters toxified by this species reflected that of this species. These results indicate that G. catenatum as well as Protogonyaulax spp. are causative organisms of PSP in Japanese waters.

INTRODUCTION

In December of 1986, bivalves were highly toxic from paralytic shellfish toxins (PSP toxins) at Senzaki Bay, Yamaguchi Prefecture, Japan. Species responsible for PSP so far in Japan are Protogonyaulax tamarensis and P. catenella. However, these species were not found in the area during the period. Instead, long-chain-forming dinoflagellates without thecal plates occurred at high cell densities. Cells from both natural waters and cultures isolated from Senzaki Bay contained PSP toxins, and their toxin profiles were similar to that of the toxic oysters.

In this report, we identify this species as Gymnodinium catenatum Graham (Dinophyceae), and describe toxin profiles from this species and from toxic oysters. This is the first report of G. catenatum and of toxification of bivalves by this species in Japan.

EXPERIMENTAL

Monitoring of G. catenatum abundance and shellfish toxicity

The monitoring station was at the head of Senzaki Bay. Seawater samples were collected between two and six times each month. Enumeration of G. catenatum was carried out microscopically after 50-fold concentration of seawater by gravity filtration using 8 µm Millipore filters. Nontoxic bivalves, Crassostrea gigas, Scapharca broughtonii, Pecten albicans, Mytilus edulis, Ruditapes philippinarum and Saxidomus purpuratus, were transplanted to 1.5 m depth at the station during January 1986. Several specimens of bivalves were collected for toxicity tests whenever the seawater sampling was done. Hepatopancreases were excised and measured for PSP toxicity by the standard mouse bioassay method [1].

Identification of G. catenatum

Cells from natural waters were fixed with glutaraldehyde. They were examined with a light microscope NIKON XF-NT21 using Nomarski interference contrast. SEM preparation consisted of putting a single chain or cell on a glass coated with poly-L-lysin hydrobromide, rinsing five times in distilled water, dehydrating in an alcohol series, replacing alcohol by iso-amyl

RED TIDES: BIOLOGY, ENVIRONMENTAL SCIENCE, AND TOXICOLOGY
Okaichi, Anderson, and Nemoto, Editors

acetate, drying with CO_2 in a critical point dryer (Hitachi HCP-2) and coating with about 200 A° of Pt-Pd in an ion spattering coater. The material was observed under SEM (Hitachi S-430). A single motile cell or chain of G. catenatum was isolated from seawater and inoculated into T1 medium [2] containing 1 µM NH_4Cl. The culture was maintained at 20°C at an illumination of 4000 lux (16:8 LD cycle). Cells in cultures were also observed by light microscopy and SEM.

Toxicity and toxin composition of G. catenatum and toxic oysters

Naturally occurring cells of G. catenatum were collected by filtering seawater successively through nylon cloths with mesh sizes of 95, 60 and 10 µm. G. catenatum cells collected in the 10 - 60 µm fraction were freeze dried. Cultures for toxin analyses were grown in 3 L conical flasks under identical conditions. Toxic specimens of the oyster Crassostrea gigas were collected from the sampling station during December 1986.

About 6 x 10^6cells of naturally occurring or cultured G. catenatum were suspended in 4 ml of water, sonicated for 2 min, and centrifuged at 1500 g for 10 min. A portion of the supernatant was added to an equal volume of 0.2 N HCl and boiled for 10 min. Toxicities of aqueous extracts and HCl treated extracts were measured by mouse bioassay [1]. The toxin composition of these extracts was analyzed using an HPLC-fluorometric analyzer of PSP toxins [3]. Whole oyster tissue was analyzed in the same manner as for the G. catenatum extracts. Alternatively, the oyster tissue extracts were prepared according to the standard method [1] by extracting the tissue with an equal volume of 0.1N HCl. The extracts obtained were added to 1N HCl to give a final HCl concentration of 0.1N HCl, and then heated in boiling water for 10 min. Toxicities of the extracts before and after the HCl treatments were measured by mouse bioassay. The toxicity was estimated from death time of mice after intraperitoneal injection, using the dose-death time table for PSP toxin [1], and expressed in mouse units (MU) where one MU represents the dose of toxin necessary to kill a 20-g male mouse (ddY strain) in 15 min.

RESULTS

Seasonal variation of G. catenatum abundance and bivalve toxicities

G. catenatum appeared in high cell densities at the beginning of December 1986. The high density of around 1000 cells/L continued for more than a month. It decreased to lower densities in January and had disappeared by the end of February 1987. Toxicities of various bivalves increased in parallel to the occurrence of G. catenatum, indicating that this species was the causative organism. Maximum toxicities of bivalves C. gigas, R. philippinarum, S. broughtonii, S. purpuratus, P. albicans and M. edulis were 1130, 1450, 2030, 360, 1500 and 2050 MU/g of hepatopancreas, respectively (Fig.1).

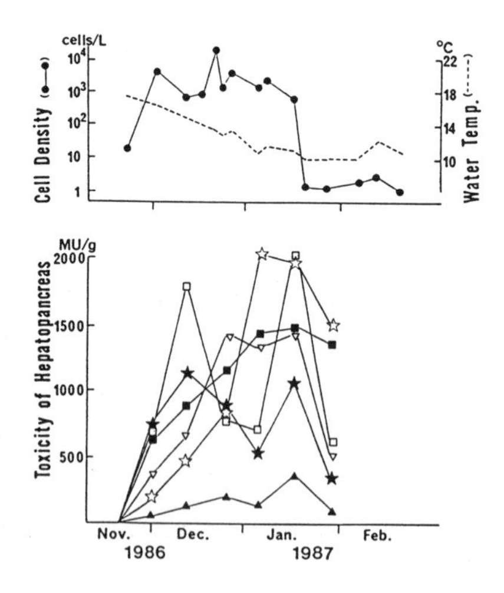

FIG. 1. Seasonal variation of G. catenatum abundance (upper) and toxicities of bivalves (lower). S. Broughtonii(☆), P. albicans(■), M. edulis(□), C. gigas(★), R. philippinarum(▽), S. purpuratus(▲).

Identification of G. catenatum

The cells were identified as G. catenatum from Morey-Gaines [4]. Cells form chains of more than 20 cells (Fig.2A). The cingulum is

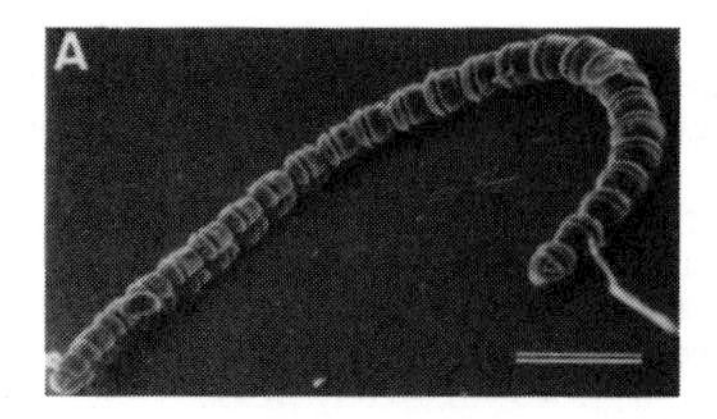

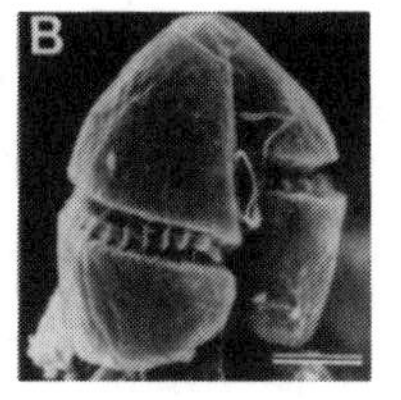

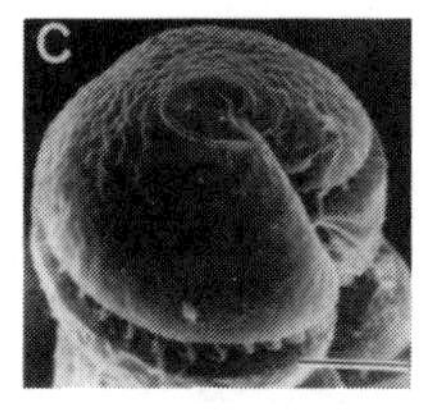

FIG. 2. Gymnodinium catenatum. A: Chain of 26 cells. The bar equals 100 µm. B: Ventral view of lead cell in the chain. The bar equals 10 µm. C: latero-apical view of lead cell in the chain. The bar equals 10 µm.

slightly pre-median, descending 2-3 times its width. The sulcus extended straight from apex to antapex (Fig.2B). The apical groove starting from near the upper end of sulcus encircled the apex (Fig.2C). The cell surface was roughly undulate and bumpy.

Toxicity and toxin composition of G. catenatum and toxic oyster

As shown in Table I, the toxicities of aqueous extracts of both naturally occurring and cultured G. catenatum much increased after treatments with HCl, indicating that most of the toxin components were N-sulfocarbamoyl derivatives of PSP toxins. It was also true for the aqueous extracts of oysters. In the case of the oyster extracts prepared according to the standard method [1], the toxicity increased about 10 times after HCl treatment, showing that the standard mouse bioassay underestimates the toxicity of shellfish toxified by G. catenatum. In the analyses of aqueous extracts from G. catenatum and oyster by HPLC-fluorometric analyzer of PS toxins [3], gonyautoxin (GTX) 8 and epiGTX 8 were predominant in all the extracts. Following HCl treatments, large peaks of GTX 3 and GTX 2 appeared in the chromatograms, indicating the presence of their N-sulfocarbamoyl derivatives, GTX 8 and epiGTX 8. In Table II, the results of the analyses are summarized. These results show that most toxin components of G. catenatum are N-sulfocarbamoyl derivatives and that the toxin profile of the oysters reflects that of G. catenatum.

TABLE I. Toxicities of oyster and G. catenatum.

Sample	Aqueous extract	HCl-treated extract	Standard extract*
Oyster tissue	7.9 MU/g	760.2 MU/g	77.5 MU/g
Cells from natural waters	No data	1.67 MU/10^4cells	-
Cultured cells	0.12 MU/10^4cells	2.13 MU/10^4cells	-

*prepared according to the standard method [1].

TABLE II. Toxin compositions of oyster and G. catenatum.

Sample	STX	nSTX	dSTX	GTX 1	2	3	4	5	6	GTX8	epiGTX8	unknowns
Oyster tissue	-	+	-	+	-	-	+	+	+	+++	+++	+
Natural cells	-	+	-	+	-	-	+	+	+	+++	+++	+
Cultured cells	-	+	±	+	-	-	+	+	+	+++	+++	++

DISCUSSION

G. catenatum is suggested to be a causative organism of PSP [5].

Recently, PSP cases coincided with the bloom of G. catenatum are reported from Spain [5] and Mexico [6]. Very recently, Oshima et al. [7] have provided the first evidence for the toxigenicity of G. catenatum in Tasmania. The present study shows the distribution of this species in Japanese waters. Oshima et al. [7] reported the predominance of N-sulfocarbamoyl derivatives of PSP toxins in Tasmanian G. catenatum as well as toxic shellfish. From these, they presented questions to the applicability of the standard mouse test in assessing potential health risks for human seafood consumers. The results of the present study show that these were also true for the Japanese strains of G. catenatum and toxic oyster. These results indicate the necessity for reinvestigation of the standard analysis method in the case of toxic shellfish caused by G. catenatum.

Onoue et al. [8] reported the PSP case in 16 patients at Senzaki Bay who ingested oysters in 1979. They analyzed the toxin composition of toxic oysters collected from the bay in 1980 and compared it with that extracted from cultures of P. catenella, which were isolated from the same bay in 1979 [9]. From the similarity of their toxin profiles, they suggested that P. catenella is the causative organism of PSP in Senzaki Bay. In the monitoring of P. catenella abundance and shellfish toxicity carried out by one of the authors (Ikeda) in 1985, shellfish toxicity increased in parallel to the occurrence of P. catenella, though the toxicity of bivalves was relatively low (maximum toxicity of pecten was 15 MU/g of hepatopancreas). However, these two parameters did not change in parallel in 1984 when the shellfish became highly toxic (maximum toxicity of mussels was 560 MU/g of hepatopancreas). Toxicity of bivalves more than 50 MU/g of hepatopancreas was maintained for about two months in that year. During the period, no P. catenella cells were observed (unpublished data). We had not paid attention to G. catenatum in the monitoring in 1984. Thus preserved phytoplankton samples collected during that period were examined again for the occurrence of G. catenatum. Although considerable amounts of cells of unarmoured species had decayed during preservation, significant amounts of G. catenatum cells were observed, suggesting that G. catenatum was the responsible species for bivalve toxicities in that year. These data suggest that G. catenatum as well as P. catenella are responsible species for toxification of bivalves in this bay.

ACKNOWLEDGEMENTS

This work was partly supported by a research fund from the Ministry of Agriculture, Forestry and Fishery in Japan. We thank Yamaguchi Prefectural Institute of Public Health for measuring the toxicities of bivalves in the monitoring survey.

REFERENCES

1. T. Kawabata in: Shokuhin Eisei Kensa Shishin (Manual for Methods of Food Sanitation Testing) Vol.2 (Japan Hygienic Association, Tokyo, 1978) pp. 240 - 244 (in Japanese).
2. T. Ogata, T. Ishimaru and M. Kodama, Marine Biol. 95, 217 - 220 (1987).
3. Y. Oshima, M. Machida, K. Sasaki, Y. Tamaoki and T. Yasumoto, Agric. Biol. Chem. 48, 1707 - 1711 (1984).
4. G. Morey-Gaines, Phycologia 21, 154 - 163 (1982).
5. S. Fraga and F.J. Sanchez in: Toxic Dinoflagellates, D.M. Anderson, A.W. White and D.G. Baden eds. (Elsevier, Amsterdam, 1985) pp.51 - 54.
6. L.D. Mee, M. Espinosa and G. Diaz, Mar. envir. Res. 19, 77 - 92 (1986).
7. Y. Oshima, M. Hasegawa, T. Yasumoto, G. Hallegraeff and S. Blackburn, Toxicon 25, 1105 - 1111 (1987).
8. Y. Onoue, T. Noguchi and K. Hashimoto, Nippon Suisan Gakkaishi 46, 1031 - 1034 (1980).
9. Y. Onoue, T. Noguchi, J. Maruyama, Y. Ueda, K. Hashimoto and T. Ikeda, Nippon Suisan Gakkaishi 47, 1347 - 1350 (1981).

VARIATIONS IN MAJOR TOXIN COMPOSITION FOR SIX CLONES OF PTYCHODISCUS BREVIS

DANIEL G. BADEN* AND CARMELO R. TOMAS**
*University of Miami Rosenstiel School of Marine and Atmospheric Science Division of Biology and Living Resources, Miami FL. 33149 U.S.A. and **Florida Department of Natural Resources Bureau of Marine Research, 100 Eighth Avenue S.E. St. Petersburg FL. 33701 U.S.A.

ABSTRACT

The toxin content of six clones of Ptychodiscus brevis was analyzed for the presence of PbTx-1, PbTx-2, and PbTx-3. The cultures included four clones established by single cell isolation of individuals from 1986 blooms in Texas and Florida coastal regions, one established from Florida in 1978, and a subculture, W53DB, of the clone originally isolated by W.B. Wilson in 1953 from Florida waters. The recent isolates were determined to be haploid while the oldest clone was previously found to be diploid. High pressure liquid chromatography analysis of toxin content, normalized on a per cell basis, indicated a wide clonal variability of the three toxin fractions. Using W53DB as the standard clone, significant differences in toxin profile were found in each of the recent clones; the most substantial variations occurring with respect to PbTx-3 concentrations. These results suggest a much wider variability in toxin content for P. brevis than was previously demonstrated from analyses of the diploid clone. The variability in toxin content from different populations exposed to dissimilar environmental conditions requires further examination.

INTRODUCTION

Studies of polyether toxins isolated from Ptychodiscus brevis, Florida'a red tide dinoflagellate, over the past decade have demonstrated the presence of eight toxic fractions which have been unified under a single nomenclature [1]. The potency of these toxins is well documented [reviewed in 2-4], with the principal locus of activity being excitable cell types [5-7].

To date, all of these studies have been conducted using toxins isolated from cultures of the original 1953 Florida isolate of W.B. Wilson. The toxin, PbTx-2 [8], was considered the predominant toxin by many groups [2-5, and references therein], and has been used for most studies. The reported amounts of each of the remaining toxins is variable. One report [9] indicates that toxin profiles are variable, depending on culture phase. The lack of other

RED TIDES: BIOLOGY, ENVIRONMENTAL SCIENCE, AND TOXICOLOGY
Okaichi, Anderson, and Nemoto, Editors

isolates has precluded studies to confirm the toxin profiles as measured in the Wilson clone. Also, new isolates would provide the opportunity to explore toxin amounts with respect to ploidy, the original Wilson isolate being diploid [10].

EXPERIMENTAL

Clonal cultures of Ptychodiscus brevis were established by single cell isolation from raw water samples from Sarasota and Gasparilla Pass, Florida and from a marina in Corpus Christi Texas. Individual cells were washed several times and placed into NH-15 medium [11]. Once cultures were established they were transferred to one L volumes and maintained at 24°C, 12:12 L:D and 50-70 uEm^{-2}s^{-1} of cool white fluorescent light. Two additional clones, 78P5 isolated from the Tampa Bay area in 1978, and W53DB originally isolated by W.B. Wilson, were also examined. For toxin analysis, eight L cultures were grown nder the same conditions as described above.

When in late logarithmic growth phase, each eight L culture was extracted using one L chloroform. Following evaporation of the chloroform solvent, the residue in each case was redissolved in 25 mL portions of 90% aqueous methanol and the solution was extracted three times with 25 mL portions of light petroleum. The combined ether fractions were back-extracted one time with 25 mL of 90% aqueous methanol, and the appropriate methanol fractions were pooled. Following flash-evaporation of the methanol solvent, residues were dissolved each in 20 mL reagent acetone, and the resulting suspensions were centrifuged at 10,000 a g for 10 min. The supernatant solutions were evaporated to small volume (ca. 0.1-0.2 mL) with a stream of nitrogen, and were applied to silica gel 4x20 cm TLC plates and chromatographed in acetone/light petroleum (30/70) and visualized with short wave UV light. Bioassays employing mosquito fish (Gambusia affinis) confirmed potency of suspect UV-absorbant fractions. Potent fractions were quantitatively removed from plates, and were rechromatographed on silica gel plates using ethyl acetate/light petroleum (50/50 for PbTx-1 and PbTx-2; 70/30 for PbTx-3) as solvent system. Potent fractions were eluted from the plates with HPLC-grade methanol, and were quantified by HPLC (C-18 reverse phase column, isocratic 85% aqueous methanol, UV detection at 215 nm). The results were expressed as averages of three 50-100 uL injections, and were reduced to per cell averages based on cell counts obtained by Coulter Counter.

RESULTS

The analysis for the three major toxic fractions of the six clones is given in Table I. Variations were observed between all clones. For the three Florida clones, the PbTx-2 fraction seemed to be the most consistent, while PbTx-1 varied between CT5D2 and the others, and PbTx-3 was elevated in the CT5D3 clone. This fraction, the greatest measured in all clones examined, was 5.5-fold greater than that observed in the Wilson clone control. The two Corpus Christi clones showed wide variations in all toxin fractions, but had the second highest values for PbTx-3 in CC6 (3.3-fold higher relative to W53DB). In all cases, PbTx-2 was more conservative and predominant than were the other toxins. Variations in toxin content within these clonal extracts is far greater than one observes upon repeated extractions of, as an example, multiple cultures of the W53 DB clone used as control. Brevetoxins 5-9 were not quantified owing to limits of detection.

TABLE I. Variation in three toxin fractions for clonal isolates of Ptychodiscus brevis

Clone	Origin	Ploidy	Toxin (pg/cell)		
			PbTx-1	PbTx-2	PbTx-3
78P5	Florida	n	1.02	8.76	-0-[a]
CT5D2	Florida	n	0.22	8.82	0.41
CT5D3	Florida	n	1.72	12.60	2.31
CC5	Texas	n	1.40	12.51	0.43
CC6	Texas	n	0.07	4.93	1.62
W53DB	Florida	2n	1.70	8.70	0.42

[a] sensitivity limit= 0.01 pg/cell.

DISCUSSION

The analysis of different clones of P. brevis serves to indicate the degree of variability which exists in this species. Examination of the data from the viewpoints of summation of total toxin, summation of toxin based on similar structural backbone, and analysis of individual amounts of each toxin in a given clone provides no obvious pattern to the toxin profiles from each respective clone. Total toxin content rnages from 6.6 to 16.6 pg/cell, and Type-1 toxin (PbTx-2 and PbTx-3) content ranges from 6.6 to 12.9 pg/cell. In all cases, PbTx-2 was found to predominate in clones, and accounted for 74-93% of the toal mg amount of toxin. PbTx-1 percentages ranged from 1-16% of total mass, and PbTx-3 accounted for 0-25% of total toxin.

That one toxin is derived from another by classical detoxification pathways has been postulated [9], and that toxin turns over rapidly in cells are two indications that P. brevis cells are remarkably active with respect to toxin metabolism and turnover. The variability in PbTx-3 concentration alone is most remarkable and we have no current experimental evidence to suggest its meaning. It is clear to us that specific labeling experiments need to be performed to answer this question with certainty.

Our finding that different clones grown under identical conditions (as identical as we can determine) provide toxin profiles which are profoundly different is significant for several reasons: (i) clones derived from different geographic localities may present different toxin profiles---which corresponds directly with the composite potency of the specific clone; (ii) clones derived from within a geographic region also appear to have variable toxin profiles---again affecting potency of blooms; (iii) the clonal variation in toxin profile may reflect the metabolic capability of the subject clone; (iv) ecological or physicochemical factors which affect the metabolic machinery for toxin production may serve to intensify or reduce the composite potency of clones or blooms; (v) clonal crosses, if possible, should expand our knowledge of the potential for toxin synthesis induction and heritability. A wider group of clonal isolates needs to be examined to confirm or refute our preliminary indications of wide variability in nature. Although the values presented within this communication appear scattered, we believe further work is needed to identify trends which may be of great importance in our understanding the phenomenon of toxic red tides.

ACKNOWLEDGEMENTS

The expert technical assistance of Laurie Roszell and Lloyd Schulman is gratefully acknowledged. This work was supported in part by the U.S. Army Medical Research and Development Command, Contract Numbers DAMD17-85-C-5171 and DAMD17-87-C-7001.

REFERENCES

1. M. Poli, T.J. Mende, and D.G. Baden, Molec. Pharmacol. 30, 129 (1986).
2. Y. Shimizu, Pure Appl. Chem. 54, 1973 (1982).
3. K.A. Steidinger and D.G. Baden in: Dinoflagellates, D. Spector, ed. (Academic Press, N.Y. 1984) pp. 201-261.
4. S. Ellis, ed. Toxicon 23, 469 (1985).
5. D.G. Baden, Int. Rev. Cytol. 82, 99 (1983).
6. J.M.C. Huang, C.H. Wu, and D.G. Baden, J. Pharmacol. Exp. Ther. 22, 615 (1984).
7. W.A. Catterall and M.A. Risk, Molec. Pharmacol. 19, 345 (1981).
8. D.G. Baden, T.J. Mende, A.M. Szmant, V.L. Trainer, and R.A. Edwards, this volume.
9. D.G. Baden, T.J. Mende and L.E. Roszell, this volume.
10. C.L. Loper, K.A. Steidinger and L.M. Walker, Trans. Amer. Micros. Soc. 99, 343 (1980).
11. D.V Aldrich and W.B. Wilson, Biol. Bull. 119, 57 (1960).

PARALYTIC SHELLFISH POISONING IN TAIWAN

D.-F. HWANG,* T. NOGUCHI,* Y. NAGASHIMA,** I-C. LIAO,*** S.-S. CHOU,**** AND K. HASHIMOTO*
*University of Tokyo, Japan; **Tokyo University of Fisheries, Japan; ***Taiwan Fisheries Research Institute, R.O.C.; ****National Health Administration, R.O.C.

ABSTRACT

Food poisoning incidents due to ingesting cultured purple clam Soletellina diphos occurred in South Taiwan in January 1986. Clam specimens were collected and assayed for lethal potency as paralytic shellfish poison (PSP). The digestive gland showed the highest potency of 2,000±1,600 (mean±S.E.) MU/g, followed by other parts (160±110 MU/g) and siphon (33±24 MU/g). HPLC and other analyses demonstrated that the responsible toxin consisted almost exclusively of gonyautoxins 1-4 (GTX_{1-4}). The potency of the clam decreased down to 6±4 MU/g digestive gland in December 1986. Subsequent surveillance for toxic purple clams showed that specimens obtained at a market were found to have a potency of 15 MU/g edible parts. The toxin was composed of GTX_{1-4}, along with some saxitoxin and neosaxitoxin.

INTRODUCTION

Serious food poisoning incidents due to ingesting the purple clam Soletellina diphos occurred in South Taiwan in January 1986. A total of about 30 victims, including two deaths, were reported. Soon after ingestion, a tingling or burning sensation appeared at the lips, with gradual progression to the neck, arms, fingertips and feet. This sensation then changed to paralysis and loss of voluntary movements. In severe cases, ataxia, aphasia, headache, thirst, nausea, and vomiting were accompanied. Two deaths occurred as a result of respiratory paralysis within 4 hr. Other victims were convalesced within 24 hr and recovered in 48 hr completely. These symptoms reminded us of those caused by paralytic shellfish poison (PSP). The causative clams were from a culture pond at Tungkang, Pintung Prefecture. In Taiwan, no purple clam-associated poisoning has so far been reported. In Japan, bivalves such as scallop Patinopecten yessoensis, mussel Mytilus edulis and oyster Crassostrea gigas, have sporadically been causing similar poisonings, in all of which the responsible toxin was PSP. The present study was undertaken in this situation. Purple clam specimens were harvested from the culture pond over one year, and assayed for lethal potency. Instrumental analyses demonstrated the responsible toxin to be PSP. On the other hand, some purple clam specimens on the market were also found to contain PSP.

MATERIALS and METHODS

Materials

In January to December 1986, specimens of purple clam Soletellina diphos were periodically collected from the culture pond at Tungkang, Pingtung Prefecture, South Taiwan, the pond where the causative purple clams were harvested. Specimens were immediately frozen with dry-ice, transported to the Laboratory of Marine Biochemistry, University of Tokyo, and kept

RED TIDES: BIOLOGY, ENVIRONMENTAL SCIENCE, AND TOXICOLOGY
Okaichi, Anderson, and Nemoto, Editors

frozen. Just before assay, ten specimens were arbitrarily taken, partially thawed and shucked. The edible part of each specimen was dissected into the digestive gland, siphon and other parts, and used to assay for lethal potency and to identify the toxin. Purple clam specimens were also obtained from the Tungkang market in March 1987, and the edible parts were immediately dissected, combined and kept frozen at -15°C for 5 months until use.

Assay for Lethal Potency

Lethal potency was assayed by the determination method for PSP [1].

Purification of Purple Clam Toxin

Another ten frozen specimens of purple clam collected on January 9, 1986, were partially thawed, and their digestive glands dissected and combined. The digestive glands (about 20 g) were homogenized for 5 min with 3 volumes of 80% ethanol adjusted to pH 2 with HCl, and centrifuged. This operation was repeated two more times. The supernatants were combined, concentrated under reduced pressure, and defatted thrice with dichloromethane. The aqueous layer was ultrafiltered through a Diaflo YM-2 membrane to cut off more than 1,000-dalton substances. The filtrate (21,000 MU) was freeze-dried, dissolved in 0.03 M acetic acid, and chromatographed on a Bio-Gel P-2 column (2.0 x 95.5 cm) using 0.03 M acetic acid as eluant. Toxic fractions were combined and freeze-dried ("toxin A", 20,000 MU, 450 MU/mg). Toxin was also partially purified from ca. 20 g edible portion of the Tungkang market specimens by essentially the same procedure as above ("toxin B", 220 MU, 20 MU/mg). Both toxins were subjected to the following analyses. Authentic GTXs, STXs, and tetrodotoxin (TTX) were prepared by the methods reported previously [2-3], and used as reference standards.

Methods of Analysis

Electrophoresis was performed for 30 min on 5 x 18 cm cellulose acetate strips (Chemetron) using 0.08 M Tris-HCl buffer (pH 8.7), under a constant current of 0.8 mA/cm width. Toxins were visualized as a greenish yellow or blue fluorescent spot under UV light at 365 nm after spraying the strip with 1% H_2O_2 and heating at 100°C for 10 min. TLC was conducted on 4 x 10 cm Whatman LHP-K high performance precoated plates with a pyridine-ethyl acetate-acetic acid-water (15:5:3:4) system. After the run, toxins were visualized as in electrophoresis. Reversed phase HPLC was carried out on an AM-314 (YMC-gel ODS S-5 resin) column, using 1-heptanesulfonic acid sodium salt as an ion-pairing reagent, with 1% methanol-0.05 M potassium phosphate (pH 7.0) for GTXs and with 20% methanol-0.05 M potassium phosphate (pH 7.0) for STXs [4]. The periodate reagent used was prepared after SULLIVAN and IWAOKA [5] with some modifications. The fluorogenic reaction was performed at 65 °C for 0.7 min. The volume ratio of column eluate to the periodate reagent in the mixing tee was 1:1. The intensity of fluorescence was measured at 388 nm with 344 nm excitation.

RESULTS

Lethal Potency of Purple Clam

Clam specimens collected from the culture pond on January 9, 1986, showed the highest lethal potency of 2,000 ± 1,600 MU (mean±S.E.)/g digestive gland, followed by other parts (160±110 MU/g) and siphon (33±24 MU/g). Purple clam specimens showed rather wide individual variations in lethal potency. Some size-dependency of lethal potency was also recognized. A specimen exhibited the highest total potency of 25,000 MU, which might

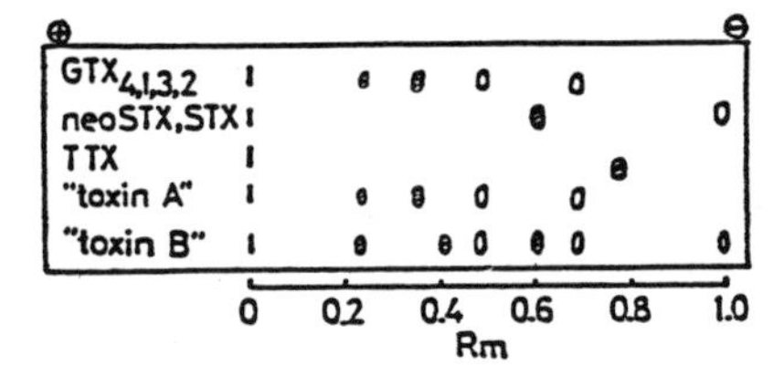

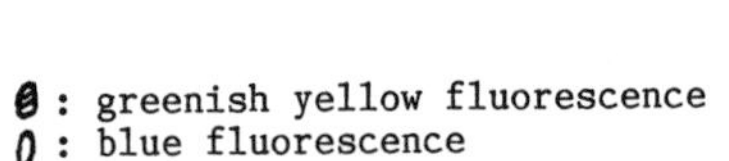
: greenish yellow fluorescence
: blue fluorescence

FIG. 1. Electrophoresis of "toxin A" and "toxin B", along with authentic PSPs and TTX.

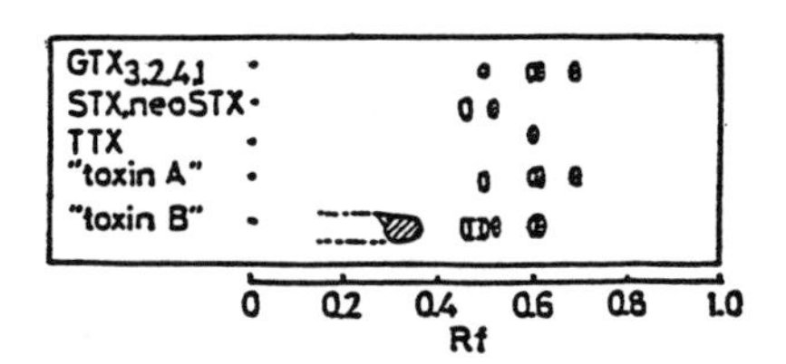

: greenish yellow fluorescence
: blue fluorescence

FIG. 2. TLC of "toxin A" and "toxin B", along with authentic PSPs and TTX.

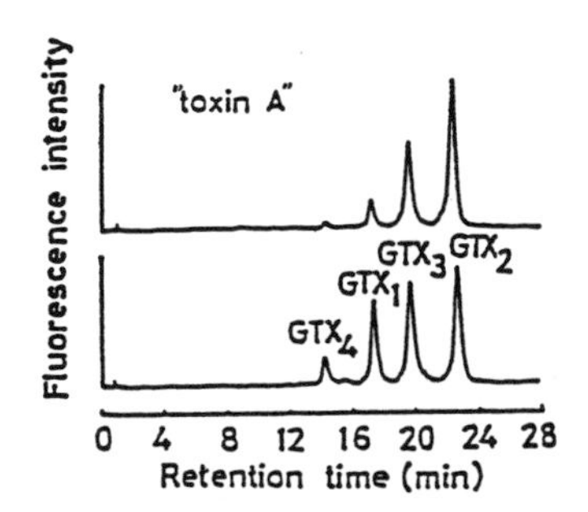

FIG. 3. HPLC for GTXs of "toxin A", along with authentic GTXs.

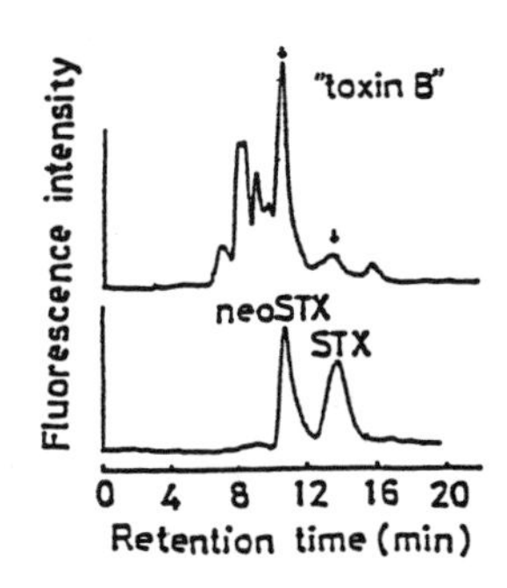

FIG. 4. HPLC for STXs of "toxin B", along with authentic STXs.

have killed at least eight adult humans if taken by mistake.

Lethal potency of the purple clam in the pond, gradually decreased over one year, resulting in December specimens with 6±4 MU/g digestive gland.

Lethal potency of the Tungkang market specimens was 15 MU/g edible part, or around 60 MU/g digestive gland as estimated.

Identification of Purple Clam Toxins

In electrophoresis, toxin A gave rise to four bands which corresponded well to GTX_{1-4}, both in the electrophoretic mobility and fluorescent color (FIG. 1). Toxin B exhibited bands which agreed well with those of GTX_{2-4}, STX and neoSTX, along with an unknown.

As illustrated in FIG. 2, toxin A gave rise to four fluorescent spots in TLC again, whose Rf values were indistinguishable from those of GTX_{1-4}. Toxin B showed six spots, five of which coincided well with GTX_{2-4}, STX and neoSTX.

In HPLC, toxin A showed four GTX peaks whose retention times (Rt) agreed well with those of GTX_{1-4} (FIG. 3). On the other hand, toxin A did not contain STX nor neoSTX at a detectable level. Toxin B afforded six peaks, four of which agreed well with GTX_{1-4} (data not shown). This toxin clearly showed STX and neoSTX peaks (FIG. 4). STXs accounted for 13% of the total potency of toxin B.

DISCUSSION

It was concluded from these results that the food poisoning incidents which occurred in South Taiwan in January 1986 were caused by PSP contained in the purple clam. The digestive gland showed the highest lethal potency and was toxic over one year. In bivalves except scallop, the toxin is fairly quickly accumulated in and eliminated from the digestive gland. In this respect, the purple clam resembles the scallop which quickly accumulates and retains the toxin in the digestive gland for a long period.

The toxin of purple clam collected from the culture pond in January 9, 1986, was composed of GTX_{1-4}, along with trace amounts of STXs. The toxin compositions of PSP have so far been reported for some bivalves such as scallop [6], mussel [7] and oyster [3]. Their toxin compositions widely differ from each other, but are commonly featured by the presence of at least a few percent of STXs. The toxin of the culture pond specimens was somewhat unique in this respect. On the other hand, the toxin of the Tungkang market specimens contained 13% STXs. The differences between toxins A and B may be intrinsic. Another possibility could not be excluded that some GTX components were converted into STXs during frozen storage [8]. In this connection, it is generally accepted that PSP is elaborated mainly by some dinoflagellate plankton. The mechanism of infestation to purple clam still remains to be elucidated.

ACKNOWLEDGEMENTS

The authors express their sincere thanks to President Dr. S.-S. Jeng, National Taiwan College of Marine Science and Technology, to Director P.-W. Yuan, Fisheries Department, Council of Agriculture, Executive Yuan, R. O. C., and Assistant Director Y.-N. Chen, Culture Department of the Association of East Asian Relations (Tokyo), for their kindness with the procurement of purple clam specimens, and for helpful discussion and encouragements during the course of this work.

REFERENCES

1. T. Kawabata, Shokuhin Eisei Kensa Shishin [Manual of Methods for Food Sanitation Testing], Vol.2 (Japan Food Hygienic Association, Tokyo 1978) p. 240.
2. T. Noguchi, J. Maruyama, Y. Ueda, K. Hashimoto, and T. Harada, Nippon Suisan Gakkaishi 47, 909-913 (1981).
3. Y. Onoue, T. Noguchi, J. Maruyama, Y. Ueda, K. Hashimoto, and T. Ikeda, Nippon Suisan Gakkaishi 47, 1347-1350 (1981).
4. Y. Nagashima, J. Maruyama, T. Noguchi, and K. Hashimoto, Nippon Suisan Gakkaishi 53, 819-823 (1987).
5. J. J. Sullivan and W. T. Iwaoka, J. Assoc. Off. Anal. Chem. 66, 297-303 (1983).
6. Y. Hashimoto, T. Noguchi, and R. Adachi, Nippon Suisan Gakkaishi 42, 671-676 (1976).
7. Y. Oshima, W. E. Fallon, Y. Shimizu, T. Noguchi, and Y. Hashimoto, Nippon Suisan Gakkaishi 42, 851-856 (1976).
8. T. Noguchi in: Toxic Phytoplankton - Occurrence, Mode of Action, and Toxins, Suisan-gaku series No. 42 (Nippon Suisan Gakkai, Tokyo 1982) pp. 88-101.

EFFECT OF WATER TEMPERATURE AND LIGHT INTENSITY ON GROWTH RATE AND TOXIN PRODUCTION OF TOXIC DINOFLAGELLATES

TAKEHIKO OGATA,* MASAAKI KODAMA,* AND TAKASHI ISHIMARU**
*Laboratory of Marine Biological Chemistry, School of Fisheries Sciences, Kitasato University, Sanriku, Iwate 022-01, Japan; **Ocean Research Institute, University of Tokyo, Nakano, Tokyo 164, Japan

ABSTRACT

The effect of water temperature and light intensity on growth rate and toxin production was examined for *Protogonyaulax catenella*, and newly found PSP producing dinoflagellates such as *P. cohorticula* and *Gymnodinium catenatum* using their unialgal cultures. The toxin content of all the species increased concomitantly with the decrease of the growth rate when the growth was controlled by temperature. However, the toxin production decreased or did not increase significantly when the growth rate decreased by lowering light intensity. These results indicate that photosynthesis is essential for the toxin production of these species.

INTRODUCTION

Several species of dinoflagellates such as *Protogonyaulax* spp. cause paralytic shellfish poisoning [1]. When these species bloom in the water, shellfish accumulate toxins by ingesting them. Therefore, not only their abundance but also the toxin content of their cells are important factors in toxification in bivalves. It has been suggested that the toxicity of *P. tamarensis* is influenced by some physiological or environmental factors [2-6]. Recently we have reported that photosynthesis is essential for toxin production of *P. tamarensis* [7].

In this report, we describe that photosynthesis is also essential for toxin production of *P. catenella*, *P. cohorticula* and *G. catenatum*.

EXPERIMENTAL

Culture of toxic dinoflagellates

Strains of *P. catenella*(OF878-1) and *G. catenatum*(Se-GC)[8] were isolated from Ofunato and Senzaki Bay, Japan, respectively. *P. cohorticula*(Chula-5)[9] was isolated from Ang Sila, Thailand in 1985. A clonal culture of each strain was made in T1 medium [7]. For culturing *G. catenatum*, 1 μM ammonium chloride was added to the medium. Culture experiments under different light intensities and water temperatures were carried out in duplicate as reported previously [7].

Measurements of growth rates and analyses of toxin contents

An aliquot of the cells was taken every day from cultures for cell counting under a microscope. The growth rate (μ_2) was estimated from cell counts during the exponential phase, according to Fukazawa *et al.* [10].

Cells harvested by centrifugation were sonicated with an equal volume of 0.2 N HCl for 2 min, and then centrifuged. The supernatant extracts were boiled for 10 min and analyzed by HPLC PSP analyzer [11]. Based on the peak height and the standard curve of each toxin component, the total molar concentration of toxin was calculated and expressed as nmol per 10^4 cells.

RED TIDES: BIOLOGY, ENVIRONMENTAL SCIENCE, AND TOXICOLOGY
Okaichi, Anderson, and Nemoto, Editors

RESULTS AND DISCUSSION

Three species were cultured under 4000 lux at 15, 20 and 25°C (P. catenella and G. catenatum) and at 20, 25 and 30 °C (P. cohorticula). Growth rates of all the species increased in proportion to increase of water temperature. As the growth rate was decreased by lowering water temperature, the toxin content of the cells increased, showing that toxin production of the cells grown under lower growth rate is high (Fig. 1, left column). It has been pointed out that toxicity of P. tamarensis and P. catenella becomes higher when the growth rate in culture of these species is low [4]. Previously, we observed the remarkable toxicity increase in P. tamarensis under lower growth rates at low water temperatures [7]. The results of the present study show that this is also true for P. catenella, P. cohorticula and G. catenatum.

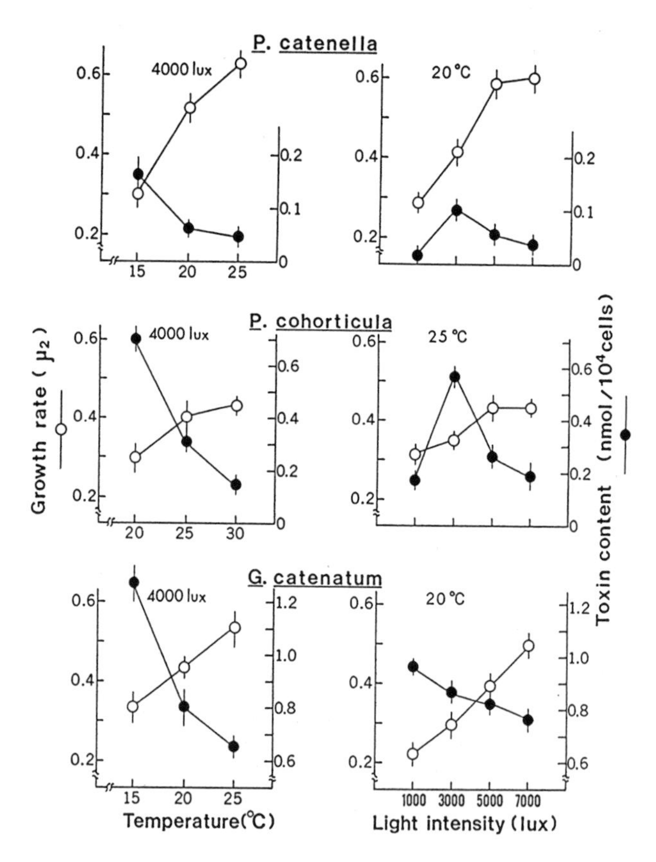

FIG. 1. Effect of water temperature and light intensity on growth rate and toxin content in dinoflagellates.

Light is essential for the growth of dinoflagellates. In a previous report [7], we observed that photosynthesis is essential for the toxin production of P. tamarensis. In order to examine the effect of light intensity on toxin production of these species, we examined the toxicity of each species grown under different light intensities of 1000, 3000, 5000 and 7000 lux at 20°C (P. catenella and G. catenatum) and at 25°C (P. cohorticula) (Fig. 1, right column). The growth rates increased with increase of light intensity within the range 1000 - 5000 lux. However, at 5000 and 7000 lux, the growth of two Protogonyaulax species was slightly inhibited, whereas growth rate of G. catenatum increased in proportion to increase of light intensity.

Toxicity of all the species increased in parallel to the decrease of growth rate under 3000 - 7000 lux. In spite of lower growth rates, however, toxin content of P. catenella and P. cohorticula grown at 1000 lux were remarkably low, suggesting the importance of photosynthesis in toxin production of these species. In the case of G. catenatum, toxin content of the cells increased proportionally to decrease of growth rate between 1000 - 7000 lux. However, the difference in toxin content was less marked when the growth rate was controlled by light intensity than by water temperature. This finding also suggests the importance of photosynthesis in toxin production in this species.

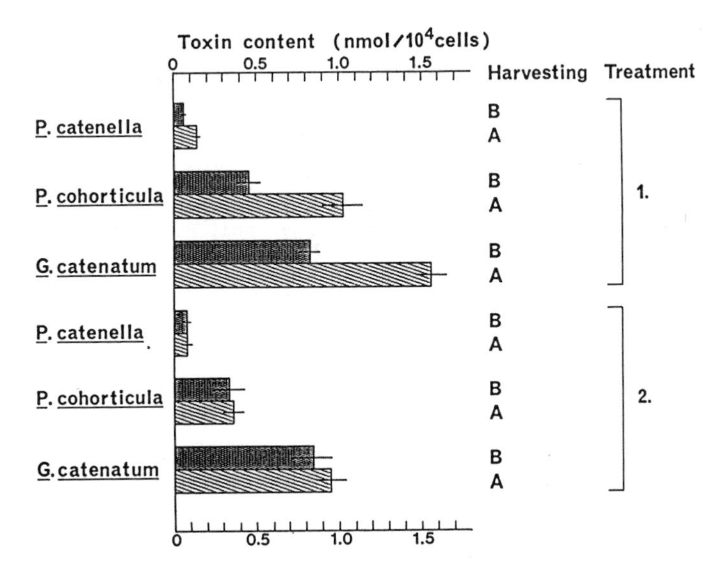

FIG. 2. Effect of slowdown of the growth rate on toxin contents in dinoflagellates. Treatment 1: Growth was stopped by lowering water temperature at the mid-exponential phase. Treatment 2: Growth was stopped by lowering light intensity at mid-exponential phase. Harvesting B: Before treatment. Harvesting A: After treatment. Error bars represent 95% confidence limits.

As described above, photosynthesis seemed to be an important factor in toxin production of these species. However, the results suggest that growth rate was related to the rate of toxin production in these species. Therefore, growth was stopped during culture by either lowering temperature or decreasing light intensity, i.e. the temperature of the culture grown under 4000 lux at 20°C (P. catenella and G. catenatum) and at 25°C (P. cohorticula) was lowered by 5°C for a period of 4 days keeping identical light intensity. Along with the temperature change, growth of all the species stopped as in a case of P. tamarensis [7]. After this treatment, toxin contents of all the species increased by roughly twice that before treatment (Fig. 2). Similar growth curves were also obtained in all the species, when the light intensity of cultures grown under the same conditions described above, was lowered from 4000 to 500 lux at the mid-exponential phase for a period of 4 days. Growth of each species was also stopped by this treatment. However, the toxin content of the cells did not increase significantly (Fig. 2). These results show that cells continue to produce toxins during the period of growth retardation when sufficient light intensity was supplied, whereas new production of toxins was suppressed under limited light supply. These facts indicate that photosynthesis is essential for toxin production of P. catenella, P. cohorticula and G. catenatum as in a case of P. tamarensis which we have reported before [7].

As described above, the toxin content of G. catenatum under 1000 lux increased because of low growth rate, whereas those of two Protogonyaulax species examined decreased in spite of lower growth rates. This suggests a difference in light demand for toxin production among these species.

REFERENCES

1. E.J.Schantz, Ann. N. Y. Acad. Sci. 479, 15-23 (1986).
2. A.Prakash, J. Fish. Res. Bd Can. 24, 1589-1606 (1967).
3. H.T.Y.Singh, Y.Oshima and T.Yasumoto, Nippon Suisan Gakkaishi 48, 1341-1343 (1982).
4. G.L.Boyer, J.J.Sullivan, R.J.Andersen, P.J.Harrison and F.J.R.Taylor, in: Toxic dinoflagellates, D.M.Anderson, A.W.White and D.G.Baden, eds. (Elsevier, New York, Amsterdam, Oxford 1985) pp. 281-286.
5. M.Kodama, Y.Fukuyo, T.Ogata, T.Igarashi, H.Kamiya and F.Matsuura, Nippon Suisan Gakkaishi 48, 567-571 (1982).
6. A.W.White, J. Phycol. 14, 475-479 (1978).
7. T.Ogata, T.Ishimaru and M.Kodama, Mar. Biol. 95, 217-220 (1987).
8. T.Ikeda, S.Matsuno, S.Sato, T.Ogata, M.Kodama, Y.Fukuyo and H. Takayama, This volume.
9. M.Kodama, T.Ogata, Y.Fukuyo, T.Ishimaru, S.Wisessang, K.Saitanu, V.Panikiyakarn and T.Piyakarnchana, Toxicon in press.
10. N.Fukazawa, T.Ishimaru, M.Takahashi and Y.Fujita, Mar. Ecol. Prog. Ser. 3, 217-222 (1980).
11. Y.Oshima, M.Machida, K.Sasaki, Y.Tamaoki and T.Yasumoto, Agric. Biol. Chem. 48, 1707-1711 (1984).

ANALYSIS OF PARALYTIC SHELLFISH POISONS, IN PARTICULAR PROTOGONYAUTOXINS, BY ION-PAIRING HIGH PERFORMANCE LIQUID CHROMATOGRAPHY

Nagashima, Y.[1], T. Noguchi[2] and K. Hashimoto[2]
1. Department of Food Science and Technology, Tokyo University of Fisheries, Tokyo 108, Japan
2. Laboratory of Marine Biochemistry, Faculty of Agriculture, University of Tokyo, Tokyo 113, Japan

ABSTRACT

A reversed-phase ion-pairing high performance liquid chromatographic method was developed to analyze paralytic shellfish poisons (PSPs), in particular protogonyautoxins (PXs).* PXs were separated from each other when 0.025M phosphate buffer (pH 7.0) containing 1mM n-decyltrimethylammonium as counterion was used as mobile phase. They were reacted with a periodate reagent and a base at 65°C, followed by treatment with chloroacetaldehyde to enhance their fluorescence intensities. The resultant fluorogenic substances were monitored at 390nm with 336nm excitation.

INTRODUCTION

Paralytic shellfish poisons (PSPs) produced by several species of dinoflagellates, are composed of more than ten components, whose structures are closely related to each other. However, their lethal potencies greatly differ from each other, ranging from 30-5,500 mouse units (MU)/mg. On mild acid-hydrolysis, "low-toxic components" (protogonyautoxins (PXs) and gonyautoxins 5 and 6) which bear 21-sulfonatecarbamoyl function are easily converted into corresponding carbamate components with an enhanced lethal potency.

In the western districts of Japan, the toxins of *Protogonyaulax catenella* are often composed mainly of low-toxic components (1). The productivity and toxin composition vary among *Protogonyaulax* spp. (2). Toxin composition, on the other hand, varies greatly with shellfish species, location and season. When fed with *P. catenella* cells containing PX_1 and PX_2 as major toxins, mussels became highly toxic. It was suggested that PX_1 and PX_2 were transformed into highly toxic analogues, since both toxins were hardly detected in the mussels (3).

In a previous paper (4), we developed a high performance liquid chromatography (HPLC) method in which gonyautoxins 1-6 (GTX_{1-6}) and saxitoxins (STXs) were separated from each other and detected with a high sensitivity. However, PX components were not separated from each other at all by the method. In this situation, we have developed an ion-pairing HPLC method using n-decyltrimethylammonium (DTMA) for this purpose.

---

* "Protogonyautoxins (PXs)" represent four PSP components (PX_{1-4}) which migrate to the anode under usual cellulose acetate membrane electrophoretic conditions. Among them, PX_1 corresponds to *epi*-GTX_8 and PX_2 to GTX_8.

RED TIDES: BIOLOGY, ENVIRONMENTAL SCIENCE, AND TOXICOLOGY
Okaichi, Anderson, and Nemoto, Editors

EXPERIMENTAL

Toxin preparations

Specimens of oyster *Crassostrea gigas* were collected at Senzaki Bay, Yamaguchi Prefecture, in January 1987, immediately frozen, transported to our laboratories, and kept frozen at below -20°C. A mixture of PX_1 (*epi*-GTX_8) and PX_2 (GTX_8) was isolated from digestive glands of the oyster specimens by the method reported previously (5), and used as reference standard. The mixture contained also appreciable amounts of $GTX_{2,3}$ which were converted from $PX_{1,2}$ during storage at around 0°C.[3]

A portion of the frozen oyster specimens was partially thawed at room temperature, and their digestive glands were excised and homogenized with two volumes of distilled water. The homogenate was subjected to ultrasonication for 30min and centrifuged at 5,000x*g* for 10min. The pellet was extracted two more times in the same manner. The supernatants were combined and defatted with dichloromethane. The aqueous layer was evaporated to remove dichloromethane and filtered through a Diaflo ultrafiltration membrane YM-2 to cut off more than 1,000-dalton substances. The filtrate was applied to a Bio-Gel P-2 column (95 x2cm i.d.) and eluted with 0.03M acetic acid. Toxic fractions were combined and used as a test sample.

HPLC system

As schematically shown in Fig. 1, the HPLC system used consisted of a Hitachi 638-50 HPLC and a Hitachi 650-10 spectrofluorometer. A silica ODS column (300x6mm i.d., YMC AM-314, Yamamura Kagaku) was used. Post-column reaction system was composed of three reagent reservoirs, three pumps and two reaction coils (7mx0.3mm i.d. and 10mx0.3mm i.d.).

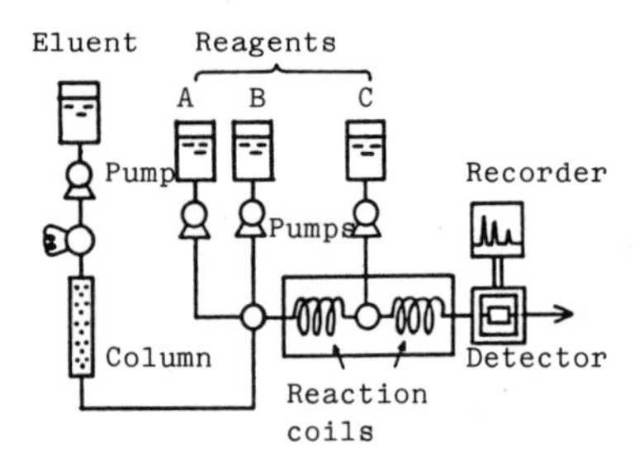

Fig. 1. Schematic diagram of HPLC system.

Mobile phase

As mobile phase for mutual separation of PX components, 0.025M phosphate buffer (pH 7.0) containing 1mM DTMA was used.

Chromatographic procedures

A sample solution was placed on the column through the injection valve, and the above phosphate buffer was isocratically pumped at a flow rate of 1m*l*/min. The eluate was first mixed with 0.05M periodic acid and a base (0.2N KOH plus 1M ammonium formate in 50% formamide), and was reacted with 1% chloroacetaldehyde in 1M citrate buffer (pH 4.0) to increase fluorescence intensity. These fluorescence reactions were conducted in a 65°C water-bath. Fluorescent substances induced were monitored at 390nm with 336nm excitation.

Chemicals

All chemicals were of analytical grade unless otherwise specified. DTMA was obtained from Tokyo Chemical Industry (Tokyo) and used without further purification.

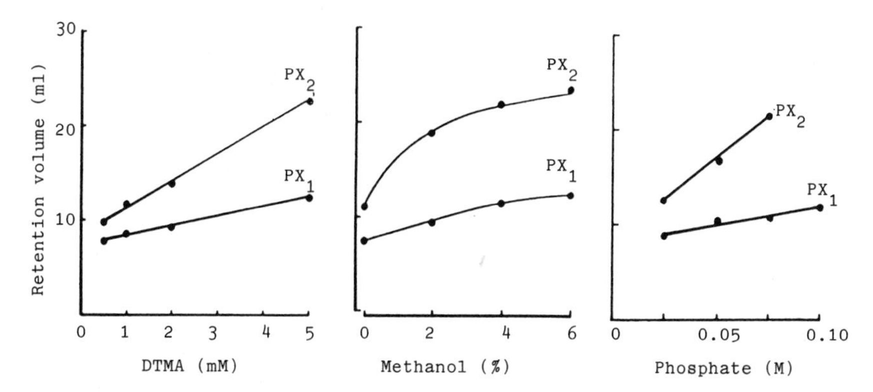

Fig. 2. Effects of n-decyltrimethylammonium, methanol and phosphate concentrations on separation of PX_1 and PX_2.

RESULTS AND DISCUSSION

HPLC behaviors of PX components were compared under various conditions in respect of phosphate concentration, pH, concentration of counterion and methanol. At first, attempts were made to separate PX components from each other using mixtures of 0.05 M phosphate buffer (pH 7.0, containing 0.5-5mM DTNA) and methanol (99:1). As shown in Fig.2 (left), both PX components tended to be separated better with an increasing DTMA concentration. At 0.5mM DTMA, however, the retention volume of PX_1 was close to that of $GTX_{2,3}$ which was not retained on the column. On the other hand, both PX peaks tended to become dull at higher concentrations of DTMA. It was judged from these results that 1mM DTMA was best for mutual separarion of PX_1 and PX_2. As shown in Fig. 2 (center), the increase of methanol concentration resulted in an apparently better separation of both components. However, chromatographic peaks, especially of PX_2, tended to become dull at higher methanol concentrations. On the other hand, PX_1 was clearly separated from PX_2 and also from $GTX_{2,3}$, even when no methanol was added. From these results, the phosphate buffer containing 1mM DTMA was used as mobile phase hereafter without adding methanol. As Fig. 2 (right) shows, the higher the phosphate concentration, the better the mutual separation of both components. However, the retention volume of each toxin, especially of PX_2, became larger, resulting in a dull peak in HPLC. Therefore, 0.025M phosphate buffer was regarded as best. The effect of pH on the separation of PXs was examined, using 1 mM DTMA-containing 0.025M phosphate buffer at several pH values. Results showed that a satisfactory separation was attained at pH 7 (data not shown). From these results, 0.025M phosphate buffer (pH 7.0) containing 1mM DTMA was selected as mobile phase for separation of PX_1 and PX_2.

As shown in Fig. 3, three peaks appeared when an authentic PSP mixture was analyzed by HPLC. The three peaks whose retention volumes were 8, 9 and 15.5ml corresponded to $GTX_{2,3}$, PX_1 and PX_2, respectively. After acid-hydrolysis, the mixture showed a single peak at a retention volume of 8ml. The hydrolyzate was composed of GTX_2 and GTX_3, as analyzed by our ion-pairing HPLC method using heptanesulfonic acid as counterion (4) (data not

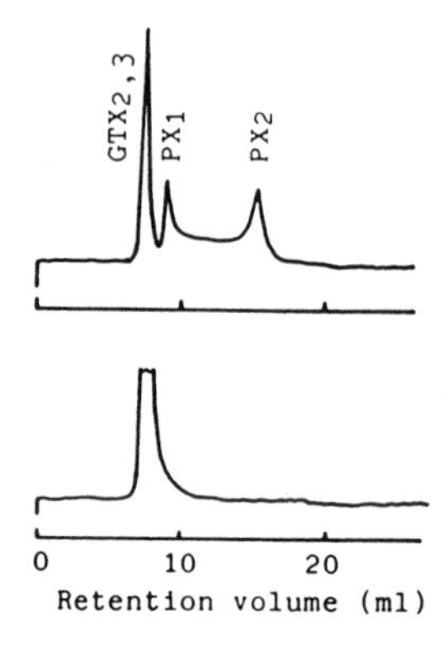

Fig. 3. HPLC of a mixture consisting of $PX_{1,2}$ and $GTX_{2,3}$ before (upper) and after (lower) acid-hydrolysis.

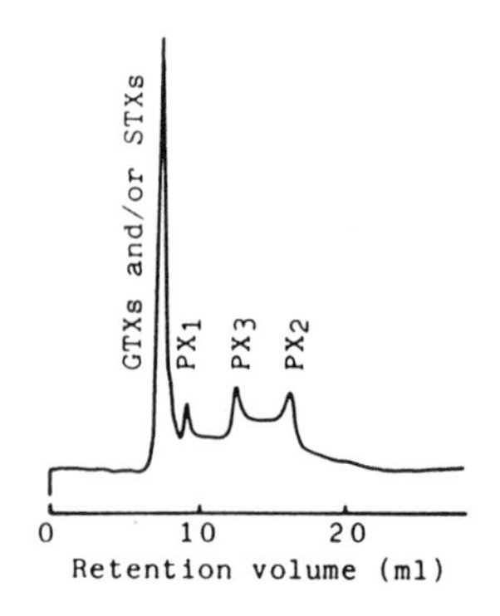

Fig. 4. HPLC of PSP fraction from oyster extract.

shown).

As shown in Fig. 4, the PSP fraction from oyster extract contained PX_1 and PX_2, along with $GTX_{2,3}$ and other PSP member. In addition, a peak appeared between both PX peaks, and was supposed to be that of PX_3 (carbamoyl-N-sulfo-11α-hydroxy neosaxitoxin sulfate), since this peak gave rise to GTX_1 on acid hydrolysis, as demonstrated by electrophoresis (data not shown).

As described above, PX components can satisfactorily be analyzed by the present method using DTMA as counterion. This method, in combination with our previous HPLC method (4), could allow us to examine the toxin composition and lethal potency of PSP-producing plankton as well as bivalves.

REFERENCES

1. Y. Oshima and T. Yasumoto, Bull. Mar. Sci. 37, 773-774 (1985).
2. G. L. Boyer, J. J. Sullivan, R. J. Anderson, F. J. R. Taylor, P.J. Harrison and A. D. Cembella, Mar. Biol. 93, 361-369 (1986).
3. T. Noguchi, J. Maruyama, Y. Onoue, K. Hashimoto and T. Ikeda, Nippon Suisan Gakkaishi 49, 499 (1983).
4. Y. Nagashima, J. Maruyama, T. Noguchi and K. Hashimoto, Nippon Suisan Gakkaishi 53, 819-823 (1987).
5. Y. Onoue, T. Noguchi, J. Maruyama, K. Hashimoto and H. Seto, J. Agric. Food Chem. 31, 420-423 (1983).

BREVETOXIN BINDING TO RAT BRAIN SYNAPTOSOMES: INHIBITION CONSTANTS OF DERIVATIVE BREVETOXINS

DANIEL G. BADEN,* THOMAS J. MENDE,** ALINA M. SZMANT,* VERA L. TRAINER,** AND RICHARD A. EDWARDS*
University of Miami *Rosenstiel School of Marine and Atmospheric Science Division of Biology and Living Resources, 4600 Rickenbacker Causeway, Miami FL. 33149 U.S.A., and **School of Medicine Department of Biochemistry and Molecular Biology, P.O. Box 016129, Miami FL. 33101 U.S.A

ABSTRACT

Florida's red tide dinoflagellate Ptychodiscus brevis produces several distinct potent materials. Radioactive brevetoxin probes were synthesized by reductive tritiation of PbTx-1 and PbTx-2. Naturally-occurring competitor brevetoxins included: the C27,28 epoxide, the C-37 O-acetate, and the C-42 primary alcohol derivative of PbTx-2; the C-43 primary alcohol derivative of PbTx-1; and PbTx-1 and PbTx-2. The more potent labeled toxin PbTx-1 exhibited the highest affinity for the specific binding site, and Ki's paralleled potency in fish bioassay. Kd's for the labeled probes were in the 1-10 nM concentration ranges. Ki's were an order of magnitude higher. PbTx-1 type toxins are more efficient at displacing probe than are PbTx-2 type probes.

INTRODUCTION

All toxins produced by Florida's red tide dinoflagellate Ptychodiscus brevis are potent in nM concentrations in a number of in vivo and in vitro systems. They are potent when injected in mice [1], in isolated axons [2],and in neuromuscular preparations [3], and induce repetitive discharges in motor endplate [4], causing an enhanced release of neurotransmitter [5]. A specific binding site for the brevetoxins located on, or proximal to, the voltage-sensitive sodium channel was postulated [6-7]. Tritiated brevetoxin PbTx-3, produced by sodium borotritiide reduction of brevetoxin PbTx-2 [8], provided a non-exchangeable tritium label of sufficient specific activity to perform binding studies in isolated excitable membrane preparations [9].

Brevetoxins have been shown to bind to Site 5 associated with voltage-sensitive sodium channels [6-7,9]. At $4^{o}C$, Rosenthal analysis of specific brevetoxin binding to rat brain synaptosomes yields a Kd of 2.9 nM and a Bmax of 6.8 pmol of toxin bound/mg of protein [9]. Previous studies have shown that labeled probe can be competitively displaced by unlabeled PbTx-3, PbTx-2, or synthetic PbTx-3 (reduced PbTx-2) but not by a nontoxic, synthetic oxidized derivative of PbTx-2 [9].

Based on competitive displacement data in the assay, and fish bioassay, we conclude that in vitro specific binding

RED TIDES: BIOLOGY, ENVIRONMENTAL SCIENCE, AND TOXICOLOGY
Okaichi, Anderson, and Nemoto, Editors

assays accurately reflect in vivo potency of the individual toxins.

EXPERIMENTAL

Cultures

Unialgal cultures of *P. brevis* were grown under continuous light (4000 lux) at 24°C in 10 L batches in NH-15 artificial seawater [10]. Cultures were harvested in stationary phase to yield sufficient quantities of the full multiplicity of brevetoxins [11].

Toxin Purification

Toxins were extracted from whole cultures using chloroform. The residue remaining after chloroform evaporation was resuspended in 90% aqueous methanol and was repeatedly extracted with petroleum ether. The methanol fraction was applied to a flash chromatography column and toxin eluted with $CHCl_3$/MeOH/acetic acid (100/10/1). The tan toxic band was applied to a silica gel thin-layer plate and was chromatographed using acetone/petroleum ether (30/70). Two toxic fractions (Rf=0.1-0.2; Rf=0.3-0.5) were separated, and each was re-chromatographed on silica gel tlc plates using ethyl acetate/petroleum ether (70/30 and 50/50, respectively). Tlc purified toxins were subjected to reverse phase HPLC (C-18 column, isocratic 85% aqueous methanol, uv detection at 215 nm) and were quantified using brevetoxin standards. Toxin structures are illustrated elsewhere in this volume [11].

Binding Assays

Synaptosomes were prepared fresh daily from rat brain [9]. Synthetic tritiated PbTx-3 preparation has been previously described [8,9]. Tritiated PbTx-7 was prepared in an identical stoichiometric fashion from PbTx-1. Quantification of tritiated toxin, and determination of specific activity was carried out using liquid scintillation techniques and appropriate quench curves, and standard brevetoxins. Competitive binding curves were generated using techniques previously described [6,7,9], and synaptosomes were rapidly centrifuged following incubation [9]. Tritium bound to pellets was quantified using liquid scintillation techniques. Nonspecific binding was measured in the presence of saturating concentrations of unlabeled PbTx-3 and was subtracted from total binding to yield specific binding. Free tritiated PbTx-3 was determined by counting directly an aliquot of supernatant solutions following centrifugation.

Potency Measurements

Gambusia affinis fish bioassays were conducted in 20 ml seawater (35 o/oo), one fish per vessel, toxin added in 0.01 ml ethanol. Each LD_{50} was determined by preparing triplicate 2-fold serial dilutions of each toxin at concentrations ranging from 1-64 nM (7 concentration plus controls, n=60). Lethality was assessed at 60 min and median lethal dose was determined [12].

RESULTS

PbTx-3 binds to Site 5 associated with voltage-sensitive sodium channels, and has a Kd of 2.9 nM and a Bmax of approximately 7 pmol/mg synaptosomal protein. Initial competitive binding studies employing PbTx-2, PbTx-3 and oxidized PbTx-2 indicated a specific displacement of tritiated toxin by the former two toxins, but not by the latter oxidized derivative. The efficiency of displacement paralleled potency in animals. Rosenthal analysis of tritiated PbTx-7 binding in the assay indicates that this material binds tighter than does tritiated PbTx-3; Kd of 0.67 nM, preliminary Bmax determination of 2.9 pmol/mg synaptosomal protein. PbTx-7 is less stable in tritiated form than is PbTx-3; the strained 5-membered lactone may open sponstaneously in protic solvent.

Using the additional natural brevetoxins we have developed specific displacement curves (n=2) which correlate well with the potency of each individual purified toxin (Fig. 1). We had sufficient PbTx-1,-2,-3, and -7 to calculate Ki's, and to correlate ED_{50} and Ki with fish potency (Table I).

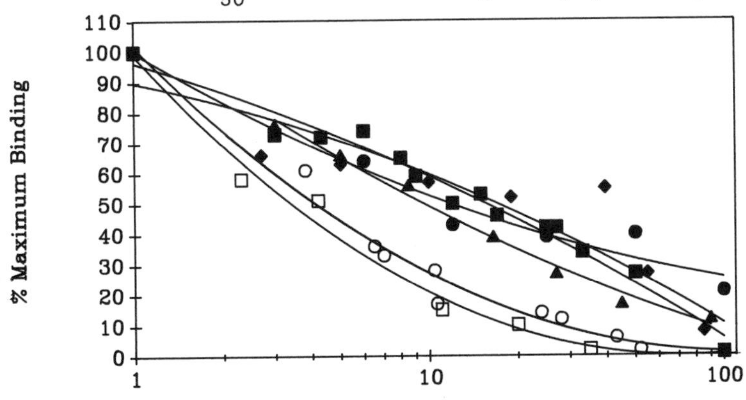

Figure 1. Specific Displacement of Tritiated PbTx-3 by Unlabeled Brevetoxins. Incubations, in the presence of 50 ug synaptosomal protein and 16 nM tritiated PbTx-3 (10.15 Ci/mmole) with increasing amounts of unlabeled PbTx-1 (□), PbTx-2 (■), PbTx-3 (●), PbTx-5 (▲), PbTx-6 (◆), or PbTx-7 (○), were for 1 hr at 4°C. Each point represents the mean of three triplicates.

DISCUSSION

T-test analysis revealed no significant difference between the PbTx-2 backbone-type toxin ED_{50}s ($p<0.01$), or between PbTx-1 and PbTx-7 ED_{50}s ($p<0.01$), but statistically significant differences were found between the two classes. Fish bioassay indicates that the two more potent ichthyotoxins were also more efficient on a molar basis in displacing brevetoxin from its specific binding site in synaptosomes. The affinities of the toxins theoretically are based on differential structural considerations involved in the binding to the site on the channel as well as lipid solubilities of

TABLE I. Comparison of ED_{50}, K_i, and LD_{50} for Synaptosome and Fish Bioassay

Toxin	Synaptosome ED_{50} (nM)	K_i	LD_{50} (nM)
PbTx-1	3.5	1.4	4.4
PbTx-7	4.1	7.1	4.9
PbTx-2	17.0	9.6	21.8
PbTx-3	12.0	9.9	10.9
PbTx-5	13.0	---	42.5
PbTx-6	32.0	---	35.0

ED_{50} are defined as the toxin conc at which 50% displacement of 16 nM tritiated PbTx-3 from sodium channels occurs. LD_{50} are determined by incubation of *Gambusia affinis* with toxin in 20 ml seawater for 60 min.

each toxin. PbTx-1 and PbTx-7 are the two most hydrophobic toxins, are the most potent, and bind with highest affinity to the site. This may be a function of the additional flexibility of this structural backbone, when compared to PbTx-2-type toxins, allowing better fit to the topography of the brevetoxin binding/active site.

The brevetoxin binding site lies in the hydrophobic portion of the channel [6,7]. Substituent character on C-42 in part determines solubility (and hence access to its site of action); the end distal to C-42 carries the active portion of each toxin. Additional brevetoxin derivatives are being synthesized to test this hypothesis.

ACKNOWLEDGEMENTS

This work was supported in part by the US Army Medical Research and Development Command, Contract Numbers DAMD17-85-C-5171 and DAMD17-87-C-7001.

REFERENCES

1. D.G. Baden and T.J. Mende, Toxicon 20, 457, (1981).
2. J.M.C. Huang, C.H. Wu and D.G. Baden, J. Pharmacol. Exp. Therapeut. 229, 615 (1984).
3. J.P. Gallagher and P. Shinnick-Gallagher, Br. J. Pharmacol. 69 367 (1980).
4. P. Shinnick-Gallagher, Br. J. Pharmacol. 69, 373, (1980).
5. J.M. Huang and C.H. Wu, this volume.
6. W.A. Catterall and M. Risk, Molec. Pharmacol. 19, 345, (1980).
7. R.G. Sharkey, E. Jover, F. Couraud, D.G. Baden, and W.A. Catterall, Molec. Pharmacol. 31, 273 (1987).
8. D.G. Baden, T.J. Mende, J. Walling and D.R. Schultz, Toxicon 22, 783, (1984).
9. M.A. Poli, T.J. Mende, and D.G. Baden, Molec. Pharmacol. 30, 129 (1986).
10. W.B. Wilson, Fla. Bd. Conserv. Prof. Pap. Ser. 7, 42 pp.(1966).
11. D.G. Baden, T.J. Mende, and L.E. Roszell, this volume.
12. C.S. Weil, Biometrics 1, 249, (1952).

INCREASES IN VISCOSITY MAY KILL FISH IN SOME BLOOMS

Ian R. JENKINSON[1]
Université de Nice, Groupe de Recherches marines, Laboratoire de Biologie et d'Ecologie marines, Parc de Valrose, 06034 Nice Cedex, France.

ABSTRACT

In both cultures of some marine flagellates, and in non-bloom seawater, viscosity is composed of a newtonian component, η_W, plus a non-newtonian component due to dissolved polymers, η_E. η_W is constant but $\eta_E = k.q^P$, where q is shear rate, k is a constant and P is a power value found to vary from 0 to -1.6. *Gyrodinium aureolum* kills fish when in excess of roughly 10,000 cm^{-3}. Trapped bubbles were seen in a bloom of 2700 cm^{-3}. This has allowed calculation of a η-q coordinate, 5.4 mPa s and 30 s^{-1} for 0.1-mm bubbles, or 140 mPa s and 5.8 s^{-1} for 0.5-mm bubbles. Based on published values for gill dimensions and pumping rate at high activity in bass and tuna of weight 0.33 to 1667 g, and assuming 0.5-mm bubbles, the energy required for gill pumping would be about 14 times normal if P is -0.3, but only 1.001 times normal if P is -1.4. High shear stress involved in pumping more viscous liquid might also tear epithelia. Whether changes in viscosity can kill fish without contributary factors depends on P. This should be investigated in red tides.

INTRODUCTION

It has been suggested that death of fish in red tides of certain species, particularly *Gyrodinium aureolum* Hulburt (north-west European Shelf population) and algae of the *Chattonella-Olisthodiscus-Hornellia* complex, may be caused by clogging of gills [15, refs in 1]. Yellowtail, *Seriola quinqueradiata*, exposed to a bloom of *Chattonella antiqua* (Hada) Ono [13] (as *Hemieutreptia antiqua* Hada) showed both lowered blood O_2 levels and behaviour characteristic of hypoxia [8], while fish exposed to blooms of *G. aureolum* also showed hypoxic-type behaviour [3]. In blooms of both species some fish died when water concentrations of O_2 were high. Bubbles occur, as if trapped, in some *G. aureolum* blooms [3,4], and at least some species of the *Chattonella* complex discharge mucus [9].

The viscosity of a *Chattonella antiqua* bloom, in which these yellowtail died, was measured using a Morton viscometer, and found to be up to twice as viscous as normal seawater [8]. From the descriptions of the Morton viscometer [5,6,8], it can be concluded that the shear rate at measurement lies between 100 and 5,000 s^{-1}.

So as to form a quantitative framework for further investigation of gill clogging in relation to red tides, published and submitted results are brought together on: bubbles trapped in *G. aureolum* blooms; viscosity variations in both phytoplankton cultures and non-bloom seawater; the relationship between polymer concentration and viscosity; and hydrodynamics of respiration in fish.

RESULTS

Phytoplankton cultures [Ref. 1]

Out of eight cultures, the three which showed the highest values of η when q was 0.15 s^{-1} were measured over a range of q. For one culture, Fig. 1a shows the variation of excess viscosity,

$$\eta_E = \eta - \eta_W \qquad \text{(Eq. 1)}$$

where η_W is the viscosity of pure water.

The cultures investigated showed constant values of P over ranges of q, where

$$\eta_E = k \,.\, q^P \qquad \text{(Eq. 2)}$$

where k and P are constants. Such power-law, shear-thinning behaviour is characteristic of many colloidal polymer solutions, and the inflexions found may result from the additive effects of different polymers. P varied from 0 to -1.3.

Rheometry of non-bloom seawater samples [Ref. 2]

Considerable variation in η_E was found between samples. Fig.1b shows the means of η_E plotted against q. The mean value of P from each sampling trip varied from -0.9 to -1.6 (Fig. 1b). Variation within and between samples indicated either considerable lumpiness (rheological heterogeneity), or that unavoidable differences in handling of the samples promoted marked changes in their rheological properties.

[1] Address for correspondance: Department of Oceanography, University College, Galway, Ireland

RED TIDES: BIOLOGY, ENVIRONMENTAL SCIENCE, AND TOXICOLOGY
Okaichi, Anderson, and Nemoto, Editors

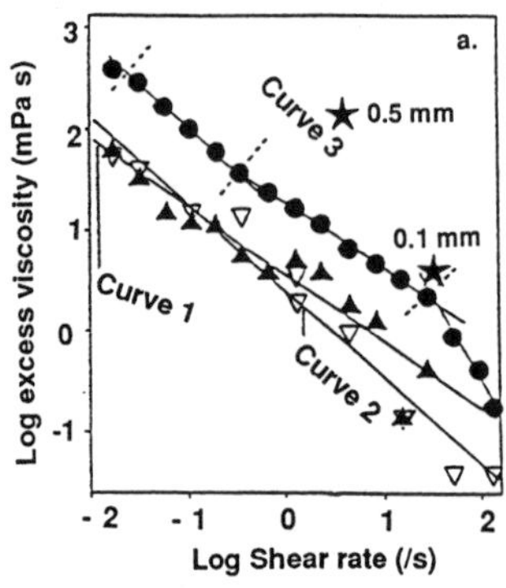

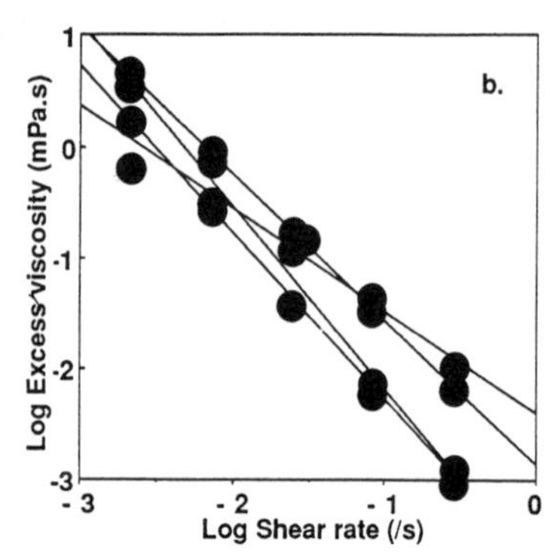

Fig. 1. *Log-log plots of excess viscosity,* η_E, *against shear rate,* q. **a.** *Culture of* Dunaliella marina, *84 cells mm*$^{-3}$. *Curves 1 and 2, less concentrated sample. Curve 3, with more settled material. Asterisks denote* η-q *intersect for bubbles [Ref. 1].* **b.** *Non-bloom sea (pooled depths 0 to 100 m) from the Rade de Villefranche, France on four different occasions. Each point represents the mean of 4 to 16 readings. Regression line shown for each occasion [Ref. 2].*

Observations on a bloom of *Gyrodinium aureolum* [Ref. 3]

Water coloured brown due to *G. aureolum* extended to a depth of 30 to 50 cm. A sample from the coloured water contained 2.7 x 10^3 cells cm^{-3} of *G. aureolum*. Where the brown water washed over rocks, barnacles and periwinkles were active. Many small bubbles were suspended in the brown layer, as if trapped. While the bubbles were not measured, I eastimate from memory that the diameter of the largest was about 0.5 mm. At a rising speed of more than 1 mm they would have cleared the brown layer in < 9 minutes, and their motion would have been noticeable.

Calculation of viscosity/shear rate intersect for the bloom

For a small, spherical particle at equilibrium in a liquid, from Stoke's law, the viscosity of the liquid,

$$\eta = \frac{2 \cdot g \cdot \rho_E \cdot r}{9 \cdot w} \quad [\text{m s}^{-1}] \quad \text{(Eq. 3)}$$

where ρ_E is the excess density of the particle, w its vertical velocity, r its radius and g is the acceleration due gravity.

The tendency of the particle to accelerate under gravity is opposed by an equal drag acting on its surface,

$$F = 4/3 \cdot \pi \cdot r^3 \cdot \rho_E \cdot g \quad [\text{kg m s}^{-2}] \quad \text{(Eq. 4)}$$

and the surface area of this spherical particle,

$$A = 4 \cdot \pi \cdot r^2 \quad [\text{m}^2] \quad \text{(Eq. 5)}$$

Combining Eqs 4 and 5, the mean shear stress over the surface of the particle,

$$\tau = \frac{F}{A} = \frac{g \cdot \rho_E \cdot r}{3} \quad [\text{kg m}^{-1}\text{ s}^{-2} \text{ or Pa}] \quad \text{(Eq. 6)}$$

and the corresponding shear rate in the water touching the particle,

$$q = \frac{\tau}{\eta} \quad [\text{s}^{-1}] \quad \text{(Eq 7)}$$

The corresponding η-q intersects for 0.5-mm bubbles (r = 2.5 x 10^{-4} m), as well as for 0.1-mm bubbles is shown in Fig. 1. The 2.6 million cell/dm^3 bloom of *G. aureolum* was thus thicker at q = 6 to 30 s^{-1} than the thickest cultures reported [1].

Flow dynamics in fish gills (ref. 7]

As fish subject to sublethal concentrations of *G. aureolum* show hyperactivity [1,8], similar to that shown in hypoxic conditions, only data for "maximum" respiratory activity will be considered here. In a shear-thinning medium, the best policy for the fish would be to maintain high inspiration pressure and thence q, as this would minimise η. Observations mentioned above suggest that the fish do this.

Published data on the dimensions of gill pores and on flow through gills have been reviewed [7] for the largemouth bass, *Micropterus salmonoides*, of sizes from 0.33 to 837g, as well as for tuna of 1667 g. The pressure across the gills is about 1 cm of water ($\approx 10^2$ kg m s^{-2}) in most fishes, but about 2 cm of water in tuna. As water flows through fish gills at only about one tenth the rate calculated from pressure and gill slit dimensions [7], instead of calculating q in gill slits from the Poiseille equation for rectangular tubes, root mean square shear rate q_{RMS}, will be estimated from energy dissipation. Energy dissipation per unit mass of water, in the gill slits,

$$\varepsilon = \frac{D \cdot V_W}{V_o} \quad [m^2\ s^{-3}] = [W\ kg^{-1}] \qquad \text{(Eq. 8)}$$

where D is the pressure difference across the gills, V_W is the mean water velocity through the gill slits and V_o is the measured volume (length x breadth x depth) of the slits. ε varies little, from 1000 $m^2\ s^{-3}$ for bass of 0.33 g to 620 for those of 837 g (mean for ten size classes 730 $m^2\ s^{-3}$), and 480 $m^2\ s^{-3}$ in tuna gills.

$$q^2 = \frac{2 \cdot \varepsilon}{7.5 \cdot \eta} \quad \text{[Ref. 12]} \qquad \text{(Eq. 9)}$$

Assuming that the reviewed studies on gill hydrodynamics were carreid out at a typical value of η of 10^{-3} Pa s [kg $m^{-1}\ s^{-1}$], the corresponding value of q is 14,000 (≈2000) s^{-1} for gill slits of bass of mass 0.33 to 837 g, and 11,000 s^{-1} for tuna gill slits.

Effects of a bubble-trapping bloom on energy of flow through fish gills

Table 1 shows the minimum possible values of k for corresponding values of P, assuming bubble diameter of either 0.1 or 0.5 mm. Also shown are the minimum values of energy dissipation in gills (for a given throughput) relative to that in newtonian, non-bloom water.

Table 1. *Minimum possible values of* **k**, *and (relative to that in non-bloom water) for "maximum" ventilation in bass (q_{RMS} in gill slits 14,000 s^{-1}) and tuna (q_{RMS} 11,000 s^{-1}), according to the maximum size of bubbles trapped in the bloom (rising rate ≤ 1 mm s^{-1}) and the value of* **P**

	P						
	-0.2	-0.3	-0.4	-0.6	-0.8	-1.0	-1.4
a. If max. bubble diameter 0.1 mm							
Value of **k** (x 10^2)	0.87	1.22	1.72	3.39	6.60	13.2	51.5
ε in gills - bass	2.29	1.70	1.38	1.11	1.031	1.009	1.0008
- tuna	2.35	1.75	1.42	1.13	1.039	1.012	1.0011
b. If max. bubble diameter 0.5 mm							
Value of **k** (x 10^2)	19.8	23.6	28.1	39.9	56.7	80.6	163
ε in gills - bass	30.3	14.4	7.17	2.23	1.27	1.058	1.0016
- tuna	31.8	15.5	7.79	2.50	1.33	1.073	1.0030

DISCUSSION

The percentage of the oxygen in inspired water extracted during passage through the gills is typically about 50% for non-tuna and 70% for tuna [7]. This does not give much margin to allow for any reduction in water throughput. Further, if throughput diminishes, so will diffusion of waste products, such as NH_4^+ and CO_2, leading to increased concentration in fish tissues.

The excess viscosity of polymer solutions depends generally on a power of the polymer concentration. For non-ionic, random-coil polymer solutions, η_E is proportional to $c^{1.4}$ to $c^{3.3}$, depending on c, where c is the polymer concentration [11]. For blooms of *Gyrodinium aureolum*, the threshold concentration for mortality to fish is 6 to 21 x 10^3 cm^{-3} [14], say, about 10^4 cm^{-3}. If the concentration of extracellular polymer present in *G. aureolum* is proportional to the concentration of the cells, and if η_E is proportional to c^2, the viscosity of blooms just lethal to most fish should be greater than that at which trapped bubbles were observed (2.7 x 10^3 cells cm^{-3}) by a factor of about $[10^4/(2.7 \times 10^3)]^2$ or ≈14. The energetic cost of irrigating gills varies from 2% to 43%, typically ≈15%, of the total metabolism in most fishes, but is only ≈1% in tuna, which ram ventilate [7]. Taking a value of 15% and assuming that the percentage of oxygen extracted does not change, the energy expended in irrigation would use all oxygen extracted when ε is increased to a factor of ≈7 times that in non-bloom water (an increase of ≈600%). The corresponding increase in ε in tuna would be 9,900%.

As shown in Table 1, considering a bloom trapping 0.5-mm bubbles with P equal to -0.8, ε in bass gill slits is 1.27 times that in non-bloom conditions (an increase of 27%). Multiplying this by 14 for an estimated "just lethal" bloom, with q in the gill slits of 14,000 s^{-1}, this gives an increase of 380%, probably not enough to kill the fish. For P equal to -0.6, the corresponding increase is 1,720%, so the fish would probably die of O_2 lack. Hence it can be predicted that if the trapped bubbles seen were 0.5 mm in diameter, and if the fish killed used ≈15% of their metabolic energy for gill irrigation, then the mean power of P over the range of q from 6 to 14,000 s^{-1}, is around -0.6 to -0.8. This value of P is typical of three phytoplankton cultures whose viscosity approached that of the bubble-trapping red tide, but higher than values found in the non-bloom conditions [2]. If the bubbles were only 0.1 mm in diameter, the corresponding value for P, over a range of q from 30 to 14,000, would be about -0.3, considerably higher than P typical of the phytoplankton cultures. To kill tuna by the same means, the bloom would have to be about 14 times more viscous, or roughly 3 to 4 times more concentrated in terms of polymer, than one able to kill other typical fish.

If P were sufficiently near zero, fish could be killed by rheological effects without the rising rate of bubbles being noticeably reduced.

The cross-gill pressure, D, is proportional to the square root of ε, and D may be limited by tissue strength, so under significantly increased viscosity the fish would have to let throughput decline, as reported for fish in *Chattonella* culture [8], with corresponding decrease in q and increase in η. Values of shear *stress*, τ, higher than normal in the gill slits may slough away protective mucus and damage underlying epithelia. The large amounts of mucus secreted from fish gills in association with sublethal blooms of *G. aureolum* and *Chattonella* [8,16] may represent efforts by the fish to replace that sloughed away. Additionally, it could be a response to damage or irritation. Sloughing of gill epithelia in trout kept in high concentrations of *G. aureolum* may indicate mechanical damage. That sloughing of epithelia occurred also in the gut, however, suggests that toxic effects may have been involved, and this has been interpreted [16] as indicating chemotoxins in *G. aureolum*. As increased viscosity, by reducing throughput, may result in increased tissue concentrations of potentially toxic waste products, evidence of a toxic pathology in association with a phytoplankton bloom should *not*, by itself, be interpreted as evidence that a chemotoxin originates from the plankton and enters the fish.

Further, some suspected gill cloggers such as *Chattonella subsalsa* Biecheler possess mucocysts [9], and the shock of being inspired with greatly increased shear stress may cause them to discharge mucus in the gill slits.

In organisms in which a much lower value of q exists at their respiratory surfaces, such as those with external gills, red tides may reduce gas and ion exchange by the following method. An increase in the thickness of the non-turbulent boundary layer may occur, associated with an increase in the size of the smallest turbulent eddies [1]. This would reduce turbulent diffusion, and increase the distance across which molecular diffusion would have to occur.

Plankton patchiness, turbulence and mucus aggregation in a bloom are all likely to result in lumpiness (rheological heterogeneity). The more viscous liquid (lumps) will pass through gill slits more slowly than the less viscous liquid (leads). One corollary to this is that the effective mean viscosity of a lumpy liquid depends on the hydrodynamics of the containing system (gill slits or a rheometer, for instance). Another is that if a chemotoxin were more concentrated in lumps than in the leads, its time in contact with the gills would be greater than if it were prefectly dispersed.

ACKNOWLEDGEMENTS

Dr. Bopiah Biddanda helped me with the style of this manuscript.

REFERENCES

1. I.R. Jenkinson, *Nature, Lond.*, **323**, 435, (1986)
2. I.R. Jenkinson, submitted to *Nature, Lond.*
3. I.R. Jenkinson and P.P. Connors, *J. Sherkin Isl.*, **1**(1), 127 (1980)
4. K.J. Jones, P. Ayres, A. M. Bullock, R.J. Roberts and P. Tett, *J. mar. biol. Ass. U.K.*, **62**, 771 (1982)
5. O. Krümmel, *Handbuch der Ozeanographie*, J. Engelhorn, Stuttgart (1907)
6. O. Krümmel and E. Ruppin, *Wiss. Meeresunters., Hydrogr. Abt., Kiel*, No. 3, 27 (1905)
7. B.L. Languille, E.D. Stevens and A. Anantaraman, *In* P.W. Webb and D. Weihs (eds), *Fish Biodynamics*, Praeger, New York, 92 (1983)
8. T. Matsusato and H. Kobayashi, *Bull. Nansei reg. Fish. Res. Lab.*, No. 7, 43 (1974)
9. J.-P. Mignot, *Protistologia*, **12**(2), 279 (1976)
10. Y. Miyake and M. Koizumi, *J. mar. Res.*, **7**, 63 (1948)
11. E.R. Morris, A.N. Cutler, S.B. Ross-Murphy and D.A. Rees, *Carbohydr. Polym.*, **1**, 5 (1981)
12. N.S. Oakey, *J. phys. Oceanogr.*, **15**, 1662 (1985)
13. C. Ono, *In* C. Ono and H. Takano, *Bull. Tokai reg. Fish. Res. Lab.*, No. 102, 93 (1980)
14. F. Partensky and A. Sournia, *Cryptogamie, Algologie*, **7**, 251 (1986)
15. G.W. Potts and J.M. Edwards, *J. mar. biol. Ass. U.K.*, **67**, 293 (1987)
16. R.J. Roberts, A.M. Bullock, M. Turner, K. Jones and P. Tett, *J. mar. biol. Ass. U.K.*, **63**, 741 (1983)

HISTOLOGICAL ALTERATIONS TO GILLS OF THE YELLOWTAIL *Seriola quinqueradiata*, FOLLOWING EXPOSURE TO THE RED TIDE SPECIES *Chattonella antiqua*

T.Toyoshima[1], M.Shimada[2], H.S.Ozaki[2], T.Okaichi[3] and T.H.Murakami[1]
1 Department of Biology, Kagawa Medical School, Kagawa 761-07, Japan
2 Department of Anatomy, Kagawa Medical School, Kagawa 761-07, Japan
3 Faculty of Agriculture, Kagawa University, Kagawa 761-07, Japan

ABSTRACT

By means of scanning electron microscopic and histochemical techniques, we examined gill filaments of yellowtails, *Seriola quinqueradiata*, which had died 25 to 90 min. after the exposure to the red tide species, *Chattonella antiqua*. The results revealed that (1) a considerable number of mucous cells were destroyed, that (2) the mucous coat disappeared, and that (3) ultrastructures on the apical surface of chloride cells conspicuously altered in the affected gill filaments. These observations imply that the osmoregulation function may be affected in the gills exposed to *C. antiqua*. Therefore, it is concluded that changes in the ion-transport function in gill filaments exposed to *C. antiqua* may be involved in edema formation, which in turn is responsible for inhibiting gas exchange across the gills. Oxygen deficiency will then cause the death of fish.

INTRODUCTION

In gills of yellowtails, *Seriola quinqueradiata*, exposed to the red tide organism, *Chattonella antiqua*, edema formation has been observed in primary and secondary lamellae. This observation strongly suggests that *C. antiqua* may have a harmful influence on osmoregulation in gill lamellae of yellowtails, which may cause death of the fish. Since, in teleosts, it is well known that chloride cells play a major role in osmotic regulation and that a mucous coat secreted by mucous cells is a barrier against external osmotic changes, we examined morphologically gill filaments of young yellowtails exposed to *C.antiqua*, focussing attention on the chloride and mucous cells.

EXPERIMENTAL

Fish and red tide species. The yellowtails, *Seriola quinqueradiata*(body weight 1080-2620 g), were obtained from the Nippon Saibai Center, Takamatsu, Kagawa Prefecture. The red tide organisms, *Chattonella antiqua*, were provided through the Red Tide Research Laboratory, Takamatsu, Kagawa Prefecture. The conditions of seawater containing *C. antiqua* with densities from 890 to 3140 cells/ml were as follows: water temperature, 22° to 25°C; salinity, 31.1 to 31.8‰; pH, 7.8 to 8.2; dissolved oxygen, 4.67 to 5.23ml/l. Those of the control seawater were: temperature, 22° to 25°C; salinity, 31.8‰; pH, 7.95 to 8.16; dissolved oxygen, 4.95 to 5.11ml/l.
Exposure to Chattonella antiqua. Two hundreds liters of seawater containing *C. antiqua* were poured into 500-l tanks, and then the fish were introduced. The tanks were aerated throughout the experiment. The fish died within 25 to 90 min. Immediately after death, gill filaments were dissected for scanning electron microscopical (SEM) histological preparations. Control gill filaments were obtained from fish kept in natural seawater.

RED TIDES: BIOLOGY, ENVIRONMENTAL SCIENCE, AND TOXICOLOGY
Okaichi, Anderson, and Nemoto, Editors

SEM and histochemical preparations. For SEM studies, the gill filaments were fixed for a minimum of 12h with 2% paraformaldehyde and 2.5% glutaraldehyde on 0.1M cacodylate-HCl buffer(pH=7.4). Since chloride cells in the gill filaments are predominantly located in the interlamellar regions of primary lamellae, the secondary lamellae of the gills were removed by tweezers and Freon spray.

For histochemical study, each specimen was prepared in the manner usual for paraffin sectioning. Sections were stained with periodic acid-Schiff reaction and alcian blue(pH=2.6).

RESULTS

Two observations were made. First, exposure to C. antiqua causes conspicuous ultrastructural alterations on the apical surface of the chloride cells of the gills(Table 1). The majority of chloride cells in normal gills showed an apical surface with numerous long cellular extensions which were about 0.3 to 0.6 μm in diameter and 50 to 200 in number per cell (Fig. 1). In the affected gills, these chloride cells with numerous long cellular extensions decreased to 31.9% and, instead, chloride cells with fewer (20 to 100 in number) and smaller (0.2 to 0.4μm) cellular extensions, which were only occasionally observed in control gill filaments, markedly increased (Table 1, Fig. 2). Further, 26.5% of the chloride cells in the affected gills exhibited a protruded apical surface with less prominent cellular extensions or a wrinkled apical surface with a few microplicae, both of which rarely occurred in normal gill filaments(Table 1, Fig. 2).

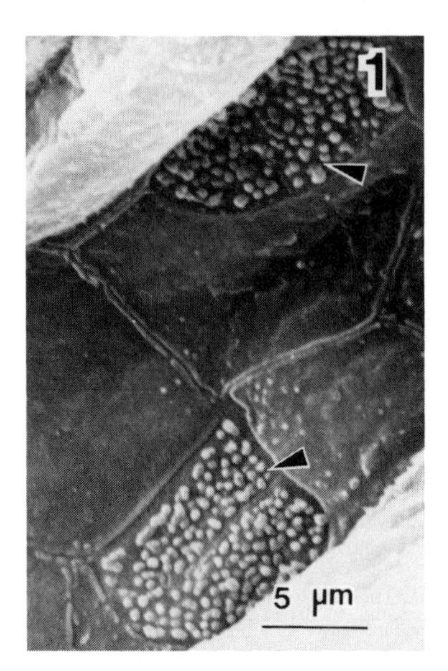

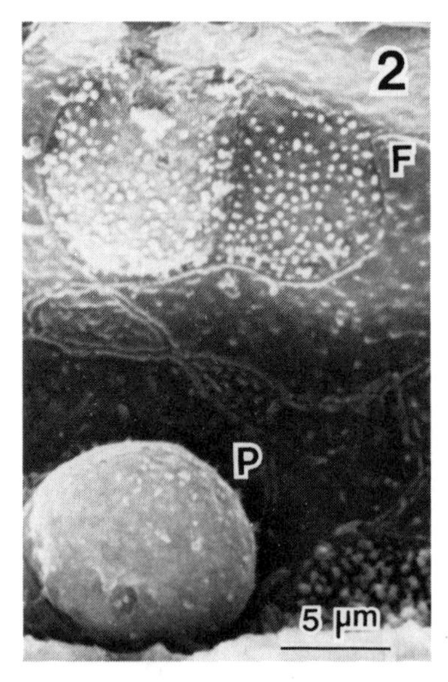

FIG. 1. Interlamellar epithelium of control gill filament. Note uniform cellular extensions on chloride cells(arrowheads).
FIG. 2. Interlamellar epithelium of gill filament exposed to artificial Chattonella antiqua red tide. It is clear that the chloride cells exhibit an apical surface with a few short cellular extensions (F) or a protruded apical surface with less prominent celluler extensions (P), instead of uniform cellular extensions.

TABLE 1. Comparison of various surface ultrastructures on chloride cells in control and exposed gill filaments

Surface morphology	Control	Affected
Apical surface with numerous long extensions	94.8%	31.9%
Apical surface with a few short extensions	3.8%	41.5%
Wrinkled apical surface	1.4%	15.1%
Protruded apical surface	0%	11.5%

The number of the chloride cells counted was 200 to 700 per fish for control(n=20) and exposed(n=14) specimens.

Second, partial destruction of mucous cells and disappearance of the mucous coat were observed in the gill filaments after exposure to the C. antiqua(Table 2, Fig. 3,4). The loss of mucous cells was seen on both afferent and efferent ridges, but its degree was somewhat more marked on afferent ridges. In many of the surviving mucous cells, vigorous mucous secretion was observed. This suggests that C. antiqua may continuously stimulate exhaustive mucous secretion by the mucous cells, leading to death of these cells.

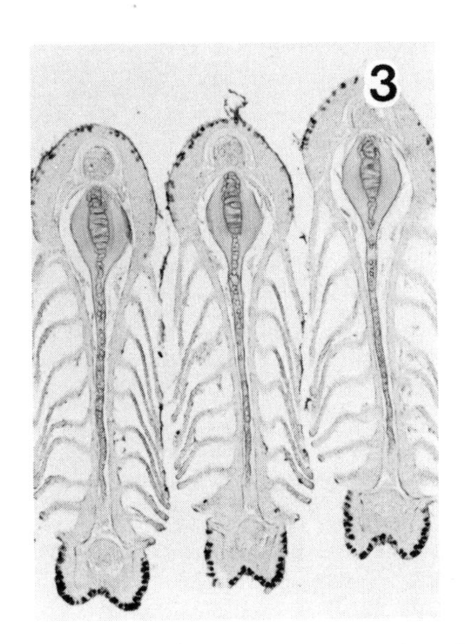

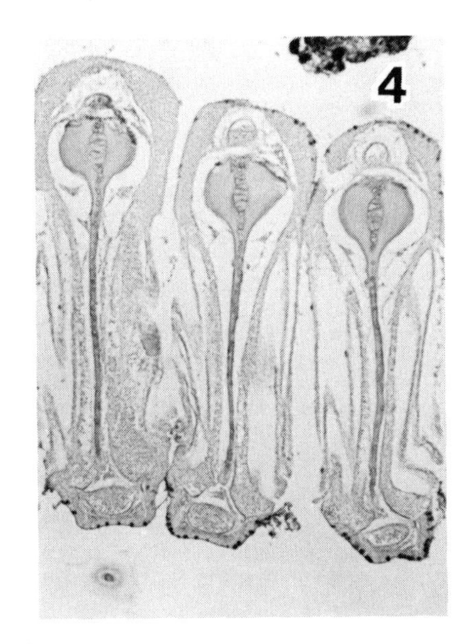

FIG. 3. Transverse section through gill primary lamella from the intact fish. A large amount of mucous cells are seen.
FIG. 4. Transverse section through gill primary lamella from the teleost exposed to sea bloom. Note a significant loss of mucous cells.

TABLE 2. Cell density (cells/100μm$\pm$ SEM) of the mucouscells on the normal and impaired gill filaments

Condition	Afferent ridge	Efferent ridge
Normal gill	13.9$\pm$ 0.7	19.5$\pm$ 0.6
Impaired gill	5.0$\pm$ 0.3	10.8$\pm$ 0.8

Each value is the mean$\pm$ SEM of 50 different gill lamellae.

DISCUSSION

Chloride cells are generally known to engage in active ion transport and are favored as a primary site of osmotic regulation in gills. Since histochemical and biochemical analyses for the ion transporting enzyme, Na-K-ATPase, were not carried out in the present study, an ion-transport function in the chloride cells with surface ultrastructures altered by the red tide organism, C. antiqua, cannot be demonstrated. However, it has recently been reported that ultrastructural modifications of chloride cells including the number of the cellular extensions very closely reflect changes in Na-K-ATPase content of gill filaments. Furthermore, the disappearance of the mucous coat may allow direct exposure to the environments hyperosmotic to the gill lamellae. Therefore, it is concluded that changes in the ion-transport function in gill filaments exposed to C. antiqua are to some extent involved with edema formation, which in turn is responsible for inhibiting gas exchange in the gills. Oxygen deficiency will then cause the death of fish.

REFERENCES

1. Dunel-Erb, S. and P. Laurent: Ultrastructures of marine teleost gill epithelia, SEM and TEM study of chloride cell apical membrane. J.Morphol. 165, 175-186 (1980)
2. Epstein, F. H., P. Silva and G. Kormanic: Role of Na-K-ATPase in chloride cell function. Am. J. Physiol. 7, 246-250 (1980)
3. Hossler, F. E.: Gill arch of the mullet, Mugi cephalus III. Rate of response to salinity change. Am. J. Physiol. 7, 160-170 (1980)
4. Kobayashi, H.: Sakana no Kokyu to Junkan. In: Respiration and circulation in fish, pp 111-124. Ed. by Japan Soc. Sci. Fish., Suisangaku series 24 Tokyo: Koseisha-Koseikaku 1978 (in Japanese)
5. Maetz, J.: Fish gills: mechanisms of salt transfer in fresh water and sea water. Phil. Trans. R. Soc. Lond. B262, 209-249 (1971)
6. Maetz, J. and M. Bornanchin: Biochemical and biophysical aspects of salt excretion by chloride cells in teleosts. Forsch. Zool. 23, 323-363 (1975)
7. Shimada, m., T. H. Murakami, A. Doi, S. Abe, T. Okaichi and M. Watanabe: A morphological and histochemical study on gill primary lamellae of the teleost, Seriola quinqueradiata, exposed to sea bloom. Acta Histochem. 15, 497-507 (1982)

RED TIDE, *CHATTONELLA ANTIQUA* REDUCES CYTOCHROME *C* FROM HORSE HEART

Shimada, M.*, Shimono, R.*, Murakami, T.H.*, Yoshimatsu, S.‡ and C. Ono‡
*Department of Anatomy and Biology, Kagawa Medical School, Kagawa 761-07 (Japan), ‡ Akashiwo Research Institute of Kagawa Prefecture, Kagawa 761-01 (Japan)

ABSTRACT

We found that the red tide, *Chattonella antiqua,* was able to reduce cytochrome *c* and this reducing ability depended on the pretreatment; 144±36 nM of cytochrome *c* /2000 cells for shaking and 235±9 nM of cytochrome *c* /2000 cells for sonication. Part of the reducing ability(60% for shaking and 13% for sonication) is thought to be due to superoxide (O_2^-) produced by the organism.

INTRODUCTION

We investigated the causative factors responsible for fish death by the red tides and showed the following(2-5) : (1)Most of the mucus goblet cells and the mucus coat disappeared from gill lamellae of fish exposed to red tide ; (2)diazo-reaction positive substances(NO_x), which may cause degeneration of the mucus of the gill lamellae, were highly concentrated in the cortex(perikaryon) of the red tide. These results suggest that when the red tide is passing between the gill lamellae, NO_x discharged from the mucocysts may act on the mucus, leading to the degeneration and subsequent removal of the mucus coat from the gill lamellae(2-5). As regards the chemical action of NO_x, it is conceivable that these substances may play a part in certain reduction processes. The present work was designed to investigate whether substances discharged from the cortex of the red tide participate in the reduction of cytochrome *c*.

MATERIALS AND METHODS

After incubation in Erd-Schreiber Modified(ESM)-culture medium (pH=8.2) for 7 days, the reducing ability of *Chattonella antiqua* was measured as follows. Firstly, one ml of the medium was transferred to a quartz-cuvette containing 100 μg of cytochrome *c* (type III from horse heart, Sigma Co.). As the reduced form of cytochrome *c* has a characteristic absorption band at 550 nm, optical density(OD) was measured with a spectrophotometer(Cary 17D, USA). A so-called silent condition (S1) OD was obtained. Next, the cuvette was vigorously shaken 20 times by hand to give ODs under shaking conditions(S2). Thirdly, the plankton was sonicated for 5 seconds:the ODs were termed OD in sonication(S3). One such experiment is displayed in figure 1a and 1b. To determine the effects on reducing ability with elapsed time after sonication, cytochrome *c* was added to the red tide at various intervals(1, 3, 7, 10, 20 minutes). As the cells have abundant chloroplasts in their perikaryon, superoxide(O_2^-) may be produced there during photosynthesis. To estimate the amount of superoxide produced, 750 units of superoxide dismutase were added before the administration of cytochrome *c*. As the plankton appears to damage fish only during summer(6), changes in reducing ability at different water temperatures were observed. After incubation for one week at 22°C, the incubation temperature was set at 26°C for one day, and then the temperature was changed to 22, 24, 28, 30°C, respectively. After a 1-hour

RED TIDES: BIOLOGY, ENVIRONMENTAL SCIENCE, AND TOXICOLOGY
Okaichi, Anderson, and Nemoto, Editors

long equilibration at each temperature, ODs were determined. Differences of oxidation-reduction potential before and after cell destruction by ice-cold condition were measured and their results are shown in figure 3.

RESULTS AND DISCUSSION

Our results show that red tide had a strong ability to reduce cytochrome *c* without any cell destruction(Table Ⅰ). About 60% of the reduction caused by the shaking(S2) depends on superoxide(Table Ⅱ), which causes degeneration of the mucus coat on the gill lamellae(3). When blood is passing through the gill lamellae the superoxide discharged from the plankton may act therefore on degeneration processes of the cell membrane of erythrocytes and on the removal of mucus coat from the gill lamellae, leading to the disturbance of branchial gas exchange. At the same time, the superoxide may also reduce cytochrome *c* of gill epithelia, resulting in morphological changes as seen before(5). About 40% of the reduction after shaking and 87% of the reduction in perikaryon were not inhibited by the addition of superoxide dismutase(Table Ⅲ). These reducing effects were very unstable and disappeared within 7 minutes after sonication(Fig. 2). As the diazo-reaction positive substances(NO_x) are highly concentrated in the cortex of red tide, there is a possibility that NO_x, especially NO^- may be partly related to the reduction processes of cytochrome *c* (4). As red tide often appears on Seto Inland Sea in Japan but only during summer(1, 6), we investigated changes of reducing ability at different water temperatures corresponding to the seasonal changes. When water temperature dropped from 26 °C to 22 °C, the reducing ability by shaking(S2) significantly declined to a trace, while no alteration was observed in ODs of perikaryon(Table Ⅲ). These data are in good agreement with field observations that the red tide is especially injurious to the fish only during summer(6). According to the measurement of oxidation-reduction potential, it was found that the cell numbers(Y) were hyperbolically related to the time(min) elapsed before complete disappearance of reducing ability(X); $Y=1600/X + 3.23$ ($r=0.9967$). Monitoring this redox potential may be useful for the supervisory system of the red tides appearing in the sea.

Table Ⅰ. Amount of cytochrome c by *Chattonella antiqua* in each condition. Cells [S3-S2] represent a mean value in the perikaryon.

	ODs/10520 cells/ml	nM of reduced cytochrome c/2000 cells
Silent(S1)	0.008 ± 0.001	72 ± 9
Shaking(S2)	0.016 ± 0.004	144 ±36
Sonication(S3)	0.042 ± 0.001	380 ± 9
Cells [S3-S2]	0.026 ± 0.001	235 ± 9

Incubation period; 7 days   Water temperature; 26 °C
Mean ± SEM are given(n=4).

Table Ⅱ. Inhibitory effects of superoxide dismutase(SDM) on reducing ability.
%Decrease=Optical density(OD) of reduced form of cytochrome c after administration of 750 units of SDM/OD of reduced cytochrome c without SDM × 100

	% decrease
Silent(S1)	65 ± 3
Shaking(S2)	60 ± 2
Sonication(S3)	27 ± 3
Cells [S3-S2]	13 ± 4

Mean ± SEM are given(n=4).

Table Ⅲ. Effects of water temperature on cytochrome c reduction.

⇦ Initial temp.⇨					
Water temp	22 ℃	24 ℃	26 ℃	28 ℃	30 ℃
S1	*0.4±0.1	1.5±0.2	1.2±0.5	0.9±0.2	1.0±0.2
S2	§0.8±0.1	2.9±0.2	3.5±0.4	*2.4±0.3	2.4±0.1
S3	*4.2±0.1	5.9±0.1	6.7±0.2	6.0±0.2	6.1±0.2
[S3-S2]	3.4±0.1	3.0±0.2	3.2±0.3	3.6±0.3	3.7±0.2

[ODs/21770 cells/ml±SEM] × 100 (n=4)
* Significant at $P<0.01$, § Significant at $P<0.001$
Note a significant decrease of optical densities(ODs) in shaking condition(S2) at 22℃ as compared with that of 26℃.

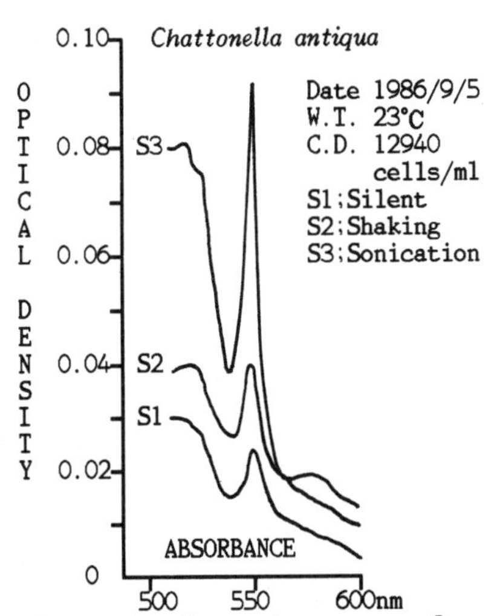

Figure 1a. Absorption curves of reduced cytochrome c. W.T.; water temperature, C.D.;cell density. Note high peaks of absorption band at 550 nm presenting an amount of reduced cytochrome c.

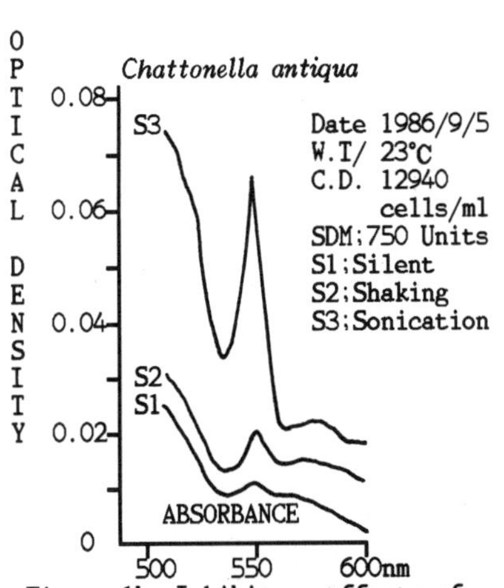

Figure 1b. Inhibitory effects of superoxide dismutase(SDM) on reducing ability. Note significant low peak of absorption band at 550 nm in each condition as compared with that of figure 1a.

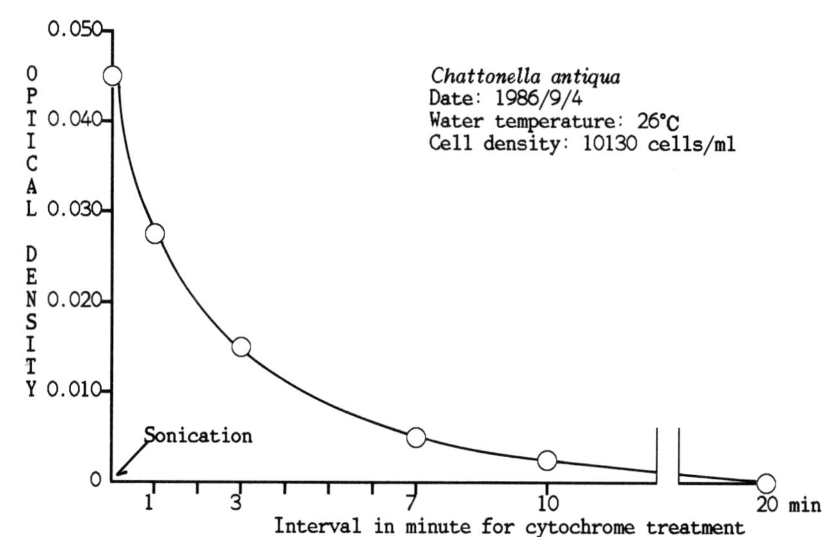

Figure 2. Decreasing effects on reducing ability with elaped time after sonication. Open circles represent a mean value from four trials. Note a trace of amounts at 7 minutes.

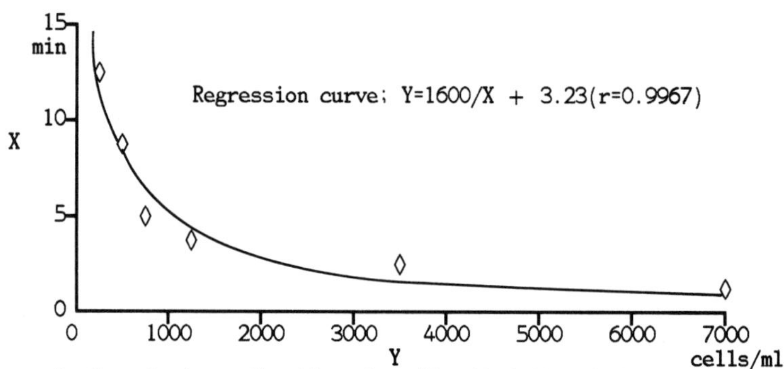

Figure 3. Correlations of cell numbers(Y;cells/ml) and the time(X;min) before complete disappearance of reducing ability.

REFERENCES

1. C.Ono & H.Takano *Bull. Tokai Reg. Fish. Res. Lab.*, 102, 93, (1980)
2. M.Shimada, T.H.Murakami, A.Doi, S.Abe, T.Okaichi & M.Watanabe *Acta Histochem, et Cytochem.* 15, 497, (1982)
3. M.Shimada, T.H.Murakami, T.Imahayashi, H.S.Ozaki & T.Okaichi *Acta Histochem. et Cytochem.* 16, 232, (1983)
4. M.Shimada, R.Shimono, T.Imahayashi, H.S.Ozaki & T.H.Murakami *Histol. and Histopathol.* 1, 327, (1986)
5. T.Toyoshima, H.S.Ozaki, M.Shimada, T.Okaichi & T.H.Murakami *Mar. Biol.* 88, 101, (1985)
6. S.Yoshimatsu & C.Ono *Bulletin of Akashiwo Res. Inst. of Kagawa Pref.* 2, 1, (1986)

ELECTROCARDIOGRAM OF A MARINE FISH, *PAGRUS MAJOR*, EXPOSED TO *CHATTONELLA MARINA*

Endo, M., R. Foscarini, and A. Kuroki

Fisheries Department, Miyazaki University, Miyazaki 889-21, Japan

ABSTRACT

Examinations of the electrocardiogram of *Pagrus major* exposed to *Chattonella marina*, a flagellate causing red tide, were made to determine the physiological effects on fish. The heart rate decreased as a result of the extention in the interval between T and P waves. The decrease in heart rate with the extended intervals between T and P waves was also recognized by the condition of decreased dissolved oxygen. Also, the decrease in the heart rate of the fish exposed to the red tide occurred while the fish were struggling. This reduction seems to be caused by the strong tension of the vagal nerve. Upon exposure to *C. marina* at high cell concentration, the heart beat at a very low frequency after 30 min. The very low heart rate is expected to seriously limit the oxygen uptake by the gill, because the cardiac output is probably very low in this situation.

INTRODUCTION

During the occurrence of red tides, fish become excited and finally die from anoxia. The cardio-physiology of fish exposed to red tides is expected to precisely reflect the influence of red tides, because of the close relationship between the blood circulation in the gill and cardiac performance. The intrinsic property of the fish heart is supposed to be primarily governed by temperature, vagal innervation, and secondarily by hormones.

Some researchers have studied fish death caused by red tide. However, there have been no reports on the cardio-physiology of fish exposed to red tide. This study examines the electrocardiogram (ECG) of a marine fish, *Pagrus major* exposed to: *Chattonella marina*; low concentration of dissolved oxygen (DO); water temperature (WT); and high pH value.

MATERIALS AND METHODS

Cultured sea bream *Pagrus major* (body weight 604-940 gm) were used. Fish were anaesthetized with MS-222 and two electrodes were implanted bipolarly 24 hrs prior to experiments by the method of Nanba *et al* (1). In the test apparatus, fish heart rate did not exhibit prominent change between the beginning and the final period of the accommodation with a coefficient of variation of 8.3% and it was not significant at 1% level (t-test). The change of environmental conditions in the test apparatus are shown in Table 1. ECG records were obtained by using a bioelectric amplifier AB-621 (Nihonkoden, Tokyo) with recorder.

RED TIDES: BIOLOGY, ENVIRONMENTAL SCIENCE, AND TOXICOLOGY
Okaichi, Anderson, and Nemoto, Editors

RESULTS

While the experimental fish were exposed to *Chattonella marina* at low concentration, they intermittently struggled. High concentration of *C. marina* induced violent swimming at the beginning of exposure with the fish remaining calm after 30 min; death occurred after 60 min. In DO, WT, and pH change tests, the fish remained quiescent.

Despite the initial large variation in heart rate between fish A and B, we observed an identical response in both fish to the respective environmental changes. Prominent reduction in heart rate was observed when the fish were exposed to *C. marina* and decreased DO and WT. Especially in the high *C. marina* concentration the heart rate markedly decreased during the initial 30 min and was very low during the last 30 min. The change in heart rate was not evident in the pH change test. The coefficient of variation of the heart rate exceeded 10% in all the tests excluding the pH increase test (Table 1).

An ECG of *P. major* was composed of P, Q, R, S, and T waves. The amount of change for every interval in each test is expressed as coefficient of variation (Table 2). The DO decrease and exposure to *C. marina* at low concentration test induced a prominent extension of TP and PP intervals. In the test with high *C. marina* concentration, the P and T waves disappeared 30 min after the exposure. In WT decrease test, all intervals except for the QRS complex were extended by the temperature change. The pH increase test induced a slight distention or a constancy in every interval.

DISCUSSION

In low concentration of *C. marina*, DO and WT tests, the heart rates were reduced according to the change of each parameter. However, analyses of ECG revealed that the reduction of heart rate was influenced by different causes. In WT decrease test, PQ, QT, and TP intervals were extended to lengthen the PP interval and thus reduced the heart rate. On the other hand, the distention of PP interval and decrease of heart rate in low concentration of *C. marina* and DO tests were due to the protracted TP intervals only. The finding in the present study confirmed the result of a previous study, that fish die of anoxia during the exposure to red tide (2).

During the last 30 min in the test of high *C. marina* concentration, the heart was beating at a very low frequency. This phenomenon seemed to limit the oxygen uptake through the gills, because the oxygen uptake is related to the amount of counter-current between the water flowing through the gill lamellae and the blood circulation of gill. Unbalanced osmoregulation and reduction of carbonic anhydrase activity in the gill, occlusion of mucous among the gill lamellae and decrease of pH in the blood have been suggested as the cause of anoxia in fish exposed to red tides. The very low heart rate observed in this study seemed to be an additional cause of death from anoxia.

The exposure to red tide in low and high concentration revealed that the decrease in heart rate occurred while fish were still struggling. Because fish heart is primarily governed by the vagal nerve (3), the decrease in the heart rate in struggling fish is probably due to the strong tonic of the

Table 1. Heart rate (beats/min, mean±s.d., n=5) of *Pagrus major* and changes in the environmental conditions

Tests		Changes of experimental conditions				CV	Other conditions
DO decrease test (ppm)		7.0	5.9	4.4	3.0		WT 25.3-26.0 ℃ pH 7.6- 7.9
Heart rate	*Fish A	146.4±10.50	135.0±2.44	98.0±6.29	62.4±2.33	29.9%	
	Fish B	104.4± 2.00	98.0±1.03	72.7±1.19	65.1±1.42	19.4%	
WT decrease test (℃)		27.6	25.3	23.1	21.1		DO 4.9- 5.4 ppm pH 7.6- 8.3
Heart rate	Fish A	107.6±0.42	108.2±1.10	96.2±1.88	83.6±1.29	10.2%	
	Fish B	132.2±6.27	131.6±3.06	115.4±1.81	98.0±2.59	11.8%	
pH increase test		8.1	8.3	8.5	8.7		DO 4.9- 5.4 ppm WT 24.2-25.3 ℃
Heart rate	Fish A	121.4±2.90	120.5±0.53	110.6±1.51	106.9±3.47	5.4%	
	Fish B	121.3±2.81	133.0±2.62	130.8±1.19	114.4±2.14	6.0%	
Low concentration red tide test[1] (min)		0	30	60	90		DO 4.4- 5.5 ppm WT 24.9-25.2 ℃ pH 8.6- 8.7
Heart rate	Fish A	119.0±0.21	103.1±2.14	77.8±0.70	65.3±0.81	25.5%	
	Fish B	130.0±0.34	120.4±1.88	104.4±4.40	102.8±4.31	10.3%	
High concentration red tide test[2] (min)		0	15	30	45		DO 4.6- 5.2 ppm WT 24.0-24.9 ℃ pH 8.8- 8.9
Heart rate	*Fish A	122.0±0.98	97.4±0.11	3.5±4.26	5.5±0.49	93.4%	

DO: dissolved oxygen, WT: water temperature, CV: coefficient variation, 1: 4,000 cells/ml *Chattonella marina*, 2: 8,000 cells/ml *C. marina*, *: A significant difference ($p>0.01$) of the heart rate was noticed between the beginning and end of the test vagal nerve which reduced the heart rate.

The vagal nerve tonic might secondarily occur as a result of anoxia induced by the exposure to *C. marina*, because asphyxia is known to strain the vagal nerve in fish (3). However, its tonic is also expected to primarily take place by visual or olfactory stimulations with exposure to *C. marina*, because it occurs with frightening stimuli (4). Furthermore, the vagal nervous system of fish might be influenced by chemical substances produced by *C. marina*.

Table 2. Coefficient of variation (%) in respective interval length of ECG of *Pagrus major* during each test

Tests		Intervals PQ	QRS	ST	QT	TP	PP
DO decrease test	Fish A	3.3	6.5	17.2	13.4	72.2	35.6
	Fish B	6.6	11.3	16.6	6.1	42.4	19.8
WT decrease test	Fish A	18.5	4.4	26.7	18.0	30.1	10.5
	Fish B	11.4	10.1	31.4	18.9	18.1	18.0
pH increase test	Fish A	3.9	7.9	10.7	5.6	16.7	4.0
	Fish B	2.3	4.3	13.3	8.9	13.7	6.2
Low concentration *C. marina*	Fish A	2.8	7.8	17.9	9.4	44.7	22.5
(4,000 cells/ml)	Fish B	11.2	10.7	8.0	7.7	49.9	16.5

DO: dissolved oxygen, WT: water temperature

REFERENCES

1. Nanba, K., S., Murachi, S. Kawamoto, and Y. Nakano, J. Fac. Fish. Anim. Husab. Hiroshima Univ., 12, 147 (1973)
2. T. Matsusato and H. Kobayashi, Bull. Nansei Reg. Fish. Res. Lab., 7, 43, (1974)
3. A.P. Farrell, Can. J. Zool., 62, 523, (1984)
4. K. Yamamori, I. Hanyu, and T. Hibiya, Bull. Japan. Soc. Sci. Fish., 37, 94, (1971)

TWO FISH-KILLING SPECIES OF COCHLODINIUM FROM HARIMA NADA, SETO INLAND SEA, JAPAN

KATSUHISA YUKI,* AND SADAAKI YOSHIMATSU**
*Matoya Oyster Research Laboratory, Matoya, Isobe-cho, Shima-gun, Mie 517-02; **Akashiwo Research Institute of Kagawa Prefecture, Yashima-higashi-machi, Takamatsu 761-01, Japan

ABSTRACT

Morphological features of two *Cochlodinium* species, *C. polykrikoides* (= *C. heterolobatum*) and *C.* sp., isolated from Harima-Nada, eastern Seto Inland Sea, are described. In laboratory tests, both these species were toxic to the juveniles of the fish *Leiognathus nuchalis*.

INTRODUCTION

Several species of the genus *Cochlodinium* are known to cause fish kills or PSP incidents in various sites of the world [1-4]. In south Harima Nada, eastern part of the Seto Inland Sea, western Japan, the routine monitoring for toxic species of the genera *Chattonella* and *Protogonyaulax* has been carried out by Kagawa Prefecture, but no information on *Cochlodinium* has been obtained except for the morphological study on the cyst and germinated motile cell of one unidentified species [5].

In 1985 we have succeeded in isolating two species of *Cochlodinium* from Harima-Nada and examined them from morphological and toxicological aspects.

MATERIALS AND METHODS

Culture of *Cochlodinium*

Two species of *Cochlodinium* were isolated from a surface seawater sample collected in Harima-Nada in September 1985. The clonal cultures of both these species were established and maintained in ESM medium [6] at 22 °C under constant fluorescent illumination at ca. 2,000 lux.

Toxicity Test

Juveniles of the fish *Leiognathus nuchalis* (35.0-40.8 mm in total length, 0.43-0.72 g in wet weight) collected in Yashima Bay, 8 km east of Takamatsu City, were used as test animal. In each test five individuals were exposed 48 hours to two liters of *Cochlodinium* culture water conducted at temperatures of ca. 22-24 °C and aerated to supply oxygen and to stir the water.

RESULTS

Morphological Features

Cochlodinium polykrikoides Margalef (FIGS. 1a-c)
Margalef, 1961, p. 76, fig. 27m.
Syn.: *Cochlodinium heterolobatum* Silva, 1967, p. 745, pl. 1, figs. A, 1-8, pl. 2, figs. 1-6, pl. 3, figs. 1-7, pl. 4, figs. 1-6, pl. 5, figs. 1-7.

Cells ellipsoidal, often forming chain of less than eight cells, 30-40 µm

RED TIDES: BIOLOGY, ENVIRONMENTAL SCIENCE, AND TOXICOLOGY
Okaichi, Anderson, and Nemoto, Editors

long, 20-30 μm wide. Cingulum descending left-hand spiral of 1.8-1.9 turns, displaced about 0.6 cell length. Sulcus making 0.8-0.9 turns before meeting distal end of cingulum, terminating at antapex, where it divides the antapex into two asymmetrical lobes. Right lobe narrower and slightly longer than left one, tongue-shaped. Apical groove beginning at anterior junction of cingulum and sulcus, counterclockwise encircling apex. Nucleus anterior in position. Chloroplasts ellipsoidal or rod-shaped, numerous. Stigma red, situated dorsally at epicone.

This species was originally reported from southern coastal waters of Puerto Rico [7]. Also recorded from Barnegat Bay, New Jersey [8] and the York River, Virginia [9]. No fisheries damage caused by this species has so far been recorded from Harima-Nada region. Although the red tide, consisting mainly of *Chattonella antiqua* and *C. polykrikoides*, has occurred in southern part of Harima-Nada in August 1986, it did not extend to the fish culturing grounds.

Cochlodinium sp. (FIGS. 2a-d)
Cells ellipsoidal, usually solitary, rarely forming short chain of less than four cells, 38-50 μm long, 25-38 μm wide. Cingulum descending in left spiral of about two turns. Sulcus making about one turn, nearly parallel to cingulum, terminating in a notch at center of antapex. Right lobe of antapex slightly obliquely depressed. Apical groove turning leftward around epicone just above cingulum, terminating at right side of its proximal end. Nucleus anterior in position. Chloroplasts yellow-brown, numerous. Stigma situated dorsally at epicone.

In late August 1985, a small-scale red tide caused by this species (400-500 cells·ml^{-1}) occurred in the northern part of Harima-Nada, where cultured yellowtails showed unusual behavior probably caused by toxin(s).

Toxicity Test

Test animals in water with *C. polykrikoides* and *C.* sp. cells exhibited mortalities of 20-40 % and 80 % respectively within 48 hours, whereas the individuals in filtered seawater (control) were unaffected (Table I). Time period before the death of animals in *C.* sp. culture water was considerably shorter than that in *C. polykrikoides* culture water. These indicate that *C.* sp. is more lethal to the present test animal than *C. polykrikoides*.

DISCUSSION

Athecate dinoflagellate tentatively called "*Cochlodinium* Type '78 Yatsushiro" [10] causing harmful red tides in southwestern coast of Kyushu Island, western Japan, seems to be identical with *C. polykrikoides*.

C. sp. is allied to *C. citron* Kofoid et Swezy in size and shape, but differs from the latter in having chloroplasts, a stigma, an apical groove in epicone (short extension of sulcus in the latter) and sulcal notch at the center of antapex (left side in the latter). *C. catenatum* Okamura also has some resemblance to *C.* sp. However, a single cell of *C. catenatum* illustrated by Kofoid & Swezy [11] possesses the cingulum with a torsion of 1.5 turns, the sulcus extending to near the apex and the nucleus situated at the center of the cell.

The present bioassay results that both *C. polykrikoides* and *C.* sp. are regarded as toxic shows the necessity for monitoring the occurrence of these toxic *Cochlodinium* species in the fish culturing grounds in Harima-Nada region in addition to toxic species of *Chattonella*.

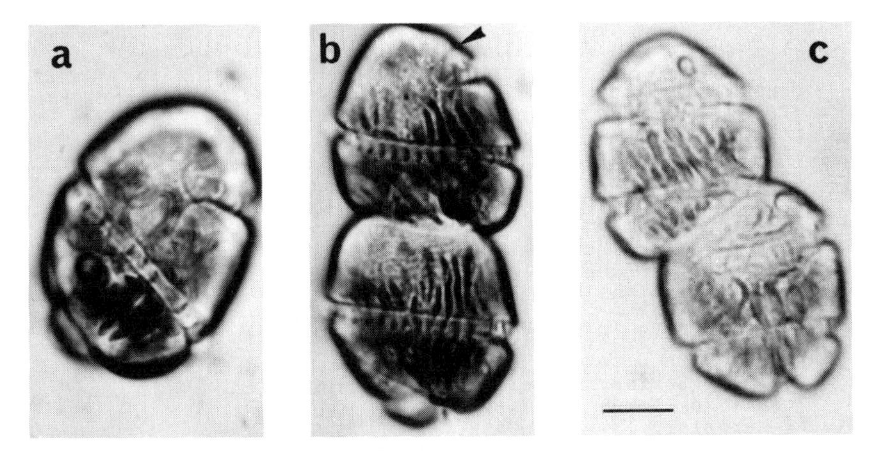

FIG. 1. Cochlodinium polykrikoides. a,b: ventral views, c: dorsal view, showing a stigma. Arrowhead indicates apical groove. Scale bar = 10 µm.

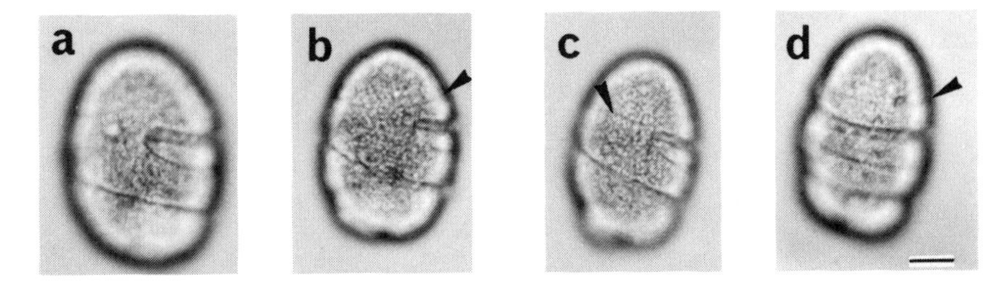

FIG. 2. Cochlodinium sp. a-c: ventral views, d: dorsal view, showing a stigma. Arrowheads indicate apical grooves. Scale bar = 10 µm.

TABLE I. Mortalities of juveniles of L. nuchalis in Cochlodinium culture water.

Species	Cochlodinium Cells·ml^{-1} 0 h	24 h	48 h	L. nuchalis Mortality(%) 24 h	48 h	Time to First Death
C. polykrikoides	6510	4820	3320	40	40	8 h
C. polykrikoides	3950	2110	1740	20	20	8 h
C. sp.	1790	180	10	80	80	3 h 15 m
Control	—	—	—	0	0	—

REFERENCES

1. K. Okamura,Rep. Fish. Inst. Japan 12, 26-41 (1916).
2. G. Reyes-Vasques, E. Ferraz-Reyes and E. Vasquez in: Toxic Dinoflagel-late blooms, D. L. Taylor and H. H. Seliger, eds. (Elsevier/North Holland, New York 1979) pp. 191-194.
3. F. J. R. Taylor in: Seafood Toxins, E. P. Ragelis, ed. (Amer. Chem. Soc., Symp. Ser. 262, Wash. D.C. 1984) pp. 77-97.
4. F. J. R. Taylor in: Toxic Dinoflagellates, D. M. Anderson, A. W. White and D. G. Baden, eds. (Elsevier/North Holland, New York 1985) pp. 11-26.
5. Y. Fukuyo in: Fundamental Studies of the Effects on the Marine Environment of the Outbreaks of Red Tides. Reports of Environmental Sciences, B148-R14-8. (Mombusho, Tokyo 1982) pp. 2050214.
6. T. Okaichi, S. Nishio and Y. Imatomi in: Toxic Phyto-plankton -- Occurence , Mode of Action, and Toxins, Japanese Society of Scientific Fisheries, ed. (Koseisha-Koseikaku, Tokyo 1982) pp. 22.34.
7. R. Margalef, Inv. Pesq. 18, 33-96 (1961).
8. E. S. Silva, J. Protozool. 14, 745-754 (1967).
9. P. L. Zubkoff, J. C. Munday, Jr., R. G. Rhodes and J. E. Warinner, III in: Toxic Dinoflagellate Blooms, D. L.Taylor and H. H. Seliger, eds. (Elsevier/North Holland, New York 1979) pp. 279-286.
10. Y. Fukuyo in: A Guide for Studies of Red Tide Organisms, Nihon Suisanshigen Hogo Kyokai, ed. (Shuwa, Tokyo 1987) pp. 332-345.
11. C. A. Kofoid and O. Swezy, Mem. Univ. Calif. 5, 1-562 (1921)

PARALYTIC SHELLFISH TOXICITY IN SHELLFISH IN HONG KONG

Lam, Catherine W.Y.[*1], M. Kodama[*2], D.K.O. Chan[*3], T. Ogata[*2], S. Sato[*2] & K.C. Ho[*1]
[*1] Environmental Protection Department, Hong Kong
[*2] School of Fisheries Sciences, Kitasato University, Japan
[*3] Department of Zoology, University of Hong Kong, Hong Kong

ABSTRACT

The PSP toxicity of popular eating shellfish from various parts of the Hong Kong territorial waters has been monitored using the mouse bioassay method since 1984, in view of the frequent red tide occurrences in local waters such as Tolo Harbour and the potential risk of public health hazards caused by toxic red tides. A very low toxicity level below 2000 mouse units per kg tissue was prevalent in the shellfish samples tested. The mean monthly toxicity level appears to follow an increasing trend with time over the past years from 1984 - 1987. The presence of saxitoxin in green-lipped mussels (Perna viridis) from Tolo Harbour has been identified by HPLC analysis. It is not known what is the origin of this baseline toxicity. No PSP causative dinoflagellates have been identified in Hong Kong waters so far. However, organisms suspected to be Protogonyaulax species have been detected recently. Furthermore, the toxicity level of shellfish in Tolo Harbour appears to follow a seasonal trend which coincided with the pattern of red tide occurrences in the marine bay.

INTRODUCTION

Red tide occurrences in Hong Kong coastal waters especially Tolo Harbour have increased over recent years [1]. So far, only two red tide incidents have been proved to be toxic by mouse bioassay test [1,2], causing fishkills [1]; contaminated shellfish and fish also contained low toxicity. The causative organisms were Noctiluca scintillans [2] and a Gymnodinium spp. [1]. The presence of well documentated PSP toxic dinoflagellates such as Protogonyaulax spp. [3] and Pyrodinium bahamense var. compressa [4] has not been recorded locally. Also, no marine food poisoning associated with red tides has been confirmed. However, in view of the potential risk of future occurrence of toxic dinoflagellates and their health hazard implications [5-9], a routine surveillance programme on PSP toxicity in shellfish has been implemented in Hong Kong as a preventive measure for toxic red tides since 1984. To further understand the situation, detailed investigations concentrating on shellfish collected from Tolo Harbour were carried out.

EXPERIMENTAL

Materials

Shellfish samples were collected monthly at random from different parts of the coastal waters in Hong Kong. From January 1984 to July 1987, 329 samples consisting of various kinds of popular eating shellfish were collected. These included green-lipped mussel (Perna viridis), fan-shell (Atrina pectinata globosa), clams (Tapes philippinarum, Marcia

RED TIDES: BIOLOGY, ENVIRONMENTAL SCIENCE, AND TOXICOLOGY
Okaichi, Anderson, and Nemoto, Editors

hiantina) and cockles (Scaphara globosa, Anadara crebricostata). Samples of green-lipped mussel and fan-shell were collected independently from Tolo Harbour at bi-weekly intervals from May 1986 to July 1987. These samples were live specimens and were dissected out from the shell and extracted immediately whenever possible; otherwise specimens were kept frozen at -20° until extracted.

Toxicity Test

The test solution was prepared by homogenising the shellfish samples with acidified 80% ethanol, defatted with chloroform and concentrated by evaporation according to Noguchi *et al*. [10]. An aliquot or its dilution was injected intra-peritoneally into mice. The toxicity was calculated from the death time of mice using dose-death time table of PSP [11] and expressed as mouse unit (MU) where one MU represents a dose to kill a 20 g mouse in 15 minutes.

Purification and Identification of Toxin

Whole tissue of green-lipped mussel (8 kg) collected from Tolo Harbour was homogenised and extracted twice with 2 volumes of boiling water containing 1% acetic acid. Combined extracts were neutralised and treated with activated charcoal. The crude toxin obtained was applied to a column of Bio-Rex 70 (H^+, 3 x 15 cm). The column was successively developed with water, 0.03 and 1.3 M acetic acid. The toxin in 1.3 M acetic acid fraction was then adsorbed to a column of Amberlite CG-50 (Na^+, 2 x 5 cm). After washing the column with 1 M Na-acetate, the toxin adsorbed was eluted with 10% acetic acid. Thus purified, the toxin was analysed by HPLC-fluorometric analyser of STX (Saxitoxin) [12].

RESULTS

Mice, into which the test solutions from almost all the shellfish samples were injected, died with signs of PSP. However, the level of toxicity was low from the public health standpoint; the highest toxicity was 1940 MU/kg whole tissue. In Fig. 1 were plotted the mean monthly toxicity levels of various shellfish collected from different areas. No distinct seasonal pattern was observed. However, a significant increasing trend of toxicity level with time ($r = 0.866$, $p < 0.01$) was apparent during the monitoring period.

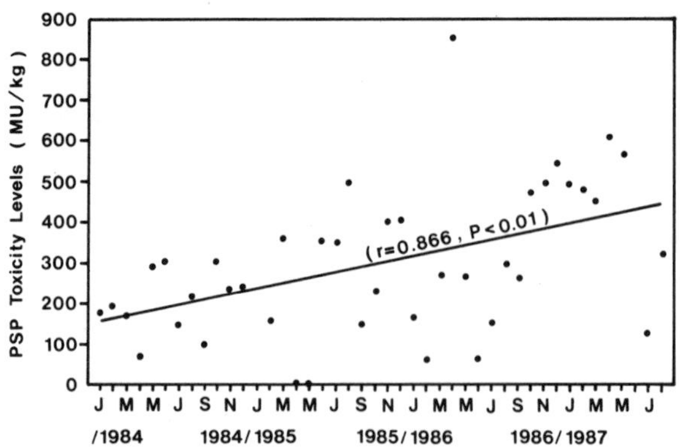

FIG. 1. Seasonal variation and increasing trend of PSP toxicity level in shellfish in Hong Kong.

Fig. 2 shows the seasonal variation of toxicities in green-lipped mussel and fan-shell in Tolo Harbour. The toxicities of both species were similar to each other and showed a more distinct seasonal pattern, exhibiting spring and autumn peaks and dropping to trace levels during the summer months of June, July and August. Such a pattern coincided with that of red tide occurrences in Tolo Harbour [13].

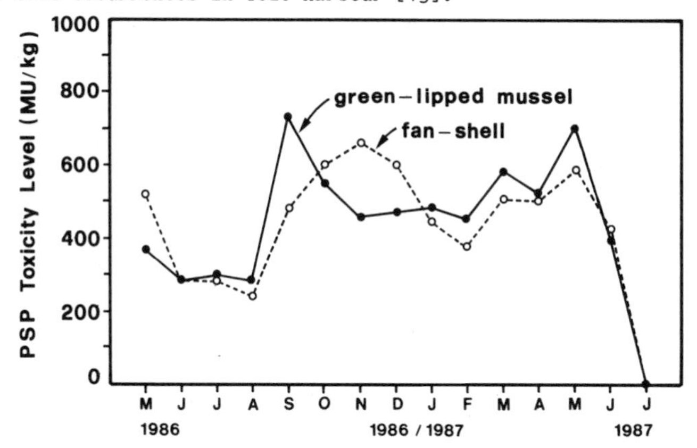

FIG. 2. Seasonal pattern of toxicities in green-lipped mussels and fan-shells in Tolo Harbour.

In order to identify the nature of this low level of toxicity, toxin in green-lipped mussel from Tolo Harbour was partially purified and analysed by the HPLC-fluorometric analyser of STX [12]. During activated charcoal treatment, about 90% of the toxicity was lost. The finally purified toxin showed a peak with identical retention time to that of STX standard (Fig. 3). The ratio of peak height to the toxicity of the toxin in mouse units well coincided with that of STX standard, indicating the presence of STX in the extract.

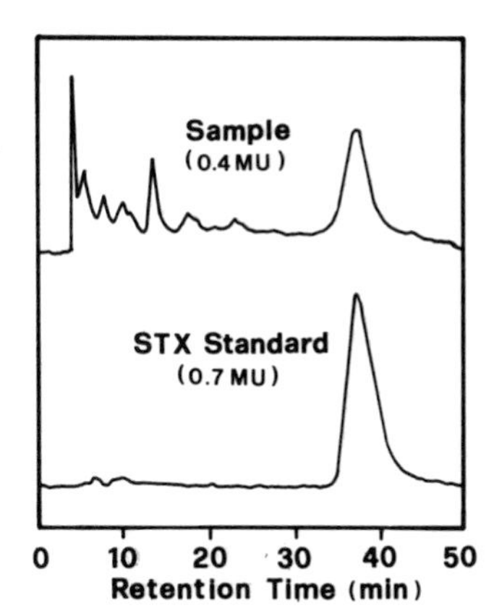

FIG. 3. Chromatographic profiles (HPLC-fluorometric analyser of STX) of toxin extract of green-lipped mussel (<u>Perna viridis</u>) from Tolo Harbour.

DISCUSSION

A low level of toxicity similar to PSP was detected in almost all the shellfish samples of various species collected from different parts of the Hong Kong coastal waters. It is noteworthy that the seasonal variation of the toxicity of shellfish in Tolo Harbour coincided with that of the red tide occurrences. However, none of the red tide causative organisms were identified as the proven toxic dinoflagellates [13] though organisms suspected to be Protogonyaulax species were detected recently in very low concentration. Saxitoxin (STX) was detected in the green-lipped mussel Perna viridis from Tolo Harbour, indicating that at least this toxin is responsible for the toxicity detected. The toxicity level in the shellfish is not significant for any health hazard implications at present. However, the presence of STX in the shellfish indicates the occurrence of some organism(s) which produce STX in the Hong Kong waters. This fact shows the potential risk of toxication of shellfish by PSP in high degree. Further studies to identify the origin of PSP and routine surveillance on the toxicity of shellfish are under progress.

ACKNOWLEDGEMENT

The Director of Environmental Protection, Hong Kong Government is acknowledged for his permission to publish this paper.

REFERENCES

1. P.R. Holmes and C.W.Y. Lam. Asian Marine Biology 2, 1 (1985).
2. B. Morton and P.R. Twentyman. Environmental Res. 4, 544 (1981).
3. E.J. Shantz in: Tetrodotoxin, Saxitoxin, and the Molecular Biology of the Sodium Channel, C.Y. Yao and S.T. Levinson, eds. (The New Academy of Science, New York 1980) pp. 15.
4. J.L. MacLean in: Toxic Dinoflagellate Blooms, Proc. 2nd Internat. Conf. on Toxic Dinoflagellate Blooms, D.L. Taylor and H.H. Seliger, eds. (Elsevier North-Holland, New York 1979) pp. 173-178.
5. D.B. Quayle. Fish. Res. Bd. of Can. Bull. 168, 1 (1969).
6. A. Prakash, J.C. Medeof and A.D. Tennant. Fish. Res. Bd. of Can. Bull. 177, 1 (1971).
7. Y. Shimizu in: Marine Natural Products, Chemical and Biological Properties (1978) pp. 1-42.
8. Y. Shimizu. Maritimes 27, 4 (1983).
9. B.W. Halstead and E.J. Shantz in: Paralytic Shellfish Poisoning. (World Health Organisation, Geneva 1984).
10. T. Noguchi, Y. Ueda, K. Hashimoto and H. Seto. Bull. of Jap. Soc. of Fish. 47(9), 1227 (1981).
11. H. Sommer and K.F. Meyer, Arch. Pathol. 24, 560 (1973).
12. Y. Oshima, M. Machida, K. Sasaki, Y. Tamaoki, T. Yasumoto. Agri. and Biol. Chem. 48(7), 1707 (1984).
13. C.W.Y. Lam, paper delivered in the International Symposium of Red Tides held in Takamatsu, Japan from 9-14 Nov 1987, Proceedings in press.

VI WORKSHOPS

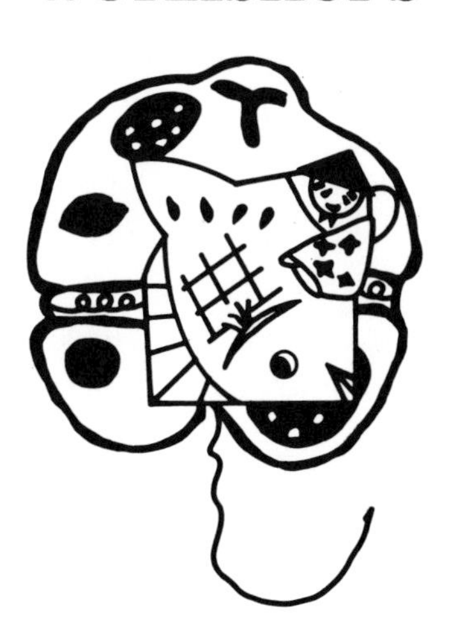

METHODS FOR MODERN DINOFLAGELLATE CYST STUDIES

K. MATSUOKA*, Y. FUKUYO**, AND D. M. ANDERSON***
*Department of Geology, Nagasaki University, 1-14 Bunkyo-Machi, Nagasaki 852, Japan; **Department of Fisheries, University of Tokyo, 1-1-1 Yayoi-Cho, Bunkyo-Ku, Tokyo 113, Japan; ***Biology Department, Woods Hole Oceanographic Inst., Woods Hole, MA 02543

INTRODUCTION

The purpose of this workshop is to provide an introduction to modern dinoflagellate cyst studies. For supporting information, the reader is referred to several detailed reviews by Wall and Dale [59], Dale [11], and Fukuyo and Matsuoka [19] (in Japanese). This manual is prepared in a course format consisting of eight sections: Introduction, Sampling, Fixation and preservation, Cleaning and concentration, Isolation of single cysts, Cyst culture, and Identification. A useful key for identifying modern dinoflagellate cysts based on shape is provided, as is a list of all known cyst-producing, living dinoflagellate species.

Definitions

The term "cyst" is used to describe a non-motile cell which lacks flagella and an ability to swim. Two types of cysts are found in dinoflagellate life cycles - the temporary cyst and the resting cyst. Definitions given by Dale [11] are repeated here, with some modifications:

Temporary Cyst: This resting cell is formed asexually when motile cells experience unfavorable conditions. When conditions become favorable again, temporary cysts quickly re-establish their ability to swim, allowing the organism to withstand temporarily adverse environments. No internal morphological changes are observed during cyst formation: cell contents round off inside a smooth and transparent wall, and the theca is quickly shed by a process called ecdysis. The stage is mainly observed in laboratory cultures, but Kita et al. [23] recorded the temporary cyst of *Goniodoma pseudogonyaulax* (= *Triadinium pseudogonyaulax*) in a rock pool and documented an important role for this stage in red tides caused by that species. Other equivalent names given to this stage are "pellicle" or "ecdysal" cysts.

Resting Cyst: This stage is occasionally formed in cultures and routinely occurs in natural plankton populations, often towards the end of a bloom. Resting cyst formation is a sexual process which may be completed within a few days under bloom conditions, after which cysts quickly become less conspicuous in the plankton and finally sink to accumulate on the bottom. This, together with the fact that only a few percent of the motile cells produce cysts probably accounts for the paucity of cyst observations in plankton records. Many cysts seem to require a mandatory resting period (6 weeks to 5 months depending on species) before they will re-establish motile populations under favorable conditions. Recent study, however, suggests that some species may not need this maturation period. Under favorable conditions, cysts can remain viable in sediments for at least 6 years. In this workshop, we use the term "cyst" to mean "resting cyst" or hypnozygote.

Significance of Cysts in Dinoflagellate Red Tides

More than 60 marine and 15 freshwater species of modern dinoflagellates are known to produce cysts (Table I). Of these species, more than

RED TIDES: BIOLOGY, ENVIRONMENTAL SCIENCE, AND TOXICOLOGY
Okaichi, Anderson, and Nemoto, Editors

Table I Dinoflagellates producing a resting cyst

MARINE SPECIES	Reference
Gymnodiniales	
Cochlodinium sp.	Fukuyo and Matsuoka (1983), Matsuoka (1985a, 1987a)
Gymnodinium breve	Walker (1982)
Gymnodinium catenatum	Anderson et al. (in press), Matsuoka (1987a)
Gyrodinium instriatum	Wall and Dale (1968), Fukuyo and Matsuoka (1983)
Gyrodinium resplendens	Dale (1983)
Gyrodinium uncatenum	Tyler et al. (1982)
Pheopolykrikos hartmannii	Fukuyo and Matsuoka (1983), Matsuoka and Fukuyo (1986), Matsuoka (1985a)
Polykrikos kofoidii	Morey-Gains and Ruse (1980), Fukuyo and Matsuoka (1983), Matsuoka (1985a)
Polykrikos schwartzii	Wall and Dale (1968a), Matsuoka (1985a)
Gonyaulacales	
Gonyaulax digitalis	Wall and Dale (1968a)
Gonyaulax polyedra	Wall and Dale (1968a), Kobayashi et al. (1981)
Gonyaulax scrippsae	Wall and Dale (1968a), Matsuoka (1982a)
Gonyaulax spinifera	Wall and Dale (1968a)
Gonyaulax cf. spinifera	Dale (1983)
Gonyaulax verior	Matsuoka et al. (in press)
Gonyaulax sp.	Dobell and Taylor (1981)
Protoceratium reticulatum	Wall and Dale (1968a)
Alexandrium monilatum	Walker and Steidinger (1979)
Protogonyaulax affinis	Fukuyo et al. (1985)
Protogonyaulax catenella	Yoshimatsu (1981), Fukuyo (1985)
Protogpnyaulax globosa	Dale (1977b)
Protogonyaulax leei	Fukuyo (unpublished data)
Protogonyaulax perviana	Fukuyo et al. (unpublished data)
Protogonyaulax tamarensis	Dale (1977b), Fukuyo (1985)
Protogonyaulax sp. OMR	Matsuoka (1987a)
Helgolandium subglobosum	von Stosch (1969b)
Pyrodinium bahamense	
var. bahamense	Wall and Dale (1969)
var. compressum	Steidinger et al. (1980)
Triadinium pseudogonyaulax	Kita et al. (1985)
Pyrophacus horologium	Wall and Dale (1971)
Pyrophacus steinii	
var. steinii	Matsuoka (1985b)
var. vancampoae	Wall and Dale (1971)
Peridiniales	
Scrippsiella trochoidea	Wall et al. (1970)
Scrippsiella sweeniae	Wall and Dale (1968b)
Ensiculifera cf. mexicana	Wall and Dale (1968b)
Ensiculifera sp.	Matsuoka and Kobayashi (unpublished data)
Cachonina hallii	von Stosch (1969a)

Table I cont.

Hetrocapsa triquetra	Braarud and Pappas (1951)
Peridinium faeroense	Dale (1977a, 1978)
Peridinium hangoei	Iwasaki (1969)
Protoperidinium avellana	Wall and Dale (1968a), Matsuoka (1984a) Lewis et al. (1984)
Protoperidinium claudicans	Wall and Dale (1968a)
Protoperidinium compressum	Wall and Dale (1968a)
Protoperidinium conicoides	Wall and Dale (1968a)
Protoperidinium conicum	Wall and Dale (1968), Fukuyo (1980) Kobayashi and Matsuoka (1984)
Protoperidinium denticulatum	Wall and Dale (1968a)
Protoperidinium divaricatum	Matsuoka et al. (1983)
Protoperidinium excentricum	Wall and Dale (1968a), Lewis et al. (1984)
Protoperidinium latissinum	Wall and Dale (1968a)
Protoperidinium leonis	Wall and Dale (1968a)
Protoperidinium minutum	Wall and Dale (1968a), Fukuyo et al. (1977)
Protoperidinium nudum	Wall and Dale (1968a)
Protoperidinium oblongum	Wall and Dale (1968a)
Protoperidinium pentagonum	Wall and Dale (1968a), Matsuoka (1982) Lewis et al. (1984)
Protoperidinium punctulatum	Wall and Dale (1968a)
Protoperidinium subinerme	Wall and Dale (1968a)
Protoperidinium thorianum	Lewis et al. (1984)
Protoperidinium cf.divergens	Dale (1983)
Protoperidinium sp.	Dale (1983)
Diplopelta parva	Matsuoka (in press)
Diplopsalis lenticula	Wall and Dale (1968a), Matsuoka (in press)
Diplopsalis lebourae	Matsuoka (in press)
Diplopsalopis orbicularis	Wall and Dale (1968a), Matsuoka (in press)
Gotoius abei	Matsuoka (in press)
Zygabikodinium lenticulatum	Wall and Dale (1968a), Matsuoka (in press)
FRESHWATER SPECIES	
Gymnodiniales	
Cystodinium bataviense (=Dinococcus oedogonii)	Pfiester and Lynch (1980)
Gymnodinium dodgei	Sarma and Shyam (1974)
Gymnodinium fungiforme	Spero and Moree (1978)
Gymnodinium pseudopalustre	von Stosch (1973)
Woloszynskia apiculata	von Stosch (1973)
Woloszynskia tylota	Bibby and Dodge (1972)
Gonyaulacales	
Ceratium carolianum	Wall and Evitt (1975)
Ceratium cornutum	Wall and Evitt (1975)
Ceratium hirundinella	Wall and Evitt (1975), Chapmann et al. (1981, 1982)
Ceratium horridum	von Stosch (1972)

Table I cont.

Peridiniales	
Peridinium cinctum	
forma ovoplanum	Pfiester (1975)
forma westii	Eren (1969)
Peridinium cunningtonii	Sako et al. (1984)
Peridinium gatunense	Pfiester (1977)
Peridinium inconspicuum	Pfiester et al. (1984), Wall and Dale (in Wall et al. 1973)
Peridinium limbatum	Evitt and Wall (1968), Pfiester and Skvarla (1980)
Peridinium lubiniensiforme	Dilwald (1937)
Peridinium penardii	Sako et al. (1987)
Peridinium volzii	Pfiester and Skvarla (1979)
Peridinium willei	Pfiester (1976)
Peridinium wisconsinense	Wall and Dale (1968a)

16 have been known to cause red tides and four are toxic. Most modern cysts are spherical, ellipsoidal to peridinioid, with or without spine-like ornaments, and range from 20-80 μm in size. As the cysts lack flagella, they float for a short period and sink to the sediments after encystment. During dormancy, cysts may be resuspended and transported in much the same way as fine silt or mud particles are.

Physiologically there are two different cyst forms observed in surface sediments: living cysts and empty cyst walls. The living cyst with fresh protoplasm can germinate under favorable conditions. The empty cyst is the wall remaining after excystment, often with a distinct opening called an archeopyle. The cyst wall of modern dinoflagellates can be calcareous or organic (sporopollenin). As the organic sporopollenin wall is highly resistant to chemical and biological attack, some cysts are not affected by harsh palynological processing with strong acids and oxidants.

SAMPLING

Planning for Sampling

Two methods have been commonly adopted for collecting surface sediments - corers and sediment traps. The sampling method should be selected according to the purpose of the cyst study. If the objective is to know when and how many cysts of a certain species are produced, sediment traps are useful. If it is to document the presence of different species or the change in a dinoflagellate community, the coring method is preferable.

Core Sampling

The core sampler is used for collecting bottom sediments. In order to get surface sediments, which include many fresh living cysts, gravity corers such as the Phleger corer (Phleger bottom sampler) and piston corers are more desirable than other bottom samplers such as dredges or grab buckets which often lose the light fluffy material at the sediment surface. A light-weight core sampler (TFO gravity corer [19]) or its equivalent can be deployed from small boats and thus is useful in the investigation of near-shore and inner bay areas.

Coring procedures are very simple. The corer is deployed from the boat and retrieved with care to avoid losing sediment. Immediately on retrieval, the top and bottom of the clear core tubing must be capped with plastic covers or some other material to prevent water leakage. The intact core can then be stored in the cold and dark until needed. When core-samplers are not available or do not work in coarser sediments, Smith McIntyre or Ekman Berg type grab samples can be used. Small 10-15 cm tubes can be inserted in the bulk sample to collect subsamples. SCUBA diving is also a useful method for obtaining undisturbed bottom cores.

Sediment Traps

Sediment traps are used to catch sinking cysts before they settle to the sea or lake floor. There are no widely-accepted procedures or designs for sediment traps, as various configurations will collect different quantities of material. Furthermore, material resuspended from the bottom and collected in a trap can complicate the interpretation of sedimentation data. Quantitative analysis of sediment trap data is thus not recommended. It is, however, possible to learn a great deal at a qualitative level about the timing and relative magnitude of cyst formation as a component of dinoflagellate population dynamics. For these purposes, sediment traps need not be elaborate or expensive. A small trap consisting of a 2-liter wide-mouth polyethylene bottle attached to a line between a surface buoy and a bottom weight will provide useful information. A vertical series of these bottles can even provide some indication of the extent of resuspension.

SEDIMENT PROCESSING

Two different processing methods have been adopted for cleaning and concentrating cysts from bulk sediment - a sieving technique that uses no chemicals and a palynological technique that use harsh chemicals. The choice of technique depends on the purpose of the study. When living cysts are needed (e.g. for germination, identification, or enumeration), the sieving technique is used. When a general survey of the cyst assemblage is desired, palynological processing can be used. One should recognize that the latter process will destroy certain fragile or non-resistant cysts that would be observed with the non-destructive sieving technique.

Sieving Procedure (Fig. 1)

1. Prepare a series of sieve of various mesh-sizes, with 250 µm being the upper sieve, 125 µm in the middle and 20 µm being the lowest sieve.

2. Mark the core tube at 2 cm intervals from the bottom.

3. Remove the plastic cover from the core tube and pipette the overlying seawater into a 50 ml vial.

4. Remove the cap from the bottom of the core tube.

5. Remove the bottom mud layer in the tube slowly by blowing at the upper end of the tube until the upper surface of the mud reaches the 2 cm mark. Sometimes it is just as easy to push the sediment out of the tube from below, using a stopper cut to be slightly smaller than the inside diameter of the core tube.

6. Put the remaining sediment into the 50 ml vial used in Step 3.

7. Rinse the inner surface of the tube with filtered seawater. Pour the rinse water into the 50 ml vial.

8. Mix all contents in the vial and pour onto the upper sieve. Rinse the vial with a small amount of filtered seawater if needed. Note that it is also possible to use an ultrasonic probe or bath to disaggregate sediment before sieving [59].

9. Wash the sediment on the upper sieve carefully with filtered seawater. The cysts and fine particles will pass through the 250 µm and 125 µm sieves and accumulate on the 20 µm sieve.

10. Transfer all the residue on the 20 µm sieve to a petri-dish.

11. Separate the cysts from the residue by squirting filtered seawater from the washing bottle. The water is injected at one side of the petri-dish in such a way that the residue is surrounded by the swirling motion of water. Cysts and other light-weight particles will be suspended in the circulating water while heavy sand particles remain at the bottom in the center of the petri-dish.

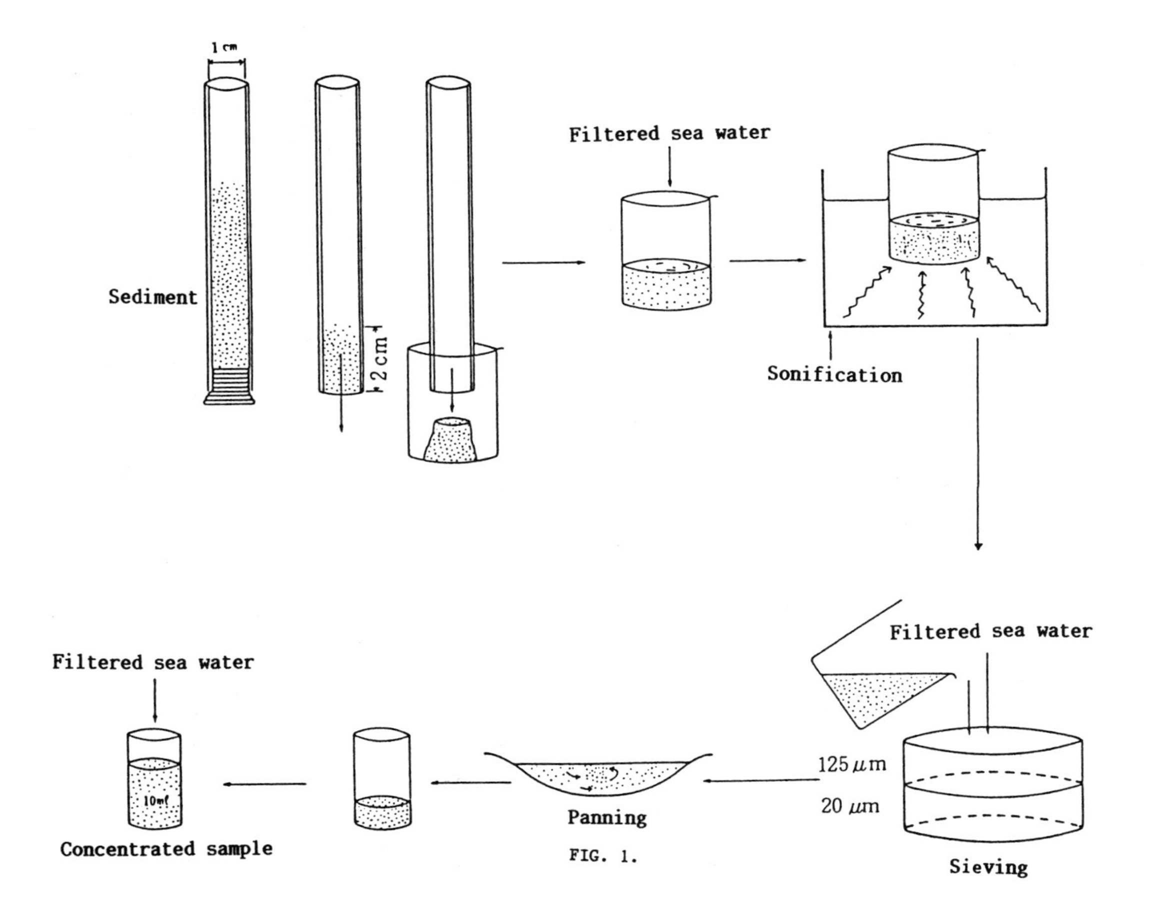
1 cm
Sediment
2 cm
Filtered sea water
Sonification
Filtered sea water
125 μm
20 μm
Sieving
Panning
Filtered sea water
10 ml
Concentrated sample

FIG. 1.

11a. Alternatively, transfer all residue on the 20 μm sieve to a watch glass. Make a water-eddy in the watch glass for separating cysts and light-weight particles from heavier sand grains (panning in the geological sense). After this process, cysts and other fine particles are concentrated at the center of the watch glass.

12. Gently pour the supernatant water with cysts suspended back onto the 20 μm sieve. The heavy sand particles are left on the petri-dish.

12a. Suck up the cysts and light-weight particles and transfer them into a test tube with a syringe. Then use a squeeze bottle to back-wash the material off the sieve. Minimize the amount of water used in order to achieve the highest concentration factor (10 ml total is useful).

13. Repeat the washing, if needed.

Palynological Technique

The technique introduced here is based on standard palynological processing. It uses several dangerous chemicals and therefore should be performed with adequate safety precautions.

Procedure

1. Put 1 ml of the original material into a 15 ml polyethylene centrifuge tube.

2. Wash with distilled water several times to remove salt.

3. Add 5% hydrochloric acid to remove calcium carbonate from calcareous nannoplankton, foraminifera and others. The calcareous cyst wall and ornaments such as on *Scrippsiella trochoidea* and *Ensiculifera* sp. are also removed at this time, but the inner organic phragma will remain.

4. Wash with distilled water.

5. Add 1% potassium hydroxide solution and warm to 70°C in a water bath for 3 mins. At higher temperature and with longer exposures, the relatively thin cyst phragma of *Protoperidinium* and *Protogonyaulax* sometimes disappear.

6. Wash with distilled water.

7. Add concentrated (25-30%) hydrofluoric acid to the tube for removing silicate materials such as sand, diatoms, silicoflagellates and others. Warm in the water bath at 70°C for 2-3 hrs. As the hydrofluoric acid is very dangerous and toxic, this processing should be conducted in a fume hood with rubber or vinyl gloves. The residue solution containing hydrofluoric acid should be neutralized with calcium carbonate.

8. Wash with distilled water.

9. When cellulose substances such as plant tissue are abundant in the sample, acetylation may be useful to remove it. The acetylation procedure is as follows:

   a. Add glacial acetic acid (CH_3COOH) to the tube.

b. After removing this chemical, add Erdtman's solution (9 parts of acetic anhydride [$(CH_3CO)_2O$] and one part of concentrated sulfuric acid [H_2SO_4]) to the tube and warm it to 70°C for 15 mins.

c. Remove the Erdtman's solution and add glacial acetic acid to the tube again.

d. Wash with distilled water.

10. Prepare a series of sieves of various mesh-sizes with 250 µm being the upper sieve, 125 µm in the middle and 20 µm at the bottom.

11. Pour all the residue onto the upper sieve. The cysts and other organic particles such as spores and pollen grains will pass through the 250 µm and 125 µm sieves and accumulate on the 20 µm sieve.

12. Wash the material on the 20 µm sieve into a vial for a final volume of 10 ml.

CYST ISOLATION

Isolation is an important technique because it must be carried out before the inoculation step in establishing cultures. It is also needed in morphological studies.

Procedure

1. A small needle (insect pin) attached to the tip of a glass tube is used to clean the background around the cysts on a slide. A Pasteur pipette drawn out under flame to give a capillary tip (50-100 um dia.) at one end and connected to a 60 cm long silicone tube at the wide end is used for micropipetting cells.

2. Take out a small amount (about 0.5 ml) of the sieved cyst sample and put it onto a large counting chamber slide. Spread the sample evenly on the slide by adding filtered seawater to 1 ml.

3. Search under the microscope for the desired cysts. Once a cyst is located, push away interfering material around the cyst with the small needle and use the Pasteur pipette and tube to suck up the cyst.

4. Transfer the cyst into culture medium. For morphological studies, the cyst is introduced into a droplet of water on a regular slide.

CYST CULTURE

Three types of culture chambers are commonly used - a glass tube, individual wells in tissue culture plates, or Palmer-Maloney slides as described in [59]. In attempts to establish axenic cultures using culture tubes, "f/10" is useful as a medium. This medium is made by adding 1 ml of the medium "f" [21] to 10 ml of sterile, filtered seawater. For Palmer-Maloney slides and tissue culture plates, sterile filtered seawater can be used without any added nutrients.

To use tissue culture plates (Corning Cell Wells No. 25820), the procedure is as follows:

1. Pipette 1 ml of sterile filtered seawater into each of the 24 wells of a tissue culture plate.

2. Inoculate one cyst into each well, cover the plate, and seal the chamber with vinyl tape to prevent evaporation.

3. Incubate the chamber at a constant temperature between 15-30°C (depending on species) with approximately 150 uE · $m^{-2}sec^{-1}$ illumination.

4. Observe the cysts daily for 2 or 3 weeks after inoculation using an inverted microscope.

5. When a cell germinates and has divided into more than 5 cells, micropipette one cell and observe morphological features under the microscope. Observation of the morphology of the empty cyst is also useful.

CYST IDENTIFICATION

The important morphological features used in identification of cysts are the shape of the cyst body and its ornaments, wall structure and color, paratabulation, and the type of archeopyle or exit opening. The archeopyle is very useful in classifying the genus and family to which cysts belong. As the opening is not visible before excystment, it is not possible to use this characteristic for identification of living cysts. Furthermore, in comparison with the morphology of motile cells of dinoflagellates, cysts are usually relatively simple, mostly spherical to peridinioid. As a result, identification of cysts based on a single morphological character is not always reliable, and other characters listed above must be examined. A diagram of archeopyle types is presented in Figure 2 and examples of species with each type are given in Table II. A key based on cyst shape is given in Table III, and a key based on shape and archeopyle type is given in Table IV.

General Character of Gymnodinalian Cysts

Shape: mostly spherical to ovoidal and sometimes ellipsoidal, with or without spinate ornaments.

Cyst wall: organic and brownish color; mostly a single layer and sometimes two layers.

Archeopyle type: cryptophylic; chasmic (slit) or tremic (hole) type [30].

Cysts of Gymnodinialean species causing red tides: *Gymnodinium breve*, *Gymnodinium instriatum*, *Pheopolykrikos hartmannii*, *Polykrikos kofoidii*, *Polykrikos schwartzii*.

Cysts of toxic Gymnodinialean species: *Gymnodinium breve* (= *Ptychodiscus brevis*), *Gymnodinium catenatum*.

General Character of Gonyaulacacean Cysts

Cyst shape: basically spherical to ellipsoidal and rarely discoidal, often without process-like ornaments.

Cyst wall: organic, colorless and sometimes transparent; usually two layers, rarely single.

Archeopyle type: mostly saphopylic; precingular, but sometimes epicystal, hypicystal or combination type.

Cysts of Gonyaulacacean species causing red tides: *Gonyaulax polyedra*, *Gonyaulax spinifera*, *Gonyaulax verior*, *Protoceratium reticulatum*, *Protogonyaulax affinis*, *Protogonyaulax peruviana*, *Triadinium pseudogonyaulax*.

Cysts of toxic Gonyaulacacean species: *Protogonyaulax catenella*, *Protogonyaulax tararensis*, *Pyrodinium bahamense* var. *bahamense*, *Pyrodinium bahamense* var. *compressum*

General Character of Peridiniacean Cysts

Cyst shape: mainly spherical, ellipsoidal, peridinioid, and rarely discoidal, mainly without process-like ornaments.

Cyst wall: mainly organic and brownish in color; sometimes calcareous; mainly a single layer and rarely two layers.

Archeopyle type: mainly saphopylic; intercalary type, and sometimes theropylic; apical, intercalary, epicystal and combination types.

Cysts of Peridiniacean species causing red tides: *Scrippsiella trochoidea*, *Cachonina hallii*, *Heterocapsa triquetra*, *Peridinium hangoei*, *Peridinium cunningtonii*.

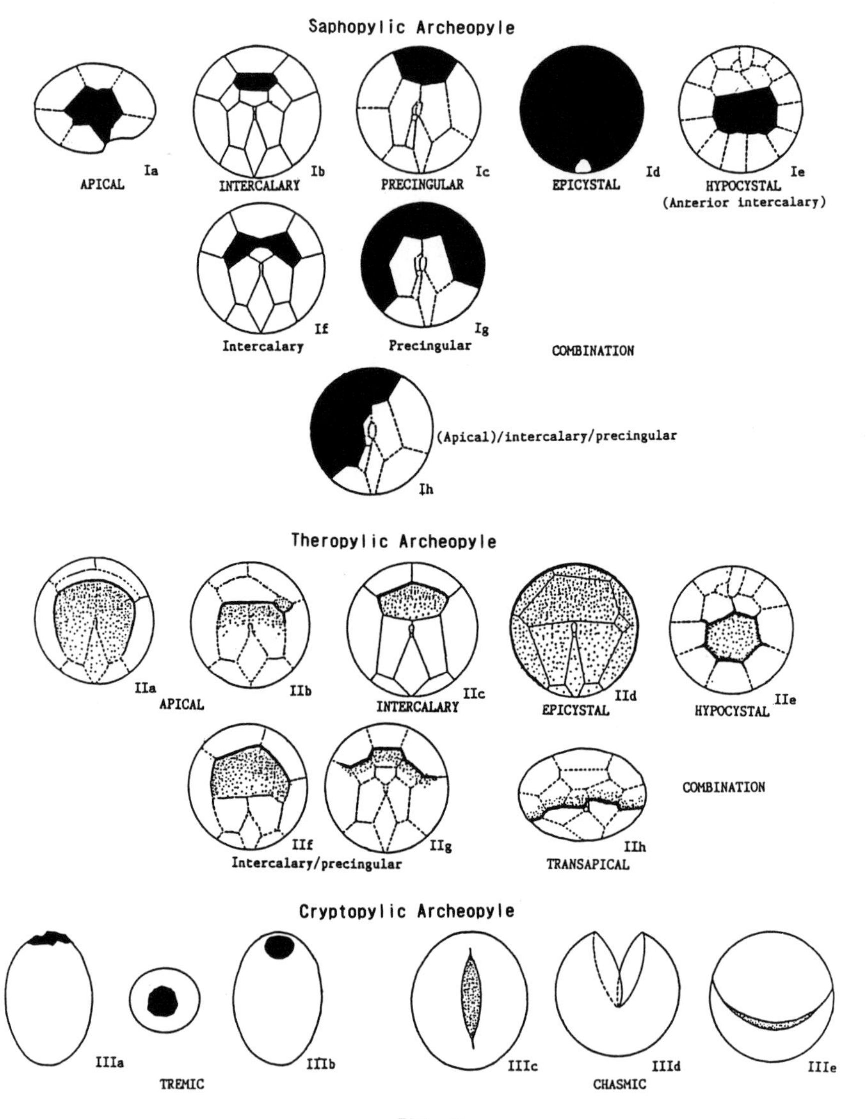
Saphopylic Archeopyle
Ia
APICAL
Ib
INTERCALARY
Ic
PRECINGULAR
Id
EPICYSTAL
Ie
HYPOCYSTAL
(Anterior intercalary)
If
Intercalary
Ig
Precingular
COMBINATION
(Apical)/intercalary/precingular
Ih
Theropylic Archeopyle
IIa
APICAL
IIb
IIc
INTERCALARY
IId
EPICYSTAL
IIe
HYPOCYSTAL
IIf
IIg
Intercalary/precingular
IIh
TRANSAPICAL
COMBINATION
Cryptopylic Archeopyle
IIIa
TREMIC
IIIb
IIIc
IIId
CHASMIC
IIIe

FIG. 2

Table II ARCHEOPYLE TYPES IN MODERN DINOFLAGELLATE CYSTS

Archeopyle type	Taxa
Ia～Ih Saphopylic type	
BASIC FORM	
Ia Apical Archeopyle	Ceratium hirundinella, Ceratium cornutum(?) Ceratium carolianum(?)
Ib Intercalary archeopyle	Brigantedinium☆, Lejeunecysta☆, Leipokatium☆ Selenopemphix☆, Stelladinium☆, Trinovantedinium☆, Xandarodinium☆, Protoperidinium latissinum, Peridinium ponticum
Ic Precingular archeopyle	Ataxiodinium☆, Impagidinium☆, Nematosphaeridium☆, Operculodinium☆, Spiniferites☆, Tectatodinium☆
Id Epicystal Archeopyle	Polysphaeridium☆
Ie Hypocystal Arcehopyle	Tuberculodinium☆
COMBINATION FORM	
If Intercalary Archeopyle	Protoperidinium (Archaeperidinium)
Ig Precingular Archeopyle	Lingulodinium☆
Ih Apical+Intercaraly+Precingular Archeopyle	Lingulodinium☆
IIa～IIh Theropylic type	
BASIC FORM	
IIa,b Apical Archeopyle	Diplopelta parva, Diplopsalis lenticula, Peridinium wisconsinense, Gonyaulax verior(?) Protogonyaulax catenella(?), Protogonyaulax tamarensis(?), Protogonyaulax leei, Protogonyaulax perviana
IIc Intercalary Archeopyle	Diplopsalis lebourae
IId Epicystal Archeopyle	Zygabikodinium lenticulatum, Dubridinium cavatum☆, Dubridinium cassiculum☆, Dubridinium ulstrum☆, Dubridinium polygonum☆, Helgolandinium subglobosum, Fragilidium heterolobum(?)
IIe Hypocystal Archeopyle	Pyrophacus horologium (?)
COMBINATIOM FORM	
IIf,g Intercalary Archeopyle	Gotius abei, Scrippsiella trochoidea Scrippsiella sweeniae, Ensiculifera cf. mexicana
Apical(?)+Intercalary Archeopyle	Diplopsalopsis orbicularis
IIh Transapical Archeopyle	Peridinium limbatum
IIIa～IIIe Cryptopylic Archeopyle	
IIIa,b Chasmic Archeopyle	Cochlodinium sp., Gymnodinium catenatum, Gymnodinium pseudopalustre, Pheopolykrikos hartmannii
IIIc,d,e Tremic Archeopyle	Polykrikos schwartzii, Polykrikos kofoidii Gyrodinium instriatum

☆:name for cyst-based taxa.

Table. III Key to modern dinoflagellate cysts based on shape

	Character	Taxon / couplet
1	Cordate in dorso-ventral view	----- *Protoperidinium oblongum*
		----- *Protoperidinium claudicans* (*Votadinium**)
1	Peridinioid in dorso-ventral view and compressed antero-posteriorly	----- *Protoperidinium conicum*
		----- *Protoperidinium nudum*
		----- *Protoperidinium subinerme* (*Selenopemphix**)
1	Roundly polygonal with hollow processes distally closed	----- *Protoperidinium divaricatum* (*Xandarodinium**)
1	Subspherical to ovoidal with well-developed parasuture	----- Gonyaulacaceae (*Impagidinium**)
1	Subspherical endophragm (inner body) with membranous periphragm	----- *Gonyaulax spinifera* complex (*Ataxiodinium**)
1	Discoidal with many short barrel-shaped processes	----- *Pyrophacus steinii* (*Tuberculodinium**)
1	Ellipsoidal with coarsely reticulate ornaments on surface	----- *Polykrikos schwartzii*
1	Ellipsoidal with shelf-like or hollow processes	----- *Polykrikos kofoidii*
1	Ovoidal and transparent phragma, sometimes with mucilaginous material	----- *Gyrodinium instriatum*
		----- *Gonyaulax verior*
1	Ellipsoidal and transparent phragma, sometimes with mucilaginous material	----- *Protogonyaulax catenella*
		----- *Protogonyaulax tamarensis*
1	Spherical and transparent phragma, sometimes with mucilaginous material	----- *Protogonyaulax leei*
		----- *Protogonyaulax perviana*
		----- *Triadinium pseudogoniaulax*
		----- *Diplopsalopsis orbicularis*
1	Spherical to ovoidal	----- 2
1	Spherical to ovoidal with processes densly distributed	----- 3
1	Peridinioid (pentagonal to stellar) in dorso-ventral view	--- 4
1	Spherical to ellipsoidal with well developed paratutures and processes	----- 5
2	Ovoidal with two phragma well adpressed (simple precingular archeopyle)	----- *Gonyaulax spinifera* complex (*Tectatodinium**)
2	Subsphaerical with two phragma well adpressed (combination precingular archeopyle)	----- *Gonyaulax spinifera* complex (*Bitectatodinium**)

Table IIIcont.

2 Spherical to subspherical with a single brouwn layer (Intercalary archeopyle) ----- Protoperidinium avellana
----- Protoperidinium denticulatum
----- Protoperidinium punctulatum
----- Gotoius abei
(Brigantedinium*)

2 Subspherical, antero-posteriolly compressed with paracingular and parasulcus ----- Zygabikodinium lenticulatum
(Dubridinium*)

2 Spherical with finely reticulate surface
----- Gymnodinium catenatum

3 Evexate to bulbous processes closed distally
----- Gonyaulax polyedra
(Lingulodinium*)

3 Slender capitate processes closed distally
----- Protoceratium reticulatum
(Operculodinium*)

3 Short denticulate to patulate processes
----- Pyrodinium bahamense
(Polysphaeridinium*)

3 Calcareous wall with acuminate to conical processes
----- Scrippsiella trochoidea

3 Brownish cyst wall with long acuminate processes
----- Protoperidinium minutum
----- Diplopelta parva

3 Brownish cyst wall with acuminate processes striated proximally
----- Pheopolykrikos hartmannii

4 Long spines developed on each corner of peridinioid shape
----- Protoperidinium compressum
(Stelladinium*)

4 Parasutural and intratabular short spines
----- Protoperidinium pentagonum
(Trinovantedinium*)

4 A single apical and two antapical horns well developed
----- Protoperidinium leonis
----- Protoperidinium latissinum
(Lejeunecysta*)

5 Parasutural furcate processes
----- Gonyaulax spinifera complex
----- Gonyaulax scrippsae
----- Gonyaulax sp.
(Spiniferites*)

5 Parasutural furcate processes connected with ectophragm
----- Gonyaulax spinifera complex
(Nematosphaeridinium*)

*: name for cyst based taxa.

Table IV

Key to modern dinoflagellate cysts based on shape and archeopyle type

1 Sahophylic archeopyle ------ 2
1 Theropylic archeopyle ------ 11
1 Cryptophylic archeopyle ---- 14
2 Epicystal archeopyle ----- Polysphaeridium
2 Hypocystal archeopyle ----- Tuberculodinium
2 Intercalary archeopyle ----- 3
2 Precingular archeopyle ----- 7
2 Combination archeopyle ----- 10
3 Cordate in dorso-ventral view ------------------------- Votadinium
3 Spehrical and brownish autophragm without ornament ----- Brigantedinium
3 Peridinioid with dome-like epicyst and two antapical horns ----- 6
3 Brownish wall with several spines ----------------------------- 4
3 Peridinioid with a single apical and two antapical horns-------- 5
4 A single apical, two antapical and a few cingular spines --- Stelladinium
4 Roundly hexagonal, antero-posteriorly compressed ----------- Selenopemphix
4 Subcircular to ellptical in dorso-ventral view with hollow spines closed distally -- Xandarodinium
5 Transparent wall with intratabular and parasutural spines - Trinovantedinium
5 Epicyst triangular in dorso-ventral view with distinct apical and antapical horns -- Lejeunecysta
6 Two small antapical horns ----------- Leipokatium
6 A single broad antapical boss ------- Selenopemphix
7 Various ornaments and processes ----- 8
7 No ornament ----- 9
8 Parasutural furcate processes and coasely reticulate ectophragm ------------------------------------ Nematosphaeropsis
8 Furcate gonal and/or processes ----- Spiniferites
8 Slender capitate processes --------- Operculodinium
9 Subspherical with thick spongy cyst wall ----- Tectatodinium
9 Spherical to ovoidal with distinct paratabulation indicated by parasutural septa --------------------------------------- Impagidinium
9 Subspherical endophragm covered with membranous periphragm -- Ataxiodinium
10 Spherical, without ornament and with operculum corresponding two precingular paraplates ----- Bitectatodinium
10 Spherical with bullbous processes and opercula corresponding four precingular and/or anterior intercalary and apical paraplates --------------------------- Lingulodinium
11 Spherical with brownish wall and acuminate processes, and apical archeopyle --------------------------- Diplopelta parva*
11 Spherical with brownish wall and apical archeopyle --------------------------- Diplopsalis lenticula*
11 Spherical with brownish wall and a simple intercalary archeopyel --------------------------- Diplopsalis lebourae*
11 Subspherical to lenticular with epicystal archeopyle ----- 12
11 Combination archopyle ----------------------------------- 13

TableIV cont.

12 Brounish wall with distinct paracingulum --- Dubridinium
Zygabikodinium lenticulatum*
12 Thin and smooth transparent wall ---------- Helgolandinium sunglobosum*
-- Fragilidium heterolobum(?)*
13 Subspherical with opercula comprising apical and anterior intercalary
paraplates -------------------------------- Diplolsalopsis orbicularis*
13 Subspherical to lenticular, with opercula consisting of two anterior
intercalary paraplates -------------------- Gotoius abei*
13 Ovoidal with calcareous wall and ornaments ----- Scrippsiella trochoidea*
13 Transapical archoepyle ------------------------ Peridinium limbatum*
14 Tremic archeopyle (hole type) ----- Polykrikos schwartzii*
---------------------------------- Gyrodinium instriatum*
14 Chasmic archeopyle (slit type) ----- Pheopolikrikos hartmannii*
---------------------------------- Gymnodinium catenatum*
------------------------------------ Gymnodinium pseudopalustre*

*: Species name for motile form.

REFERENCES

1. A.M. Anderson and F.M. Morel, Est. Coast. Mar. Sci. 8, 279-293 (1979).
2. D.M. Anderson, S.W. Chisholm, and C.J. Watras, Mar. Biol. 76, 179-189 (1983).
3. D.M. Anderson, D. Jacobson, I. Bravo, and J.H. Wrenn, J. Phycol. In Press.
4. B.T. Bibby and J.D. Dodge, Br. phycol. J. 7, 85-100 (1972).
5. T. Braarud and I. Pappas, Lebour. Vid.-Akad. Avh. I. M. -N. Kgl. 2, 1-23 (1951).
6. D.V. Chapmann, D. Livingstone, and J.D. Dodge, Br. phycol. J. 16, 183-194 (1981).
7. D.V. Chapmann, J.D. Dodge, and S.I. Heaney, J. Phycol. 18, 121-129 (1982).
8. B. Dale, Br. phycol. J. 12, 241-253 (1977a).
9. B. Dale, Sarsia 63, 29-34 (1977b).
10. B. Dale, Palynology 2, 187-193 (1978).
11. B. Dale in: Survival Strategies of the Algae, G.A. Fryxell, ed. (Cambridge Univ. Press, Cambridge 1983), pp. 69-144.
12. K. Dilwald, Flora 132, 174-192 (1937).
13. P.E.R. Dobell and F.J.R. Taylor, Palynology 5, 99-106 (1981).
14. J. Eren, Verh. Internat. Verein. Limnol. 17, 1013-1016 (1969).
15. Y. Fukuyo in: The Working Party on Taxonomy in the Akashiwo Kenkyukai, Sheet No. 59 (1980).
16. Y. Fukuyo, Bull. Mar. Res. 37, 529-537 (1985).
17. Y. Fukuyo, J. Kittaka, and R. Hirano, Bull. Plankton Soc. Japan 24, 11-18 (1977).
18. Y. Fukuyo and K. Matsuoka, Marine Environmental Sciences, 389-410 (1983).
19. Y. Fukuyo and K. Matsuoka in: A Guide for Studies of Red Tide Organisms, Japan Fisheries Resource Conservation Association, ed. (Shuwa, Tokyo 1987), pp. 85-101.
20. Y. Fukuyo, K. Yoshida, and H. Inoue in: Toxic Dinoflagellates, D.M. Anderson et al., eds. (Elsevier, Amsterdam 1985), pp. 11-25.
21. R.R.L. Guillard and J.H. Ryther, Gran. Can. J. Microbiol. 8, 229-239 (1962).
22. H. Iwasaki, Bull. Plankton Soc. Japan 16. 132-139 (1969).
23. T. Kita, Y. Fukuyo, H. Tokuda, and R. Hirano, Bull. Mar. Sci. 37, 643-651 (1985).
24. S. Kobayashi, S. Iizuka and K. Matsuoka, Bull. Plankton Soc. Japan, 28, 53-57 (1981).
25. S. Kobayashi and K. Matsuoka, J. Phycol. 32, 53-57 (1984).
26. J. Lewis, J.D. Dodge, and P. Tett, J. Micropalaeontol. 3, 25-34 (1984).
27. K. Matsuoka in: The Working Party on Taxonomy in the Akashiwo Kenkyukai, Sheet No. 108 (1982).
28. K. Matsuoka, Bull. Fac., Liberal Arts, Nagasaki Univ. 25, 37-47 (1984a).
29. K. Matsuoka in: The Working Party on Taxonomy in the Akashiwo Kenkyukai, Sheet No. 166 (1984b)
30. K. Matsuoka, Rev. Palaeoboton. Palynol. 44, 217-231 (1985a).
31. K. Matsuoka, Trans. Proc. Palaeont. Soc. Japan, N.S. 140, 240-263 (1985b)
32. K. Matsuoka in: A Guide for Studies of Red Tide Organisms, Japan Fisheries Resource Conservation Association, ed. (Shuwo, Tokyo 1987a), pp. 399-476.
33. K. Matsuoka, Bull. Fac. Liberal Arts, Nagasaki Univ. 28, 35-123 (1987b).
34. K. Matsuoka, Rev. Palaeoboton. Palynol. In press.
35. K. Matsuoka and Y. Fukuyo, J. Plankton Res. 8, 811-818 (1986).

36. K. Matsuoka, Y. Fukuyo, and D.M. Anderson, In press.
37. K. Matsuoka, S. Kobayashi, and S. Iizuka, Rev. Palaeobotan. Palynol. 38, 109-118 (1982).
38. G. Morey-Gaines and R.H. Ruse, Phycologia 19, 230-232 (1980).
39. L.A. Pfiester, J. Phycol. 11, 259-265 (1975).
40. L.A. Pfiester, J. Phycol. 12, 234-238.
41. L.A. Pfiester, J. Phycol. 13, 95-95.
42. L.A. Pfiester and R.A. Lynch, Phycologia 18, 13-18 (1980).
43. L.A. Pfiester and J.J. Skvarla, J. Phycol. 18, 13-18 (1979).
44. L.A. Pfiester and J.J. Skvarla, Am. J. Bot. 67, 955-958 (1980).
45. L.A. Pfiester, P. Timpano. J.J. Skvarla and J.R. Holt, Am. J. Bot. 71, 1121-1127 (1984).
46. Sako, Y., Y. Ishida, H. Kadota and Y. Hata, Bull. Japan. Soc. Sic. fish. 50, 743-750 (1984).
47. Sako, Y., Y.Ishida, T. Nishijima and Y. Hata, Bull. Japan. Soc. Sci. Fish. 53, 267-272 (1987).
48. W.A.S. Sarjenat, T. Lacalli, and G. Gains, Micropaleontology 33, 1-36 (1987).
49. Y.S.R.K. Sarma and R. Shyam, Br. phycol. J. 9, 21-29 (1974).
50. H.J. Spero and M.D. Moree, J. Phycol. 17, 43-51 (1978).
51. K.A. Steidinger, L.S. Tester and F.J.R. Taylor, Phycologia 19, 329-337 (1980).
52. H.A. von Stosch, Helgolander wiss. Meeresunt. 19, 558-568 (1969a).
53. H.A. von Stosch, Helgolander wiss. Meeresunt. 19, 569-577 (1969b).
54. H.A. von Stosch, Soc. Bot. Fr. Mem. 1972, 201-202 (1972).
55. H.A. von Stosch, Br. phycol. J. 8, 105-134 (1973).
56. T. Takeuchi in: Toxic Dinoflagellate - Implication in Shellfish Poisoning, Y. Fukuyo, ed. (Koseisha-Koseikaku, Tokyo 1985), pp. 98-108.
57. M.A. Tyler, D.W. Coats, and D.M. Anderson, Mar. Ecol. Prog. Ser. 7, 163-178 (1982).
58. L.M. Walker, Trans. Am. Microsc. Sci. 101, 287-293 (1982).
59. L.M. Walker and K.A. Steidinger, J. Phycol. 15, 312-315 (1979).
60. D. Wall and B. Dale, Micropaleontology 14, 265-304 (1968).
61. D. Wall and B. Dale, J. Phycol. 5, 140-149 (1969).
62. D. Wall and B. Dale, J. Phycol. 7, 221-235 (1971).
63. D. Wall, B. Dale, and K. Harada, Micropaleontology 19, 18-31 (1973).
64. D. Wall and W.R. Evitt, Micropaleontology 21, 18-31 (1975).
65. D. Wall, R.R.L. Guillard, B. Dale, E. Swift, and N. Watabe, Phycologia 9, 151-156 (1970).
66. S. Yoshimatsu, Bull. Plankton Soc. Japan 28, 131-139 (1981).

INSTRUMENTAL ANALYSES OF DINOFLAGELLATE TOXINS

TAKESHI HASUMOTO*, YASUKATSU OSHIMA*, MICHIO MURATA* and SACHIO NISHIO**
*Faculty of Agriculture, Tohoku University, Tsutsumidori, Sendai, 980, Japan
**Shikoku Women's University, Ojin-cho, Tokushima, 771-11, Japan

A highly sensitive and specific analytical method was developed for the determination of okadaic acid (OA) and dinophysistoxin-1 (DTX1), the principle toxins responsible for diarrhetic shellfish poisoning. The toxins, which bear a carboxylic group, were esterified with 9-anthryldiazomethane and the resultant fluorescent derivatives were determined by high performance liquid chromatography (HPLC) after clean-up on a disposable cartridge column (Sep-pak Silica, Waters). Satisfactory separation of the toxin derivatives was achieved on a reverse phase column (Develosil ODS) using acetonitrile-methanol-water (8:1:1) as a moble phase. The excitation and emission wavelengths were set at 365 and 412 nm, respectively. The minimum detection level was about 1 pmol. The high sensitivity of the method enabled us not only to confirm the toxigenicity of *Dinophysis* spp. but also to determine the levels and profiles of the toxins with a small number of cells (100-1000) collected under a microscope. Thus the method provides an efficient means to test the toxicigenicity of species incapable of growing in laboratories. The method is also useful in distinguishing different strains of *Prorocentrum lima* by toxin profiles.

The fluoremetric HPLC method for determination of paralytic shellfish toxins was also improved. The method is capable of analyzing toxin profiles using only a few hundred dinoflagellate cells and thus saves the time and labor needed to prepare cultures. Another advantage of the instrumental assay is the easiness with which it determines sulfamate toxins, which are difficult to analyze by mouse bioassay due to their instability and low potency.

Details of these methods are to be published elsewhere.

Published 1989 by Elsevier Science Publishing Co., Inc.
RED TIDES: BIOLOGY, ENVIRONMENTAL SCIENCE, AND TOXICOLOGY
Okaichi, Anderson, and Nemoto, Editors

INDEX

D

E

O

P